Control and Nonlinear Dynamics on Energy Conversion Systems

Control and Nonlinear Dynamics on Energy Conversion Systems

Special Issue Editors

Herbert Ho-Ching Iu
Abdelali El Aroudi

MDPI • Basel • Beijing • Wuhan • Barcelona • Belgrade

MDPI

Special Issue Editors
Herbert Ho-Ching Iu
The University of Western Australia
Australia

Abdelali El Aroudi
Universitat Rovira i Virgili
Spain

Editorial Office
MDPI
St. Alban-Anlage 66
4052 Basel, Switzerland

This is a reprint of articles from the Special Issue published online in the open access journal *Energies* (ISSN 1996-1073) from 2018 to 2019 (available at: https://www.mdpi.com/journal/energies/special_issues/energy_conversion)

For citation purposes, cite each article independently as indicated on the article page online and as indicated below:

LastName, A.A.; LastName, B.B.; LastName, C.C. Article Title. *Journal Name* **Year**, *Article Number*, Page Range.

ISBN 978-3-03921-110-4 (Pbk)
ISBN 978-3-03921-111-1 (PDF)

Contents

About the Special Issue Editors

Herbert Ho-Ching Iu received a B.Eng. (Hons) degree in electrical and electronic engineering from the University of Hong Kong, Hong Kong, in 1997. He received a Ph.D. degree in Electronic and Information Engineering from the Hong Kong Polytechnic University, Hong Kong, in 2000. In 2002, he joined the School of Electrical, Electronic and Computer Engineering of the University of Western Australia where he is currently a Professor. His research interests include power electronics, renewable energy, nonlinear dynamics, current sensing techniques, and memristive systems. He won the IET Power Electronics Premium Award, the IET Generation, Transmission and Distribution Premium Award, UWA Vice-Chancellor Mid-Career Research Award, and the IEEE PES Western Australian Chapter Outstanding Engineer Award in 2012, 2014, 2014 and 2015, respectively. He currently serves as Editor for IEEE Transactions on Smart Grids, Associate Editor for *IEEE Transactions on Power Electronics, IEEE Transactions on Network Science and Engineering, IEEE Transactions on Circuits and Systems-II and IEEE Access*.

Abdelali El Aroudi received a graduate degree in physical science from Faculté des sciences, Université Abdelmalek Essaadi, Tetouan, Morocco, in 1995, and a Ph.D. degree (hons) in applied physical science from Universitat Politécnica de Catalunya, Barcelona, Spain in 2000. During the period 1999–2001 he was a Visiting Professor at the Department of Electronics, Electrical Engineering and Automatic Control, Technical School of Universitat Rovira i Virgili (URV), Tarragona, Spain, where he became an Associate Professor in 2001 and a full-time tenure Associate Professor in 2005. His research interests are in the field of the structure and control of power conditioning systems for autonomous systems, power factor correction, renewable energy applications, stability problems, nonlinear phenomena, and bifurcation control. He was a Guest Editor of the *IEEE Journal on Emerging and Selected Topics on Circuits and Systems*, the Special Issue on the Design of Energy-Efficient Distributed Power Generation Systems (2015), Guest Editor of the *IEEE Transactions on Circuits and Systems II* (2018) and Guest Editor of *Energies* (2018, 2019). He currently serves as Associate Editor in *IET Power Electronics, IET Circuits, Systems* and *Devices and IET Electronics Letters*.

energies

MDPI

Article

A New Bridgeless High Step-up Voltage Gain PFC Converter with Reduced Conduction Losses and Low Voltage Stress

Xiang Lin [1], Faqiang Wang [1,2,*] and Herbert H. C. Iu [2]

1   State Key Laboratory of Electrical Insulation and Power Equipment, School of Electrical Engineering, Xi'an Jiaotong University, Xi'an 710049, China; linxiangjob@163.com
2   School of Electrical, Electronic and Computer Engineering, The University of Western Australia, Crawley W.A. 6009, Australia; herbert.iu@uwa.edu.au
*   Correspondence: faqwang@mail.xjtu.edu.cn; Tel.: +29-82668630-218

Received: 24 August 2018; Accepted: 1 October 2018; Published: 2 October 2018

Abstract: Bridgeless power factor correction (PFC) converters have a reduced number of semiconductors in the current flowing path, contributing to low conduction losses. In this paper, a new bridgeless high step-up voltage gain PFC converter is proposed, analyzed and validated for high voltage applications. Compared to its conventional counterpart, the input rectifier bridge in the proposed bridgeless PFC converter is completely eliminated. As a result, its conduction losses are reduced. Also, the current flowing through the power switches in the proposed bridgeless PFC converter is only half of the current flowing through the rectifier diodes in its conventional counterpart, therefore, the conduction losses can be further improved. Moreover, in the proposed bridgeless PFC converter, not only the voltage stress of power switches is lower than the output voltage, but the voltage stress of the output diodes is lower than the conventional counterpart. In addition, this proposed bridgeless PFC converter features a simple circuit structure and high PFC performance. Finally, the proposed bridgeless PFC converter is analyzed and designed in the discontinuous conduction mode (DCM). The simulation results are presented to verify the effectiveness of the proposed bridgeless PFC converter.

Keywords: bridgeless converter; discontinuous conduction mode (DCM); high step-up voltage gain; power factor correction (PFC)

---

1. Introduction

In the past decades, AC-DC converters have been widely used in numerous power electronic equipment supplied by the power grid in order to obtain the DC voltage. For the passive AC-DC rectifier, the input current harmonics are large, which is very harmful for the power grid and other power electronic equipment. In order to alleviate the input current harmonics and satisfy the rigorous input current harmonic standards, for instance, the IEC 61000-3-2 criterion, the active power factor correction (PFC) converter has become a popular and effective method to shape the input current waveform and achieve the near unity power factor (PF) in the power supplies. For single-phase power supplies, the boost topology is the most popular option as the PFC pre-regulator, by reason of its simple circuit structure and high PFC performance [1–3]. Unfortunately, the boost topology cannot achieve a very high voltage gain in practical applications, because the extremely high duty cycle is unpractical. Therefore, in some high voltage applications, for example, X-ray medical/industry equipment, HVDC system insulator testing, electrostatic precipitators and high voltage battery charger, the boost PFC converter is a poor candidate, especially for the universal line [4,5].

For outputting high voltage, many conventional high step-up voltage gain PFC converters have been studied in the past decade [6–16]. Based on the Cockcroft–Walton (CW) structure, some high step-up voltage gain PFC converters were proposed in [6–10]. In [6], a three-stage CW PFC converter was proposed. This converter can achieve a high output voltage and a high PFC performance. In [7], a transformerless hybrid boost and CW PFC converter was presented. By adding the CW voltage multiplier (VM) stages, high output voltage and high power factor are obtained. A single-phase single-stage high step-up matrix PFC converter using CW-VM was proposed in [8]. By combining a four bidirectional-switch matrix converter and the CW-VM, a high step-up voltage gain is achieved. Based on [6], a more comprehensive analysis and validation were presented in [9]. Based on [9], an improved high step-up voltage gain PFC converter with soft-switching characteristic was introduced in [10]. Besides the CW structure, some efforts focused on the switched-capacitor PFC topology to produce the high output voltage [11–13]. In [11], a family of high-voltage gain hybrid switched-capacitor PFC converters were proposed and validated, which can achieve a high output voltage and good PFC performance. A high voltage gain PFC converter based on a hybrid boost DC-DC converter was presented in [12]. By integrating boost topology and the switched-capacitor voltage doubler, a high output voltage and nearly unity PF are produced. In [13], a hybrid single ended primary inductor converter (SEPIC) PFC converter using switched-capacitor voltage doubler was proposed, which also owns a high voltage gain and a good PFC performance. Other new PFC converters can also achieve a high voltage gain [14–16]. In [14], a single-stage boost PFC converter with zero current switching (ZCS) characteristic was proposed and studied, which has a high voltage gain. In [15], a modified SEPIC PFC converter with a high voltage gain was proposed. A family of ZCS isolated high voltage gain PFC converters were proposed in [16]. Also, many high step-up voltage gain DC-DC converters have been studied in [17–24]. These DC-DC topologies can also be applied as the PFC converters for the high voltage applications. However, all the PFC converters, as mentioned above [6–24], are the conventional PFC type. The rectifier bridge is necessary for them, and their topology structures are more complex.

Compared to the conventional PFC converters, the bridgeless PFC converters possess the merits of low conduction losses and higher efficiency. That is because the input rectifier diodes of the bridgeless PFC converters are reduced, leading to a less number of semiconductors in the current-flowing path [2]. In order to improve efficiency, some bridgeless PFC converters with high output voltage are proposed in [25–27]. In [25], a bridgeless Cuk PFC converter was proposed for high voltage battery charger. In [26], a bridgeless modified SEPIC PFC converter was proposed with extended voltage gain. Two bridgeless hybrid boost PFC converters using the switched-capacitor structure were presented in [27]. All these bridgeless PFC converters can be applied for the high voltage applications, and their efficiency are improved compared to their conventional counterparts.

Based on the conventional high step-up voltage gain PFC converter shown in Figure 1, which was first proposed in [18] as the DC-DC converter, a new bridgeless high step-up voltage gain PFC converter with improved efficiency shown in Figure 2 is proposed for high voltage applications. By reducing the number of semiconductors in the current-flowing path and reducing the current stress of semiconductors, the proposed bridgeless PFC converter can achieve a reduced conduction losses and a higher efficiency compared to its conventional counterpart. Besides, the proposed bridgeless PFC converter owns a lower voltage stress of output diodes than the conventional one. The high PF and low total harmonic distortion (THD) are also obtained in the proposed bridgeless PFC converter. In addition, the proposed bridgeless PFC converter features a very simple circuit structure, contributing to cost and power density. The discontinuous conduction mode (DCM) is utilized with the merits of zero current turned on in the power switches, zero current turned off in the diodes, nature current-sharing ability and a simple control method. As a result, the proposed bridgeless PFC converter is more suitable for the high voltage applications than its conventional counterpart.

The operation principle of the proposed bridgeless PFC converter is discussed in Section 2. A detailed theoretical analysis and design guideline is presented in Section 3. The validation by the simulated results is shown in Section 4, followed by the conclusions in Section 5.

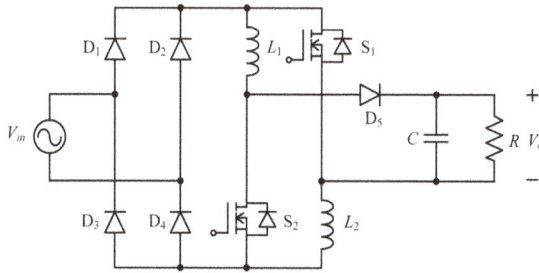

Figure 1. The conventional high step-up voltage gain PFC converter.

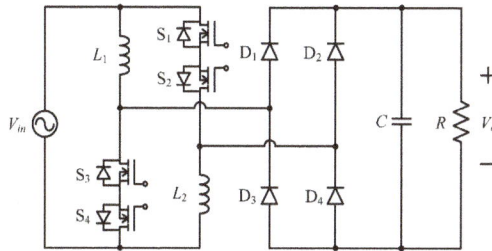

Figure 2. The proposed bridgeless high step-up voltage gain PFC converter.

2. Operation Principle

This proposed bridgeless high step-up voltage gain PFC converter uses two bidirectional switches in series with two same level inductors. Each bidirectional switch is constructed by two anti-series power switches. It should be noted that the two power switches in one bidirectional switch have the common source terminal, which can simplify the drive circuit. Simultaneously, an output bridge including D_1, D_2, D_3 and D_4 which are fast-recovery diodes is used to obtain a high DC output voltage in the proposed bridgeless PFC converter, while only D_5 is the fast-recovery diode in its conventional counterpart. The proposed bridgeless PFC converter is designed to operate in DCM. Thereby, it has three operation modes during one switching period. The detailed operation modes of one switching period in the positive line cycle are presented in Figure 3. Since the proposed bridgeless PFC converter is symmetrical, the operation modes in negative line cycle are similar to the modes in positive line cycle. Its key time-domain waveforms are exhibited in Figure 4.

Mode I shown in Figure 3a: when the power switches S_1 and S_3 are turned on, the input sinusoidal source v_{in} charges the two inductors L_1 and L_2, simultaneously, through the power switches S_2 and S_4. The output bulk capacitor maintains the output voltage v_o. In each branch of the proposed bridgeless PFC converter, only two semiconductors consisting by two power switches are active, while three semiconductors are active in one branch of the conventional counterpart. In this mode, the inductor currents satisfy:

$$v_{in} = L_1 \frac{di_{L1}}{dt} = L_2 \frac{di_{L2}}{dt} \tag{1}$$

Mode II shown in Figure 3b: when all the power switches are turned off, the input source and the two inductors releases energies to the load. Only two output fast-recovery diodes conduct in this mode, while three semiconductors including two slow-recovery diodes and one fast-recovery diode conduct in the corresponding conventional counterpart. In this mode, the inductor currents satisfy:

$$\frac{v_{in} - v_o}{2} = L_1 \frac{di_{L1}}{dt} = L_2 \frac{di_{L2}}{dt} \tag{2}$$

3

Mode III shown in Figure 3c: all the semiconductors are in the off state. The inductor currents are zero. The output bulk capacitor C maintains the output voltage.

Figure 4 presents the key waveforms of duty cycle D, inductor current i_{L1}, i_{L2}, input current i_{in}, and the voltage v_{S1}, v_{S3}, v_{D1}, v_{D4} across the semiconductors in the positive line cycle. From this figure, the inductor current i_{L1}, i_{L2} are equal to each other. When the power switches are turned on, the inductor current are half of the input current i_{in}. When the power switches are turned off, the inductor current are same with the input current. The maximum voltage across the power switches and the output diodes are $(v_{in} + v_o)/2$ in the positive line cycle. It should be noted that the duty cycle D equals to $(t_2 - t_1)/T_S$, where T_S is the switching period.

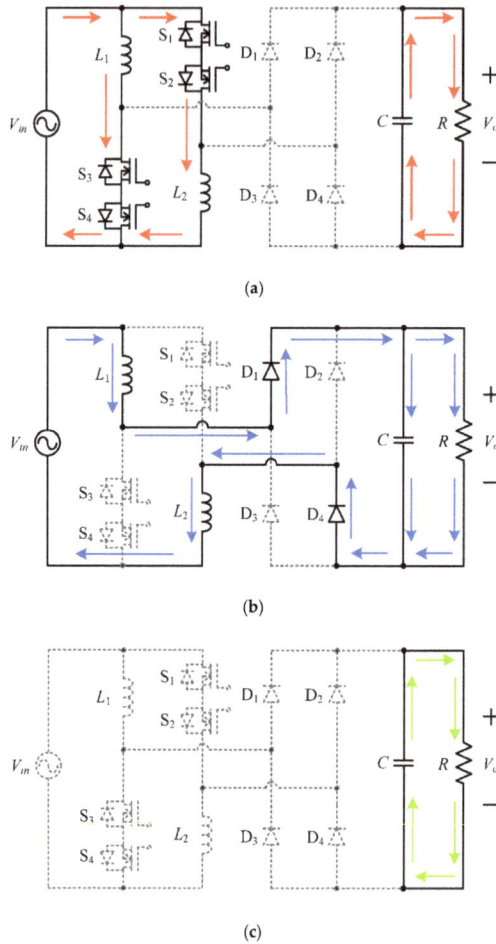

(a)

(b)

(c)

Figure 3. The operation modes of the proposed bridgeless high step-up voltage gain PFC converter in the positive line cycle: (**a**) mode I; (**b**) mode II and (**c**) mode III.

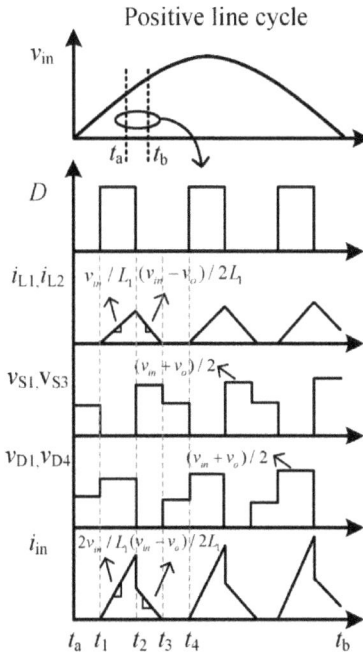

Figure 4. The key time-domain waveforms of the proposed bridgeless high step-up voltage gain PFC converter.

3. Theoretical Analysis

The detailed theoretical analysis and designed consideration in DCM are presented in this subsection. First of all, some ideal assumptions are provided to simplify the analysis. Notably, the theoretical analysis is made in one positive line cycle. These assumptions are shown as follows:

- The switching frequency f_s is much higher than the line frequency. Thus, the input voltage is constant during one switching period.
- The capacitance of the bulk capacitor is large enough. Thereby, the output voltage is ideal constant.
- All the components are ideal without losses.
- The input voltage is ideally sinusoidal.

3.1. The Voltage Conversion Ratio M

Appling the voltage-second balance principle to the inductor L_1, the voltage conversion ratio M is derived as follows:

$$M = \frac{v_o}{v_m} = \frac{2D + D_x}{D_x} \times \sin\theta \tag{3}$$

where v_m is the amplitude of the sinusoidal input voltage v_{in}, θ is the angle of the input voltage v_{in}, and D_x is equal to $(t_3 - t_2)/T_S$.

Based on (3), the relationship between the duty cycle D and D_x can be expressed as:

$$D_x = 2D \times \frac{\sin\theta}{M - \sin\theta} \tag{4}$$

In addition, the peak inductor current $i_{L1\text{-}peak}$ in one switching period is:

$$i_{L1-peak} = \frac{DT_S}{L_1} \times v_m \sin\theta \tag{5}$$

Due to the power balance between input power and output power, we can get:

$$\frac{1}{\pi}\int_0^\pi \frac{1}{2} \times i_{L1-peak} \times (2D+D_x) \times v_{in}d\theta = \frac{v_o^2}{R} \tag{6}$$

Substituting (4) and (5) into (6), the relationship of the voltage conversion ratio M and duty cycle D is derived as follows:

$$M = D \times \sqrt{\frac{\beta}{\pi K}} \tag{7}$$

where the dimensionless conduction parameter K is:

$$K = \frac{2L_1}{RT_S} \tag{8}$$

and the parameter β is

$$\beta = \int_0^\pi \left(\frac{2M}{M-\sin\theta}\right) \times \sin^2\theta d\theta \tag{9}$$

The relationship of the voltage conversion ratio M and duty cycle D is presented in Figure 5. From this figure, one can see that the voltage conversion ratio M increases with the lower parameter K. Compared to the conventional boost PFC converter, the voltage conversion ratio M of the proposed bridgeless PFC converter is much higher. Therefore, the proposed bridgeless PFC converter is more suitable for the high voltage applications.

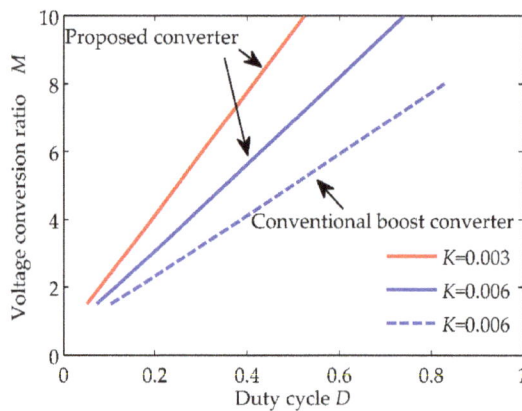

Figure 5. The relationship of the voltage conversion ratio M and duty cycle D.

3.2. The Operation Conditon for DCM

In order to operate in DCM, the operation condition must satisfy as follows:

$$D + D_x < 1 \tag{10}$$

Substituting (3) and (7) into (10), the operation condition for DCM is derived as:

$$K < \frac{\beta}{\pi} \times \frac{1}{M^2} \times \left(\frac{M-\sin\theta}{M+\sin\theta}\right)^2 \tag{11}$$

The proposed bridgeless PFC converter is designed to operate in DCM totally. Therefore, the inductor currents should be discontinuous at the peak point in the line cycle. Thus, the simplified operation condition for DCM is:

$$K < \frac{\beta}{\pi} \times \frac{1}{M^2} \times \left(\frac{M-1}{M+1} \right)^2 \tag{12}$$

Figure 6 draws the operation boundary between the DCM and the continuous conduction mode (CCM). From this figure, the operation boundary is higher at the low voltage conversion ratio. However, for the universal line, the voltage conversion ratio is different under different input voltage. Hence, the key parameter K must be designed at the lowest input voltage.

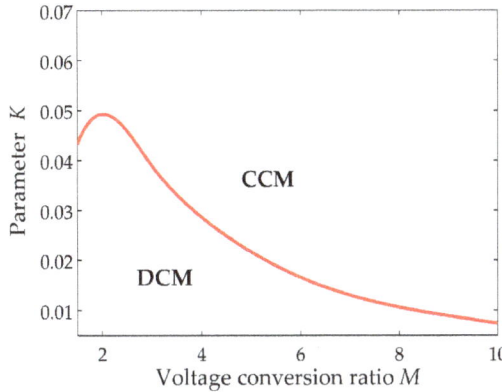

Figure 6. The operation boundary between DCM and CCM.

3.3. The Voltage Stress and Current Stress

The voltage stress of semiconductors in the proposed bridgeless PFC converter and in its conventional bridge counterpart are shown in Table 1. From this table, the voltage stress of power switch in the proposed bridgeless PFC converter is same with its conventional bridge converter, and it is lower than the output voltage. The voltage stress of fast-recovery diode in the proposed bridgeless PFC converter is lower than that in the conventional bridge converter. Therefore, the lower rated diode can be used in the proposed bridgeless PFC converter. It is beneficial to improve cost and losses. In addition, no slow-recovery diode is used in the proposed bridgeless PFC converter, while four slow-recovery diodes as the input bridge are used in its conventional bridge counterpart, and their voltage stress is v_m.

Table 1. The voltage stress of semiconductors.

	Proposed Bridgeless PFC Converter	Conventional Bridge PFC Converter
Power switch	$(v_m + v_0)/2$	$(v_m + v_0)/2$
Fast-recovery diode	v_0	$v_m + v_0$
Slow-recovery diode	-	v_m

The root-mean-square (RMS) current i_{S1-rms} of power switch in one switching period is shown as follows:

$$i_{S1-rms} = \frac{v_{in}DT_S}{L_1} \sqrt{\frac{D}{3}} \tag{13}$$

The averaged current $i_{D1\text{-}avg}$ of output diode in one switching period is derived as follows:

$$i_{D1-avg} = \frac{v_{in}DD_xT_S}{2L_1} \qquad (14)$$

3.4. The Conduction Losses

In this subsection, the conduction losses of semiconductors are calculated. The detail derivations in one positive line cycle are exhibited as follows:

$$P_{S1} = \frac{1}{\pi}\int_0^\pi \left(\frac{v_{in}DT_S}{L_1}\right)^2 \times \frac{D}{3} \times R_{on}d\theta \qquad (15)$$

$$P_{D1} = \frac{1}{\pi}\int_0^\pi \frac{v_{in}DD_xT_S}{2L_1} \times V_F d\theta \qquad (16)$$

where R_{on} is the conduction resistance of the power switch and V_F is the forward voltage of diodes.

Under the operation condition v_{in} = 220 V_{rms}/50 Hz, v_o = 800 V, f_s = 30 kHz and P_o = 500 W, the conduction losses of semiconductors are calculated. It should be noted that the parameters R_{on} and V_F are chosen from the datasheet of the selected components. The conduction losses of semiconductors of the proposed bridgeless PFC converter and its conventional counterpart are presented in Figure 7. From this figure, it can be found that the total conduction losses of semiconductors in the proposed bridgeless PFC converter is much lower than its conventional bridge counterpart. The conduction losses of power switches in the proposed bridgeless PFC converter are higher, while it has no conduction losses of input rectifier diodes.

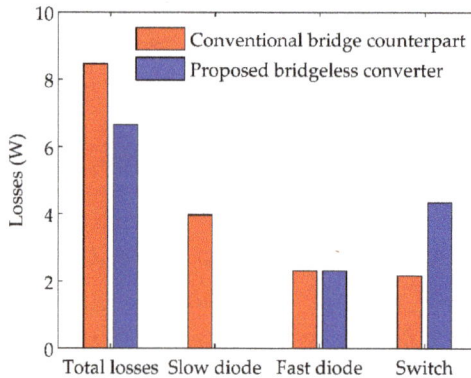

Figure 7. The calculated conduction losses of semiconductors.

3.5. The Control Principle

This proposed bridgeless PFC converter is designed in DCM. The DCM possesses the merit of a naturally current-sharping ability, which contributes to a simple control method. Thereby, the voltage control loop is applied in order to obtain the constant DC output voltage. The control principle is displayed in Figure 8. From this figure, the controller mainly contains one compensator, one PWM generator and four drivers. It should be noted that the four power switches in the proposed bridgeless PFC converter can be driven by one same control signal, which simplifies the controller, significantly. Notably, the signal V_{g1}, V_{g2}, V_{g3} and V_{g4} drive the power switches S_1, S_2, S_3 and S_4, respectively.

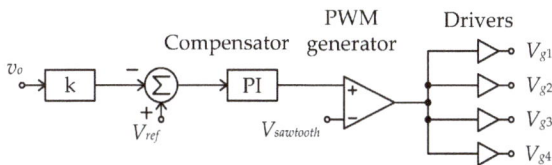

Figure 8. The control diagram of the proposed bridgeless high step-up voltage gain PFC converter.

4. Simulation Results

The effectiveness of the proposed bridgeless PFC converter is validated in the SIMetrix/SIMPLIS (version 8.00, company SIMetrix Technologies Ltd., Thatcham, UK) environment. The simulation program with integrated circuit emphasis (SPICE) models of practical components are employed in this simulation. The key operation parameters of the proposed bridgeless PFC converter is v_{in} = universal line 95–265 V_{rms}, v_o = 800 V, f_s = 30 kHz and P_o = 500 W. The selected components are shown in Table 2. Considering the voltage stress, current stress and safety margin, the SPP17N80C3 (company Infineon, GER) with R_{on} = 0.29 Ω and V_{DS} = 800 V is chosen as the power switches. The MUR490 (company On Semiconductor, Phoenix, AZ, USA) with V_F = 1.85 V and V_D = 900 V is chosen as the fast-recovery diodes in the proposed bridgeless PFC converter. Since the voltage stress of the fast-recovery diode in the conventional bridge counterpart is up to around 1200 V, which is much larger than the voltage stress 800 V of the fast-recovery diode in the proposed bridgeless PFC converter, we have to choose two series MUR490 as the fast-recovery diode in the conventional bridge counterpart. In the conventional bridge converter, 8EWS08 (company International Rectifier, El Segundo, CA, USA) with V_F = 1 V is used as the input rectifier diodes.

Table 2. The selected components.

	Proposed Bridgeless PFC Converter	**Conventional Bridge PFC Converter**
Power switches	SPP17N80C3	SPP17N80C3
Fast-recovery diodes	MUR490	MUR490
Slow-recovery diodes	—	8EWS08
Output capacitor	200μF	200μF
Inductors	200μH	200μH

The input current after the input LC filter at the typical input line is displayed in Figure 9. From this figure, the input current is shaped to be almost sinusoidal at the typical low line 110 V_{rms} and the typical high line 220 V_{rms}. Thereby, it is validated that the proposed bridgeless PFC converter owns a good current-shaping ability. Figure 10 presents the key time-domain waveforms of the proposed bridgeless PFC converter. It can be figure out that the simulated waveforms are in agreement with the theoretical analysis. The key waveforms also validate that the proposed bridgeless PFC converter operates in DCM.

Figure 11 presents the simulated PF and THD under the universal line. From this figure, one can see that nearly unity PF is achieved and the THD is low under the universal line. The high PF and low THD validate that the proposed bridgeless PFC converter owns a good PFC performance.

The simulated efficiency of the proposed bridgeless PFC converter and its conventional bridge counterpart under the universal line is shown in Figure 12. From this figure, it is clear that the efficiency of the proposed bridgeless PFC converter is higher than its conventional bridge counterpart, due to the reduced semiconductors and the reduced current. Also, the efficiency of other state of the art high step-up voltage gain converter in [12] is simulated. Under the same operation parameters and components, the efficiency of the converter in [12] is 97.42% at the typical line V_{in} = 220 V_{rms}, while the efficiency of the proposed bridgeless PFC converter can reach up to 98.78% at the typical line V_{in} = 220 V_{rms}. Therefore, the proposed bridgeless PFC converter is more suitable for the practical application.

Figure 13 displays the simulated input current harmonics compared with the IEC 61000-3-2 class D limits. From this figure, the input current harmonics of the proposed bridgeless PFC converter are much lower than the IEC 61000-3-2 class D limits under both the typical low line and high line. Namely, the proposed bridgeless PFC converter can easily satisfy the international harmonic standards, which is very beneficial to practical application.

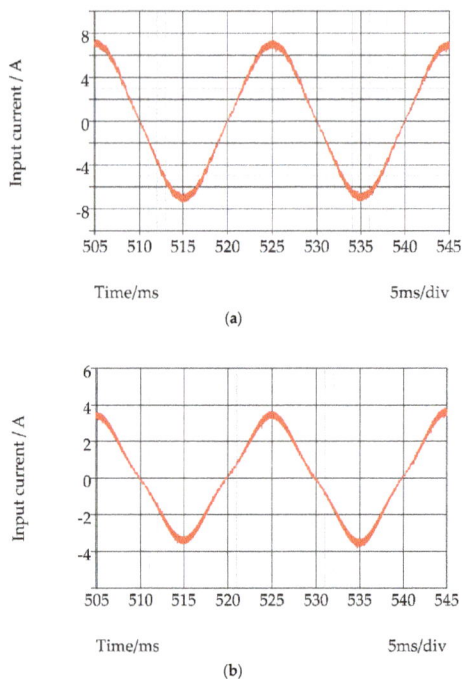

(a)

(b)

Figure 9. The input current waveforms after the input LC filter: (a) v_{in} = 110 V$_{rms}$; (b) v_{in} = 220 V$_{rms}$.

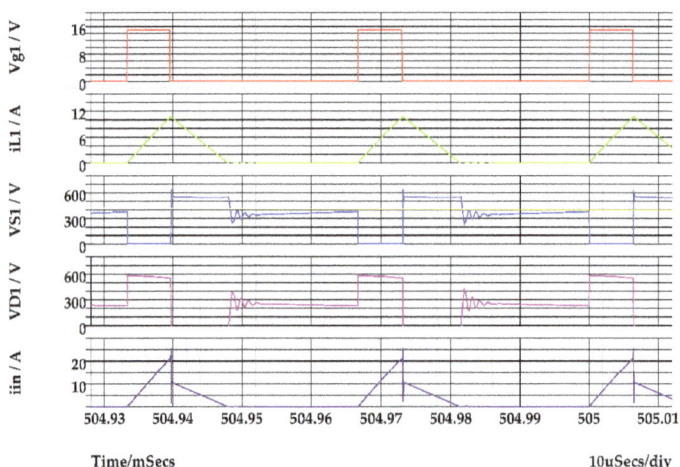

Figure 10. The key time-domain waveforms at v_{in} = 220 V$_{rms}$.

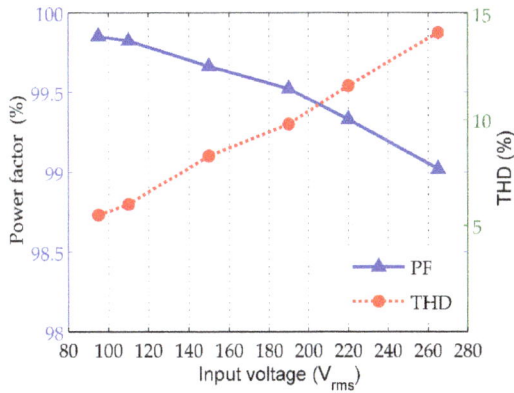

Figure 11. The simulated PF and THD of the proposed bridgeless high step-up gain PFC converter.

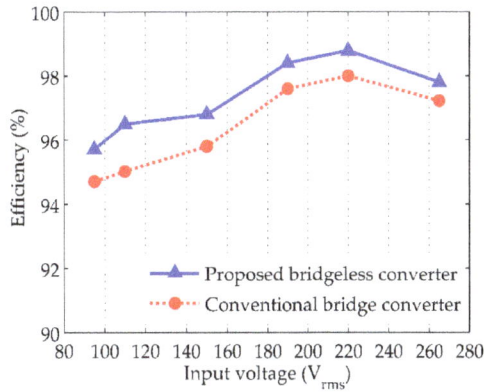

Figure 12. The simulated efficiency of the proposed bridgeless high step-up voltage gain PFC converter and its conventional bridge counterpart.

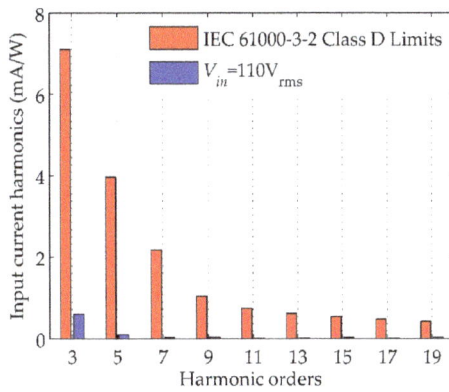

(a)

Figure 13. *Cont.*

11

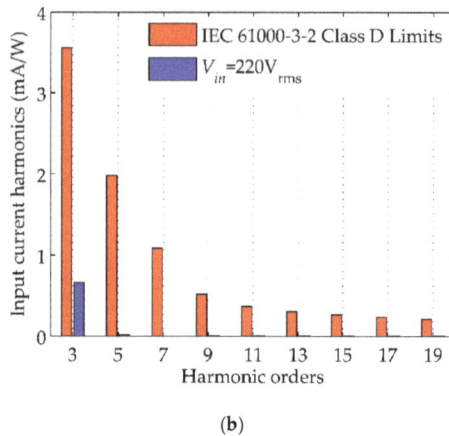

(b)

Figure 13. The simulated input current harmonics of the proposed bridgeless high step-up voltage gain PFC converter compared with the IEC 61000-3-2 class D limits: (**a**) v_{in} = 110 V$_{rms}$; (**b**) v_{in} = 220 V$_{rms}$.

5. Conclusions

A new bridgeless high step-up voltage gain PFC converter with low conduction losses and low voltage stresses for high voltage applications is proposed, analyzed and verified in this paper. The theoretical analysis and design consideration in DCM are presented. The simulated results validate that the proposed bridgeless PFC converter has a higher efficiency than its conventional bridge counterpart. Moreover, the proposed bridgeless PFC converter can achieve a very high PF and low THD, and it can easily satisfy the IEC 61000-3-2 class D limits, thereby, the proposed bridgeless PFC converter is a competitive option for the high voltage applications.

Author Contributions: X.L. conceived, validated and wrote the manuscript. F.W. and H.H.C.I. participated in the research plan development and revised the manuscript. All authors contributed to the manuscript.

Funding: This work was supported in part by the National Natural Science Foundation of China under Grant 51377124 and Grant 51521065, in part by the New Star of Youth Science and Technology of Shaanxi Province under Grant 2016KJXX-40, and in part by the China Scholarship Council under Grant 201706285022.

Conflicts of Interest: The authors declare no conflict of interest.

References

1. Salmon, J.C. Circuit topologies for single-phase voltage-doubler boost rectifiers. *IEEE Trans. Power Electron.* **1993**, *8*, 521–529. [CrossRef]
2. Huber, L.; Jang, Y.; Jovanović, M.M. Performance evaluation of bridgeless PFC boost rectifiers. *IEEE Trans. Power Electron.* **2008**, *23*, 1381–1390. [CrossRef]
3. Muhammad, K.S.B.; Lu, D.D. ZCS bridgeless boost PFC rectifier using only two active switches. *IEEE Trans. Ind. Electron.* **2015**, *62*, 2795–2806. [CrossRef]
4. Bellar, M.D.; Watanabe, E.H.; Mesquita, A.C. Analysis of the dynamic and steady-state performance of Cockcroft-Walton cascade rectifiers. *IEEE Trans. Power Electron.* **1992**, *7*, 526–534. [CrossRef]
5. Momayyezan, M.; Hredzak, B.; Agelidis, V.G. Integrated reconfigurable converter topology for high-voltage battery systems. *IEEE Trans. Power Electron.* **2016**, *31*, 1968–1979. [CrossRef]
6. Young, C.M.; Chen, M.H. A novel single-phase AC to high voltage DC converter based on Cockcroft-Walton cascade rectifier. In Proceedings of the IEEE 2009 International Conference on Power Electronics and Drive Systems, Taipei, Taiwan, 2–5 November 2009; pp. 822–826.

7. Young, C.M.; Chen, M.H.; Chen, H.L.; Chao, J.Y.; Ko, C.C. Transformerless single-stage high step-up AC-DC converter based on symmetrical Cockcroft-Walton voltage multiplier with PFC. In Proceedings of the IEEE Ninth International Conference on Power Electronics and Drive Systems, Singapore, 5–8 December 2011; pp. 191–196.

8. Young, C.M.; Chen, M.H.; Yeh, S.H.; Yuo, K.H. A single-phase single-stage high step-up AC-DC matrix converter based on Cockcroft-Walton voltage multiplier with PFC. *IEEE Trans. Power Electron.* **2012**, *27*, 4894–4905. [CrossRef]

9. Young, C.M.; Chen, M.H.; Ko, C.C. High power factor transformerless single-stage single-phase ac to high-voltage dc converter with voltage multiplier. *IET Power Electron.* **2012**, *5*, 149–157. [CrossRef]

10. Young, C.M.; Wu, S.F.; Chen, M.H.; Chen, S.J. Single-phase AC to high-voltage DC converter with soft-switching and diode-capacitor voltage multiplier. *IET Power Electron.* **2014**, *7*, 1704–1713. [CrossRef]

11. Cortez, D.F.; Barbi, I. A family of high-voltage gain single-phase hybrid switched-capacitor PFC rectifiers. *IEEE Trans. Power Electron.* **2015**, *30*, 4189–4198. [CrossRef]

12. Cortez, D.F.; Maccarini, M.C.; Mussa, S.A.; Barbi, I. High static gain single-phase PFC based on a hybrid boost converter. *Int. J. Electron.* **2017**, *104*, 821–839. [CrossRef]

13. Costa, P.J.S.; Font, C.H.I.; Lazzarin, T.B. Single-phase hybrid switched-capacitor voltage-doubler SEPIC PFC rectifiers. *IEEE Trans. Power Electron.* **2018**, *33*, 5118–5130. [CrossRef]

14. Chen, J.F.; Chen, R.Y.; Liang, T.J. Study and implementation of a single-stage current-fed boost PFC converter with ZCS for high voltage applications. *IEEE Trans. Power Electron.* **2008**, *23*, 379–386. [CrossRef]

15. Melo, P.F.D.; Gules, R.; Romaneli, E.F.R.; Annunziato, R.C. A modified SEPIC converter for high-power-factor rectifier and universal input voltage applications. *IEEE Trans. Power Electron.* **2010**, *25*, 310–321. [CrossRef]

16. Li, C.; Xu, D. Family of enhanced ZCS single-stage single-phase isolated AC-DC converter for high-power high-voltage DC supply. *IEEE Trans. Ind. Electron.* **2017**, *64*, 3629–3639. [CrossRef]

17. Wai, R.J.; Lin, C.Y.; Duan, R.Y.; Chang, Y.R. High efficiency DC-DC converter with high voltage gain and reduced switch stress. *IEEE Trans. Ind. Electron.* **2007**, *54*, 354–364. [CrossRef]

18. Yang, L.S.; Liang, T.J.; Chen, J.F. Transformerless DC-DC converters with high step-up voltage gain. *IEEE Trans. Ind. Electron.* **2009**, *56*, 3144–3152. [CrossRef]

19. Qian, W.; Cao, D.; Cintron-Rivera, J.G.; Gebben, M.; Wey, D.; Peng, F.Z. A swithed-capacitor DC-DC converter with high voltage gain and reduced component rating and count. *IEEE Trans. Ind. Appl.* **2012**, *48*, 1397–1406. [CrossRef]

20. Tofoli, F.L.; Oliveira, D.S.; Torrico-Bascope, R.P.; Alcazar, Y.J.A. Novel nonisolated high-voltage gain DC-DC converters based on 3SSC and VMC. *IEEE Trans. Power Electron.* **2012**, *27*, 3897–3907. [CrossRef]

21. Hu, X.; Gong, C. A high voltage gain DC-DC converter integrating coupled-inductor and diode-capacitor techniques. *IEEE Trans. Power Electron.* **2014**, *29*, 789–800. [CrossRef]

22. Silveira, G.C.; Tofoli, F.L.; Bezerra, L.D.S.; Torrico-Bascope, R.P. A nonisolated DC-DC boost converter with high voltage gain and balanced output voltage. *IEEE Trans. Ind. Electron.* **2014**, *61*, 6739–6746. [CrossRef]

23. Cao, Y.; Samavatian, V.; Kaskani, K.; Eshraghi, H. A novel nonisolated ultra-high-voltage-gain DC-DC converter with low voltage stress. *IEEE Trans. Ind. Electron.* **2017**, *64*, 2809–2819. [CrossRef]

24. Siwakoti, Y.P.; Blaabjerg, F. Single switch nonisolated ultra-step-up DC-DC converter with an integrated coupled inductor for high boost applications. *IEEE Trans. Power Electron.* **2017**, *32*, 8544–8558. [CrossRef]

25. Al-Kaabi, A.K.; Fardoun, A.A.; Ismail, E.H. Bridgeless high voltage battery charger PFC rectifier. *Renew. Energy* **2013**, *56*, 24–31. [CrossRef]

26. Gabri, A.M.A.; Fardoun, A.A.; Ismail, E.H. Bridgeless PFC-modified SEPIC rectifier with extended gain for universal input voltage applications. *IEEE Trans. Power Electron.* **2015**, *30*. [CrossRef]

27. Dias, J.C.; Lazzarin, T.B. A family of voltage-multiplier unidirectional single-phase hybrid boost PFC rectifiers. *IEEE Trans. Ind. Electron.* **2018**, *65*, 232–241. [CrossRef]

energies

MDPI

Article

Analysis of Nonlinear Dynamics of a Quadratic Boost Converter Used for Maximum Power Point Tracking in a Grid-Interlinked PV System

Abdelali El Aroudi [1,*], Mohamed Al-Numay [2], Germain Garcia [3], Khalifa Al Hossani [4], Naji Al Sayari [4] and Angel Cid-Pastor [1]

[1]   Departament d Enginyeria Electrònica, Universitat Rovira i Virgili, Elèctrica i Automàtica, Av. Paisos Catalans, No. 26, 43007 Tarragona, Spain; angel.cid@urv.cat
[2]   Department of Electrical Engineering, College of Engineering, King Saud University, P.O. Box 800, Riyadh 11421, Saudi Arabia; alnumay@ksu.edu.sa
[3]   Laboratoire d'Analuse et Architecture des Systèmes, Centre Nationale de Recherche Scientifique (LAAS-CNRS), Institut National des Sciences Appliquées (INSA), 7 Avenue du Colonel Roche, 31077 Toulouse, France; garcia@laas.fr
[4]   Department of Electrical and Computer Engineering, Khalifa University of Science and Technology, Abu Dhabi, UAE; khalhosani@pi.AC.ae (K.A.H.); nalsayari@pi.AC.ae (N.A.S.)
*   Correspondence: abdelali.elaroudi@urv.cat; Tel.: +34-977558522

Received: 30 September 2018; Accepted: 14 December 2018; Published: 25 December 2018

Abstract: In this paper, the nonlinear dynamics of a PV-fed high-voltage-gain single-switch quadratic boost converter loaded by a grid-interlinked DC-AC inverter is explored in its parameter space. The control of the input port of the converter is designed using a resistive control approach ensuring stability at the slow time-scale. However, time-domain simulations, performed on a full-order circuit-level switched model implemented in PSIM© software, show that at relatively high irradiance levels, the system may exhibit undesired subharmonic instabilities at the fast time-scale. A model of the system is derived, and a closed-form expression is used for locating the subharmonic instability boundary in terms of parameters of different nature. The theoretical results are in remarkable agreement with the numerical simulations and experimental measurements using a laboratory prototype. The modeling method proposed and the results obtained can help in guiding the design of power conditioning converters for solar PV systems, as well as other similar structures for energy conversion systems.

Keywords: DC-DC converters; quadratic boost; maximum power point tracking (MPPT); nonlinear dynamics; subharmonic oscillations; photovoltaic (PV)

---

1. Introduction

Electrical power grids feature many changes in their paradigm since they are no longer based only on coal-fired power stations [1]. The production of electrical energy in many countries is also based on renewable energy resources such as solar photovoltaic (PV) arrays, wind turbines, and batteries, forming nano-and micro-grids [1]. In particular, solar PV technology is considered as one of the most environmentally-friendly energy sources since it generates electricity with almost zero emissions while requiring low maintenance efforts. Despite the relatively high cost, the reduced number of installed capacities, the damaging effect of the temperature on their efficiency, as well as the need for cooling techniques [2], PV modules remain the most important renewable energy sources that can meet the power requirements of residential applications. This explains the increasing demand of PV array installation in homes and small companies in both grid-connected and in stand-alone operation modes.

PV modules are nonlinear energy sources with a maximum power point (MPP) voltage ranging from 15 V–40 V. Hence, a major challenge that needs to be addressed, if string-connected modules are to be avoided, is to take the low voltage at the output of the PV source and convert it into a much higher voltage level such as the standard 380 V DC-link voltage. This requires a DC-DC converter with a high-voltage-gain as a power interface between the PV source and the DC-AC inverter. The conventional canonical boost converter cannot be used in this case because the maximum conversion gain that can attain this converter is limited by parasitic resistances in the switching devices and the reactive components [3]. Typically, to deal with this problem, several PV modules are connected in series to obtain a sufficiently high voltage at the input of the DC-DC converter, hence not requiring an extremely high value of the duty cycle. However, series connection of PV modules has the inconvenient of undertaking shadowing effects that reduce the power production [4]. To overcome this drawback, module integrated converters (MICs) featuring distributed maximum power point tracking (MPPT) are used [5]. Such a PV system composed of a PV source with a DC-DC power electronics converter loaded by a DC-AC inverter is called a microinverter [6].

Because of the independent operation of each PV module in the microinverter approach, this has other advantages such as modularity, increased reliability, long life-time and better efficiency. In the microinverter or in the MIC approach, DC-DC converters with a high voltage conversion ratio are used as a first stage to perform the maximum power extraction.

MIC converters in a DC microgrid can be connected to the common DC-link voltage (DC bus) through the output of the *m* different branches, each one consisting of a PV module connected to a high-voltage-gain DC-DC converter, as depicted in Figure 1. A back-up storage battery is also connected to the main DC bus through a bidirectional DC-DC converter. In a real application, the number of branches in Figure 1 will be fixed according to the rated power. In microinverter applications, a number between two and twelve branches can be used, the rated power being between 170 W and 1 kW approximately. In some PV applications, a high-voltage-gain of about twenty is needed in each branch. This is the case of converting the voltage of a single PV module of about 18 V to the standard voltage of a DC bus of 380 V. The conventional canonical boost converter cannot be used for this kind of applications since, due to the losses, this converter cannot provide a voltage conversion gain higher than six.

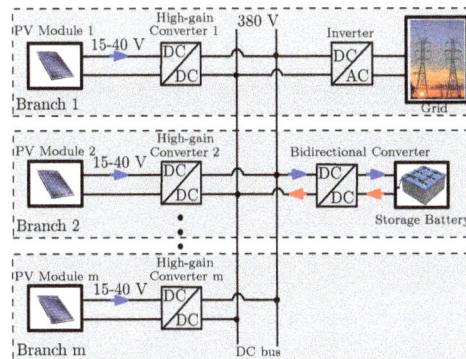

Figure 1. A model of a PV-based DC microgrid equipped with high-voltage-gain MICs.

The quadratic boost converter is an interesting topology for this kind of applications because it is a transformer-less circuit using only one active switch [7]. Its conversion ratio is ideally a quadratic function of the duty cycle allowing a larger gain than the conventional boost converter. Therefore, it could be a low cost and efficient solution capable of achieving a high-voltage-gain with a relatively low control complexity [8]. Recently, this topology has attracted the interest of many researchers in different power electronics applications such as in power factor correction [9], in fuel cell energy processing [10], in PV systems [11], [12] and in DC microgrids [13].

The quadratic boost converter is a high-order nonlinear and complex system with a large number of parameters. The optimization of its performances in terms of these parameters requires accurate models to be used, in particular when subharmonic oscillation is of concern. The design of the controller of the DC stage in a PV system is accomplished based on a linearized model in a suitable operating point. However, this operating point is constantly changing in a PV system, and the design of the controller is usually performed based on the lowest irradiance level [14]. Nevertheless, this approach does not take into account the possibility of subharmonic oscillation, which takes place precisely for high levels of irradiance as will be shown later in this paper.

Recently, much effort has been devoted to the study of nonlinear behavior such as subharmonic oscillation and other complex phenomena [15], [16] and is still attracting the interest of researchers even for simple converter topologies such as the buck converter [17] and the boost converter [18] with ideal constant input voltage and resistive load. In PV applications of switched mode power converters, the PV source is nonlinear and the output voltage is either controlled by the DC-AC inverter or fixed by a storage element such as a battery. The control objectives and functionalities of the DC side are also different since MPPT is usually performed at the input port [19]. As a consequence, all the well known features of DC-DC converters with constant voltage source, resistive load and under output voltage control are no more valid in the case of a DC-DC converter used in a PV system. For instance, it is well known that boost and boost-derived topologies are non-minimum phase systems when the controlled variable is the output voltage. This is not the case for the same converters with the input voltage as a control variable.

So far, the results concerning nonlinear dynamics in general and subharmonic oscillation in particular, in switching converters when supplied by nonlinear source, are sparse and limited. For instance, nonlinear dynamics was explored in [20,21] for a boost converter for PV applications. In [20], the nonlinear dynamics of a boost converter supplied from a PV source and loaded by a resistive load was investigated. In [21], a current-mode controlled boost DC-DC converter charging a battery from a PV panel was considered, and its dynamics was analyzed using the switched model of the converter and the nonlinear model of the PV generator.

The design of DC-DC switching power electronics converters in PV applications still requires a comprehensive knowledge about suitable ways of their accurate modeling and stability analysis, particularly, in the presence of parametric variations, nonlinear energy sources and loads. To accurately predict the dynamic behavior of a switching converter, appropriate modeling approaches, taking into account the switching action, must be used. Usually, the prediction of subharmonic instability has been addressed numerically by discrete time-modeling [15,16] or Floquet theory [22].

The relevant performance metrics for any power converter used in PV systems include MPPT, fast transient response under the constantly varying voltage/current reference due to the MPPT and low sensitivity to load and other parameter disturbances. The success in achieving these metrics can only be guaranteed by avoiding all kind of instability. In particular, subharmonic oscillation has many jeopardizing effects on the performances of the power converter such as increased ripple in the state variables and stresses in the switching devices and it could even make a PV system to operate out of the MPP [23]. Therefore, in this particular application, it is very important to dispose of accurate mathematical tools to predict this phenomenon.

The determination of critical system parameters for stable operation of switching converters in PV applications has had a growing interest recently [24,25]. Most of past works focused on low frequency (slow time-scale) behavior of these systems based on their averaged models. The slow time-scale instability problems can be avoided by using a Loss-Free-Resistor (LFR) [26] approach also known as resistive control [24]. However, although the low frequency instability could be guaranteed with this control, subharmonic oscillation may still occur.

The main purpose of the present paper is to present a methodology which is applicable to any single-switch converter topology either in PV systems or in other similar applications where nonlinearities can take place either in the energy source or in any other system parameter. The main contributions of this study are:

- Development of a methodology to accurately predict subharmonic oscillation in switching converters used for MPPT for PV applications considering the nonlinearity of the PV energy source and the saturability of the inductors.
- Analytical and experimental determination of subharmonic oscillation boundaries in terms of relevant system parameters of different nature.

The remainder of this paper is organized as follows: In Section 2, the system description and its modeling are presented. The controller design the DC-DC quadratic boost converter when used for MPPT is described in Section 3. A closed-loop state-space switched model of the system is presented in Section 4. Using numerical simulations from the detailed and complete switched model including the PV-fed DC-DC quadratic converter, a DC-AC H-bridge inverter and an extremum seeking MPPT controller, it is shown in Section 5 that the system may exhibit complex nonlinear phenomena in the form of subharmonic oscillation when the irradiance level increases. In Section 6, a stability analysis is performed and the observed phenomenon is studied in the light of Floquet theory. In the same section, an analytical expression for accurately locating the boundary of this phenomenon is presented. In Section 7, results obtained from this mathematical expression are validated by numerical computer simulations and experimental measurements. Finally, concluding remarks of this study are given in the last section.

2. System Description and its Mathematical Modeling

2.1. Operation Principle

The schematic diagram of a DC-DC quadratic boost converter fed by a PV generator and loaded by a DC-AC grid-connected inverter is shown in Figure 2. In this kind of applications, the input voltage is controlled using the switch of the DC-DC stage [27–29] while the output DC-link voltage is regulated by acting on the switches of the DC-AC inverter. As the solar irradiation S or the temperature Θ change during the operation, the voltage/current of the PV module is adjusted to correspond to the maximum available power. Here, the input port of the DC-DC side is controlled using a resistive control approach for the quadratic boost converter defining the appropriate conductance to match the MPP. This approach is known in the literature as Loss-Free-Resistor (LFR) [26] and it makes the controlled port of the converter to behave like a virtual resistance in average. To achieve this, the reference i_{ref} for the input current is generated proportionally to the input voltage v_{pv}, i.e, $i_{ref} = G_{mpp}v_{pv}$. The proportionality factor $g^* = G_{mpp}$ is a conductance provided by an MPPT controller. The error between the inductor current and the generated reference is controlled by type-II average controller in such a way that the inductor current tightly tracks its reference hence imposing the LFR behavior. The activation of the switch S is carried out as follows: the output v_{con} of the type-II controller is connected to the inverting pin of the comparator whereas a sawtooth signal $v_{ramp} = V_M(t/T) \bmod 1$ is applied to the non inverting pin. The output of the comparator is applied to the reset input of a set-reset (SR) latch and a periodic clock signal is connected to its set input in such a way that the switch S is ON at the beginning of each switching cycle and is turned OFF whenever $v_{con} = v_{ramp}$. The state of the diodes D_1 and D_3 are complementary to that of the switch S while that of D_2 is the same as that of S.

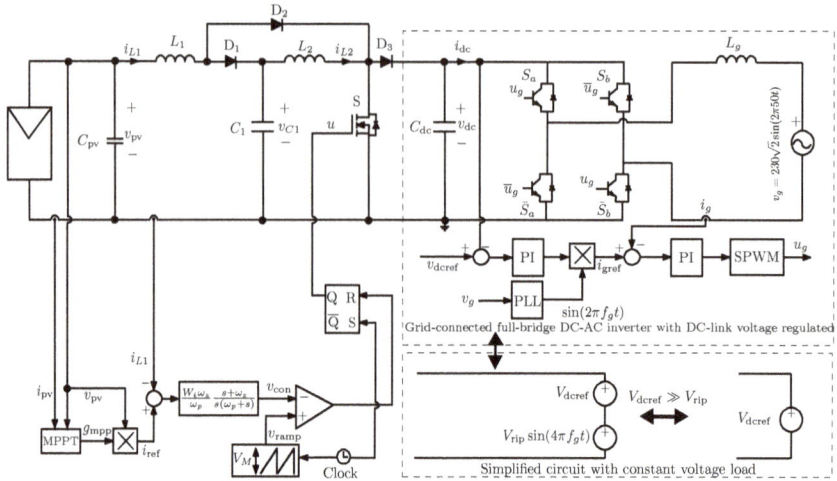

Figure 2. Two-stage grid connected PV system with a quadratic boost converter in the DC-DC stage.

Remark 1. *For making the steady-state conductance of the input-stage to match the one corresponding to the MPP, the inductor current has been used instead of the PV current in the synthesis of the LFR. This is because in steady-state, their average values are identical. However, from stability and performance point of view, it is better to use the inductor current which contains both the PV current and the capacitor current. The latter introduces suitable damping and speed-up the system response as detailed in [28].*

2.2. The Nonlinear Model of a PV Generator

The PV generators have a nonlinear characteristic changing with the temperature Θ and irradiation S. Their $i - v$ characteristic equation can be found in many references in the literature. A comparison between the different models are presented in [30]. The single diode model, shown in Figure 3, is one of the most widely used since it has a good compromise between simplicity and accuracy. The equation of this model can be written as follows [31]:

$$
i_{pv} = I_{pv} - I_s \left(e^{\frac{v_{pv} + R_s i_{pv}}{AV_t}} - 1 \right) - \frac{v_{pv} + R_s i_{pv}}{R_p},
\tag{1}
$$

where i_{pv} and v_{pv} are, respectively, the current and voltage of the PV module, I_{pv} and I_s are the photogenerated and saturation currents respectively, $V_t = N_s K\theta/q$ is the thermal voltage, A is the diode ideality constant, K is Boltzmann constant, q is the charge of the electron, Θ is the PV module temperature and N_s is the number of the series-connected cells. The photogenerated current I_{pv} depends on the irradiance S and temperature Θ according to the following equation:

$$
I_{pv} = I_{sc}\frac{S}{S_n} + C_\Theta(\Theta - \Theta_n),
\tag{2}
$$

where I_{sc} is the short circuit current, Θ_n and S_n are the nominal temperature and irradiance respectively and C_Θ is the temperature coefficient. Practical PV generators have a series resistance R_s and a parallel resistance R_p. These parameters can be ignored for simplicity.

2.3. The PV Generator Model Close to the MPP

A PV generator has mainly three working regions. Namely, a constant current region where the generator works as a current source, a constant voltage region where the generator works as a voltage source and a maximum power point region where the power drawn from the generator is the optimal one. For a large part of its $i - v$ curve, the PV generator can be considered as a constant current source. However, since the system desired operation is the MPP, this generator can be better linearized by expanding its nonlinear model as a Taylor series and ignoring high-order terms. Therefore, the $i - v$ equation of the PV model can be approximated by the following linear Norton equivalent model:

$$i_{pv} \approx I_{mpp} + \frac{\partial i_{pv}}{\partial v_{pv}}(v_{pv} - V_{mpp}) = I_{mpp} + G_{pN}(v_{pv} - V_{mpp}). \tag{3}$$

where $G_{pN} = \partial i_{pv}/\partial v_{pv}$ is the equivalent Norton conductance. In contrast to the ideal current source mode, this linearization reveals correctly the effect of the parameters that arise due to the nonlinear nature of the generator such as its dynamic Norton equivalent conductance G_{pN} and its Norton equivalent current i_{pN} that vary with the weather conditions. From (3), and making the PV voltage v_{pv} zero, the equivalent Norton current i_{pN} is as follows:

$$i_{pN} = I_{mpp} - G_{pN}V_{mpp}, \tag{4}$$

The equivalent conductance G_{pN} can be obtained by differentiating (1) which by using the implicit function theorem results in the following expression:

$$G_{pN} = -\frac{AV_t + R_p I_s e^{\frac{V_{mpp} + R_s I_{mpp}}{AV_t}}}{AV_t(R_p + R_s) + R_p R_s I_s e^{\frac{V_{mpp} + R_s I_{pv}}{AV_t}}} \tag{5}$$

where I_{mpp} and V_{mpp} are the generator current and voltage at the MPP. Based on the data provided in [32], the used PV generator has an open circuit voltage around 22 V under nominal conditions. Its internal parameters are depicted in Table 1 being its nominal power of 85 W. It is worth noting that the input voltage of the used PV module varies between 0 and the open circuit voltage with an optimum MPP value of about 18 V at nominal weather conditions.

Figure 4 shows its $i - v$ curve together with its linearized approximation close to the MPP for $S = 1000$ W/m^2 and $\Theta = 25$ °C. The corresponding load line of the optimum value of the conductance $G_{mpp} = g^* = 0.2524$ S is also shown in the same figure.

Table 1. Parameters of the PV module.

Parameter	Value
Number of cells N_s	36
Standard light intensity S_n	1000 W/m^2
Ref temperature Θ_n	25 °C
Series resistance R_s	0.005 Ω
Parallel resistance R_p	1000 Ω
Short circuit current I_{sc}	5 A
Saturation current I_0	1.16×10^{-8} A
Band energy E_g	1.12
Ideality factor A	1.2
Temperature coefficient C_Θ	0.00325 A/°C

A PV generator has a single operating point where the power $P = i_{pv}v_{pv}$ reaches its maximum value P_{max}. The values of the current I_{mpp} and the voltage V_{mpp} at this point correspond to a particular load resistance. Its corresponding inductance $G_{mpp} = g^*$ is equal to I_{mpp}/V_{mpp}. Hence, this generator can operate at the MPP by appropriately selecting that conductance whose load line intersects the $i - v$ curve of the PV generator at the MPP.

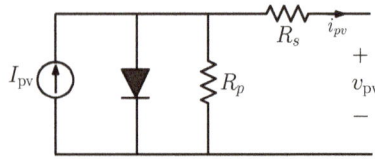

Figure 3. The single-diode five-parameter equivalent circuit diagram of the PV generator according to (1) [31].

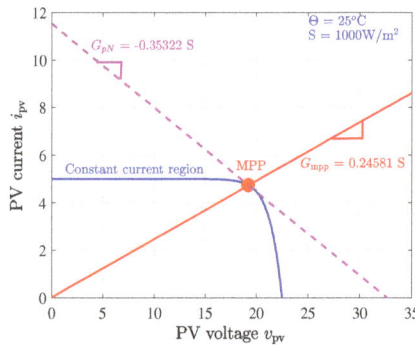

Figure 4. The BP585PV module $i - v$ characteristic and its linear approximation (dashed) at the MPP for $S = 1000$ W/m^2 and $\Theta = 25$ °C. The load line of the optimum conductance $G_{mpp} = 0.2524$ S and the Norton equivalent conductance $G_{pN} = 0.35322$ S are also shown.

2.4. Modeling of the DC-AC Inverter

The DC-AC inverter stage is responsible for injecting a sinusoidal grid current i_g in phase with the grid voltage $v_g = V_g \sin(2\pi f_g t)$. For this, a two-loop control strategy is used where the outer DC-link voltage controller provides the reference grid current amplitude I_{gref} for the inner current controller. This amplitude is multiplied by a sinusoidal signal synchronized with the grid voltage v_g, using a phase-locked loop (PLL), to obtain the time varying current reference $i_{gref} = I_{gref} \sin(2\pi f_g t)$. The current controller is conventionally a PI regulator that aims to make the grid current i_g to accurately track i_{gref} hence making the reactive power as close as possible to zero. This outer loop regulates the DC-link voltage by varying the current reference amplitude. A low-pass filter with a cut-off frequency at the grid frequency is also usually added to the PI voltage controller with the aim to reduce the harmonic distortion introduced by second harmonic of the grid frequency. The output of the current controller is fed to a Sinusoidal Pulse Width Modulator (SPWM). The output of this modulator generates the driving signal u_g of the DC-AC H-bridge. The study presented in this paper is constrained to the DC-DC stage assuming a quasi steady-state operation of the DC-AC inverter. This is an accurate assumption provided that the grid voltage v_g and the grid current i_g vary much slower than the variables at the DC-DC stage. The state-space model describing the dynamical behavior of the DC-AC inverter can be written in the following form:

$$\frac{dv_{dc}}{dt} = \frac{i_{L2}}{C_{dc}}(1 - u) - \frac{(2u_g - 1)i_g}{C_{dc}}, \tag{6}$$

$$\frac{di_g}{dt} = (2u_g - 1)\frac{v_{dc}}{L_g} - \frac{v_g}{L_g}. \tag{7}$$

A simple steady-state analysis based on a power balance reveals that the DC-link voltage can be approximated by:

$$v_{dc} \approx V_{dcref} + V_{rip}\sin(4\pi f_g t), \tag{8}$$

where V_{rip} is the amplitude of the ripple at the double frequency of the grid which can be expressed as follows [33]:

$$V_{rip} = \frac{\eta P_{pv}}{4\pi f_g C_{dc} V_{dcref}}, \tag{9}$$

η is the efficiency of the DC stage, f_g is the grid frequency, C_{dc} is the DC-link capacitance and V_{dcref} is the desired DC-link voltage. For a well designed inverter, one has $V_{dcref} \gg V_{rip}$. Moreover, the switching frequency is much higher than the grid frequency and therefore, the DC-link voltage can be considered constant at the switching time-scale. This is a widely used assumption in two-stage PV systems when the design of the DC-DC stage is of concern [24,25,28].

2.5. Dynamic Modeling of the Quadratic Boost Regulator Powered by a PV Generator

In PV systems, the input voltage of the DC-DC converter is controlled, not its output voltage. Therefore, it is modeled and analyzed as a current-fed converter. If the Norton equivalent model of the PV generator is used and the DC-link voltage ripple is neglected, the circuit configurations of the quadratic boost converter corresponding to the two different switch states are the ones depicted in Figure 5a,b.

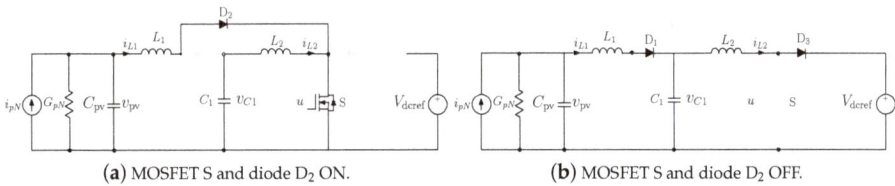

(**a**) MOSFET S and diode D_2 ON.   (**b**) MOSFET S and diode D_2 OFF.

Figure 5. The two simplified equivalent circuit configurations of the system of Figure 2 for the different switch S states where the PV generator is substituted by its linearized Norton equivalent and the grid-interlinked inverter is substituted by a constant DC voltage.

The application of Kirchhoff's laws to the circuit, after substituting the nonlinear PV generator by its Norton equivalent model, leads to the following set of differential equations describing the quadratic boost converter dynamical behavior:

$$\frac{dv_{pv}}{dt} = \frac{i_{pN}}{C_{pv}} - \frac{G_{pN}v_{pv}}{C_{pv}} - \frac{i_{L1}}{C_{pv}}, \tag{10}$$

$$\frac{di_{L1}}{dt} = \frac{v_{pv}}{L_1} - \frac{v_{C1}}{L_1}(1-u), \tag{11}$$

$$\frac{di_{L2}}{dt} = \frac{v_{C1}}{L_2} - \frac{V_{dcref}}{L_2}(1-u), \tag{12}$$

$$\frac{dv_{C1}}{dt} = \frac{i_{L1}}{C_1}(1-u) - \frac{i_{L2}}{C_1}, \tag{13}$$

where L_1 and L_2 are the inductances of the input and intermediate inductors, C_{pv} and C_1 are the capacitances of the input and the intermediate capacitors. All other parameters and variables that appear in (10)–(13) are shown in Figure 2. By applying a net volt-second balance [3], the following expressions are obtained relating the average steady-state values of the state variables to the operating duty cycle D:

$$I_{L1} = i_{pN} - G_{pN}V_{mpp}, \; I_{L2} = (1-D)I_{L1}, \tag{14}$$
$$V_{C1} = V_{dcref}(1-D), \quad V_{pv} = V_{mpp} = V_{dcref}(1-D)^2. \tag{15}$$

From (15), it can be observed that for a fixed value of D, the main advantage of the quadratic boost converter is that the voltage conversion gain defined as V_{dcref}/V_{pv} is the square of the conversion ratio corresponding to the canonical boost converter. According to (15), D is related to the PV generator average voltage $V_{pv} = V_{mpp}$ and the average output voltage V_{dcref} by the following expression:

$$D(S,\Theta) = 1 - \sqrt{\frac{V_{mpp}(S,\Theta)}{V_{dcref}}}. \tag{16}$$

For a slowly-varying output voltage, the quasi-steady-state duty cycle D is a function of the climatic conditions, and it is constrained by (16) with V_{mpp} as a function of the temperature Θ and the irradiance S.

2.6. Modeling the Input Port Controller

Since a dynamic controller is used for controlling the input port of the quadratic boost converter, its corresponding state equations are needed to complete the system model. The transfer function of the type-II controller is as follows:

$$H_i(s) = \frac{W_i \omega_p}{\omega_z} \frac{s + \omega_z}{s(s + \omega_p)}, \tag{17}$$

where W_i is the integrator gain, ω_z is the cut-off frequency of the controller zero and ω_p is the cut-off frequency of its pole. Let $W_p = (\omega_p - \omega_z)W_i/\omega_z$. A partial fraction decomposition of the transfer function defined in (17) lead to the following equivalent form which is suitable to be converted to a state space representation [34]:

$$H_i(s) = \frac{W_i}{s} + \frac{W_p}{s + \omega_p}, \tag{18}$$

Figure 6 shows an equivalent block diagram of the type-II controller where its corresponding state variables are represented together with their weighting factors in the feedback loop. From this block diagram, the time-domain state equations corresponding to the previous Laplace domain transfer function can be expressed as follows:

$$\frac{dv_p}{dt} = -\omega_p v_p + G_{mpp}v_{pv} - i_{L1}, \tag{19}$$
$$\frac{dv_i}{dt} = G_{mpp}v_{pv} - i_{L1}. \tag{20}$$

where v_p and $v_i := \int(G_{mpp}v_{pv} - i_{L1})dt$ are the state variables corresponding to the type-II controller [35].

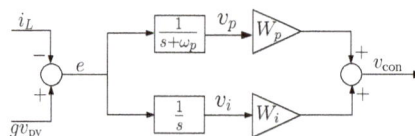

Figure 6. Equivalent block diagram of a type-II controller.

2.7. The State-Space Switched Model of the Quadratic Boost Converter

The model of the quadratic boost converter given in (10)–(13) can be written in the following matrix form:

$$\dot{\mathbf{x}}_p = \mathbf{A}_{p1}\mathbf{x}_p + \mathbf{B}_{p1}\mathbf{w}_p \text{ if } u = 1 \tag{21}$$

$$\dot{\mathbf{x}}_p = \mathbf{A}_{p0}\mathbf{x}_p + \mathbf{B}_{p0}\mathbf{w}_p \text{ if } u = 0 \tag{22}$$

$$e = G_{mpp}v_{pv} - i_{L1} := \mathbf{C}_p^\mathsf{T}\mathbf{x}_p \tag{23}$$

where $\mathbf{x}_p = (v_{pv}, i_{L1}, i_{L2}, v_{C1})^\mathsf{T}$ is the vector of the state variables of the converter and $\mathbf{A}_{pu}$ and $\mathbf{B}_{pu}$, $u = 1, 0$, are the state and input matrices corresponding to the different switch states. According to (10)–(13), the matrices $\mathbf{A}_{pu}$ and $\mathbf{B}_{pu}$ for $u = 1$ and $u = 0$, and the external input parameters vector $\mathbf{w}_p$ are as follows:

$$\mathbf{A}_{p1} = \begin{pmatrix} -\dfrac{G_{pN}}{C_{pN}} & -\dfrac{1}{C_{pv}} & 0 & 0 \\ \dfrac{1}{L_1} & 0 & 0 & 0 \\ 0 & 0 & 0 & \dfrac{1}{L_2} \\ 0 & 0 & -\dfrac{1}{C_1} & 0 \end{pmatrix}, \ \mathbf{A}_{p0} = \begin{pmatrix} -\dfrac{G_{pN}}{C_{pN}} & -\dfrac{1}{C_{pv}} & 0 & 0 \\ \dfrac{1}{L_1} & 0 & 0 & 0 \\ 0 & 0 & 0 & \dfrac{1}{L_2} \\ 0 & 0 & \dfrac{1}{C_1} & -\dfrac{1}{C_1} & 0 \end{pmatrix}, \tag{24}$$

$$\mathbf{B}_{p1} = \begin{pmatrix} \dfrac{1}{C_{pv}} & 0 \\ 0 & 0 \\ 0 & -\dfrac{1}{L_2} \\ 0 & 0 \end{pmatrix}, \ \mathbf{B}_{p0} = \begin{pmatrix} \dfrac{1}{C_{pv}} & 0 \\ 0 & 0 \\ 0 & 0 \\ 0 & 0 \end{pmatrix}, \ \mathbf{w}_p = \begin{pmatrix} i_{pN} \\ V_{dcref} \end{pmatrix}. \tag{25}$$

3. Small-Signal Model of the DC-DC Quadratic Boost Converter and Its Input Controller Design

The design of the controller in a switching converter is conventionally based on a small-signal averaged model, which can be obtained from (10)–(13) after substituting the control signal u by its duty cycle d and performing a perturbation and linearization close to the operating point of the converter.

The averaged small-signal model of the quadratic boost power stage can be expressed in the state-space form $\dot{\tilde{\mathbf{x}}}_p = \mathbf{A}\tilde{\mathbf{x}} + \mathbf{B}\tilde{d}$, where $\tilde{\ }$ stands for a small-signal variation, $\mathbf{A} = \mathbf{A}_{p1}D + \mathbf{A}_{p0}(1-D)$ and $\mathbf{B} = (\mathbf{A}_{p1} - \mathbf{A}_{p0})\mathbf{x}_{av} + \mathbf{B}_{p1} - \mathbf{B}_{p0}$ and $\mathbf{x}_{av} = -\mathbf{A}^{-1}(\mathbf{B}_{p1}D + \mathbf{B}_{p0}(1-D))$. Selecting the output represented by the small-signal error signal $\tilde{e} = \tilde{i}_{L1} - G_{mpp}\tilde{v}_{pv}$ and using the Laplace transform, the small-signal transfer functions can be straightforwardly obtained using the well-known formula $\tilde{e}(s) = \mathbf{C}_p^\mathsf{T}(s\mathbf{I} - \mathbf{A})^{-1}\mathbf{B}\tilde{d}$, where $\mathbf{C}_p^\mathsf{T} = (G_{mpp} \ -1 \ 0 \ 0)$ and $\mathbf{I}$ is a 4×4 identity matrix. Hence, the d-to-e transfer function can be expressed as follows:

$$H_p(s) = \mathbf{C}_p^\mathsf{T}(s\mathbf{I} - \mathbf{A})^{-1}\mathbf{B} \tag{26}$$

The zeros can be obtained by solving for s the equation $\mathbf{C}_p^\mathsf{T}(s\mathbf{I} - \mathbf{A})^{-1}\mathbf{B} = 0$. In doing so and after some algebra taking into account (14)–(15), the following expressions for the zeros are obtained:

$$z_1 = -\frac{G_{pN} + G_{mpp}}{C_p}, \tag{27}$$

$$z_2 = \frac{-I_{mpp}}{2C_1 V_{dcref}} + j\frac{\sqrt{8C_1/L_2 V_{dcref}^2 - I_{L1}^2}}{2C_1 V_{dcref}} \tag{28}$$

$$z_3 = \frac{-I_{mpp}}{2C_1 V_{dcref}} - j\frac{\sqrt{8C_1/L_2 V_{dcref}^2 - I_{L1}^2}}{2C_1 V_{dcref}} \tag{29}$$

Note that in addition to the left half plane zero z_1, which also exists in the small-signal model of the canonical boost converter with input current feedback, an extra complex conjugate zeros pair appears in the small-signal model of the quadratic boost converter. Note also that because $8C_1/L_2 V_{dcref}^2 - I_{L1}^2 > 0$, the extra complex conjugate zeros are located in the left half side of the complex plane, and therefore, the input controlled quadratic boost converter is a minimum phase system. This is also the case of the boost converter with input voltage feedback [36]. On the other hand, the poles can be obtained by solving for s the equation $\det(s\mathbf{I} - \mathbf{A}_{ss}) = 0$, i.e.,

$$s^4 + a_3 s^3 + a_2 s^2 + a_1 s + a_0 = 0 \tag{30}$$

where the coefficients a_3, a_2, a_1, and a_0 are given by the following expressions:

$$a_3 = \frac{G_{pN}}{C_p}, \quad a_2 = \frac{C_p L_1 + L_2(C_1 + C_p(1-D)^2)}{C_1 C_p L_1 L_2}, \quad a_1 = \frac{G_{pN}(L_1 + L_2(1-D)^2)}{C_1 C_p L_1 L_2}, \quad a_0 = \frac{1}{C_1 C_p L_1 L_2}. \tag{31}$$

It is worth noting that the desired working point of the PV source is the MPP characterized by a Norton equivalent conductance $G_{pN} \neq 0$. In this case, according to Routh-Hurwitz criterion, all the poles of the quadratic boost converter are located in the left half side of the complex plane. However, if under any circumstance, such as at startup or during a transient, the PV source works in the constant current region characterized by a zero Norton equivalent conductance, the quadratic boost converter will exhibit two pairs of purely imaginary complex conjugate poles that can lead to undamped low frequency oscillation. With an appropriate control design, such oscillation will disappear as soon as the system reaches the operation in the MPP mode forced by the MPPT controller.

Using the previously-obtained small-signal model, the input port controller design can be performed by appropriately selecting the required performances in terms of settling time, crossover frequency, and stability phase margin. With this averaged small-signal approach, the controller is designed for the lowest irradiance level [14]. Figure 7 shows the crossover frequency f_c and the phase margin φ_m of the model of the quadratic boost converter under the type-II input port controller when the irradiance is varied in the range (500, 1000) W/m². According to the small-signal averaged model, as the irradiance level is increased, the crossover frequency f_c increases at the expense of a decrease of the phase margin φ_m. Despite this, according to the same model, the system remains stable and exhibits a sufficient phase margin above 40° and an infinite gain margin for the whole range of the varied parameter. The gain margin is infinite because the total loop gain presents six stable poles (four from the power stage and two from the controller) and four stable zeros (three from the power stage and one from the controller), and the asymptotic behavior at high frequencies is similar to a minimum phase continuous-time second order system whose phase never crosses −180 degrees; therefore, the gain can be increased as much as possible without destabilizing the system. However, the values of the gain and the phase obtained from the small-signal average model are different from the actual phase of the switched system in the vicinity of the Nyquist frequency, as was recently reported in [37]. Indeed, it will be shown later using accurate discrete-time modeling that the system exhibits instability in the form of subharmonic oscillation for values of irradiance larger than approximately 820 W/m² with the fixed values of parameters shown in Tables 1–3.

Table 2. The parameters used for the DC-AC inverter.

Parameter	Value
Inductance L_g	20 mH
DC-link capacitance C_{DC}	47 μF
Grid frequency f_g	50 Hz
PWM switching frequency f_s	50 kHz
RMS value of the grid voltage	230 V
Proportional gain (current) k_{ip}	1 Ω
Integral gain (current) k_{ii}	20 krad/s
Cut-off frequency of the filter (current controller)	50 Hz
Proportional gain (voltage) k_{vp}	0.019
Integral gain (voltage) k_{vi}	0.51 rad/s

Table 3. The parameter values used for the quadratic boost converter.

L_1 (μH)	L_2 (mH)	C_1, C_{pv}, C_{dc}(μF)	V_M (V)	V_g (V)	V_{dcref} (V)	ω_p, ω_z, W_i (krad/s)	f_s (kHz)
120–138	3.5–5.5	10, 10, 47	variable	$230\sqrt{2}$	380	50π, 1, 1	50

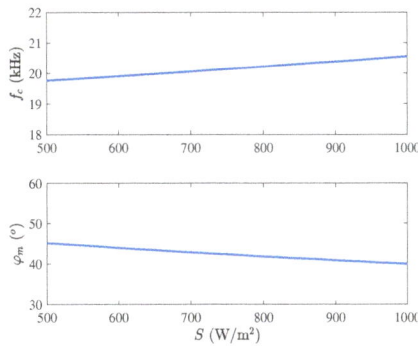

Figure 7. The crossover frequency f_c (top) and the phase margin φ_m (bottom) of the small-signal model of the quadratic boost converter with the input voltage control for different values of the irradiance S between 500 W/m² ($P_{max} \approx 42$ W) and 1000 W/m² ($P_{max} \approx 85$ W). V_M =4 V. $\Theta = 25\,°C$.

4. The Complete State-Space Switched Model of the Closed-Loop Quadratic Boost Regulator

The complete model of the quadratic boost regulator is obtained by including the state variables corresponding to the input port controller. This model can be written in the following augmented matrix form:

$$\dot{x} = A_1 x + B_1 w \text{ if } u = 1, \tag{32}$$

$$\dot{x} = A_0 x + B_0 w \text{ if } u = 0, \tag{33}$$

$$\dot{v}_i = e = G_{mpp} v_{pv} - i_{L1}. \tag{34}$$

where $x = (v_{pv}, i_{L1}, i_{L2}, v_{C1}, v_p)^\top$ is the augmented vector of state variables, $A_0 \in \mathbb{R}^{5\times5}$, $A_1 \in \mathbb{R}^{5\times5}$, $B_0 \in \mathbb{R}^{5\times2}$ and $B_1 \in \mathbb{R}^{5\times2}$ are the augmented system state matrices taking into account the state variables of the power stage and the controller and excluding the state variable corresponding to the integral action and $w = (i_{pN}, V_{dcref})^\top$ is the vector of the external parameters supposed to be constant within a switching cycle. To avoid matrix singularity problems in computer computations and to start with a well-posed mathematical problem, the state variable v_i was excluded from the rest of state variables in the vector $\mathbf{x}$ [35]. According to (10)–(13) and (19), the matrices $\mathbf{A}_u$ and $\mathbf{B}_u$ and the input vector $\mathbf{w}$ for $u = 1$ and $u = 0$ are as follows:

$$\mathbf{A}_1 = \begin{pmatrix} -\dfrac{G_{pN}}{C_{pN}} & -\dfrac{1}{C_{pv}} & 0 & 0 & 0 \\ \dfrac{1}{L_1} & 0 & 0 & 0 & 0 \\ 0 & 0 & 0 & \dfrac{1}{L_2} & 0 \\ 0 & 0 & -\dfrac{1}{C_1} & 0 & 0 \\ G_{mpp} & -1 & 0 & 0 & -\omega_p \end{pmatrix}, \quad \mathbf{A}_0 = \begin{pmatrix} -\dfrac{G_{pN}}{C_{pN}} & -\dfrac{1}{C_{pv}} & 0 & 0 & 0 \\ \dfrac{1}{L_1} & 0 & 0 & 0 & 0 \\ 0 & 0 & 0 & \dfrac{1}{L_2} & -\dfrac{1}{L_1} \\ 0 & \dfrac{1}{C_1} & -\dfrac{1}{C_1} & 0 & 0 \\ G_{mpp} & -1 & 0 & 0 & -\omega_p \end{pmatrix}, \quad (35)$$

$$\mathbf{B}_1 = \begin{pmatrix} \dfrac{1}{C_{pv}} & 0 \\ 0 & 0 \\ 0 & -\dfrac{1}{L_2} \\ 0 & 0 \\ 0 & 0 \end{pmatrix}, \quad \mathbf{B}_0 = \begin{pmatrix} \dfrac{1}{C_{pv}} & 0 \\ 0 & 0 \\ 0 & 0 \\ 0 & 0 \\ 0 & 0 \end{pmatrix}, \quad \mathbf{w} = \begin{pmatrix} i_{pN} \\ V_{dcref} \end{pmatrix}. \quad (36)$$

5. A Glimpse at the Solar PV System Behavior from Its Complete Mathematical Model

Let us take a quick glimpse at some of the typical operating dynamic behaviors of the system in terms of different parameter values. The numerical simulations are performed using PSIM© software using the detailed switched model of the complete system consisting of the DC-DC quadratic boost converter performing MPPT and interlinked to the grid-connected DC-AC inverter as depicted in Figure 2. The nonlinear PV panel model is implemented using the physical model of the solar module in the renewable energy package of PSIM©. The set of parameter values shown in Table 3 is used for the quadratic boost converter, those in Table 1 for the PV module, and the ones in Table 2 for the DC-AC inverter. The inductance values were selected to guarantee continuous conduction mode (CCM), and the capacitance values were chosen to get acceptable voltage ripple amplitudes. The compensator zero $\omega_z = 1$ krad/s was placed in such a way to damp partially one of the complex conjugate poles pair resonant effect. The low-pass filter pole ω_p was placed at one half the switching frequency. An extremum seeking algorithm was used for performing MPPT [38,39].

5.1. System Startup and Steady-State Response

The response of the complete system starting from zero initial conditions is depicted in Figure 8. It can be seen from the plots that after an initial transient, the state variables and the control signals of the system reached their desired periodic steady-state. The extracted power also converged to its MPP value.

Figure 9a illustrates the response of the system to a change in the irradiance level from 500 W/m² ($P_{max} \approx 42$ W) to 1000 W/m² ($P_{max} \approx 85$ W). In that figure, the waveforms of the control signals v_{ramp} and v_{con}, the instantaneous power P, its reference value P_{max} are depicted. The DC link voltage and the grid current in the AC side are also shown in the same figure. A detailed view of the ramp modulator, the control signal and the inductor currents at DC-DC stage is shown in Figure 10 where it can be observed that desired periodic operation (stable) takes place for $S = 500$ W/m² while nonlinear phenomena in the form of subharmonic oscillation is exhibited for $S = 1000$ W/m². It is worth noting that the dynamical behavior and the stability at the AC side is not affected by the subharmonic oscillation at the DC side as can be observed in Figure 9b. Moreover, the grid current i_g exhibits a low total harmonic distortion of about 2% as calculated by PSIM© software.

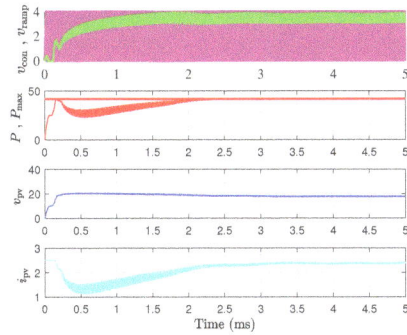

Figure 8. The startup response of the quadratic boost converter with a nonlinear PV source under MPPT control $S = 500\,\text{W/m}^2$, $V_M = 4\,\text{V}$. $\Theta = 25\,^\circ\text{C}$.

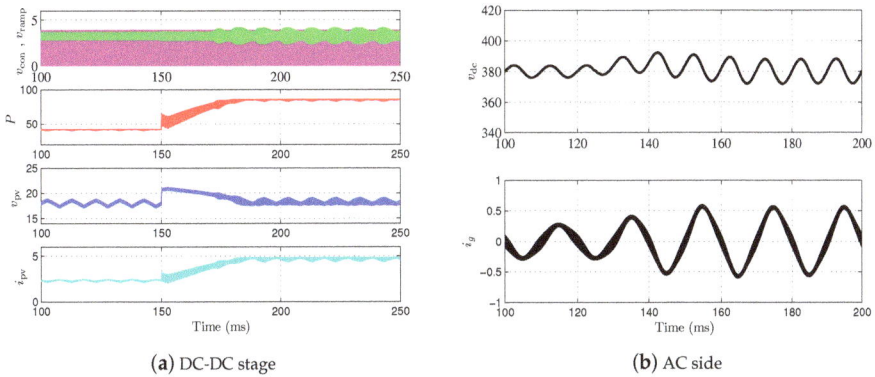

(**a**) DC-DC stage

(**b**) AC side

Figure 9. The simulated PV system response to a change at $t = 150$ ms in the irradiance level from $500\,\text{W/m}^2$ ($P_{\max} \approx 42$ W) to $1000\,\text{W/m}^2$ ($P_{\max} \approx 85$ W). $V_M = 4\,\text{V}$. $\Theta = 25\,^\circ\text{C}$.

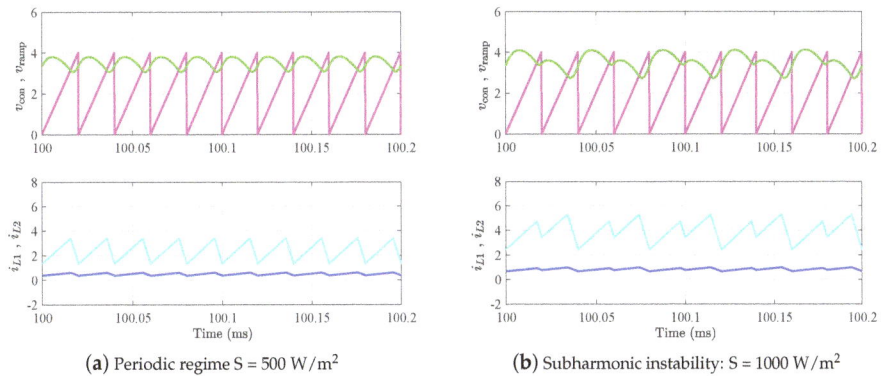

(**a**) Periodic regime S = 500 W/m²

(**b**) Subharmonic instability: S = 1000 W/m²

Figure 10. Close view of the ramp signal v_{ramp}, the control signal $v_{\text{con},}$, and the inductor currents i_{L1} and I_{L2} at the DC-DC stage.

5.2. Bifurcation Diagram of the PV System by Varying the Irradiance Level

In order to understand the mechanisms of how the subharmonic oscillation takes place, a bifurcation diagram for the system is plotted by considering the irradiance S as a bifurcation parameter which is varied within the range (500, 1000) W/m^2. This bifurcation diagram is obtained by sampling the vector of state variables $\mathbf{x}(t)$ at the switching period rate, thus yielding $\mathbf{x}(nT)$, $n = 0, 1 \ldots 100 \times 10^3$. The last 100 samples are considered as steady-state and the corresponding inductor current samples $i_{L1}(nT)$ are plotted in terms of the bifurcation parameter. Two bifurcation diagrams were computed and the results are shown in Figure 11. In the first diagram, a constant value g^* of the conductance was used for simplicity. In the second one, the dynamic conductance G_{mpp} provided by the extremum seeking MPPT controller was used. As can be observed, the system undergoes a period doubling at $S \approx 836 \text{ W}/\text{m}^2$, which explains the observed subharmonic oscillation in Figures 9 and 10 for $S = 1000 \text{ W}/\text{m}^2$. Note that the dynamics of the MPPT controller slightly alters the location of the bifurcation boundary, improving the stability at the fast time-scale for larger irradiance values. Such a stabilizing effect of a periodic time-varying signal in a switching converter has been already reported in previous works such as [40].

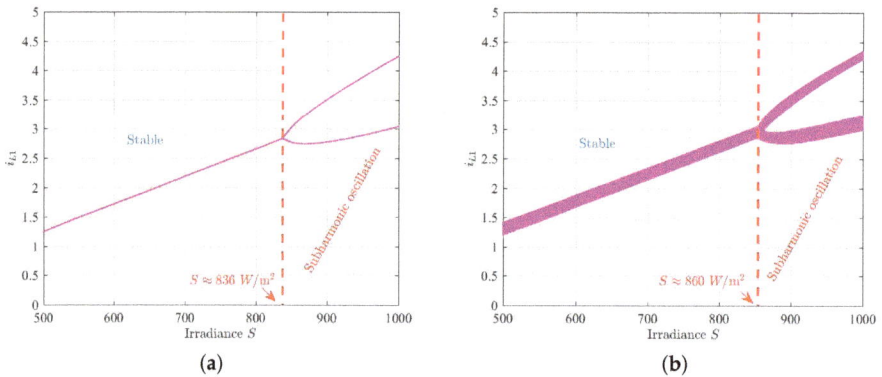

Figure 11. The bifurcation diagram of the quadratic boost regulator with a nonlinear PV source under extremum seeking MPPT control for regulating the input voltage taking the irradiance S as a bifurcation parameter. (**a**) With the exact theoretical conductance g^* and (**b**) with the conductance G_{mpp} provided by the extremum seeking MPPT. $V_M = 4$ V. $\Theta = 25\,°\text{C}$.

6. Stability Analysis of Periodic Orbits and Subharmonic Oscillation Boundary

6.1. Stability Analysis of Periodic Orbits

The switching from the ON to the OFF phase takes place whenever the ramp modulator signal v_{ramp} and the control signal $v_{\text{con}} := W_p v_p + W_i v_i$ intersect, i.e, whenever the following equality holds:

$$W_i v_i(d_n T) + \mathbf{K}^\mathsf{T} \mathbf{x}(d_n T) - v_{\text{ramp}}(d_n T) = 0, \tag{37}$$

where $\mathbf{K} = (0, 0, 0, 0, W_p)^\mathsf{T}$ is the vector of feedback gains and d_n is the discrete-time the duty cycle during the nth switching cycle. The steady-state value D of d_n is imposed by the output DC-link voltage V_{dcref} and the MPP voltage V_{mpp}. Therefore, for a fixed DC-link voltage V_{dcref}, the steady-state duty cycle D is a function of the climatic conditions, and it is constrained by (16) with V_{mpp} as a function of the temperature Θ and the irradiance S.

To perform a stability analysis of the system, Floquet theory is used and therefore the monodromy matrix $\mathbf{M}$ is first obtained. Let $\mathbf{x}(DT) = (\mathbf{I} - \boldsymbol{\Phi})^{-1}\boldsymbol{\Psi}$ be the steady-state value of $\mathbf{x}(t)$ at time instant DT, where $\boldsymbol{\Phi} = \boldsymbol{\Phi}_1\boldsymbol{\Phi}_0$, $\boldsymbol{\Phi}_1 = e^{\mathbf{A}_1 DT}$, $\boldsymbol{\Phi}_0 = e^{\mathbf{A}_0(1-D)T}$, $\boldsymbol{\Psi}_1 = (e^{\mathbf{A}_1 DT} - \mathbf{I})^{-1}\mathbf{B}\mathbf{w}$, $\boldsymbol{\Psi}_0 = (e^{\mathbf{A}_0(1-D)T} - \mathbf{I})^{-1}\mathbf{B}\mathbf{w}$,

$\Psi = \Phi_1\Psi_0 + \Psi_1$. Let $m_a = V_M/T$ be the slope of the ramp-modulating signal, where V_M is its peak-to-peak value. Let $\mathbf{m}_1(\mathbf{x}(t)) = \mathbf{A}_1\mathbf{x}(t) + \mathbf{B}_1\mathbf{w}$ and $\mathbf{m}_0(\mathbf{x}(t)) = \mathbf{A}_0\mathbf{x}(t) + \mathbf{B}_0\mathbf{w}$. Then, the monodromy matrix can be expressed as follows [22]:

$$\mathbf{M} = \Phi_0\mathbf{S}\Phi_1, \tag{38}$$

where $\mathbf{S}$ is the saltation matrix given by:

$$\mathbf{S} = \mathbf{I} + \frac{(\mathbf{m}_0(\mathbf{x}(DT)) - \mathbf{m}_1(\mathbf{x}(DT)))\mathbf{K}^\mathsf{T}}{W_i v_i(DT) + \mathbf{K}^\mathsf{T}\mathbf{m}_1(\mathbf{x}(DT)) - m_a}. \tag{39}$$

Once the MPP voltage is obtained by maximizing the PV power, the steady-state duty cycle D is determined according to (16). The expression of $v_i(DT)$ that appears in (39) can be obtained from (37) in steady-state:

$$v_i(DT) = \frac{1}{W_i}(\mathbf{K}^\mathsf{T}\mathbf{x}(DT) - m_a DT) \tag{40}$$

The study is done by using the set of parameter values of Table 3 for the quadratic boost converter and those shown in Table 1 for the PV module. First, $\mathbf{x}(DT)$ and $\mathbf{x}(0)$ are calculated, and the stability of the system is checked by observing the location of the eigenvalues of the monodromy matrix in the complex plane. Figure 12a shows the loci of these eigenvalues when the irradiance S is varied in the range $(500, 1000)$ W/m² for $V_M = 4$ V. It can be observed that as the irradiance is increased above a critical value of $S \approx 820$ W/m², the system undergoes a period doubling because one eigenvalue of the monodromy matrix leaves the unit disk from the point $(-1,0)$. This explains the exhibition of the subharmonic oscillation observed previously in the time-domain waveforms of Figures 9a and 10b and in the bifurcation diagrams of Figure 11. Note that the critical value predicted by the eigenvalues of the monodromy matrix is very close to the one predicted by the bifurcation diagram in Figure 11a. In turn, by fixing the irradiance S and the varying the amplitude V_M of the ramp voltage v_{ramp}, the same phenomenon is observed when V_M is decreased. The variation of other parameters also leads to the exhibition of the same phenomenon whenever the operation in CCM is guaranteed.

Remark 2. *It can be observed that when the parameter values vary, only the eigenvalues of the monodromy located at the real axis move, while the complex conjugate ones remain practically constant and are maintained inside the unit disk. Therefore, the system does not undergo a slow-scale instability. This is due to the imposition of the LFR behavior at the input port of the converter, as already mentioned before. This is particularly important for a PV system since the optimum conductance G_{mpp} is constantly changed by the MPPT controller and the damping of the undesired oscillations caused by this change is better than in other control strategies, such as in [14,27].*

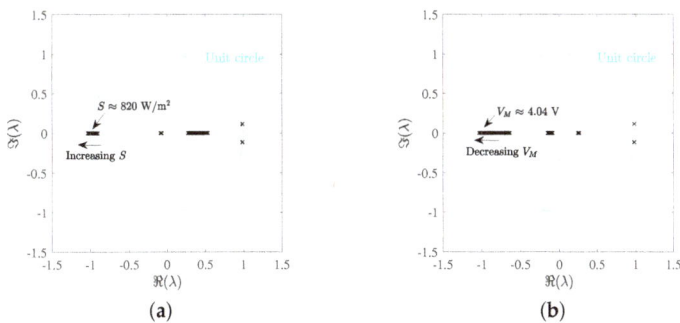

Figure 12. Monodromy matrix eigenvalues' loci for (**a**) the irradiance $S \in (500, 1000)$ W/m², $V_M = 4$ V, $\Theta = 25\,^\circ$C, and (**b**) the ramp peak-to-peak amplitude $V_M \in (4, 5)$ V, $S = 1000$ W/m², $\Theta = 25\,^\circ$C.

6.2. Analytical Determination of the Subharmonic Instability Boundaries

It was demonstrated in [35] that at the onset of subharmonic instability for a single-switch DC-DC regulator working in CCM, the following equality holds (The sign convention of the feedback coefficients has been adapted from [35]):

$$m_a = \mathbf{K}^\mathsf{T}(\mathbf{I} + \mathbf{\Phi})^{-1}\mathbf{\Phi}_1(\mathbf{m}_1(\mathbf{x}(0)) + \mathbf{m}_0(\mathbf{x}(0))) + m_i, \tag{41}$$

where $\mathbf{x}(0) = (\mathbf{I} - \overline{\mathbf{\Phi}})^{-1}\overline{\mathbf{\Psi}}$, $\overline{\mathbf{\Phi}} = \mathbf{\Phi}_0\mathbf{\Phi}_1$, $\overline{\mathbf{\Psi}} = \mathbf{\Phi}_0\mathbf{\Psi}_1 + \mathbf{\Psi}_0$, and $m_i = W_i(G_{mpp}v_{vp}(DT) - i_{L1}(DT))$. The terms $v_{vp}(DT)$ and $i_{L1}(DT)$ can be extracted from $\mathbf{x}(DT)$ defined previously. The theoretical results from expression (41) will be presented together with those corresponding to computer simulations and experimental results.

7. Validation of the Theoretical Results by Using Numerical Simulations and Experimental Results

To verify the theoretical and the time-domain simulation results, a DC-DC quadratic boost prototype was designed and implemented (Figure 13). In order to simplify the experimental setup and to obtain repeatable experiments, the PV emulator was used rather than a real PV generator. The main conclusions can be translated to real PV modules under the same weather conditions. An electronic active load was programmed in constant voltage mode and was connected at the output of the quadratic boost regulator with a type-II controller at the input side. A bank of capacitors of 28.2 mF was connected between the converter and the active load to fix the output voltage.

The inductances have been built in-house and had the same nominal values as the ones used in the numerical simulations presented previously, i.e., $L_1^* = 138\ \mu\text{H}$ and $L_2^* = 5.5\ \text{mH}$. The input capacitor of 10 μF was a metallized polyester capacitor (MKT) technology, and its rated voltage was 63 V. The intermediate and output capacitors of 10 μF were metalized polypropylene film technology (MKP), and their rated voltage was 560 V. The power MOSFET (SIHG22N60E-GE3), with a rated voltage of 600 V, was used as a controlled switch of the quadratic boost regulator. The silicon carbide Schottky diodes (C3D10065A CREE) with a maximum reverse voltage VRRM voltage of 650 V were the diodes. The current sensing was performed by means of shunt resistors of 20 mΩ. Operational amplifiers MC33078 were used to amplify the sensed current. The analog multiplier (AD633JNZ) was used to obtain the reference current. The current error is processed by a PI controller with a tunable proportional gain. The output of the PI controller was followed by a low-pass filter hence obtaining the type-II controller. Like in the numerical simulations, the cut-off frequency of the low-pas filter was at one half the switching frequency (25 kHz). Note that a type-II controller is equivalent to a PI compensator cascaded with a low-pass filter. The same switching logic used in numerical simulations was used in the experimental prototype.

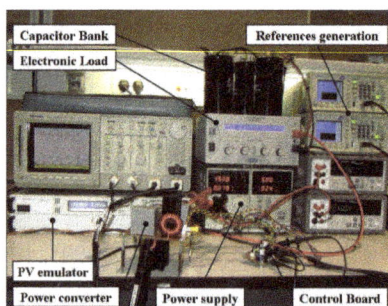

Figure 13. A picture of the experimental setup where the quadratic boost converter, the PV emulator, and the electronic load are used to obtain the experimental results.

7.1. Experimental Test 1

To validate the numerical simulations experimentally, first, the experimental system response corresponding to Figure 9 was obtained from the laboratory prototype, and the results are depicted in Figure 14. The step change in the irradiance level was from 500 W/m²–1000 W/m². First, for $S = 500$ W/m², the system worked in the stable periodic regime. For $S = 1000$ W/m², the subharmonic oscillation was exhibited. As can be observed, a close agreement between the numerical simulations in Figure 9 and the experimental measurements in Figure 14 was obtained.

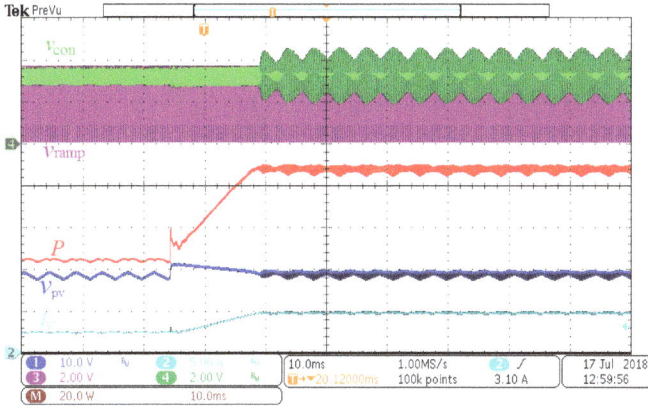

Figure 14. The experimental PV system response due to a change of step type in the irradiance level from 500 W/m²–1000 W/m² as in Figure 9. $V_M = 4$ V.

To validate the previous methodology, the ramp signal amplitude V_M was fixed in a relatively large value and then decreased till observing subharmonic instability at the oscilloscope screen, and the critical value of the ramp amplitude was recorded for several values of the operating duty cycle D in the range (0.2, 0.8). The duty cycle was varied by sweeping the active load voltage while maintaining the operation of the system at the MPP by selecting the suitable value of the conductance g^* to be equal to the optimum value $G_{mpp} = I_{mpp}/V_{mpp}$. Figure 15 shows the subharmonic instability boundary in the plane (D, V_M) obtained from (41) (dashed curve) using the values of inductances corresponding to no loading conditions and by experimental measurements ($\star$). A small discrepancy between the results can be observed. For instance, for $V_{dcref} = 380$ V, i.e, $D = 0.7824$, the critical value of the ramp voltage amplitude from the theoretical expression was $V_M \approx 4.8$ V, while the one from the experimental measurements was $V_M \approx 5.2$ V. This mismatching between the theoretical and the experimental results can be attributed to many parasitic factors and non-modeled effects. However, it was observed that partial saturation of the inductors and the drop of their inductance values with the operating currents [41], is the main factor. Next, the saturability of the inductors will be taken into account. The variation of the inductance values versus their operating DC currents was experimentally determined.

An LCR meter and a current source, both controlled by a LabView© software program, were used to measure the values of the inductances for different current levels. The experimental data obtained and a regression analysis based on least squared error revealed that in the range of current values used, the following linear expressions, relating the inductances L_1 and L_2 and their currents, can be used:

$$L_1 \approx L_1^* - \sigma_1 I_{L1}, \quad L_2 \approx L_2^* - \sigma_2 I_{L2}, \tag{42}$$

where $L_1^* = 138$ µH and $L_2^* = 5.5$ mH are the inductance values under no load condition, $\sigma_1 = 3$ µH/A, $\sigma_2 = 1.2$ mH/A, and I_{L1} and I_{L2} are given by (14). The previous equations were used in both the theoretical expression (41) and in the numerical results. The theoretical results from (41) are depicted in Figure 15 in the solid curve and those from numerical simulations using PSIM© software are indicated by △. After taking into account the inductances drop with the inductor current, a remarkable agreement among the experimental, theoretical and numerical results was obtained.

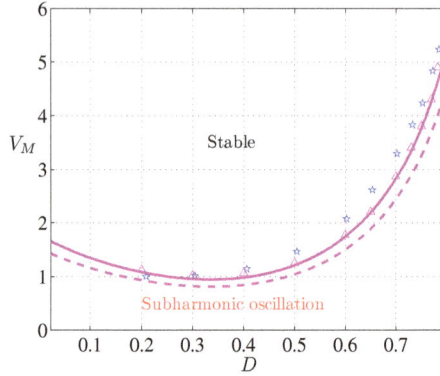

Figure 15. The stability boundary in the (D, V_M) parameter space from the theoretical expression (41) by using fixed values of the inductances $L_1^* = 138$ μH and $L_2^* = 5.5$ mH (dashed curve), by updating the inductances L_1 and L_2 values according to (42) (solid curve and △) and experimentally (⋆).

The waveforms of the inductor currents i_{L1} and i_{L2} at both sides of the subharmonic instability boundary are represented in Figure 16 together with the ramp signal and the control voltage. By comparing the waveforms in this figure and those in Figure 10, one can observe a good agreement between the measured and the simulated system dynamics.

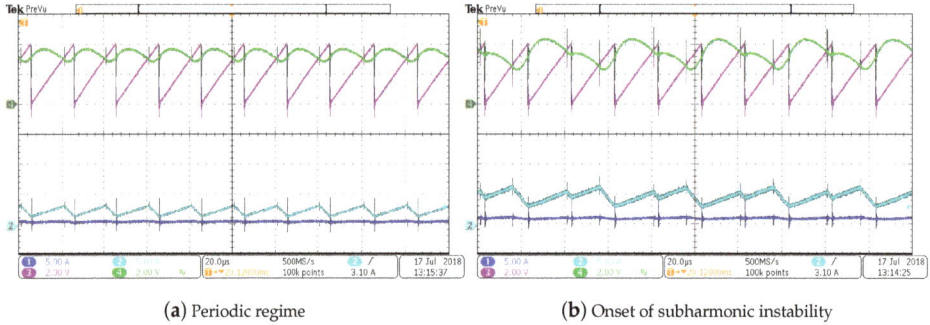

(**a**) Periodic regime (**b**) Onset of subharmonic instability

Figure 16. Experimental waveforms of the quadratic boost converter fed by a PV generator before (S = 500 W/m²) and after (S = 750 W/m²) subharmonic oscillation takes place. $V_M = 4$ V. Θ = 25 °C. Other parameters' values are from Table 3.

7.2. Experimental Test 2

In this test, the output voltage was fixed at $V_{dcref} = 380$ V, the ramp signal peak-to-peak value was fixed at $V_M = 4$ V, and the dynamics of the quadratic boost converter was explored by varying parameters corresponding to temperature and irradiance. Figure 17 shows the subharmonic instability boundary in the plane (Θ, S) obtained from (41) while maintaining the PV emulator at its MPP. The four parameters needed to define the PV curve in this emulator were adjusted to different values to correspond to a temperature variation between 10 °C and 70 °C. The stability boundary is depicted in Figure 17. In this figure, the theoretical boundary obtained using (41) with fixed values of the inductances represented by the thin curve (upper) and experimental measurements are depicted. A significant mismatching can be observed between the results obtained by using the mathematical expression (41) with fixed values of inductances and the experimental measurements. The subharmonic instability boundary from numerical simulations ($\triangle$) and from (41) (thick curve) by updating the inductances values according to (42) in both cases is also shown in Figure 17. Taking into account the inductances' variation with the operating current, the agreement between the results is remarkable.

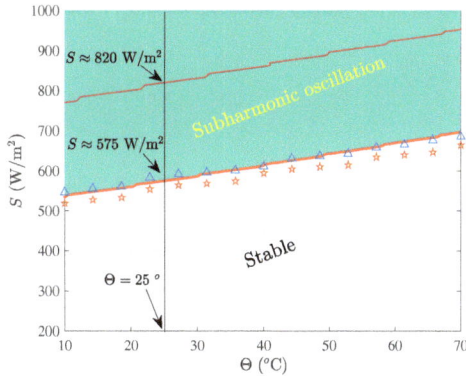

Figure 17. The subharmonic instability boundary in the parameter plane (Θ, S). The results are obtained from the theoretical expression (41) with L_1 and L_2 fixed (thin curve), with L_1 and L_2 varied according to (42) (thick curve), from computer simulations performed on the switched model with L_1 and L_2 varied according to (42) ($\triangle$), and experimentally ($\star$). $V_M = 4$ V.

8. Conclusions

This study has shown that a high-voltage-gain DC-DC quadratic boost power converter connected to a grid-interlinked inverter in a PV system may undergo subharmonic instability when parameters such as those relayed to climatic conditions, loading and control circuit vary. The boundary of this instability has been located accurately using an analytical expression. Experimental tests, carried out using a laboratory prototype and numerical simulations from the switched model of the system, have been used to validate the theoretical derivations. The study provides a methodology for control-oriented modeling, nonlinear analysis and analytical determination of subharmonic instability boundary of energy conversion circuits used in PV systems. The presented methodology could help in tuning the different parameter values in order to avoid the undesired subharmonic oscillation, particularly as nonlinearity and/or parameter variations can be taken into account in the approach used. From a design perspective, the average small-signal model of the system can be used to achieve the desired performances in terms of stability phase margin, crossover frequency and settling time. However, subharmonic instability cannot be predicted by using this approach. Then, as a second step in the design, one should take into account the boundary condition given in this study to avoid problems related to subharmonic instability. In particular, the switching regulator control parameters such as the amplitude of the ramp modulator or the gain of the controller can be tuned according to the operating point in order to avoid the jeopardizing effects of such instability problems. These parameters must be tuned based on the highest irradiance level.

Author Contributions: Conceptualization, A.E.A.; Methodology, A.E.A. and G.G.; Software, A.E.A.; Validation, A.C.-P.; Formal Analysis, A.E.A. and G.G.; Investigation, A.E.A.; Resources, A.C.-P.; Data Curation, M.A.-N.; Writing—Original Draft Preparation, A.E.A.; Writing—Review & Editing, K.A.H. and N.A.S.; Visualization, N.A.S.; Supervision, A.C.-P.; Funding Acquisition, A.C.-P. and M.A.-N.

Funding: This research was funded by the Spanish Agencia Estatal de Investigación (AEI) and the Fondo Europeo de Desarrollo Regional (FEDER) under grant DPI2017-84572-C2-1-R (AEI/FEDER, UE). Abdelali El Aroudi and Mohamed Al-Numay extend their appreciation to the International Scientific Partnership Program ISPP at King Saud University for funding this work through ISPP# 00102.

Acknowledgments: The authors would like to thank Reham Haroun for obtaining some of the experimental results.

Conflicts of Interest: The authors declare no conflict of interest.

Abbreviations

The following abbreviations are used in this manuscript:

PWM	Pulse width modulation
CCM	Continuous conduction mode
LFR	Loss-free-resistor
MPP	Maximum power point
MPPT	Maximum power point tracking
PV	Photovoltaic
SPWM	Sinusoidal pulse width modulation

References

1. Fahimi, B.; Kwasinski, A.; Davoudi, A.; Balog, R.S.; Kiani, M. Powering a more electrified planet. *IEEE Power Energy Mag.* **2011**, 54–64. Available online: https://www.ieee-pes.org/images/files/pdf/2012-pe-smart-grid-compendium.pdf (accessed on 28 December 2018).
2. Grubišić-Čabo, F.; Nizetić, S.; Giuseppe Marco, T. Photovoltaic panels: A review of the cooling techniques. *Trans. Famena* **2016**, *40*, 63–74.
3. Erickson, R.; Maksimovic, D. *Fundamentals of Power Electronics*, 2nd ed.; Kluwer Academic/Plenum Publishers: New York, NY, USA, 2001.
4. Alajmi, B.N.; Ahmed, K.H.; Finney, S.J.; Williams, B.W. A maximum power point tracking technique for partially shaded photovoltaic systems in microgrids. *IEEE Trans. Ind. Electron.* **2013**, *60*, 1596–1606. [CrossRef]
5. Sahan, B.; Vergara, A.N.; Henze, N.; Engler, A.; Zacharias, P. A single-stage PV module integrated converter based on a low-Power current-source inverter. *IEEE Trans. Ind. Electron.* **2008**, *55*, 2602–2609. [CrossRef]
6. Xiao, W. *Photovoltaic Power System: Modelling, Design and Control*; John Wiley & Sons: Hoboken, NJ, USA, 2017.
7. Wijeratne, D.S.; Moschopoulos, G. Quadratic power conversion for power electronics: Principles and circuits. *IEEE Trans. Circuits Syst. I Regul. Pap.* **2012**, *59*, 426–438. [CrossRef]
8. Lopez-Santos, O.; Martinez-Salamero, L.; Garcia, G.; Valderrama-Blavi, H.; Sierra-Polanco, T. Robust sliding-mode control design for a voltage regulated quadratic Boost Converter. *IEEE Trans. Power Electron.* **2015**, *30*, 2313–2327. [CrossRef]
9. Chen, Z.; Yang, P.; Zhou, G.; Xu, J.; Chen, Z. Variable duty cycle control for quadratic boost PFC converter. *IEEE Trans. Ind. Electron.* **2016**, *63*, 4222–4232. [CrossRef]
10. D. Langarica-Cordoba, L.; Diaz-Saldierna, H.; Leyva-Ramos, J. Fuel-cell energy processing using a quadratic boost converter for high conversion ratios. In Proceedings of the IEEE 6th International Symposium on Power Electronics for Distributed Generation Systems (PEDG), Aachen, Germany, 22–25 June 2015; pp. 1–7.
11. Deivasundari, P.S.; Uma, G.; Poovizhi, R. Analysis and experimental verification of Hopf bifurcation in a solar photovoltaic powered hysteresis current-controlled cascaded-boost converter. *IET Power Electron.* **2013**, *6*, 763–773. [CrossRef]
12. El Aroudi, A. Prediction of subharmonic oscillation in a PV-fed quadratic boost converter with nonlinear inductors. In Proceedings of the IEEE International Symposium on Circuits and Systems (ISCAS), Florence, Italy, 27–30 May 2018; pp. 1–5.
13. Valderrama-Blavi, H.; Bosque, J.M.; Guinjoan, F.; Marroyo, L.; Martinez-Salamero, L. Power adaptor device for domestic DC microgrids based on commercial MPPT inverters. *IEEE Trans. Ind. Electron.* **2013**, *60*, 1191–1203. [CrossRef]

14. Femia, N.; Petrone, G.; Spagnuolo, G.; Vitelli, M. Optimization of perturb and observe maximum power point tracking method. *IEEE Trans. Power Electron.* **2005**, *20*, 963–973. [CrossRef]

15. Banerjee, S.; Verghese, G.C. (Eds.) *Nonlinear Phenomena in Power Electronics: Attractors, Bifurcations Chaos, and Nonlinear Control*; IEEE Press: New York, NY, USA, 2001.

16. Tse, C.K. *Complex Behavior of Switching Power Converters*; CRC Press: New York, NY, USA, 2003.

17. Cheng, L.; Ki, W.H.; Yang, F.; Mok, P.K.T.; Jing, X. Predicting subharmonic oscillation of voltage-mode switching converters using a circuit-oriented geometrical approach. *IEEE Trans. Circuits Syst. I Regul. Pap.* **2017**, *64*, 717–730. [CrossRef]

18. Lu, W.; Li, S.; Chen, W. Current-ripple compensation control technique for switching power converters. *IEEE Trans. Ind. Electron.* **2018**, *65*, 4197–4206. [CrossRef]

19. Islam, H.; Mekhilef, S.; Shah, N.B.M.; Soon, T.K.; Seyedmahmousian, M.; Horan, B.; Stojcevski, A. Performance Evaluation of maximum power point tracking approaches and photovoltaic systems. *Energies* **2018**, *11*, 365. [CrossRef]

20. Abusorrah, A.; Al-Hindawi, M.M.; Al-Turki, Y.; Mandal, K.; Giaouris, D.; Banerjee, S.; Voutetakis, S.; Papadopoulou, S. Stability of a boost converter fed from photovoltaic source. *Sol. Energy* **2013**, *98*, 458–471. [CrossRef]

21. Al-Hindawi, M.; Abusorrah, A.; Al-Turki, Y.; Giaouris, D.; Mandal, K.; Banerjee, S. Nonlinear dynamics and bifurcation analysis of a boost converter for battery charging in photovoltaic applications. *Int. J. Bifurc. Chaos* **2014**, *24*, 1450142. [CrossRef]

22. Giaouris, D.; Banerjee, S.; Zahawi, B.; Pickert, V. Stability analysis of the continuous-conduction-mode buck converter via Filippov's method. *IEEE Trans. Circuits Syst. I Regul. Pap.* **2008**, *55*, 1084–1096. [CrossRef]

23. El Aroudi, A. Out of Maximum Power Point of a PV system because of subharmonic oscillations. In Proceedings of the International Symposium on Nonlinear Theory and Its Applications, NOLTA2017, Cancún, Mexico, 4–7 December 2017.

24. Lee, J.H.; Bae, H.S.; Cho, B.H. Resistive control for a photovoltaic battery charging system using a microcontroller. *IEEE Trans. Ind. Electron.* **2008**, *55*, 2767–2775. [CrossRef]

25. Valencia, P.A.O.; Ramos-Paja, C.A. Sliding-mode controller for maximum power point tracking in grid-connected photovoltaic systems. *Energies* **2015**, *8*, 12363–12387. [CrossRef]

26. Shmilovitz, D. On the control of photovoltaic maximum power point tracker via output parameters. *IEE Proc. Electr. Power Appl.* **2005**, *152*, 239–248. [CrossRef]

27. Xiao, W.; Ozog, N.; Dunford, W.G. Topology study of photovoltaic interface for maximum power point tracking. *IEEE Trans. Ind. Electron.* **2007**, *54*, 1696–1704. [CrossRef]

28. Bianconi, E.; Calvente, J.; Giral, R.; Mamarelis, E.; Petrone, G.; Ramos-Paja, G.C.A.; Spagnuolo, G.; Vitelli, M.M. A fast current-based MPPT technique employing sliding mode control. *IEEE Trans. Ind. Electron.* **2012**, *60*, 1168–1178. [CrossRef]

29. Huang, L.; Qiu, D.; Xie, F.; Chen, Y.; Zhang, B. Modeling and stability analysis of a single-phase two-stage grid-connected photovoltaic system. *Energies* **2017**, *10*, 2176. [CrossRef]

30. Rodrigues, E.M.G.; Godina, R.; Marzband, M.; Pouresmaeil, E. Simulation and Comparison of Mathematical Models of PV Cells with Growing Levels of Complexity. *Energies* **2018**, *11*, 2902. [CrossRef]

31. Villalva, M.; Gazoli, J.; Filho, E. Comprehensive approach to modeling and simulation of photovoltaic arrays. *IEEE Trans. Power Electron.* **2009**, *24*, 1198–1208. [CrossRef]

32. BP Solar BP585 Datasheet. Available online: http://www.electricsystems.co.nz/documents/BPSolar85w.pdf (accessed on 19 December 2018).

33. Krein, P.T.; Balog, R.S.; Mirjafari, M. Minimum energy and capacitance requirements for single-phase inverters and rectifiers using a ripple port. *IEEE Trans. Power Electron.* **2012**, *27*, 4690–4698. [CrossRef]

34. El Aroudi, A. A new approach for accurate prediction of subharmonic oscillation in switching regulators—Part II: Case studies. *IEEE Trans. Power Electron.* **2017**, *32*, 5835–5849. [CrossRef]

35. El Aroudi, A. A new approach for accurate prediction of Subharmonic oscillation in switching regulators—Part I: Mathematical derivations. *IEEE Trans. Power Electron.* **2017**, *32*, 5651–5665. [CrossRef]

36. Xiao, W.; Dunford, W.G.; Palmer, P.R.; Capel, A. Regulation of photovoltaic voltage. *IEEE Trans. Ind. Electron.* **2007**, *54*, 1365–1374. [CrossRef]

37. Al-Turki, Y.; El Aroudi, A.; Mandal, K.; Giaouris, D.; Abusorrah, A.; Al Hindawi, M.; Banerjee, S. Nonaveraged control-oriented modeling and relative stability analysis of DC-DC switching converters. *Int. J. Circuit Theory Appl.* **2018**, *46*, 565–580. [CrossRef]

38. Haroun, R.; El Aroudi, A.; Cid-Pastor, A.; Garcia, G.; Olalla, C.; Martinez-Salamero, L. Impedance matching in photovoltaic systems using cascaded boost converters and sliding-mode control. *IEEE Trans. Power Electron.* **2015**, *30*, 3185–3199. [CrossRef]

39. Leyva, R.; Alonso, C.; Queinnec, I.; Cid-Pastor, A.; Lagrange, D.; Martinez-Salamero, L. MPPT of photovoltaic systems using extremum-seeking control. *IEEE Trans. Aerosp. Electron. Syst.* **2006**, *42*, 249–258. [CrossRef]

40. Zhou, Y.; Tse, C.K.; Qiu, S.S.; Lau, F.C.M. Applying resonant parametric perturbation to control chaos in the buck DC/DC converter with phase shift and frequency mismatch considerations. *Int. J. Bifurc. Chaos Appl. Sci. Eng.* **2003**, *13*, 3459–3471. [CrossRef]

41. Di Capua, G.; Femia, N. A novel method to predict the real operation of ferrite inductors with moderate saturation in switching power supplies applications. *IEEE Trans. Power Electron.* **2016**, *31*, 2456–2464. [CrossRef]

Article

Control of a DC-DC Buck Converter through Contraction Techniques

David Angulo-Garcia [1],*, Fabiola Angulo [2], Gustavo Osorio [2] and Gerard Olivar [3]

[1] Grupo de Modelado Computacional–Dinámica y Complejidad de Sistemas, Instituto de Matemáticas Aplicadas, Universidad de Cartagena, Carrera 6 # 36–100, Cartagena de Indias 130001, Bolívar, Colombia
[2] Departamento de Ingeniería Eléctrica, Electrónica y Computación, Percepción y Control Inteligente–Bloque Q, Universidad Nacional de Colombia–Sede Manizales, Facultad de Ingeniería y Arquitectura, Campus La Nubia, Manizales 170003, Colombia; fangulog@unal.edu.co (F.A.); gaosoriol@unal.edu.co (G.O.)
[3] Departamento de Matemáticas, Percepción y Control Inteligente–Bloque W, Universidad Nacional de Colombia–Sede Manizales, Facultad de Ciencias Exactas y Naturales, Campus La Nubia, Manizales 170003, Colombia; golivart@unal.edu.co
* Correspondence: dangulog@unicartagena.edu.co

Received: 20 September 2018; Accepted: 25 October 2018; Published: 8 November 2018

Abstract: Reliable and robust control of power converters is a key issue in the performance of numerous technological devices. In this paper we show a design technique for the control of a DC-DC buck converter with a switching technique that guarantees both good performance and global stability. We show that making use of the contraction theorem in the Jordan canonical form of the buck converter, it is possible to find a switching surface that guarantees stability but it is incapable of rejecting load perturbations. To overcome this, we expand the system to include the dynamics of the voltage error and we demonstrate that the same design procedure is not only able to stabilize the system to the desired operation point but also to reject load, input voltage, and reference voltage perturbations.

Keywords: DC-DC buck converter; contraction analysis; global stability; matrix norm

1. Introduction

Many industrial and residential applications use voltage regulation with DC-DC power converters; such applications include fuel cells [1], photovoltaic sources [2,3], control of DC motors [4], lighting appliances [5], computer power supplies [6], and many others. Power converters transform a non regulated voltage/current source (DC or AC) into a regulated voltage/current output, which can be either larger or smaller than the non regulated input. Usually, the underlying structures in these devices are the so-called buck (step-down), boost (step-up), buck-boost (step down-step up), flyback, Ćuk, to mention few, depending on the type of application [7,8]. DC-DC power converters show both fast speed and capability of managing high power if needed [9]. More than 90% of the total amount of power supply in the world is processed through power converters [10]. For this reason, a precise control of these converters is a critical factor and therefore a vast amount of literature has been devoted to their control. For instance, PID-based schemes [11], Fuzzy PID control [12], robust controllers [13], predictive control [14], sliding mode control [15], and a controller based on a modified pulse-adjustment of the PWM [16], just to mention few.

The DC-DC buck power converter supplies a lower voltage than the input voltage and is one of the most widely studied power converters: Some recent applications include battery chargers [17], hybrid electric vehicles [18], quadropter's control [19], among others. The underlying topology of the buck converter is non-smooth, meaning that it switches back and forth according to a control signal, between an ON and OFF state, to guarantee a required output voltage. Some examples of control techniques applied to the buck converter include zero voltage control technique [20], fractional derivative control [21], controller based on active ramp tracking [22], and fuzzy PID controllers [23].

Even though the effectiveness of these control actions is out of doubt, usually the design of such controllers are based either on the averaged version of the system which effectively disregards the non-smoothness; or via linearization. This is because the effect of the nonlinearity is not always entirely understood and therefore the system can only be analyzed in the vicinity of the operation point (see [24] for a review on stability methods). This may result in undesired effects such as destabilization when the system is far from the operation point and limits the range of operation in which the DC-DC converter can work. This is because linearizing the system can only assure local stability, and the region of attraction is usually unknown.

Recently, a novel method to design an asymptotically globally stable controller for switched systems has been introduced in the literature [25,26] following the ideas of contraction theory, also used in [27]. Inspired by these papers, where some illustrative cases were developed in a few academic examples with limited application into the physical realm, the aim of this paper is to use the novel concepts of contraction theory on switched systems to design a switched controller that guarantees asymptotic global stability on the buck DC-DC power converter. With this purpose, the paper is organized as follows: In Section 2 we present some preliminary concepts needed for the development of the paper, specifically on linear transformations, matrix measure, Filippov systems, and contraction theory. After, in Section 3, the buck power converter is presented as well as its principle of operation. In Section 4, a controller based on contraction theory is designed and tested for the buck power converter. As the system is not robust, in Section 5 we develop a modified control action that uses the principles of integral control which shows robustness preserving global stability. We conclude this paper with some remarks and future perspectives.

2. Mathematical Methods

In this section we present some standard theory on linear systems (see [28]), matrix measures [29–31] and contraction theory applied to stability of switched systems [25,26] Most of the material can be found in the cited documents and references therein.

2.1. Linear Transformations

Let us consider the piece-wise linear system (PWLS) given by

$$\dot{\mathbf{x}} = A\mathbf{x} + Bu \,, \tag{1}$$

where $u \in \{u_1, u_2\}$ and it commutes between booth values depending on the value of the switching surface $h(\mathbf{x}) = 0$. A is a Hurwitz matrix, the pair (A, B) is controllable and all eigenvalues are distinct but not necessarily real. Then, there exists a real matrix P which transforms the original system into a canonical form, so called the Jordan form, in the following way:

$$A_J = P^{-1}AP \qquad B_J = P^{-1}B \,. \tag{2}$$

The transformation matrix can be constructed as follows: For each real eigenvalue, its corresponding eigenvector is computed and assigned to one column of the matrix P. For every pair of complex eigenvalues their corresponding complex eigenvectors are computed but only one of them is used to construct two column vectors of the matrix P. The first one is composed by the real parts of the complex eignevector, while the other one is composed by the imaginary parts of the same eigenvector. For example, in a system with one real eigenvector $\mathbf{v}_1$ and two complex conjugate $\mathbf{v}_2$ and $\hat{\mathbf{v}}_2$ the matrix P takes the form

$$P = [\mathbf{v}_1 \quad \mathrm{Re}(\mathbf{v}_2) \quad \mathrm{Im}(\mathbf{v}_2)] \,. \tag{3}$$

Using this transformation matrix we obtain that every real eigenvalue ($\lambda_j = \eta_j$) produces a column in the matrix A_J with the eigenvalue in the corresponding diagonal element with other elements equal to zero. Every complex pair of eigenvalues ($\lambda_{k,k+1} = \alpha_k \pm \beta_k$) instead, generates a 2 × 2 block in the Jordan matrix such that the diagonal part corresponds to the real part of the eigenvalues and the other positions correspond to the positive and negative imaginay part: Other elements are zero. A general example of this Jordan form is:

$$
A_J = \begin{bmatrix}
\eta_1 & 0 & 0 & 0 & 0 & 0 & \cdots \\
0 & \ddots & 0 & 0 & 0 & 0 & \cdots \\
0 & \cdots & \alpha_k & -\beta_k & 0 & 0 & \cdots \\
0 & \cdots & \beta_k & \alpha_k & 0 & 0 & \cdots \\
0 & \cdots & 0 & 0 & \alpha_{k+2} & -\beta_{k+2} & \cdots \\
0 & \cdots & 0 & 0 & \beta_{k+2} & \alpha_{k+2} & \cdots \\
\vdots & \vdots & \vdots & \vdots & \vdots & \vdots & \ddots
\end{bmatrix}.
\tag{4}
$$

2.2. Matrix Measure

The norm-2 induced measure of a matrix A is defined as:

$$
\mu_2(A) = \lambda_{\max}[A' + A]/2,
\tag{5}
$$

where $\lambda_{max}[\cdot]$ is the largest eigenvalue and A' is the transpose of A. It is possible to verify that, if the matrix A in (1) is Hurwitz, then $\mu_2(A_J)$ is always negative. This is an important issue in the stability analysis performed in this paper.

2.3. Contraction Analysis for Filippov Systems

Another way to define the system (1) is as a bimodal Filippov system

$$
\dot{x} = \begin{cases} F^+(x) & \text{if } x \in S^+ \\ F^-(x) & \text{if } x \in S^-. \end{cases}
\tag{6}
$$

where

$$
F^+(x) = Ax + Bu_1 \quad \text{and} \quad F^-(x) = Ax + Bu_2
$$

being

$$
S^+ = \{x \in \mathcal{U} : h(x) > 0\} \quad \text{and} \quad S^- = \{x \in \mathcal{U} : h(x) < 0\}.
$$

Here, $h : \mathcal{U} \to R$ is a smooth function called switching function and the surface Σ defined as

$$
\Sigma = \{x \in \mathcal{U} : h(x) = 0\}
\tag{7}
$$

is called the switching surface.

According to [25,26], the bimodal Filippov system (6) is incrementally exponentially stable in a so-called K-reachable set $\mathcal{C} \subseteq \mathcal{U}$ with convergence rate $r = \min\{r_1, r_2\}$, if there exists some norm in $\mathcal{C}$ with associated measure μ such that for some positive constants r_1, r_2

$$
\mu\left(\frac{\partial F^+(x)}{\partial x}\right) \leq -r_1 \qquad \forall x \in \overline{S^+},
$$
$$
\mu\left(\frac{\partial F^-(x)}{\partial x}\right) \leq -r_2 \qquad \forall x \in \overline{S^-},
\tag{8}
$$

and

$$\mu((F^+(\mathbf{x}) - F^-(\mathbf{x})) \cdot \nabla h(\mathbf{x})) = 0 \qquad \forall \mathbf{x} \in \Sigma, \tag{9}$$

where $\overline{S^+}$ and $\overline{S^-}$ represent the closures of the sets S^+ and S^- respectively.

If system (6) is incrementally exponentially stable then there exist constants $k \geq 1$ and $\lambda > 0$ such that

$$|\mathbf{x}(t) - \mathbf{y}(t)| \leq k e^{-\lambda(t-t_0)} |\mathbf{x}(0) - \mathbf{y}(0)| \; \forall t \geq t_0 \; \forall \mathbf{x}(0), \mathbf{y}(0) \in \mathcal{C}$$

where $\mathbf{x}(t)$ and $\mathbf{y}(t)$ are solutions of the system. Thus we can establish global stability properties for system (1). Making use of the previous concepts, we will design an hybrid control for a buck power converter that guarantees not only global stability, but is also robust to different disturbances.

3. The Buck Power Converter

The scheme of a buck power converter is depicted in Figure 1. The equations describing this dynamical system in Continuous Conduction Mode CCM (see [10,32,33]) are

$$\begin{pmatrix} \dot{v} \\ \dot{i} \end{pmatrix} = \begin{pmatrix} -\frac{1}{RC} & \frac{1}{C} \\ -\frac{1}{L} & 0 \end{pmatrix} \begin{pmatrix} v \\ i \end{pmatrix} + \begin{pmatrix} 0 \\ \frac{E}{L} \end{pmatrix} u \tag{10}$$

where R is the load resistance, C is the capacitor's capacitance, L is the coil's inductance, and E is the voltage provided by the power source. The state variable v corresponds to the voltage across the capacitor and i quantifies the current flowing through the inductor. The control signal u takes values in the discrete set $\{0, 1\}$. When $u = 0$ the switch is opened and the power source (input voltage) does not feed the system. In this case, the load is being fed by the capacitor and the inductor. For simplicity we will perform a first transformation which maps the original system (10) into a dimensionless framework by means of the following similarity transformation $x = M^{-1}(v\ i)'$, where

$$M = \begin{pmatrix} E & 0 \\ 0 & \frac{E}{\sqrt{L/C}} \end{pmatrix} \tag{11}$$

Also we perform a normalization of the time as $\tau = t/\sqrt{LC}$, such that a new and unique parameter $\gamma = \frac{1}{R}\sqrt{\frac{L}{C}}$ holds the information of the parameters in the system. Therefore we can rewrite the equations as:

$$\begin{pmatrix} \dot{x}_1 \\ \dot{x}_2 \end{pmatrix} = \begin{pmatrix} -\gamma & 1 \\ -1 & 0 \end{pmatrix} \begin{pmatrix} x_1 \\ x_2 \end{pmatrix} + \begin{pmatrix} 0 \\ 1 \end{pmatrix} u \tag{12}$$

or in a compact form as $\dot{x} = Ax + Bu$.

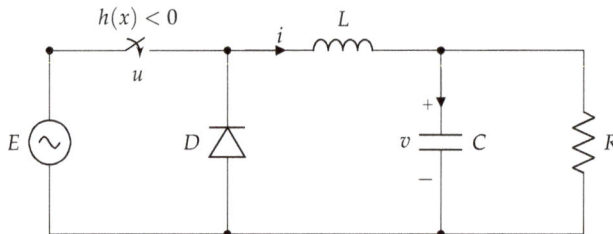

Figure 1. Schematic diagram of a buck power converter.

With the aim of designing the controller, it is necessary to transform the system to the Jordan normal form. As the pair (A, B) is controllable, then as outlined in Section 2.1, there exists a transformation matrix P given by

$$P = \begin{pmatrix} \gamma/2 & \rho \\ 1 & 0 \end{pmatrix} \tag{13}$$

which transforms the system into:

$$\begin{pmatrix} \dot{z}_1 \\ \dot{z}_2 \end{pmatrix} = \begin{pmatrix} -\gamma/2 & -\rho \\ \rho & -\gamma/2 \end{pmatrix} \begin{pmatrix} z_1 \\ z_2 \end{pmatrix} + \begin{pmatrix} 1 \\ -\gamma/(2\rho) \end{pmatrix} u \tag{14}$$

where $z = P^{-1}x$ and we have used $\rho := \rho(\gamma) = \sqrt{4 - \gamma^2}/2$. These equations are noted in a compact form as $\dot{z} = A_J z + B_J u$.

4. Application to 2D-Case

4.1. Controller Design

Using the contraction theorem outlined in Section 2.3, we can establish that the converter operating with the switched signal control u will be stable if the following two conditions are satisfied:

$$a)\ \mu_2(A_J) < -r_1\ , \forall z$$
$$b)\ \mu_2(B_J \cdot \nabla h(z)) = 0, \forall z \in h(z) = 0\ . \tag{15}$$

One can easily show that $\mu_2(A_J) = -\gamma/2$, hence condition (a) is always met as γ is always positive. Then, considering $h(z)$ as a linear function of the states $h(z) = (h_1\ h_2) \cdot (z_1\ z_2)'$, the condition (b) can be written as:

$$\mu_2 \left(\begin{pmatrix} 1 \\ -\gamma/(2\rho) \end{pmatrix} \cdot (h_1\ h_2) \right) = 0\ . \tag{16}$$

It is possible to demonstrate (see Appendix A) that the following choice of $h(z)$:

$$h(z) = h_1 z_1 + \frac{B_J(2)}{B_J(1)} h_1 z_2 = (h_1 - h_1\gamma/(2\rho)) \cdot (z_1\ z_2)' := h_z \cdot z\ , \tag{17}$$

where $B_J(i)$ is the $i - th$ row element in B_J, fulfills condition (16) if the pairs $\{B_J(i), h(i)\}$ have opposite signs. Then, according to the signs of $B_J(i)$, it is necessary to choose $h_1 < 0$ and $h_2 > 0$. In this way, the matrix from which the maximum eigenvalue needs to be calculated according to Equation (5), has one null eigenvalue and the other one can be computed as $\lambda_2 = h_1/\rho^2$, which is smaller than zero. Since the switching surface has been calculated in the canonical space, this result needs to be transformed back into the dimensionless state variables through $x = Pz$. The switching manifold is then obtained as $h(x) = h_z \cdot P^{-1} \cdot x$, or equivalently:

$$h(x) = (h_1\ h_1(1 + (\gamma/(2\rho))^2)) \cdot (x_1\ x_2)' := h_x \cdot x \tag{18}$$

Of course, the term h_x correspond the vector in the normal direction of the switching surface. With the aim of simplifying the calculations we normalize such vector such that $|h_x| = 1$. Moreover, we need to subtract the reference values to the states to ensure the regulation to the operation point.

$$h(x) = \left(-\frac{\gamma}{\sqrt{4+\gamma^2}}\quad \frac{2}{\sqrt{4+\gamma^2}} \right) \cdot (x_1 - \bar{x}_{1ref}\quad x_2 - \bar{x}_{2ref})' = 0\ . \tag{19}$$

Finally, we can define the switching manifold in terms of the original state variables $(x_1 \ x_2) = M^{-1}(v \ i)'$ leading to:

$$h(v,i) = \left(-\frac{\gamma}{E\sqrt{4+\gamma^2}} \quad \frac{2\sqrt{L/C}}{E\sqrt{4+\gamma^2}} \right) \cdot (v - \bar{v}_{ref} \quad i - \bar{i}_{ref})' = 0. \tag{20}$$

It is worth noticing that neither h_1 nor h_2 appear in the calculations. On the one hand h_2 is parametrized via h_1 (see Appendix A), on the other hand h_1 disappear via the normalization, reducing effectively two degrees of freedom. Also, for the sake of simplifying the calculations we have considered a switching function with zero offset. Introducing the offset in this function, which amounts to perform a translation of the switching surface, does not change any of the stability criteria that we are presenting here and can indeed be employed as a further degree of freedom.

4.2. Simulation Results

The design methodology described so far is independent of the parameters. However, in order to show the numerical behavior in a realistic set up we will use the following set of parameters for numerical computations: $L = 2$ mH, $C = 40$ μF, $E = 40$ V, and $\bar{v}_{ref} = 32$ V. This range of input/output operation can be found, for instance, in solar panel arrays feeding a battery charger through a buck converter. In our particular numerical example we will assume a load $R = 20\,\Omega$. The desired current reference can be assumed to be $\bar{i}_{ref} = \bar{v}_{ref}/R = 1.6$ A. With this, $\gamma \approx 0.35$ and $\rho \approx 0.98$. Also, as electronic devices cannot switch with infinite speed, it is necessary to implement a hysteresis band for simulating the change in the position of the MOSFET. We have designed this band in such a way that the switching time is close to 175 μs. The size of the hysteresis band is an important issue because its width also determines the size of the chattering in the voltage variable. Under these assumptions, Equation (20) takes the following values:

$$h(v,i) = \left(-4.4 \times 10^{-3} \quad 0.1741 \right) \cdot (v - \bar{v}_{ref} \quad i - \bar{i}_{ref})' \pm 0.02 \tag{21}$$

In Figure 2 we show the performance of the designed control. In particular, in Figure 2a the time trace of the voltage v is depicted in response to a drastic change in the reference output voltage $\bar{v}_{ref}$. During the first 30 ms, where the system is subject to $\bar{v}_{ref} = 32$ V (top dashed line), the output voltage reaches the steady state close to 5.7 ms, with no overshoot and the maximum error in steady state is lower than 0.6% (see inset). After 30 ms, the reference voltage is changed to $\bar{v}_{ref} = 16$ V (bottom dashed line) and the system is able to track the change and stabilize to the new value of output voltage. In Figure 2b,c we plot the orbit in the (v, i) space during the steady state for the two references used in panel A) of the same figure. From this, one can observe that indeed the equilibrium value $(\bar{v}_{ref}, \bar{i}_{ref})$ is reached through the continuous rippling of the orbit around the equilibrium point (red symbol).

We also tested the robustness of the control to changes in the load. In Figure 3 is depicted the time trace of the voltage in this scenario. Following a similar procedure as in Figure 2, after 30 ms, a change in the resistance from $R = 20\,\Omega$ to $R = 15\,\Omega$ (10% difference) is applied. From this figure it is possible to see that the system drifts away from the reference output $\bar{v}_{ref} = 32$ V (dashed line), producing a steady state error of around 18% (see inset).

So far, the controller designed with contraction theory has been successful to operate in a desired way and reject disturbances in the output voltage. However, when a disturbance in the load is presented (a common situation in power converters) the system loses the ability to follow the desired output voltage, indicating that the controller is not robust. To solve this problem, we extend the proposed controller based on the idea of an integral control action.

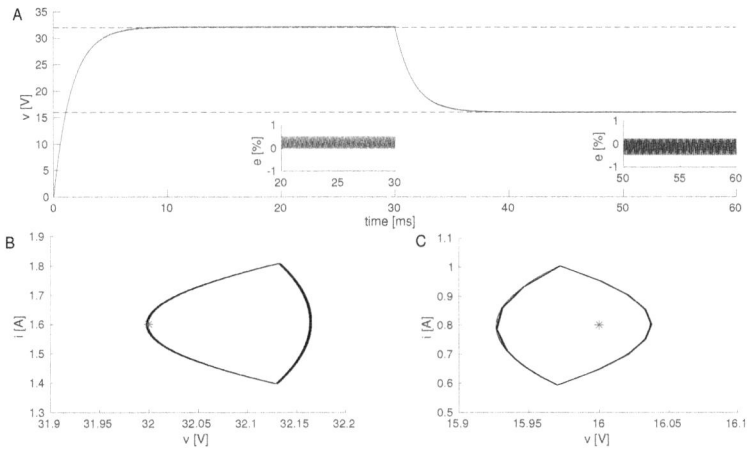

Figure 2. (**A**) Time trace of the voltage in the capacitor v. During the first 30 ms a $\bar{v}_{ref} = 32$ V is used, after this a drastic change to $\bar{v}_{ref} = 16$ V is applied (depicted in the dashed lines). The time trace of the steady state percentage error is also depicted in the insets for both values of $\bar{v}_{ref}$; (**B**) Phase representation of the steady state for $\bar{v}_{ref} = 32$; (**C**) $\bar{v}_{ref} = 16$, with the equilibrium point indicated by the red star. Simulations were performed using MATLAB® with a fourth order Runge-Kutta algorithm with variable step and event detection to identify collisions with the hysteresis band. Steady state was considered after 20 ms of simulation time. Initial conditions were chosen as $(v, i) = (0, 0)$. Other parameters as in the main text.

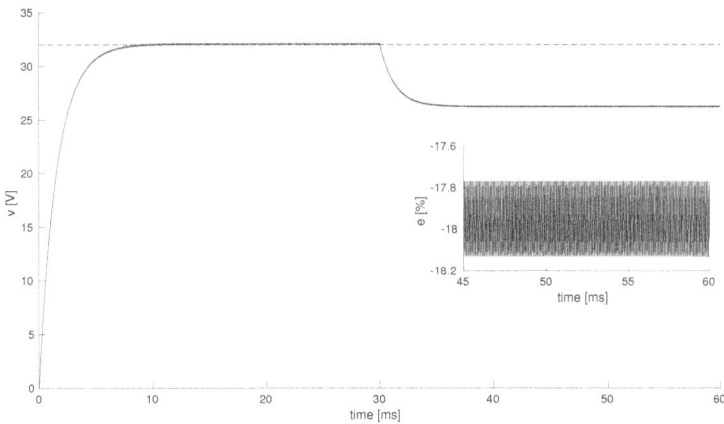

Figure 3. Time response of the capacitor's voltage v. During the first 30 ms, the value of the resistor is set to $R = 20\,\Omega$, after this the load is changed to $R = 18\,\Omega$. Inset: Steady state percentage error (considered 15 ms after the presentation of the disturbance). The desired output is plot with the dashed line. Other details as in Figure 2.

5. Application to 3D-Case

5.1. Controller Design Based on a Modified Integral Control Action

In control theory it is known that perturbations are better rejected by a PI controller; however, in this case, adding a PI controller implies to add a pole in the origin of the system which prevents us from applying contraction theorem. Then, with the aim of enhancing the robustness of the controlled system, we will modify the control action in such a way that it introduces the dynamics of the error. To do so we introduce a new state variable x_3 in the dimensionless system in the following way: $\dot{x}_3 = e - \delta x_3$, with $e = \tilde{x}_{1ref} - x_1$ defined as the output error and δ as the time constant of x_3. As $x_1 = v/E$, then $\tilde{x}_{1ref} = \bar{v}_{ref}/E$. Under these assumptions, the system takes the following form:

$$\begin{pmatrix} \dot{x}_1 \\ \dot{x}_2 \\ \dot{x}_3 \end{pmatrix} = \begin{pmatrix} -\gamma & 1 & 0 \\ -1 & 0 & 0 \\ -1 & 0 & -\delta \end{pmatrix} \begin{pmatrix} x_1 \\ x_2 \\ x_3 \end{pmatrix} + \begin{pmatrix} 0 \\ 1 \\ 0 \end{pmatrix} u + \begin{pmatrix} 0 \\ 0 \\ 1 \end{pmatrix} \tilde{x}_{1ref} \tag{22}$$

with

$$u = \begin{cases} 1 & \text{if } h(\mathbf{x}) \leq 0 \\ 0 & \text{otherwise.} \end{cases} \tag{23}$$

or in compact form $\dot{\mathbf{x}} = A\mathbf{x} + B u + Q\tilde{x}_{1ref}$. The aim of the term $-\delta$ appearing in position $\{3,3\}$ in the matrix A is to stabilize the system allowing us to apply the contraction theorem. In this way, the value of δ must be very small to avoid high steady state error. As the pair (A, B) is controllable we then proceed to apply the general theory with a new consideration: In the construction of the matrix P we will take into account the norm of the eigenvectors $\mathbf{v}_i$, which will allow us to gain more degrees of freedom in the system to tune the controller. Indeed this is not an issue when obtaining the canonical form A_J as the operation $P^{-1}AP$ cancels out any norm that may have been considered. However, the transformed matrix B_J, which is critical for the stability conditions Equation (15), may depend on the chosen modules of the eigenvectors. To take this into account, we need to include in Equation (3) the magnitude of the eigenvectors via the scaling factors c_1 and c_2 as follows:

$$P = [c_1 \mathbf{v}_1 \quad c_2 \text{Re}(\mathbf{v}_2) \quad c_2 \text{Im}(\mathbf{v}_2)]. \tag{24}$$

The general form of the transformation matrix can then be written as

$$P = \begin{pmatrix} 0 & c_2(\gamma - 2\delta)/2 & -c_2\rho \\ 0 & c_2(2 - \gamma\delta)/2 & -c_2\rho\delta \\ c_1 & c_2 & 0 \end{pmatrix} \tag{25}$$

which leads to the transformed system $\dot{\mathbf{z}} = A_J \mathbf{z} + B_J u + Q_J \tilde{x}_{1ref}$, where

$$A_J = \begin{pmatrix} -\delta & 0 & 0 \\ 0 & -\gamma/2 & \rho \\ 0 & -\rho & -\gamma/2 \end{pmatrix} \tag{26}$$

$$B_J = \begin{pmatrix} -1/(c_1(\delta^2 - \gamma\delta + 1)) \\ 1/(c_2(\delta^2 - \gamma\delta + 1)) \\ -(2\delta - \gamma)/(2c_2\rho(\delta^2 - \gamma\delta + 1)) \end{pmatrix}$$

$$Q_J = \begin{pmatrix} 1 & 0 & 0 \end{pmatrix}'$$

The purpose will be again to find a switching function $h(\mathbf{z}) = h_1 z_1 + h_2 z_2 + h_3 z_3$ that meets the conditions of global stability in Equation (15) in the transformed space. One can easily verify that, provided that $\delta < \gamma/2$, $\mu(A_J) = -\gamma/2$, fulfilling condition (a). Moreover, one of the eigenvalues of $B_J \cdot \nabla h(\mathbf{z})$ is always 0 due to the fact that the matrix is constructed using only two linearly independent vectors (see Appendix A.2). Also, following a similar procedure as in the 2D case, choosing the following switching function:

$$h(\mathbf{z}) = h_1 z_1 + \frac{B_J(2)}{B_J(1)} h_1 z_2 + \frac{B_J(3)}{B_J(1)} h_1 z_3, \tag{27}$$

the condition (b) in Equation (9) is always guaranteed if, for every pair $\{B_J(i), h(i)\}$, its elements have opposite signs and the signs of c_1 and h_1 are equal (see Appendix A). From this, the switching surface in the canonical space is:

$$h(\mathbf{z}) = \left(h_1 \quad -h_1 \frac{c_1}{c_2} \quad h_1 \frac{c_1(2\delta - \gamma)}{2c_2\rho} \right) \cdot (z_1 \ z_2 \ z_3)' := h_z \cdot \mathbf{z} \tag{28}$$

It is worth noticing that for the 2D case, considering arbitrary norms for the eigenvectors does not have an effect in the possible switching functions, in contrast to the extended system. This is because there is only one constant associated to that norm (two complex eigenvalues). Another important aspect is that the plane defined in Equation (28) depends on the ratio c_1/c_2 and not on their individual values which effectively reduces one degree of freedom in the tunning parameters of the hybrid controller based on the integral action. As in the previous case, the next steps in the design are (i) apply the transformation to the dimensionless variables; (ii) normalize by the norm of the resulting orthogonal vector to the switching surface in the $\mathbf{x}$ space, i.e., $|h_z \cdot P^{-1}|$; and (iii) transform back to the original buck converter states variables (v, i) via the matrix M (recall that the similarity transformation is $x = M^{-1}(v \ i)'$). It is important to notice that for the 3D system, the matrix M is not unique, as we don't know the exact mapping between the extended variable x_3 and its counterpart in the real system y. We can assume without loss of generality and preserving the idea of the integral action, that the mapping between x_3 and y is given by a scaling factor, which after some algebra can be demonstrated to be $x_3 = y/(E\sqrt{LC})$. This results preserves the information of the error defined by $\bar{v}_{ref}$. The similarity transformation matrix is then given by:

$$M = \begin{pmatrix} E & 0 & 0 \\ 0 & E/\sqrt{L/C} & 0 \\ 0 & 0 & E\sqrt{LC} \end{pmatrix}. \tag{29}$$

We will avoid displaying the rather long expression of performing the aforementioned steps, but they can be summarized in the operation:

$$h(v, i, y) = \frac{h_z \cdot P^{-1}}{|h_z \cdot P^{-1}|} \cdot M^{-1}(v \ i \ y)'. \tag{30}$$

The system finally reads in its original variables as:

$$\begin{pmatrix} \dot{v} \\ \dot{i} \\ \dot{y} \end{pmatrix} = \begin{pmatrix} -\frac{1}{RC} & \frac{1}{C} & 0 \\ -\frac{1}{L} & 0 & 0 \\ -1 & 0 & -\frac{\delta}{\sqrt{LC}} \end{pmatrix} \begin{pmatrix} v \\ i \\ y \end{pmatrix} + \begin{pmatrix} 0 \\ E/L \\ 0 \end{pmatrix} u + \begin{pmatrix} 0 \\ 0 \\ 1 \end{pmatrix} \bar{v}_{ref} \tag{31}$$

with

$$u = \begin{cases} 1 & \text{if } h(v, i, y) \leq 0 \\ 0 & \text{otherwise.} \end{cases} \tag{32}$$

Simulation Results

From Equations (30) and (31), the resulting controlled system can be tuned via two parameters, namely the time constant of the extended variable δ, and the ratio c_1/c_2 of the norm of the eigenvectors associated with matrix A. To tune these parameters, we performed an optimization routine which explored several possible combinations of parameters δ and c_1/c_2 in a wide range of values. Following an heuristic approximation we chose the values which met some desired criteria, namely small overshoot and small settling time. From this analysis we concluded that a sufficiently small value of δ is necessary in order for the steady state error to be small. Also, as c_1/c_2 is decreased, the system evolves faster but produces large overshoots; conversely, increasing the ratio reduces the overshoot but slows down the system. A good performance was achieved by choosing $\delta = 1 \times 10^{-4}$ and $c_1/c_2 = 9$. With these choices, the numerical values for the switching surface are:

$$h(v,i,y) = (-4.3 \times 10^{-3} \quad 0.1741 \quad -1.03) \cdot (v \quad i \quad y)' \pm 0.05 \, , \tag{33}$$

where we have set the hysteresis to a value that meets the MOSFET switching frequency criterion as in the previous section. It can be noted that this controller does not require any information about current reference as in 2D-case.

The results of the 3D system behavior and its ability to reject disturbances in the reference voltage are depicted in Figure 4. In this figure, a reference voltage of $\bar{v}_{ref} = 32$ V is applied during the first 40 ms of the simulation, after this, the reference voltage is drastically decreased by a 50%, i.e., $\bar{v}_{ref} = 16$ V and the system is allowed to evolve during 40ms more. From Figure 4a it is possible to deduce that, in the 3D system, the controller is also able to regulate with a settling time of ≈ 10 ms and a steady state error smaller than 1%. Not only this, but also the control is robust against disturbances in the reference output value. Panel B,C of the same figure show the orbit exhibited by the system in the steady state before and after the disturbance, which clearly evolves in the neighborhood of the equilibrium value $(\bar{v}_{ref}, \bar{i}_{ref})$ (red star).

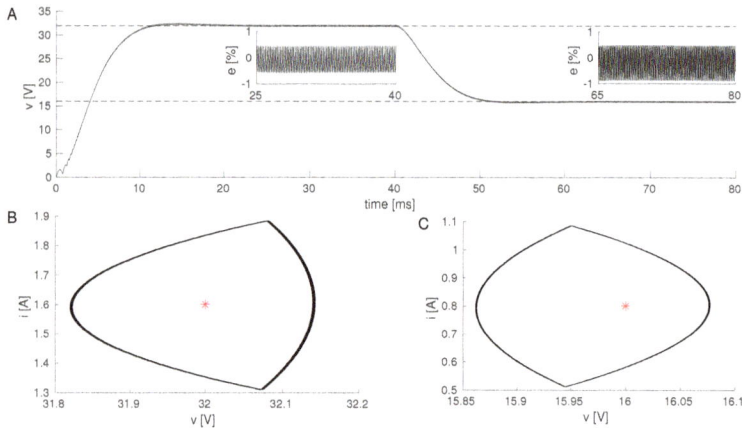

Figure 4. (a) Time trace of the voltage in the capacitor v. During the first 40 ms a $\bar{v}_{ref} = 32$ V is used, after this a drastic change to $\bar{v}_{ref} = 16$ V is applied (depicted in the dashed lines). The time trace of the steady state percentage error is also depicted in the insets for both values of $\bar{v}_{ref}$; (b) Phase representation of the steady state for $\bar{v}_{ref} = 32$; (c) $\bar{v}_{ref} = 16$, with the equilibrium point indicated by the red star. This results were obtained by making $c_1/c_2 = 9$ and $\delta = 1 \times 10^{-4}$. Steady state was considered after 25 ms of transient dynamics. Other parameters as in the main text and Figure 2.

We also tested the capability of the system to reject disturbances both in the load R and the input voltage E. To do so we simulated a similar set-up to the one described for the 2D system. In particular we evolved the unperturbed system during 40 ms to achieve a steady state, and immediately after the perturbation is presented. For Figure 5a the perturbation is induced as a sudden change in the load from $R = 20\,\Omega$ to $R = 15\,\Omega$ (25% change). As depicted in the main figure of the panel and its inset, the system recovers to the reference voltage $\bar{v}_{ref} = 32$ V (dashed line) with a percentage error smaller than 1%. A similar scenario is plotted in Figure 5b, in this case the perturbation is presented as a change in the input voltage from $E = 40$ V to $E = 50$ V. Even though the perturbation in the input corresponds to a 25% change, the system barely moves from its steady state, and the perturbation only induces a slight increase in the error. This small error, which is never larger than 1%, rapidly returns to the steady value after 10 ms. (see inset). An important aspect of the controller design is that the first two elements in the normal vector of the switching surface in Equation (33), are exactly those of Equation (21) for the 2D case, where neither the norm of the vectors nor δ were involved. Hence, the effect of c_1/c_2 and δ are only exhibited in the third term.

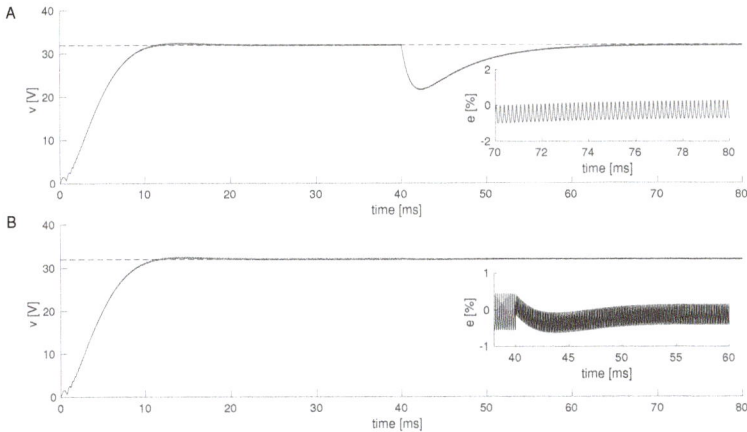

Figure 5. (**a**) Time response of the capacitor's voltage v. During the first 40ms, the value of the resistor is set to $R = 20\,\Omega$, after this the load is changed to $R = 15\,\Omega$. Inset: Steady state percentage error (considered 25 ms after the presentation of the disturbance). The desired output is plot with the dashed line; (**b**) Same as (**A**) for a disturbance in the input voltage E. During the first 40 ms $E = 40$ V, after this it is changed to $E = 50$ V. In this panel, the inset shows the percentage error during the first 20 ms after the presentation of the input disturbance. Other details as in Figure 2.

So far we have numerically analyzed the controller for a particular design of a buck converter determined by the values of the parameters R, L, C, and E and v_{ref}. However, as we demonstrated in Section 5, the methodology proposed here is general. To demonstrate this generality we performed simulations of our proposed controller under different values of parameters L and C which preserved the value of $\gamma \approx 0.3536$. To do so, we fixed the value of $R = 20\,\Omega$. Then the inductance was varied in the range $L = [20\,\mu\text{H}\ 10\,\text{mH}]$ and the capacitance was automatically set to $C = L/50$ F. According to the normalization of the buck converter, this amounts to change the time scale $\sqrt{LC}$ in which the power converter evolves, while keeping the dynamical behavior invariant (recall that the actual dynamics of the normalized system only depend on the value of γ). The results of this analysis are reported in Figure 6. In particular, Figure 6a shows how the settling time changes when varying the time scale $\sqrt{LC}$. Not surprisingly, increasing values of $\sqrt{LC}$ produce a linearly increase also in the settling time, since the effect of the former is to stretch and compress the time in which the system evolves. The inset in this panel shows also how the switching frequency increases as the evolution of the system is faster (decreasing values of $\sqrt{LC}$). This behavior emerges since we have kept the hysteresis band $\epsilon = \pm 0.05$, which highlights the need to use faster switches when trying to achieve faster dynamics. Finally we calculated also the average steady state error and the overshoot for each one of the $\sqrt{LC}$ values (see Figure 6b,c). On the one hand, the steady state error was always kept below 0.3%, regardless the time scale of the system. On the other hand, overshoot didn't change significantly which supports again the claim that the dynamical behavior remains unchanged under this particular design criteria.

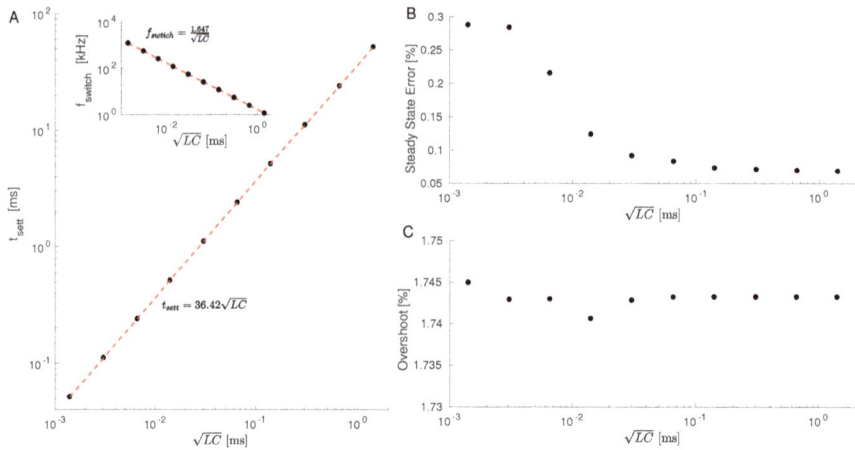

Figure 6. (A) Black symbols: Settling time of the buck converter with different values of $\sqrt{LC}$ and fixed $R = 20\,\Omega$, preserving the value of $\gamma = 0.3536$. Red line: Linear fit of the simulation points. For this plot, we chose inductance values in the range $L = [20\,\mu\text{H}\ 10\,\text{mH}]$ and a capacitance $C = L/50$ F. Settling time was calculated as the time it takes to the system to evolve from $(v,\ i,\ y) = (0,0,0)$ to the point in which the error doesn't leave the $\pm 2\%$ band. Inset: Switching frequency resulting from the hysteresis band in Equation (33) for the different values of $\sqrt{LC}$ reported in the main figure (black symbols). Red line depicts the fitted function reported in the text box; **(B)** Average steady state error calculated as the mean value of the error after the settling time; **(C)** percentage overshoot for the values of $\sqrt{LC}$ reported in panel **(A)**. For each simulation point, the system is evolved during a time span of $T = 200\sqrt{LC}$. Other details as in Figure 2.

6. Conclusions and Future Work

In this paper we developed a switched control action for the buck power converter that guarantees global asymptotic stability, by applying recent results from contraction analysis. To do so, we took advantage of the Jordan canonical form of the system to fulfill the conditions of global stability resulting from contraction analysis, which wouldn't have been met in the original form of the system. At first, we applied the design to the original 2D buck converter model where the controller presented good performance and robustness to voltage reference changes; however, as the load varied, regulation was lost. To overcome this issue, we extended the 2D-system to take into account the dynamics of the error inspired by the disturbance-rejection effect of a PI controller. With this design, the controlled system showed robustness to several types of disturbances including load and input voltage changes.

Although the 3D system is robust, it comes with the price of increasing the settling time respect to the 2D design. To overcome this issue, one can make use of a different buck converter design (different capacitance and/or inductance) to achieve the desired time-scale of the dynamics (which is mainly driven by the factor $\sqrt{LC}$) and then design the controller according to our methodology. It shall be noticed that other control techniques designed for the buck converter may show better performance in terms of efficiency, however, it is important to stress that the method outlined in this paper is not only simple in its implementation (design based on hysteresis band) but also quite general. This is because it is not based on the linearized version of the system but on the nonlinear form, such that the resulting controller is globally stable, a feature that cannot usually be guaranteed using linearization. Indeed we have numerically tested the globally stability property by performing extensive simulations for different initial conditions in the (v, i, y) space. These tests showed convergence for all the simulations.

Throughout this paper we have analyzed and designed the controller assuming that the current flowing through the inductor is always positive, a topology known as Continuous Conduction Mode (CCM). Depending on the value of the load and disturbances in it, the buck converter can enter in

Discontinuous Conduction Mode (DCM), where the current through the inductor is zero. The control design that we have provided in this paper only takes into consideration the dynamical behavior of the buck converter in CCM. Considering also DCM implies adding a further topology to the system (vector field) which implies the study of contraction in multimodal filippov systems. This issue is indeed a current topic of research in the field of applied mathematics.

Finally, whether the approach presented here can be applied to other power converters such as the boost, is currently an open problem. This is because not every single system can be easily approached by contraction theory and other standard tools for stability analysis might be the best option in these cases.

Author Contributions: Conceptualization was made by F.A. and D.A.-G.; Supervision and Project Administration were made by F.A., G.O. (Gustavo Osorio) and G.O. (Gerard Olivar); Software and Validation and Funding Acquisition by D.A.-G., F.A. and G.O. (Gustavo Osorio); Visualization made by D.A.-G. and G.O. (Gustavo Osorio). All authors contributed to Formal Analysis, Investigation and Writing.

Funding: F.A. and G.O. (Gustavo Osorio) were supported by Universidad Nacional de Colombia, Manizales, Project 31492 from Vicerrectoría de Investigación, DIMA, and COLCIENCIAS under Contract FP44842-052-2016. D.A.-G. was supported by Universidad de Cartagena through the "Vicerrectoría de Investigaciones" under contract No. 085-2018. G.O. (Gerard Olivar) was supported by Universidad Nacional de Colombia, Manizales, Project 35467 from Vicerrectoría de Investigación, DIMA, and also by COLCIENCIAS under Contract FP44842-022-2017.

Conflicts of Interest: The authors declare no conflict of interest. The founding sponsors had no role in the design of the study; in the collection, analyses, or interpretation of data; in the writing of the manuscript, and in the decision to publish the results.

Appendix A

In this appendix we prove that, for a particular selection of the constants h_i, $\mu_2(B \cdot \nabla h(\mathbf{x})) := \mu_2(B \cdot h) \leq 0$.

Appendix A.1. Matrix Measure for a 2D System

Without loss of generality, we can consider two vectors $B = [b_1 \ b_2]'$ and $h = [h_1 \ h_2]$. The matrix C is formed as $C = B \cdot h$ and matrix N is defined as

$$N = C + C'.$$

The measure of this matrix must be equal to zero over the switching surface to meet the theorem in [25,26]. Then:

$$\mu_2(C) = \lambda_{\max}[N]/2 = \lambda_{\max}[N] = 0 \tag{A1}$$

This condition is equivalent to the matrix N being negative semidefinite, or the matrix $-N$ being positive semidefinite, i.e.,

$$\lambda_{\min}[-N] = 0$$

An extensive discussion about positive definiteness can be found in [28,29]. Then, the conditions associated with the eigenvalues can be computed using theory of positive definite matrices which states that in a symmetric matrix all its eigenvalues are greater than zero if and only if all its principal minors are positive. This matrix is called positive definite. A matrix is positive semidefinite if all its eigenvalues are greater than or equal to zero. On the other hand, a matrix N is negative semidefinite if $-N$ is positive semidefinite.

In this way, the matrix N is given by:

$$N = \begin{pmatrix} 2b_1h_1 & b_1h_2 + b_2h_1 \\ b_1h_2 + b_2h_1 & 2b_2h_2 \end{pmatrix}. \tag{A2}$$

To fulfill the condition to be negative semidefinite, we have to check that all principal minors of $-N$ are greater than or equal to zero ($\Delta_k \geq 0$)

First order principal minors. The matrix has two first order principal minors which are:

$$M_1^1(-N) := \Delta_{11}(-N) = -N_{11} = -2b_1 h_1 \geq 0 \tag{A3}$$

and

$$M_1^2(-N) := \Delta_{12}(-N) = -N_{22} = -2b_2 h_2 \geq 0 \tag{A4}$$

As it can be seen, the only condition is that the pairs $\{b_i, h_i\}$ have opposite signs.

Second order principal minors. This system has only one second order principal minor which is computed as:

$$\Delta_2^1(-N) = \det \begin{pmatrix} -2b_1 h_1 & -b_1 h_2 - b_2 h_1 \\ -b_1 h_2 - b_2 h_1 & -2b_2 h_2 \end{pmatrix} \geq 0$$

From this inequality is obtained:

$$b_1 h_2 = b_2 h_1 \tag{A5}$$

Supposing h_1 as a free parameter to tune, it is obtained that:

$$h_2 = \frac{b_2}{b_1} h_1 \tag{A6}$$

Replacing (A6) in (A4) it can be seen that independently of the value and sign of b_2 the inequality is satisfied.

The proof is complete.

Appendix A.2. Matrix Measure for 3D System

Following the ideas of previous section, N is given by:

$$N = \begin{pmatrix} 2b_1 h_1 & b_1 h_2 + b_2 h_1 & b_1 h_3 + b_3 h_1 \\ b_1 h_2 + b_2 h_1 & 2b_2 h_2 & b_2 h_3 + b_3 h_2 \\ b_1 h_3 + b_3 h_1 & b_2 h_3 + b_3 h_2 & 2b_3 h_3 \end{pmatrix} \tag{A7}$$

To fulfill the condition to be negative semidefinite, we have to check that all principal minors of $-N$ are greater than or equal to zero ($\Delta_k \geq 0$)

First order principal minors. Here, there are three first order principal minors, they are:

$$M_1^1(-N) := \Delta_{11}(-N) = -N_{11} = -2b_1 h_1 \geq 0 \tag{A8}$$

$$M_1^2(-N) := \Delta_{12}(-N) = -N_{22} = -2b_2 h_2 \geq 0 \tag{A9}$$

and

$$M_1^3(-N) := \Delta_{13}(-N) = -N_{33} = -2b_3 h_3 \geq 0 \tag{A10}$$

As it can be seen, the only condition is that the pairs $\{b_i, h_i\}$ have opposite signs.

Second order principal minor. In this case, there are three second order principal minors. The first one is:

$$\Delta_2^1(-N) = \det \begin{pmatrix} -2b_2 h_2 & -b_2 h_3 - b_3 h_2 \\ -b_2 h_3 - b_3 h_2 & -2b_3 h_3 \end{pmatrix} \geq 0$$

After some computations the following equation is obtained.

$$b_2 h_3 = b_3 h_2 \tag{A11}$$

Other second order principal minor is given by:

$$\Delta_2^2(-N) = det \begin{pmatrix} -2b_1h_1 & -b_1h_3 - b_3h_1 \\ -b_1h_3 - b_3h_1 & -2b_3h_3 \end{pmatrix} \geq 0$$

As in previous case, it is obtained:

$$b_1h_3 = b_3h_1 \tag{A12}$$

The last second order principal minor is computed as:

$$\Delta_2^3(-N) = det \begin{pmatrix} -2b_1h_1 & -b_1h_2 - b_2h_1 \\ -b_1h_2 - b_2h_1 & -2b_2h_2 \end{pmatrix} \geq 0$$

and in a similar way it is obtained:

$$b_1h_2 = b_2h_1 \tag{A13}$$

Taking into account these three inequalities and considering h_1 as a free parameter to tune, it is obtained from (A13)

$$h_2 = \frac{b_2}{b_1}h_1$$

From (A12)

$$h_3 = \frac{b_3}{b_1}h_1$$

To finally prove from (A11) that h_3 takes the same value as already given. Replacing these values in expressions (A8) to (A10), the equalities are still preserved regardless of the value and sign of constants b_i.

Third order principal minor. As matrix N is obtained from two vectors, its range cannot be greater than two, then its third order principal minor namely

$$\Delta_3^1(-N) = det(-N) = 0$$

The proof is complete.

References

1. Chen, S.M.; Wang, C.Y.; Liang, T.J. A novel sinusoidal boost-flyback CCM/DCM DC-DC converter. In Proceedings of the 2014 IEEE Applied Power Electronics Conference and Exposition–APEC, Fort Worth, TX, USA, 16–20 March 2014; pp. 3512–3516.
2. Tseng, K.C.; Lin, J.T.; Cheng, C.A. An Integrated Derived Boost-Flyback Converter for fuel cell hybrid electric vehicles. In Proceedings of the 2013 1st International Future Energy Electronics Conference (IFEEC), Tainan, Taiwan, 3–6 November 2013; pp. 283–287.
3. Siouane, S.; Jovanovic, S.; Poure, P. Service Continuity of PV Synchronous Buck/Buck-Boost Converter with Energy Storage. *Energies* **2018**, *11*, 1369. [CrossRef]
4. Ortigoza, R.S.; Rodriguez, V.H.G.; Marquez, E.H.; Ponce, M.; Sanchez, J.R.G.; Juarez, J.N.A.; Ortigoza, G.S.; Perez, J.H. A Trajectory Tracking Control for a Boost Converter-Inverter-DC Motor Combination. *IEEE Lat. Am. Trans.* **2018**, *16*, 1008–1014. [CrossRef]
5. Fernando, W.A.; Lu, D.D. Bi-directional converter for interfacing appliances with HFAC enabled power distribution systems in critical applications. In Proceedings of the 2017 20th International Conference on Electrical Machines and Systems (ICEMS), Sydney, Australia, 11–14 August 2017; pp. 1–6. [CrossRef]
6. Singh, S.; Singh, B.; Bhuvaneswari, G.; Bist, V. A Power Quality Improved Bridgeless Converter-Based Computer Power Supply. *IEEE Trans. Ind. Appl.* **2016**, *52*, 4385–4394. [CrossRef]

Energies **2018**, 11, 3086

7. Undeland, T.M.; Robbins, W.P.; Mohan, N. *Power Electronics: Converters, Applications, and Design*; John Wiley and Sons: New York, NY, USA, 2003.
8. Erickson, R.W.; Maksimović, D. *Fundamentals of Power Electronics*; Springer: Berlin, Germany, 2001.
9. Mueller, J.A.; Kimball, J. Modeling Dual Active Bridge Converters in DC Distribution Systems. *IEEE Trans. Power Electron.* **2018**. [CrossRef]
10. Banerjee, S.; Verghese, G.C. *Nonlinear Phenomena in Power Electronics: Attractors, Bifurcations, Chaos, and Nonlinear Control*; IEEE Press: New York, NY, USA, 2001.
11. Cheng, C.H.; Cheng, P.J.; Xie, M.J. Current sharing of paralleled DC–DC converters using GA-based PID controllers. *Expert Syst. Appl.* **2010**, 37, 733–740. [CrossRef]
12. Guo, L.; Hung, J.Y.; Nelms, R.M. Evaluation of DSP-Based PID and Fuzzy Controllers for DC-DC Converters. *IEEE Trans. Ind. Electron.* **2009**, 56, 2237–2248.
13. Rodríguez-Licea, M.A.; Pérez-Pinal, F.J.; Nuñez-Perez, J.C.; Herrera-Ramírez, C.A. Nonlinear Robust Control for Low Voltage Direct-Current Residential Microgrids with Constant Power Loads. *Energies* **2018**, 11, 1130. [CrossRef]
14. Aguilera, R.P.; Quevedo, D.E. Predictive Control of Power Converters: Designs With Guaranteed Performance. *IEEE Trans. Ind. Inf.* **2015**, 11, 53–63. [CrossRef]
15. Alsmadi, Y.M.; Utkin, V.; Haj-ahmed, M.A.; Xu, L. Sliding mode control of power converters: DC/DC converters. *Int. J. Control* **2017**. [CrossRef]
16. Khaligh, A.; Rahimi, A.M.; Emadi, A. Modified Pulse-Adjustment Technique to Control DC/DC Converters Driving Variable Constant-Power Loads. *IEEE Trans. Ind. Electron.* **2008**, 55, 1133–1146. [CrossRef]
17. Gabian, G.; Gamble, J.; Blalock, B.; Costinett, D. Hybrid buck converter optimization and comparison for smart phone integrated battery chargers. In Proceedings of the 2018 IEEE Applied Power Electronics Conference and Exposition (APEC), San Antonio, TX, USA, 4–8 March 2018; pp. 2148–2154.
18. Lai, C.; Cheng, Y.; Hsieh, M.; Lin, Y. Development of a Bidirectional DC/DC Converter With Dual-Battery Energy Storage for Hybrid Electric Vehicle System. *IEEE Trans. Veh. Technol.* **2018**, 67, 1036–1052. [CrossRef]
19. Lukmana, M.A.; Nurhadi, H. Preliminary study on Unmanned Aerial Vehicle (UAV) Quadcopter using PID controller. In Proceedings of the 2015 International Conference on Advanced Mechatronics, Intelligent Manufacture, and Industrial Automation (ICAMIMIA), Surabaya, Indonesia, 15–17 October 2015; pp. 34–37. [CrossRef]
20. Chiang, C.; Chen, C. Zero-Voltage-Switching Control for a PWM Buck Converter Under DCM/CCM Boundary. *IEEE Trans. Power Electron.* **2009**, 24, 2120–212. [CrossRef]
21. Calderón, A.; Vinagre, B.; Feliu, V. Fractional order control strategies for power electronic buck converters. *Signal Process.* **2006**, 86, 2803–2819. [CrossRef]
22. Suh, J.; Seok, J.; Kong, B. A Fast Response PWM Buck Converter with Active Ramp Tracking Control in Load Transient period. *IEEE Trans. Circuits Syst. II Express Briefs* **2018**. [CrossRef]
23. Chang, C.; Yuan, Y.; Jiang, T.; Zhou, Z. Field programmable gate array implementation of a single-input fuzzy proportional-integral-derivative controller for DC-DC buck converters. *IET Power Electron.* **2016**, 9, 1259–1266. [CrossRef]
24. El Aroudi, A.; Giaouris, D.; Iu, H.H.C.; Hiskens, I.A. A review on stability analysis methods for switching mode power converters. *IEEE J. Emerg. Sel. Top. Circuits Syst.* **2015**, 5, 302–315. [CrossRef]
25. Fiore, D.; Hogan, S.J.; di Bernardo, M. Contraction analysis of switched systems via regularization. *Automatica* **2016**, 73, 279–288. [CrossRef]
26. di Bernardo, M.; Fiore, D. Switching control for incremental stabilization of nonlinear systems via contraction theory. In Proceedings of the 2016 European Control Conference (ECC), Aalborg, Denmark, 29 June–1 July 2016; pp. 2054–2059. [CrossRef]
27. Lohmiller, W.; Slotine, J.J.E. On Contraction Analysis for Non-linear Systems. *Automatica* **1998**, 34, 683–696. [CrossRef]
28. Chen, C.T. *Linear Systems Theory and dEsign*; Oxford University Press: Oxford, UK, 1999.
29. Vidyasagar, M. *Nonlinear Systems Analysis*; Prentice Hall: Englewood Cliffs, NJ, USA, 1993.
30. Slotine, J.J.E.; Li, W. *Applied Nonlinear Control*; Prentice Hall: Englewood Cliffs, NJ, USA, 1991; Volume 199.
31. Khalil, H.K. *Nonlinear Control*; Pearson: New York, NY, USA, 2015.

32. Deane, J.H.; Hamill, D.C. Analysis, simulation and experimental study of chaos in the buck converter. In Proceedings of the PESC'90 Record 21st Annual IEEE Power Electronics Specialists Conference, San Antonio, TX, USA, 11–14 June 1990; pp. 491–498.
33. Fossas, E.; Olivar, G. Study of chaos in the buck converter. *IEEE Trans. Circuits Syst. I Fundam. Theory Appl.* **1996**, *43*, 13–25. [CrossRef]

energies

MDPI

Article

Efficiency Optimization of a Variable Bus Voltage DC Microgrid

David García Elvira *, Hugo Valderrama Blaví *, Àngel Cid Pastor and Luis Martínez Salamero

Department of Electrical, Electronic, and Automatic Control Engineering, Universitat Rovira i Virgili, 43007 Tarragona, Spain; angel.cid@urv.cat (À.C.P.); luis.martinez@urv.cat (L.M.S.)
* Correspondence: david.garciae@urv.cat (D.G.E.); hugo.valderrama@urv.cat (H.V.B.);
    Tel.: +34-977-297-051 (D.G.E.); +34-977-558-523 (H.V.B.)

Received: 28 September 2018; Accepted: 5 November 2018; Published: 8 November 2018

Abstract: A variable bus voltage DC microgrid (MG) is simulated in Simulink for optimization purposes. It is initially controlled with a Voltage Event Control (VEC) algorithm supplemented with a State of Charge Event Control (SOCEC) algorithm. This control determines the power generated/consumed by each element of the MG based on bus voltage and battery State of Charge (SOC) values. Two supplementary strategies are proposed and evaluated to improve the DC-DC converters' efficiency. First, bus voltage optimization control: a centralized Energy Management System (EMS) manages the battery power in order to make the bus voltage follow the optimal voltage reference. Second, online optimization of switching frequency: local drivers operate each converter at its optimal switching frequency. The two proposed optimization strategies have been verified in the simulations.

Keywords: DC micro grid; efficiency optimization; variable bus voltage MG; variable switching frequency DC-DC converters; centralized vs. decentralized control; local vs. global optimization

1. Introduction

Electric power transmission network's topology is being rethought and reformulated nowadays. Ecological, social and economic perspectives recommend moving towards grid decentralization. According to [1], distributed generation provides a range of benefits, including:

- Generation, transmission, and distribution capacity investments deferral.
- Ancillary services.
- Environmental emissions benefits.
- System losses reduction.
- Energy production savings.
- Reliability enhancement.

Development of microgrids is part of the new grid model. A microgrid (MG) consist of a number of interconnected and coordinated elements: generator(s), load(s), energy storage(s) and electrical grid. In DC microgrids, all these elements are connected to a common DC bus through individual DC-DC converters. Power management of the different elements of a MG is necessary to guarantee that the system operates always stable and that, whenever possible, Renewable Energy Sources (RES) operate at their Maximum Power Point (MPP) and critical loads are supplied.

Adequate management of the power of MG's elements is a key to accomplishing stable operation, improving energy efficiency, extending battery lifetime and achieving maximum economic yield.

MG stable operation require to coordinately control all the MG's elements. Control algorithms for microgrids, following [2], can be divided into three categories from the communication perspective:

- Decentralized control: Digital Communication Links (DCL) do not exist and power lines are used as the only channel of communication. It is generally based on the interpretation of the voltage in the common DC bus.
- Centralized control: Data from distributed units are collected in a centralized aggregator, processed and feedback commands are sent back to them via DCLs.
- Distributed control: DCLs are implemented between units and coordinated control strategies are processed locally.

Droop control [3,4] is the basic diagram for decentralized control. In DC microgrids, primary droop control achieves power sharing among the parallel connected sources and bidirectional DC-DC converters. The control of each converter locally determines the power that it must perform according to a linear control law based on the bus voltage. Droop control changes the power reference of the sources' converters as the bus voltage varies due to variations in load or generation. For example, starting from a stable operation point, if load increases, bus voltage tends to decrease. This makes the decentralized control system increase the power supplied by each source according to its particular linear control law. Larger sources contribute with more power thanks to the different droop coefficients for different units. Bus voltage can be restored to its initial value implementing secondary and tertiary droop control. However, only primary droop control is utilized in the MG studied in this paper.

Besides stable operation, more advanced control strategies enable to improve the MG performance. Most MG management optimization studies focus on finding out the optimal economic dispatch. Some examples are summarized next. In [3], an optimization problem is formulated to achieve load sharing minimizing fuel and operation costs. In [5], a multi–objective optimization function is utilized to balance the tradeoff between maximizing the MG revenue and minimizing the MG operation cost, including penalties for bid deviation, renewable energy curtailment, and involuntary load shedding. In [6], a genetic algorithm solves an optimization problem to minimize the instantaneous (no forecasting is considered) MG total operation cost, taking into account real-time pricing of electricity from the utility grid. In [7], a genetic algorithm is implemented to minimize the daily net cost of the Battery Energy Storage System (BESS) scheduling. In [8], a genetic algorithm schedule is used to minimize economic operation cost, including demand response price policy in the model. In [9], a linear optimization problem is solved to determine the day ahead and intraday markets bids that maximize the overall profits of a photovoltaic plant with BESS, considering battery aging, incomes and penalties due to provision of ancillary services and forecast uncertainty.

This paper presents a DC microgrid managed in a stable manner by an Events–Based Control System, and proposes two strategies for efficiency optimization, rather than economic optimization, which has been more frequently addressed. The two proposed strategies are:

- Bus Voltage Optimization Control (BVOC).
- Online Optimization of Switching Frequency (OOSF).

The objective of both strategies is to minimize the DC-DC converters' losses. The proposed optimization algorithms are compatible with economic optimization strategies.

To the best of the authors' knowledge, online bus voltage optimization of a DC microgrid has never been addressed. The most similar study to BVOC optimization found is [10], which determines optimal bus voltages for residential and commercial DC MGs applications. In this case, though, the optimal voltages are constant, so it provides a design criterion rather than an online optimization algorithm. In [11,12], bus voltage control is addressed with different approaches, namely, double loop PI control and droop coefficients optimization, but considering a fixed and predetermined bus voltage level in both cases.

Switching frequency optimization of individual DC-DC converters has been previously modeled in [13–16] and implemented online in [17–19].

After providing basic background on microgrids' management and optimization in this introduction, the sections below are organized as follows. In Section 2, a simulated variable bus voltage MG is described, as well as its existing control algorithm based on bus voltage and State of Charge events. Afterwards, the efficiency curves of the MG's converters are analyzed in detail in Section 3. In sections four and five, the aforementioned efficiency optimization strategies, Bus Voltage Optimization Control and Online Optimization of Switching Frequency, are explained thoroughly. In Section 6, the MG is simulated implementing both BVOC and OOSF, and their performance and dynamic evolution are evaluated. Section 7, finally, contains the summarization and discussion of the results.

2. Description of the MG Studied

The MG studied is simulated in Simulink [20] and it is based on the MG described in [21]. The MG's elements are listed in Table 1.

Table 1. MG's elements.

Sources	PV: Photovoltaic Field	WT: Wind Turbine	FC: Fuel Cell
Loads	LOAD: Residential profile DC load	EZ: Electrolyzer	
Bidirectional Units	BESS: Battery Energy Storage System	INV: Bidirectional inverter	

Figure 1 shows the MG's elements interconnected in a variable voltage DC bus through DC-DC converters and the topology chosen for each of them. All the MG's converters are controlled as power sources. Converters with nominal power over 3 kW have been designed with IGBTs while those with lower nominal power have been designed with MOSFETs. It has been imposed an inductor current ripple lower than 20% in the worst-case scenario: nominal operating conditions and minimum switching frequency (3 kHz for IGBT converters and 20 kHz for MOSFET converters). The transistors operate hard switching in continuous conduction mode. Averaged models of the converters' losses based on small-signal analysis have been employed. The converters are modeled considering conduction and switching losses, following [22,23]. Average current values are used, since the error is small thanks to the low current ripple.

The existing MG's control system consists of a decentralized Voltage Event Control (VEC) in every unit, supplemented with a State of Charge Event Control (SOCEC) in FC, EZ and INV. This initial control system will be denominated Events-Based Control System (E-BCS = VEC + SOCEC). VEC is a primary droop control using no secondary or tertiary droop control to stabilize bus voltage.

Each unit follows its correspondent linear control law based on the bus voltage (v_{bus}). FC, EZ and INV also follow their additional control laws based on the battery State of Charge (SOC).

The E-CBS determines a coefficient c_i that multiplies the absolute value of the available/demanded/nominal power of the source/load/bidirectional unit i. The coefficient c_i value is the saturated sum of VEC coefficient c_{VEC}^i plus SOCEC coefficient c_{SOCEC}^i, as shown in Equation (1). The coefficients c_{VEC}^i and c_{SOCEC}^i are defined in Figure 2. The power reference P_i imposed in the converter i is calculated as in Equation (2). The sign convention is: P_i is positive if the power is coming into the DC bus and negative is power is coming out:

$$P_i = \begin{cases} \min(1, \max(0, (c_{VEC}^i(v_{bus}) + c_{SOCEC}^i(SOC))) & \text{sources}: i = PV, WT, FC \\ \min(0, \max(-1, (c_{VEC}^i(v_{bus}) + c_{SOCEC}^i(SOC))) & \text{loads}: i = LOAD, EZ \\ \min(1, \max(-1, (c_{VEC}^i(v_{bus}) + c_{SOCEC}^i(SOC))) & \text{bidirectional units}: i = BESS, INV \end{cases}, \quad (1)$$

$$P_i = \begin{cases} P_{PV} = c_{PV}|P_{MPP\ PV}| \\ P_{WT} = c_{WT}|P_{MPP\ WT}| \\ P_{FC} = c_{FC}|P_{nom\ FC}| \\ P_{LOAD} = c_{LOAD}|P_{demand}| \\ P_{EZ} = c_{EZ}|P_{nom\ EZ}| \\ P_{BESS} = c_{BESS}|P_{nom\ BESS}| \\ P_{INV} = c_{INV}|P_{nom\ INV}| \end{cases} , \tag{2}$$

$P_{MPP\ i}$ is the maximum power that a RES can generate at a given moment; $P_{nom\ i}$ is the nominal power of the source/load/bidirectional unit i; and P_{demand} is the load's electricity demand.

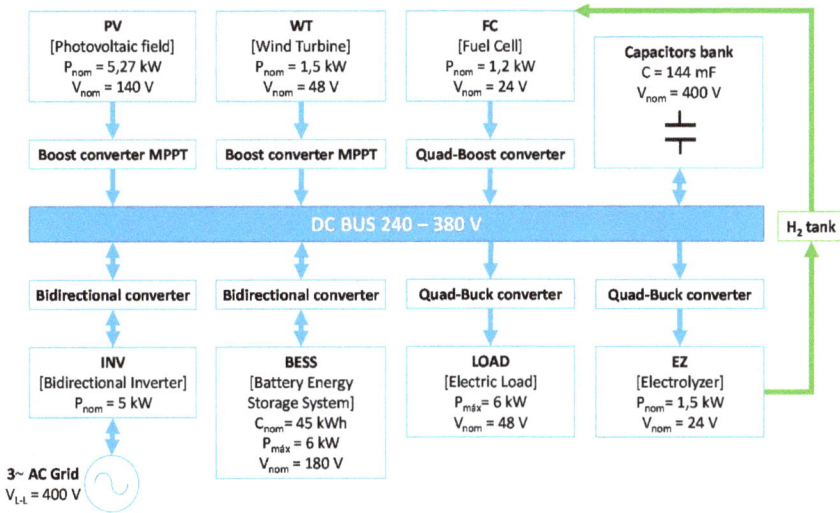

Figure 1. MG architecture. The MG is composed of a DC bus in which the following elements are connected: two Renewable Energy Sources (RES) (PV and WT), a controllable source (FC), a capacitors bank, a bidirectional grid-inverter (INV), a battery system (BESS), critical loads (LOAD), and a controllable load (EZ).

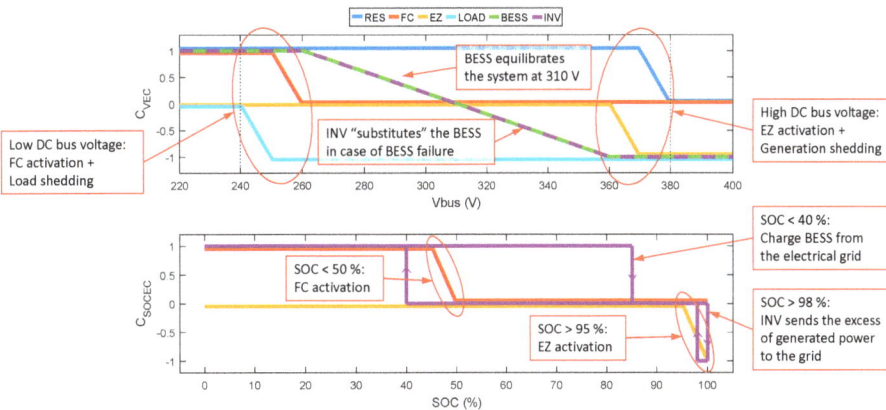

Figure 2. E-BCS control laws: VEC in the graph above and SOCEC in the graph below.

Voltage limitations are established in 240 V $\leq v_{bus} \leq$ 380 V. The lower limit is chosen such that v_{bus} always remains above the maximum PV voltage (190 V approx.) and the maximum BESS voltage

(220.5 V). This permits to discard the use of buck-boost and bidirectional buck-boost converters, which are less efficient than the finally selected PV's boost converter and BESS' bidirectional converter (see Figure 1). The higher voltage limit is the result of applying a safety margin with respect to the maximum voltage of the capacitors bank (400 V). A 20 V safety margin is sufficient thanks to the robustness of E-CBS maintaining voltage peaks lower than this value.

Voltage Event Control algorithm manages power balance, in order to keep the system stable. Power from Renewable Energy Sources (RES) (i.e., PV and WT) is limited when v_{bus} is too high (over 370 V). LOAD power is limited when v_{bus} is too low (under 250 V)—this can only happen if the MG operates in islanded mode and with fully discharged battery. BESS power tends to equilibrate the system at the medium voltage level (i.e., 310 V). FC and EZ are utilized as emergency support source/load when v_{bus} approaches the lower/higher voltage limit.

FC, EZ and INV also incorporate State of Charge Event Control: an independent power control based on battery SOC that, on the one hand, avoids performing deep discharges and, on the other, avoids missing RES generation when SOC $\approx$ 100%. Hydrogen production/consumption are activated when the SOC reaches high (95%)/low (50%) values. The inverter will follow the BESS control law in case of BESS failure or when SOC exceeds 98%; and will deliver full power to the DC bus when SOC drops down to 40%. All the linear control laws apply simultaneously, except the INV VEC control law (dashed line), which only applies in case of BESS failure.

MG's operation controlled by E-BCS is taken as the reference to assess Bus Voltage Optimization Control (BVOC) and Online Optimization of Switching Frequency (OOSF) performances. The power that each source/load/bidirectional converter has to deliver is determined by E-BCS. When BVOC is implemented, it substitutes the E-BCS control (only) in the BESS' converter. The rest of the converters maintain their VEC and SOCEC control laws, except for FC and EZ VEC control laws, which only regulate FC and EZ powers in some specific cases. BVOC controller calculates BESS power as the power that makes the bus voltage reach its optimal value. OOSF, in its turn, does not affect the power reference of any of the MG's converters—OOSF locally optimizes the switching frequency of the converters in which it is applied, while their power references remain unaltered. Thus, the power references in all the MG's converters are imposed by E-BCS, except when BVOC is implemented, that it modifies the BESS' converter power reference.

3. Analysis of the Energy Efficiency Curves of the DC-DC Converters

Figure 3 shows the efficiency curves of two representative converters of the seven MG's DC-DC converters (LOAD's and BESS' converters) operating at different switching frequencies and bus voltage values. In this section, these curves are analyzed to anticipate the order of magnitude of the efficiency improvement that the two proposed optimization strategies BVOC and OOSF will produce.

The efficiency curves of the LOAD's and BESS' converters (which are quadratic buck and bidirectional converters, respectively, as can be seen in Figure 1) have been chosen because they serve to illustrate general issues that occur in all the MG's converters, namely:

- First, there is little room for energy efficiency improvement for both optimization strategies. The efficiency curves reveal small efficiency differences within the studied range of possible v_{bus} and f_{sw} values. Efficiency increases are limited to a fraction of a percentage point.
- Second, the optimal bus voltage is different for individual converters. Thus, a change in v_{bus} leading to catch up with the overall optimal bus voltage v_{optim}, causes some of the MG's converters to operate more efficiently but, also, it causes some other converters to operate less efficiently. The overall effect is that BVOC achieves efficiency improvements, yet low ones for this reason.

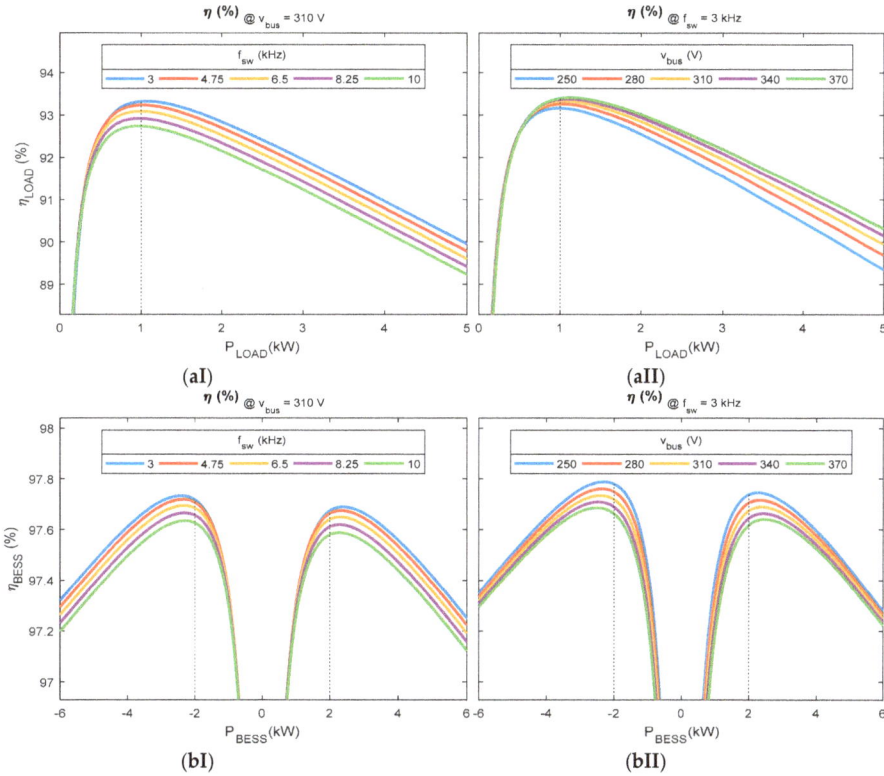

Figure 3. Efficiency curves of LOAD's and BESS' converters at different switching frequencies (I) and bus voltages (II): (**a**) Quadratic buck converter interfacing with LOAD; (**b**) Bidirectional converter interfacing with BESS.

Figure 3(aI) shows the efficiency curves of the LOAD's quadratic buck converter for five values of switching frequency, linearly distributed within the range 3–10 kHz, at a constant bus voltage of 310 V. The Online Optimization of Switching Frequency algorithm applies to each converter individually, so they all will operate at their optimal switching frequency (the highest curve in the graph). OOSF controls the gate driver of the transistors to operate them at optimal switching frequency (i.e., 3 kHz, blue curve, for most P_{LOAD} values). The efficiency of the LOAD converter increases steeply from $P_{LOAD} = 0$ kW to 1 kW approx. In this first part of the curve, it is possible to obtain significant efficiency improvements in relative terms, but with low impact in absolute terms. In the second part of the curve, from $P_{LOAD} = 1$ kW to 5 kW, the maximum possible efficiency improvement remains constant with a value of 0.75%. Figure 3(bI) is essentially the same graph but with the efficiency curves of the BESS's bidirectional DC-DC converter. The left half of the graph ($P_{BESS} < 0$ kW) represents the buck operation mode; the right part represents the boost operation mode. The turning points where the slope changes sharply occur at $P_{BESS} = \pm 2$ kW approx. The maximum possible efficiency improvement for high P_{BESS} is as low as 0.12% in this case.

Figure 3(aII) shows the efficiency curves of the LOAD quadratic buck converter for five bus voltage levels, linearly distributed within the range 250–370 V, at a constant switching frequency of 3 kHz. BVOC regulates the duty cycle of the transistors in order to operate the MG at optimal bus voltage, which minimizes the losses of the set of DC-DC converters taken as a whole. This optimization algorithm is applied globally, so, in general, global optimum will not coincide with local optimums. In this curve, the efficiency of the LOAD converter increases steeply also from $P_{LOAD} = 0$ kW to 1 kW approx. For power greater than 1 kW, the maximum possible efficiency improvement steadily increases from 0.23% at 1 kW to 0.96% at 5 kW. Figure 3(bII) is again the same graph but representing the BESS' bidirectional DC-DC converter. The maximum possible efficiency improvement for high values of P_{BESS} is as low as 0.06% in this case.

Figure 3(aII,bII) show that, in general, local optimums for different MG's units occur at opposed v_{bus} increments: for example, if $P_{LOAD} > 1$ kW, raising v_{bus} makes the efficiency of LOAD's converter increase but, at the same time, it makes the efficiency of BESS' converter decrease (and vice versa), regardless of what the P_{BESS} is. Optimum bus voltage v_{optim} reference calculated by BVOC is the trade-off that minimizes the sum of power losses in all the converters. Hence, not all of them will operate simultaneously at their local optimum point in general.

4. Bus Voltage Optimization

This first optimization approach, Bus Voltage Optimization Control (BVOC), makes use of the distinctive varying DC bus voltage characteristic of the MG to online improve the efficiency of the set of MG's converters taken as a whole, for all the possible power flows.

The power of the MG's sources, loads and bidirectional units continuously varies so individual optimization of the converters at their rated power is not an adequate approach in MG applications. Global optimization, considering all the possible power flows (i.e., all the possible operation points of the converters) is conducted next.

BVOC controls the BESS power substituting the VEC control law (shown in Figure 2) in the BESS' converter. RES's, LOAD's and INV's converters maintain their E-BCS control laws. FC's and EZ's converters maintain their SOCEC control laws; but their VEC control laws only apply if one of the following three conditions is met: BVOC controller failure, energy shortage or energy excess.

BVOC keeps the bus voltage at its optimal value in every moment. The calculation variable V_{BUS} is used in the system of equations of the BVOC optimization problem (in capital letters to differentiate between V_{BUS}, this calculation variable, and v_{bus}, which represents the bus voltage not as a calculation variable, but as the real, measurable, physical magnitude). The optimal bus voltage v_{optim} is defined here as the value of the calculation variable V_{BUS} that minimizes the sum of the converters' losses ($P_{loss}^{MG}(V_{BUS})$) for the instantaneous power flows among the MG's units. A centralized Energy Management System (EMS) calculates v_{optim} online every 100 ms, evaluating a V_{BUS}-dependent deterministic model of P_{loss}^{MG} explained below. With the reference v_{optim}, a PI controller determines the power that the BESS must perform to make v_{bus} catch up with v_{optim}.

From now on, this optimization strategy will be denoted by BVOC$_{BESS}$, to highlight that this control algorithm applies only to the BESS converter, while the remaining converters maintain the E-BCS presented in the previous section.

The EMS gathers sensor readings of switching frequencies, and input and output voltages and currents from all the converters. With them, the EMS evaluates an online loss models of all the converters. The Matlab [20] function *fminbnd* is used to solve the nonlinear optimization problem formulated in Equations (3) and (4). It returns the value of the calculation variable V_{BUS} that is a minimizer of the sum of the converters' losses (P_{loss}^{MG}), within the interval in which both load and RES remain unconstrained, i.e., 250 V–370 V:

$$v_{optim} = \min\left[P_{loss}^{MG}(V_{BUS})\right]_{f_{sw}^i, P^i, v^i = csts.} \quad \text{with } 250 \text{ V} \leq V_{BUS} \leq 370 \text{ V}, \quad (3)$$

$$P_{loss}^{MG}(V_{BUS}) = \sum_i \left[P_{loss}^i(V_{BUS})\right]_{f_{sw}^i, P^i, v^i = csts.} \quad \text{with } i = \text{PV, WT, FC, LOAD, EZ, BESS, INV}, \quad (4)$$

where P_{loss}^{MG} is the sum of the V_{BUS}—dependent expressions of power loss (P_{loss}^i) in the converters of each unit i, given their measured switching frequencies (f_{sw}^i), powers (P^i) and voltages (v^i). Measured values get updated every 100 ms. Notice that P_{loss}^{MG} and P_{loss}^i are dependent on V_{BUS}.

In order to clarify how the EMS calculates v_{optim}, an example is provided: the power loss equations of one of the DC-DC converters, which is one of the seven addends in Equation (4). PV boost converter's power loss ($P_{loss}^{PV}(V_{BUS})$) is calculated evaluating the Equations (5)–(7). Note that the remaining P_{loss}^i expressions (see Appendix A) are analogous and must be added up to P_{loss}^{MG} according to Equation (4), to calculate the total power loss as a function of V_{BUS} and obtain the v_{optim} that minimizes it:

$$\eta = \frac{v_{out}i_{out}}{v_{in}i_{in}}, \tag{5}$$

$$D = \frac{(V_{BUS} - v_{in}\cdot\eta)}{V_{BUS}}, \tag{6}$$

$$P_{loss}^{PV}(V_{BUS}) = \underbrace{\left\{ R_L i_{in}^2 + K\left(\frac{\Delta B(V_{BUS}, f_{sw})}{2}\right)^\beta f_{sw}^\alpha \right\}}_{inductor} + \underbrace{\left\{ (1-D)(R_d i_{in}^2 + V_f i_{in}) \right\}}_{diode}$$
$$+ \underbrace{\left\{ D(R_d' i_{in}^2 + V_f' i_{in}) + (0.5 V_{BUS} i_{in} t_{sw} f_{sw}) \right\}}_{IGBT}, \tag{7}$$

where η is the converter's efficiency; v_{out}, i_{out}, v_{in} and i_{in} are the measured output and input voltages and currents (note that v_{out} is the measured value of v_{bus}); D is the transistor's duty cycle (the value of η in Equation (6) is considered equal to that of Equation (5) during each time step; the value of η gets corrected every time step as the measured v_{out} gets updated, and so does the value of D); R_L is the inductor resistance; ΔB is the flux density ripple calculated using the Faraday's Law, see [24]; K, α, β are the characteristic parameters of the inductor's magnetic core; f_{sw} is the actual switching frequency; R_d, V_f are the resistance and the forward voltage drop of the diode; R_d', V_f' are the resistance and the forward voltage drop of the IGBT; and t_{sw} is the sum of transition turn–on and turn–off times of the IGBT. Notice that P_{loss}^{PV}, D, and ΔB are dependent on V_{BUS}. The full system of equations and list of parameters is provided in the Appendix A.

BVOC$_{BESS}$ achieves two main goals: first, it balances the power of sources, loads and inverter and, second, it controls the bus voltage. For a better understanding of how MG powers are balanced and bus voltage is controlled thanks to BVOC$_{BESS}$, Equation (8) clarifies how the power balance is achieved in the MG:

$$P_{BESS}(t) + P_C(t) = -\sum_j P_j(t), \quad \text{with } j = \text{PV, WT, FC, LOAD, EZ, INV,} \tag{8}$$

where t represents any time, P_C is the capacitor's bank power, P_{BESS} is the BESS power, and $\sum_j P_j$ is the resultant power of the rest of the MG's units. Positive sign is assigned to the power flows coming into the DC bus, and negative to the power coming out.

Equation (8) shows that the power of the two storage elements, BESS and capacitor's bank, always equal to the sum of the powers coming in the DC bus, but with opposite sign. If net power from sources, loads and inverter is positive (power is being injected to the DC bus), then $P_{BESS} + P_C$ must be negative and, thus, behave as a load (withdrawing power from the DC bus) to balance the system, and vice versa. BVOC$_{BESS}$ keeps P_{BESS} very close (or equal) to $(-\sum_j P_j)$ at every time, thus, forcing P_C to be very low (few or zero watts). This is how MG powers are balanced.

If v_{bus} is already equal to v_{optim}, P_{BESS} will be exactly equal to $(-\sum_j P_j)$ to maintain that voltage level constant. Else, P_{BESS} will differ from $(-\sum_j P_j)$ in a very low power. This (very low) extra power from the BESS is determined by BVOC$_{BESS}$ to make v_{bus} catch up with v_{optim}. The (very low) extra BESS power is absorbed by the capacitor's bank, which regulates the bus voltage. Agreeing to the aforementioned sign criterion, positive P_C discharges the capacitors bank (bus voltage decreases), injecting power to the DC bus; negative P_C takes power from the DC bus to charge the capacitors bank (increasing the bus voltage). P_C must be kept small to avoid too rapid v_{bus} changes that could produce an oscillating behavior. This is how bus voltage stabilization at optimal value is accomplished.

As previously mentioned, there are three abnormal operation conditions in which VEC would be in charge of achieving power balance instead of BVOC$_{BESS}$:

- BVOC$_{BESS}$ controller failure
- Energy shortage: high LOAD + low RES + fully discharged battery + failure in bidirectional inverter
- Energy excess: low LOAD + high RES + fully charged battery + failure in bidirectional inverter

In these three cases (and only in these three cases), VEC will trigger FC and EZ. If v_{bus} eventually surpasses 370 V or goes below 250 V, VEC will shed generation or loads, respectively, to help to achieve power balance. The MG operation in such cases is as follows. When the bus voltage descends below 260 V (ascends over 360 V), E-BCS activates FC (EZ), providing (consuming) surplus power. If this is not enough and bus voltage leaves the interval 250 V–370 V, then, then VEC would perform load shedding or generation shedding, accomplishing power balance and limiting the bus voltage to a minimum of 240 V and a maximum of 380 V.

5. Switching Frequency Optimization

An alternative and complementary approach to improve the converter's efficiency is the Online Optimization of Switching Frequencies (OOSF). The converters are allowed to vary the switching frequency within a predefined interval: depending on whether the converter's transistor is an IGBT (3 kHz $\leq f_{sw} \leq$ 10 kHz) or if it is a MOSFET (20 kHz $\leq f_{sw} \leq$ 100 kHz).

The calculation variable F_{SW} is used in the equations of the OOSF optimization problems (in capital letters to differentiate between F_{SW}, this calculation variable, and f_{sw}, which represents the switching frequency not as a calculation variable, but as the real, measurable, physical magnitude). The optimal switching frequency value for converter i (f^i_{optim}) is calculated every 100 ms as the value of the calculation variable F_{SW} that minimizes its individual loss power expression (P^i_{loss}), given the i converter's measured power (P^i) and voltage (v^i) (see Equation (9)):

$$f^i_{optim} = \min \left[P^i_{loss}(F_{SW}) \right]_{v_{bus}, P^i, v^i = csts.} \quad \text{with} \quad \begin{cases} \text{MOSFET}: 20 \text{ kHz} \leq f_{sw} \leq 100 \text{ kHz} \\ \text{IGBT}: 3 \text{ kHz} \leq f_{sw} \leq 10 \text{ kHz} \end{cases} . \tag{9}$$

Notice that P^i_{loss} is dependent on F_{SW}, and that in this case v_{bus} is not a calculation variable but a measured value.

The OOSF optimization is independent for each converter. Optimization function *fminbnd* is used again to obtain the value f^i_{optim}. Calculation of f^{PV}_{optim} is shown in Equations (10) and (11) to provide an example (again of PV converter) of optimal switching frequency calculation. Equations (10) and (11) are the same as (6) and (7), but with F_{SW} being the calculation variable instead of v_{bus} because now v_{bus} is considered constant (v_{out}):

$$D = \frac{(v_{out} - v_{in} \cdot \eta)}{v_{out}}, \tag{10}$$

$$P^{PV}_{loss}(F_{SW}) = \left\{ R_L i^2_{IN} + K \left(\frac{\Delta B(v_{out}, F_{SW})}{2} \right)^\beta F^\alpha_{SW} \right\}_{inductor} + \left\{ (1 - D)(R_d i^2_{in} + V_f i_{in}) \right\}_{diode} + \left\{ D(R'_d i^2_{in} + V'_f i_{in}) + (0.5 v_{out} i_{in} t_{sw} F_{SW}) \right\}_{IGBT}. \tag{11}$$

Notice that P^{PV}_{loss} and ΔB are dependent on F_{SW}.

Unlike in BVOC$_{BESS}$, bus voltage is considered here as a constant, equal to its measured value v_{out}, which gets updated every time step. Thus, Equation (6) is different from (10) because, although they both express the same relation between D, v_{bus}, v_{in} and η: bus voltage is a variable (V_{BUS}) in Equation (6), but it is a constant ($v_{bus} = v_{out}$) in Equation (10).

Figure 4 is a flow chart that shows how E-BCS, BVOC$_{BESS}$ and OOSF can be executed simultaneously. BVOC$_{BESS}$ and OOSF do not directly interfere with each other because they are independent. Nevertheless, they do interfere with each other indirectly. Indeed, BVOC$_{BESS}$ affects the values of the BESS power (P_{BESS}), the SOC and v_{bus}, which will be measured in the next time step, and will affect then to the OOSF algorithm (see Equations (10) and (11)). And the other way round,

OOSF produces changes in f_{sw}^i for every converter i which, in its turn, will affect BVOC$_{BESS}$ when their measured values are fed to the EMS in the next time step (see Equation (7)).

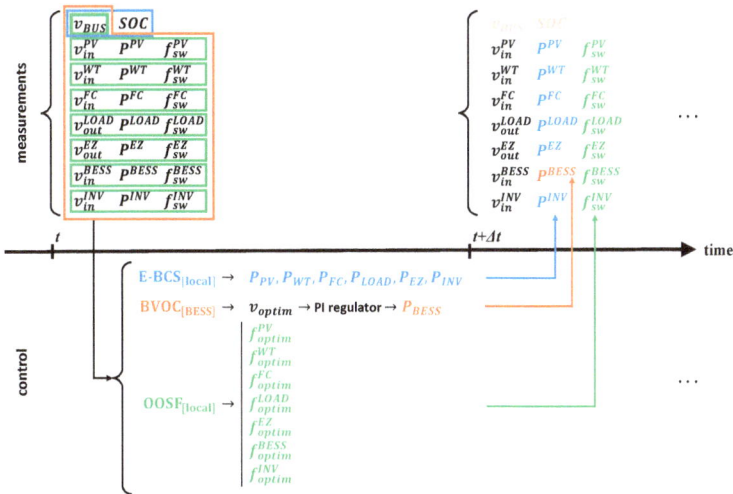

Figure 4. MG management flow chart describing the process of simultaneous E-BCS (in blue), BVOC$_{BESS}$ (in orange) and OOSF (in green). Each time step, say in time t, the E-BCS local controller of each converter i receives the measured value of v_{bus} and SOC and, from them, determines the correspondent P_i. The BVOC$_{BESS}$ centralized controller (i.e., the BESS controller) receives all the MG's measured values and, from them, calculates v_{optim}; a PI regulator determines then P_{BESS} from this reference. The OOSF local controller of each converter i receives the measured value of v_{in}^i, v_{out}^i (note that one of these two is invariably equal to v_{bus}. for all converters i), P^i, and f_{sw}^i and, from them, calculates the optimal $f_{sw\ optim}^i$. The calculated values P_i and $f_{sw\ optim}^i$ by the three algorithms are then imposed and, thus, they will coincide with the measured values in the next time step $t + \Delta t$. Measured v_{bus} and SOC appear in light orange in $t + \Delta t$. because, although they are not directly determined by BVOC$_{BESS}$, they are highly influenced by P_{BESS}.

6. Simulation Results

The MG operation has been simulated over seven days, in ten different scenarios S0 to S9. This section presents the results in three sub-sections. In the first sub-section, the quantitative results of efficiency improvement are provided for the ten scenarios considered. In the second sub-section, the dynamic response of the MG is graphically represented and explicated. Finally, in the third sub-section, the control parameters of BVOC$_{BESS}$ and OOSF are changed to evaluate how they affect the results.

6.1. Efficiency Increase

The global energy efficiency (η_{global}) is calculated to quantify the efficiency increase ($\Delta\eta_{global}$) achieved by each strategy with respect to the reference case (E-BCS). System $\boxed{MG}$ (in green dotted line in Figure 5) is defined as that composed of the DC bus, the capacitors bank, and all the DC-DC converters.

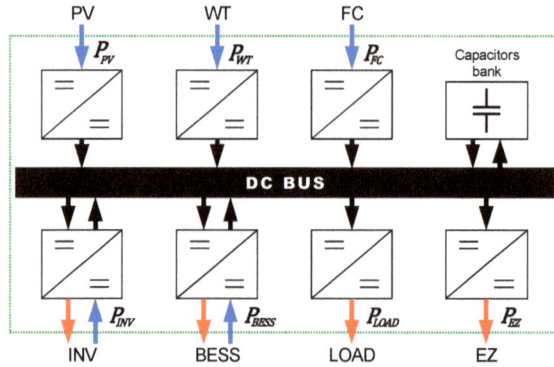

Figure 5. Possible power flows in the MG. The green rectangle defines the system $\boxed{MG}$.

This system $\boxed{MG}$ is taken as a black box in which there are energy inputs and outputs. Global efficiency η_{global} is defined in Equation (12) as the quotient of the energy coming out ($E_{out}^{\boxed{MG}}$) divided by energy coming in ($E_{in}^{\boxed{MG}}$):

$$S_{BESS} := P_{BESS} > 0 \quad \text{[BESS works as an energy source (logical variable)],} \tag{12}$$

$$S_{INV} := P_{INV} > 0 \quad \text{[INV works as an energy source (logical variable)],} \tag{13}$$

$$\eta_{global}(t) = \frac{E_{out}^{\boxed{MG}}(t)}{E_{in}^{\boxed{MG}}(t)} = \frac{\int_0^{t_{end}} abs(P_{LOAD} + P_{EZ} + S_{BESS}^* \cdot P_{BESS} + S_{INV}^* \cdot P_{INV}) \, dt}{\int_0^{t_{end}} (P_{PV} + P_{WT} + P_{FC} + S_{BESS} \cdot P_{BESS} + S_{INV} \cdot P_{INV}) \, dt}, \tag{14}$$

where P_i represents the power injected/withdrawn to/from the $\boxed{MG}$ system by the converter i. The logical variables S_{BESS} and S_{INV} are defined to distinguish when the BESS and the INV converters work as sources and when they work as loads (the superscript * indicates negation). The integration limits are zero (the beginning of the simulation), and t_{end} (the end of the simulation $t_{end} = 7 \times 24 = 168$ h).

Ten simulations, corresponding to ten different scenarios, are performed. In the first simulation, S0, several elements have been simulated to fail in order to verify the robustness of the control system. The simulations S1 to S9, on the contrary, represent different scenarios of normal operation (i.e., no failures). In each simulation, different coefficients multiply the base RES generation and base LOAD consumption profiles. Table 2 summarizes the differences between the nine simulations S1 to S9:

Table 2. Characteristics of the ten simulation scenarios.

RES & LOAD Energies	Failures Simulation	Low Energy			Intermediate Energy			High Energy		
	S0	S1	S2	S3	S4	S5	S6	S7	S8	S9
E_{RES} (kWh)	161.2		37.7			113.1			188.5	
$\frac{E_{LOAD}}{E_{RES}}$	1.20	0.75	1.00	1.33	0.75	1.00	1.33	0.75	1.00	1.33

For these ten simulations, base RES generation profile has been estimated based on meteorological data (solar irradiance, temperature and wind speed) from the Servei Meteorològic de Catalunya [25], and base LOAD profile corresponds to a residential electricity consumption profile. A time step of 100 ms is selected, coinciding with the refresh rate of both BVOC$_{BESS}$ and OOSF controllers. The switching frequency ranges are 20–100 kHz for IGBTs and 3–10 kHz for MOSFETs, as indicated previously in Equation (9). In each of the ten scenarios, the MG is simulated with four different control schemes:

- E-BCS
- E-BCS + BVOCBESS + OOSF
- E-BCS + BVOCBESS
- E-BCS + OOSF

BVOC$_{BESS}$ controls the BESS' converter and OOSF is applied in all the MG's converters. Global efficiency results obtained in the simulations are shown in Table 3:

Table 3. MG global efficiency improvement under four different control strategies, in ten different scenarios S0–S9, over seven days operation.

Control Strategy	Failures Simulation	No Failures Simulation								
		Low Energy			Intermediate Energy			High Energy		
	S0	S1 E_{LOAD} $< E_{RES}$	S2 E_{LOAD} $= E_{RES}$	S3 E_{LOAD} $> E_{RES}$	S4 E_{LOAD} $< E_{RES}$	S5 E_{LOAD} $= E_{RES}$	S6 E_{LOAD} $> E_{RES}$	S7 E_{LOAD} $< E_{RES}$	S8 E_{LOAD} $= E_{RES}$	S9 E_{LOAD} $> E_{RES}$
Control					η: Efficiency					
E-CBS (reference)	86.184%	77.948%	79.504%	80.841%	86.126%	86.990%	87.295%	87.250%	87.140%	87.339%
Optimization					$\Delta\eta$: Efficiency improvement over reference E-CBS					
BVOC$_{BESS}$ + OOSF	0.608%	1.006%	0.580%	0.441%	0.648%	0.343%	0.329%	0.571%	0.4776%	0.458%
BVOC$_{BESS}$	0.146%	0.481%	0.359%	0.270%	0.164%	0.136%	0.131%	0.137%	0.1707%	0.255%
OOSF	0.444%	0.723%	0.360%	0.280%	0.559%	0.241%	0.205%	0.446%	0.3100%	0.190%

S0 represents the operation of the MG scheduling failures in RES, BESS and INV. This permits to verify that the optimized control system responds correctly to failures and rapid changes. S0 is analyzed in the next sub-section. S1, S2 and S3 represent the MG operation when the power of all the elements is low (compared to their nominal power). These three simulations correspond to increasing ratios E_{LOAD}/E_{RES}: 0.75, 1.00 and 1.33. This ratios scheme is repeated for intermediate (S4–S6) and high powers (S7–S9).

Each of the 40 simulations performed has taken an average of 15 min to complete, using MatlabR2017a–Simulink in Intel® Core™ i7-6700 CPU @ 4.40 GHz, 16 GB RAM. Figure 6 represents graphically the results of Table 3:

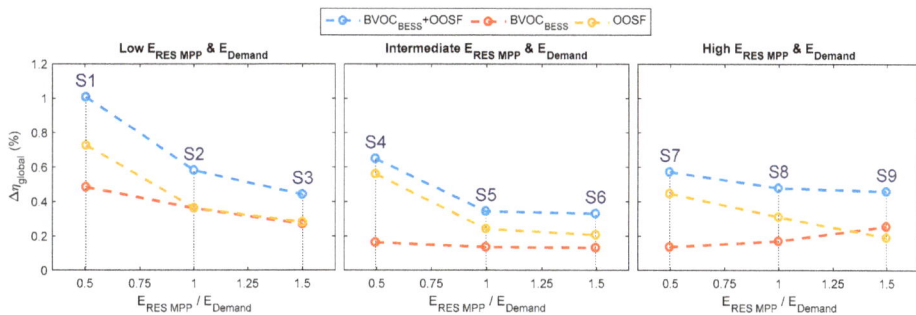

Figure 6. Energy efficiency improvement achieved by simultaneous BVOC$_{BESS}$ + OOSF optimization strategies, only BVOC$_{BESS}$, and only OOSF in different scenarios.

The results of the simulations reveal that simultaneously applying both BVOC$_{BESS}$ and OOSF strategies, result in a $\Delta\eta_{global}^{BVOC_{BESS}+OOSF}$ (blue in Figure 6) ranging from 0.329% in the worst scenario (S6) to 1.006% in the best case scenario (S1). Applying only BVOC$_{BESS}$ (orange) while switching the transistors at fixed frequencies (MOSFETs at 50 kHz and IGBTs at 4 kHz), the $\Delta\eta_{global}^{BVOC_{BESS}}$ ranges from 0.131% (S6) to 0.481% (S1). And last, applying only OOSF (yellow), letting the bus voltage evolve according to VEC and SOCEC, the achieved $\Delta\eta_{global}^{OOSF}$ ranges from 0.190% (S9) to 0.723% (S1).

These results are coherent with the efficiency curves information provided in Figure 3. There is a general tendency towards higher $\Delta\eta_{global}$ for lower powers, and vice versa, due to the characteristic shape of the efficiency curves. In the steep part of the curves (which corresponds to low power values), higher efficiency improvements are achievable, while for powers greater than maximum–efficiency power, efficiency optimization performance is limited due to the small differences existing in $\Delta\eta_{global}$ for different values of f_{sw} and v_{bus}.

Efficiency improvements accomplished by OOSF strategy do consistently show this effect in all the simulated scenarios: in all the cases, $\Delta\eta_{global}^{OOSF}$ is higher for lower power flows. Hence, for a given RES profile, $\Delta\eta_{global}^{OOSF}$ is higher for lower LOAD profiles. Indeed, the optimization performance of OOSF verifies $\Delta\eta_{global}^{OOSF}(S1) > \Delta\eta_{global}^{BOOSF}(S2) > \Delta\eta_{global}^{BOOSF}(S3)$ (left graph of Figure 6). Similarly, for a given ratio $E_{RES\ MPP}/E_{Demand}$, $\Delta\eta_{global}^{OOSF}$ is higher for lower power flows: $\Delta\eta_{global}^{OOSF}(S1) > \Delta\eta_{global}^{OOSF}(S4) > \Delta\eta_{global}^{OOSF}(S7)$ (first scenario of each of the three graphs in Figure 6).

On the other hand, BVOC$_{BESS}$ performance does not show this regularity as it highly depends on the specific combination of active converters performing the instantaneous power flows. For example, $\Delta\eta_{global}^{BVOC_{BESS}}$ would be high in a moment when only WT and LOAD were active (WT directly feeding the LOAD) because the individual optimal bus voltage is 370 V for both converters, but $\Delta\eta_{global}^{BVOC_{BESS}}$ would be much lower if the same power came from BESS instead of WT because individual optimal bus voltage is 250 V for the BESS converter and 370 V for the LOAD converter. The global optimal bus voltage in this last case would be the trade-off which minimizes the sum of power losses in both converters. This dependence on the particular MG's power flows evolution is the cause for $\Delta\eta_{global}^{BVOC_{BESS}}$ irregular results.

The efficiency improvements achieved with the two proposed optimization strategies, BVOS$_{BESS}$ and OOSF, are small but still interesting. Even though the MG's efficiency increases range only from 0.3 to 1.0% (approx.), they are achieved by introducing changes in the discrete control algorithms of the DC-DC converters. Thus, they require (almost) no extra investment. Implementing both optimization strategies produces net energy (and economic) savings in all possible operating conditions. Nevertheless, BVOC$_{BESS}$ and OOSF optimization are best suited for microgrids applications which require low power flows compared to the nominal power of the converters and/or microgrids in which all the MG's converters coincide in the same optimal bus voltage value v_{optim}.

Some application examples that meet these requirements can be Uninterruptible Power Supply (UPS) systems (in which the BESS operates at near zero power to maintain the battery floating), and remote Base Transceiver Stations (RBT) and weather stations, because their BESS are normally significantly oversized so that they can sustain the operation of the station for several days in case of generation shortage. In both cases, the BESS converter normally operates well below its nominal power and can benefit from OOSF high performance for low power. Furthermore, if the individual optimal bus voltage values of the MG's converters coincide, BVOC$_{BESS}$ performance can be more significant than in the MG studied in this paper.

For other microgrids applications in which higher powers are involved and with a diversity of individual v_{optim} for different MG's converters, BVOC$_{BESS}$ might be discarded in the sake of simplicity and robustness, considering that it is difficult to implement (it performs complex calculations and requires the utilization of an EMS that communicates with all the MG's converters), and that $\Delta\eta_{global}^{BVOC_{BESS}}$ is small for intermediate and high power flows.

6.2. Dynamic Response

The simulation S0 is represented in Figure 7 and analyzed in this sub-section. The MG has been simulated implementing both BVOC$_{BESS}$ (in the BESS' converter) and OOSF (in all the MG's converters) simultaneously. In order to test MG's robustness and verify that operation is sustained even in very adverse circumstances, several failures have been simulated:

- From $t = 24$ h to $t = 48$ h, all the renewable energy generation (PV and WT) is switched off.

- From $t = 96$ h to $t = 144$ h, the BESS is disconnected.
- From $t = 120$ h to $t = 144$ h, the INV is disconnected.

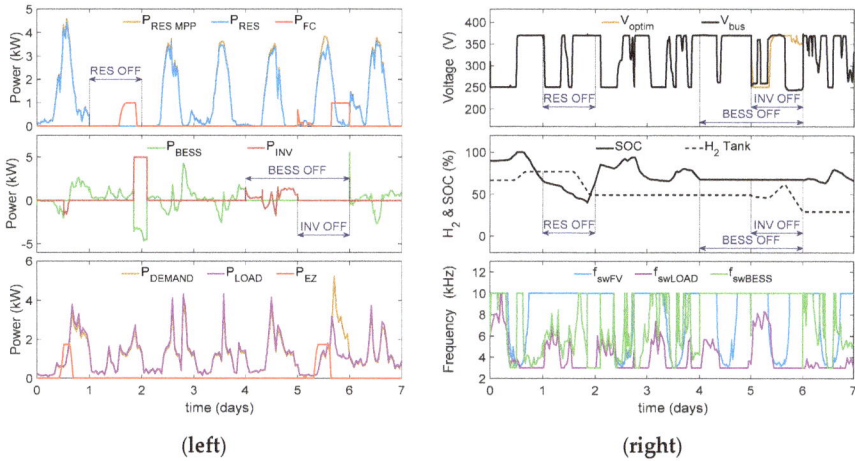

(left) (right)

Figure 7. MG simulation over 7 days with the optimized control system: E-BCS + BVOC$_{BESS}$ + OOSF. From top to bottom, the graphs in the **left** column represent: (1) Sources power: available RES power, actual RES power injected in the DC bus, and FC power injected in the DC bus; (2) BESS and INV power; (3) Loads power: critical loads electricity demand, actual LOAD power consumed from the DC bus, and EZ power consumed from the DC bus. The graphs in the **right** column represent: (1) Optimum bus voltage reference and actual bus voltage evolution; (2) SOC and hydrogen tank level evolution; (3) Optimized switching frequencies of three converters: PV, LOAD and BESS.

At the beginning of the first day, generation is higher than consumption and the surplus energy is used to charge the BESS until it reaches 100% SOC. EZ and INV are then activated. EZ leverages high generation to produce H_2. INV injects the excess of RES generation power to the grid. During the evening, RES generation decreases and LOAD increases, so the BESS begins to discharge to provide power.

The second day, RES remain disconnected and the battery keeps feeding the LOAD. Battery SOC descends to low values, triggering the activation of FC and INV. First, FC delivers emergency power to the DC bus (consuming H_2) and since SOC still keeps descending down to 40%, INV starts consuming full power from the grid to feed the loads and charge the battery up to 85%.

The third and fourth days, the BESS gets charged when RES power is greater than LOAD, and gets discharged when the opposite occurs. This is normal operation.

The fifth and sixth days, BESS remains disconnected. The control system permits to maintain MG stability, making the INV assume the power profile that would otherwise correspond to the BESS. The sixth day, INV (together with BESS) remains disconnected (this represents a BESS controller fault), leaving the MG not only operating in islanded mode, but also with no possibility of utilizing any bidirectional unit. Thus, generated power must constantly equal consumed power. VEC manages unidirectional units to balance the power being injected to and being consumed from the DC bus. The last day, the MG comes back to normal operation.

The system proves to be robust even in the face of severe failures. No undesired downtimes have occurred. The top right graph shows bus voltage evolution under BVOC$_{BESS}$: the optimal voltage reference (v_{optim}) is imposed, and BESS power is controlled such that bus voltage (v_{bus}) follows that reference—except when both BESS and INV are down. The bottom right graph shows the result of OOSF: the switching frequencies constantly vary to match the optimal frequency of every converter.

In all cases, the optimal switching frequency turns out to be high for low values of power, i.e., when core losses prevail.

6.3. Influence of Control Parameters in Optimization Performance

In this last sub-section, the parameters of BVOC$_{BESS}$ and OOSF are modified to evaluate how they affect the efficiency improvement achieved. The parameters studied are:

- Δt_{BVOC}: Refresh rate of BVOC$_{BESS}$ controller
- Δt_{OOSF}: Refresh rate of OOSF controllers
- Switching frequency optimization intervals

For this study, the operation of the MG has been simulated over a period of only 24 h because the simulation time step has been reduced from previous 100 ms to 10 ms. Executing the simulations over a seven days period would have required excessive computing resources and time. Each of the executed simulations (of 24 h of MG's operation) has taken an average of 15 minutes to complete. The results are presented in Table 4:

Table 4. Influence of control parameters on optimization performance.

	Reference Case Scenario S0	Modified Control Parameters				
		Faster BVOC	Slower BVOC	Wider F_{SW} Range	Faster OOSF	Slower OOSF
Control Strategy	Δt_{BVOC} = 100 ms Δt_{OOSF} = 100 ms $F_{SW\ MOSFET}$ ∈ 20–100 kHz $F_{SW\ IGBT}$ ∈ 3–10 kHz	Δt_{BVOC} 10 ms	Δt_{BVOC} 5 s	$F_{SW\ MOSFET}$ 10–200 kHz $F_{SW\ IGBT}$ 0.5–50 kHz	Δt_{OOSF} 10 ms	Δt_{OOSF} 5 s
E-BCS	η: Efficiency = 85.969%					
Optimization	$\Delta\eta$: Efficiency improvement over E-BCS	$\Delta\eta$: Efficiency improvement over reference case				
BVOC$_{BESS}$ + OOSF	1.029%	3.8×10^{-4}%	−0.009%	0.293%	1×10^{-6}%	-4×10^{-6}%
BVOC$_{BESS}$	0.216%	0.001%	−0.014%	-	-	-
OOSF	0.768%	-	-	0.265%	1×10^{-6}%	-3×10^{-6}%

Table 4 results indicate that BVOC$_{BESS}$ and OOSF performance can be slightly improved to some extent if the control parameters are chosen properly:

- Δt_{BVOC}: The refresh rate of BVOC$_{BESS}$ controller has very limited influence in overall BVOC$_{BESS}$ performance for the studied values. The capacitor's bank smooths the bus voltage dynamic so it is slow compared to the evaluated Δt_{BVOC}. Calculating and updating the optimal bus voltage reference, v_{optim}, more frequently does not significantly improve the efficiency of BVOC$_{BESS}$ algorithm.
- Switching frequency ranges: If the range of possible switching frequencies is widened, efficiency improvements close to 0.3% can be obtained. However, it is important to mention that a new optimization problem arises here to select the most appropriate frequency limits. The trade-off between OOSF efficiency improvement, passive filters size and electromagnetic interferences restriction must be established.
- Δt_{OOSF}: The variations in OOSF performance are insignificant. Again, the power in all the MG's units evolves very slowly compared to the evaluated Δt_{OOSF}.

Summarizing the results, the control parameters were already satisfactory. Changing the refresh rate of BVOC$_{BESS}$ and OOSF controllers produce nearly zero improvements. Only by widening the range of switching frequency window some appreciable efficiency increases might be obtained. Nevertheless, this has important drawbacks (e.g., increased current ripple) that might require re-designing the MG's converters.

7. Conclusions

A self–managed DC MG has been described. Events-Based Control System (E-BCS), which is constituted of decentralized Voltage Event Control (VEC) and State of Charge Event Control (SOCEC), has been the starting point to evaluate two strategies to optimize the efficiency of the MG's converters. E-BCS determines the power of each unit based on the bus voltage (v_{bus}) and on the battery SOC. These laws prevent that v_{bus} and SOC reach extreme values by activating loads and deactivating sources when v_{bus} or SOC get too high, and deactivating loads and activating sources when v_{bus} or SOC get too low.

The two proposed optimization strategies have been explained and simulated. Both of them increase the efficiency of the MG's DC-DC converters online, in any possible operation point. They utilize with two different approaches to minimize the power losses of the MG's converters. The nonlinear optimization problems that they solve are formulated based on small–signal averaged deterministic loss models of the converters. The two strategies are independent from each other and can be implemented simultaneously. They are also compatible with economic optimization strategies.

Representative efficiency curves of two of the MG's DC-DC converters at different switching frequencies and bus voltage values have been analyzed with the purpose of estimating the efficiency improvement that the optimization strategies can produce.

- Bus Voltage Optimization Control (BVOC$_{BESS}$): The BESS power is controlled such that it fulfills two purposes: first, to balance the power of sources, loads and inverter and, second, to make the system work at the (varying) optimal bus voltage v_{optim}. The value of v_{optim} is calculated online as the minimizer of the sum of power losses in the MG's DC-DC converters. Thus, this optimization applies to the whole set of converters, not to any particular individual converter.
- Online Optimization of Switching Frequency (OOSF): The gate drivers' switching frequency f_{sw} of each DC-DC converter is optimized online to improve its energy efficiency.

MG operation has been simulated over seven days in ten different scenarios representing different electricity generation and consumption profiles. The resulting control system (E-BCS + BVOC$_{BESS}$ + OOSF) manages the MG in a stable manner and is capable of overcoming abnormal difficulties, such as the simultaneous failure of all the bidirectional units. The two optimization strategies result in low (yet interesting) efficiency improvements: $\Delta\eta_{global}^{BVOC_{BESS}+OOSF} = 0.329\text{–}1.006\%$ (depending on the scenario considered); $\Delta\eta_{global}^{BVOC_{BESS}} = 0.131\text{–}0.481\%$; $\Delta\eta_{global}^{OOSF} = 0.190\text{–}0.723\%$.

Even though the efficiency improvements are low, they are based in discrete control algorithms so they require (almost) no extra investment. Best results occur when the MG's sources/loads deliver/consume low powers.

Two main reasons explain the low results. First and foremost, the converter's efficiency curves reveal small variations within the range of variation of v_{bus} and f_{sw}. Second, the optimal bus voltage is different for individual converters causing the BVOC$_{BESS}$ algorithm to calculate the v_{optim} as the trade–off that minimizes the overall power losses.

BVOC$_{BESS}$ and OOSF optimization strategies are appropriate for microgrids applications which require low power flows compared to the nominal power of the converters and/or microgrids in which all the MG's converters coincide in a single optimal bus voltage value v_{optim}. Possible application examples that meet these requirements include Uninterruptible Power Supply (UPS) systems and remote Base Transceiver Stations (RBT) and weather stations.

Author Contributions: Conceptualization, D.G.E., H.V.B., À.C.P. and L.M.S.; Data curation, D.G.E.; Formal analysis, D.G.E.; Funding acquisition, H.V.B. and À.C.P.; Investigation, D.G.E.; Methodology, D.G.E.; Project administration, H.V.B. and À.C.P.; Resources, H.V.B., À.C.P. and L.M.S.; Software, D.G.E.; Supervision, L.M.S.; Visualization, D.G.E.; Writing—original draft, D.G.E.; Writing—review & editing, D.G.E., H.V.B. and À.C.P.

Funding: This research was funded by the Spanish Ministry of Economy and Competitiveness under grants BES-2016-077460, DPI2015-67292-R and DPI2017-84572-C2-1-R.

Conflicts of Interest: The authors declare no conflict of interest.

Appendix A

This Appendix contains the complete system of equations describing the operation of the converters and their energy efficiency. Equations (A1)–(A37) are utilized to model the DC-DC converters in the simulation of the MG operation.

Equations (A1)–(A37) are also used to solve the BVOC$_{BESS}$ optimization problem considered in Equation (3) (substituting the physical magnitude v_{bus} by the calculation variable V_{BUS}) and the OOSF optimization problems considered in Equation (9) (substituting the physical magnitude f_{sw} by the calculation variable F_{SW}).

Only the variables and parameters appearing here for the first time will be described.

Appendix A.1 General Equations

The following equations apply to all the DC-DC converters present in the MG:

$$v_{out}i_{out} = v_{in}i_{in}\eta, \tag{A1}$$

$$i_{out} = i_{in} \cdot f(D), \tag{A2}$$

$$v_{out} = \frac{\eta v_{in}}{f(D)}, \tag{A3}$$

$$\Delta B = \frac{v_L D}{N A_C f_{sw}}, \tag{A4}$$

where v_L is the voltage across the inductor during the ON time of the duty cycle, N is the number of turns, A_C is the cross-sectional area of the inductor's core.

Appendix A.2 General Parameters

The following parameters apply to all the DC-DC converters present in the MG:

$$K = \begin{cases} 170.17 & for\ f_{sw} \leq 5\ kHz \\ 170.17 - 124.69 \cdot \frac{f_{sw}-5}{10} & for\ 5\ kHz \leq f_{sw}(kHz) \leq 10\ kHz \\ 45.48 & for\ f_{sw} > 15\ kHz \end{cases}, \tag{A5}$$

$$\alpha = \begin{cases} 1.03 & for\ f_{sw} \leq 5\ kHz \\ 1.03 + 0.43 \cdot \frac{f_{sw}-5}{10} & for\ 5\ kHz \leq f_{sw}(kHz) \leq 10\ kHz \\ 1.46 & for\ f_{sw} > 15\ kHz \end{cases}, \tag{A6}$$

$$\beta = 1.774. \tag{A7}$$

Appendix A.3 Converters Models: Parameters

Table A1. Converters parameters.

Device	PV	WT	FC	LOAD	EZ	BESS	INV
Inductors	R_L 120 mΩ N 183 A_C 9.87 cm² V_C 407 cm³	R_L 140 mΩ N 245 A_C 9.87 cm² V_C 407 cm³	R_{L1} 25 mΩ N_1 72 A_{C1} 6.78 cm² V_{C1} 220 cm³ R_{L2} 300 mΩ N_2 219 A_{C2} 6.78 cm² V_{C2} 220 cm³	R_{L1} 110 mΩ N_1 166 A_{C1} 9.87 cm² V_{C1} 407 cm³ R_{L2} 25 mΩ N_2 65 A_{C2} 9.87 cm² V_{C2} 407 cm³	R_{L1} 230 mΩ N_1 210 A_{C1} 6.78 cm² V_{C1} 220 cm³ R_{L2} 2 mΩ N_2 59 A_{C2} 6.78 cm² V_{C2} 220 cm³	R_L 130 mΩ N 189 A_C 9.87 cm² V_C 407 cm³	R_L 110 mΩ N 222 A_C 1.76 cm² V_C 34.5 cm³
Transistors	R'_d 80 mΩ V_f 1.75 V t_{sw} 135 ns (f_{sw}) 4 kHz	R'_d 7.9 mΩ t_{sw} 78 ns (f_{sw}) 50 kHz	R'_d 63 mΩ t_{sw} 106 ns (f_{sw}) 50 kHz	R'_d 9.8 mΩ V_f 2.0 V t_{sw} 287 ns (f_{sw}) 4 kHz	R'_d 22 mΩ t_{sw} 174 ns (f_{sw}) 50 kHz	R'_{d1} 15 mΩ V_{f1} 2.5 V t_{sw1} 131 ns (f_{sw1}) 4 kHz R'_{d2} 15 mΩ V_{f2} 2.5 V t_{sw2} 131 ns (f_{sw1}) 4 kHz	R'_{d1} 3.3 mΩ V_{f1} 1.55 V t_{sw1} 197 ns (f_{sw1}) 4 kHz R'_{d2} 3.3 mΩ V_{f2} 1.55 V t_{sw2} 197 ns (f_{sw1}) 4 kHz
Diodes	R_d 23.2 mΩ V_f 1.7 V	R_{d1} 11.3 mΩ V_{f1} 1.8 V	R_{d1} 2.2 mΩ V_{f1} 0.83 V R_{d2} 11.3 mΩ V_{f2} 1.5 V R_{d3} 150 mΩ V_{f3} 1.4 V	R_{d1} 11.2 mΩ V_{f1} 1.5 V R_{d2} 11.2 mΩ V_{f2} 1.5 V R_{d3} 1.6 mΩ V_{f3} 0.89 V	R_{d1} 39 mΩ V_{f1} 1.3 V R_{d2} 11.2 mΩ V_{f2} 1.5 V R_{d3} 3 mΩ V_{f3} 0.74 V	–	–
P.S.	P_{ctrl} 10 W	P_{ctrl} 5 W	P_{ctrl} 5 W	P_{ctrl} 10 W	P_{ctrl} 5 W	P_{ctrl} 10 W	P_{ctrl} 10 W

P.S. stands for power supply of the driver, and (f_{sw}) is the fixed frequency value when E-BCS only, or E-BCS + BVOC, are utilized.

Appendix A.4 Converter Models: Equations

The following equations are employed in the simulation to model the DC-DC converters operation. They are executed iteratively until the efficiency error is less than 0.1%. These equations are also utilized to formulate the optimization problems of BVOC$_{BESS}$ (substituting the physical magnitude v_{bus} by the calculation variable V_{BUS}) and OOSF (substituting the physical magnitude f_{sw} by the calculation variable F_{SW}):

Table A2. Equations utilized to model the converters' power losses.

PV converter (Boost) [IGBT]	$D = \frac{(v_{bus} - v_{in} \cdot \eta)}{v_{bus}}$	(A8)
	$v_L = v_{in}$	(A9)
	$P_{loss}^{PV} = \{R_L i_{in}^2 + K(\Delta B/2)^\beta f_{sw}^\alpha\}_{inductor} + \{(1-D)(R_d i_{in}^2 + V_f i_{in})\}_{diode}$ $+ \{D(R_d' i_{in}^2 + V_f' i_{in}) + (0.5 v_{bus} i_{in} i_{sw} f_{sw})\}_{IGBT} + P_{ctrl}$	(A10)
WT converter (Boost) [MOSFET]	$D = \frac{(v_{bus} - v_{in} \cdot \eta)}{v_{bus}}$	(A11)
	$v_L = v_{in}$	(A12)
	$P_{loss}^{WT} = \{R_L i_{in}^2 + K(\Delta B/2)^\beta f_{sw}^\alpha\}_{inductor} + \{(1-D)(R_d i_{in}^2 + V_f i_{in})\}_{diode}$ $+ \{DR_d' i_{in}^2 + (0.5 v_{bus} i_{in} i_{sw} f_{sw})\}_{MOSFET} + P_{ctrl}$	(A13)
FC converter (Quad. boost) [MOSFET]	$D = 1 - \sqrt{\frac{v_{in} \cdot \eta}{v_{bus}}}$	(A14)
	$v_{L1} = v_{in}$	(A15)
	$v_{L2} = \frac{v_{in}}{1-D}$	(A16)
	$P_{loss}^{FC} = \left\{R_{L1} i_{in}^2 + K(\Delta B_{L1}/2)^\beta f_{sw}^\alpha\right\}_{inductor1} + \left\{R_{L2}(1-D) i_{in}^2 + K(\Delta B_{L2}/2)^\beta f_{sw}^\alpha\right\}_{inductor2}$ $+ \left\{(1-D)\left(R_{d1} i_{in}^2 + V_{f1} i_{in}\right)\right\}_{diode1} + \left\{D\left(R_{d2} i_{in}^2 + V_{f2} i_{in}\right)\right\}_{diode2}$ $+ \left\{(1-D)\left(R_{d3} i_{out}^2 + V_{f3} i_{out}\right)\right\}_{diode3}$ $+ \left\{DR_d'(D i_{in})^2 + (0.5 v_{bus} D i_{in} i_{sw} f_{sw})\right\}_{MOSFET} + P_{ctrl}$	(A17)
LOAD converter (Quad. buck) [IGBT]	$D = \sqrt{\frac{v_{out}}{v_{bus} \cdot \eta}}$	(A18)
	$v_{L1} = (1-D) v_{in}$	(A19)
	$v_{L2} = \frac{(1-D) v_{out}}{D}$	(A20)
	$P_{loss}^{LOAD} = \left\{R_{L1}(D i_{out})^2 + K(\Delta B_{L1}/2)^\beta f_{sw}^\alpha\right\}_{inductor1} + \left\{R_{L2} i_{out}^2 + K(\Delta B_{L2}/2)^\beta f_{sw}^\alpha\right\}_{inductor2}$ $+ \left\{D\left(R_{d1}(D i_{out})^2 + V_{f1} D i_{out}\right)\right\}_{diode1}$ $+ \left\{(1-D)\left(R_{d2}((1-D) i_{out})^2 + V_{f2}(1-D) i_{out}\right)\right\}_{diode2}$ $+ \left\{D\left(R_{d3} i_{out}^2 + V_{f3} i_{out}\right)\right\}_{diode3}$ $+ \left\{D\left(R_d' i_{out}^2 + V_f' i_{out}\right) + (0.5(1+D) v_{bus} i_{out} i_{sw} f_{sw})\right\}_{IGBT} + P_{ctrl}$	(A21)
EZ converter (Quad. buck) [MOSFET]	$D = \sqrt{\frac{v_{out}}{v_{bus} \cdot \eta}}$	(A22)
	$v_{L1} = (1-D) v_{in}$	(A23)
	$v_{L2} = \frac{(1-D) v_{out}}{D}$	(A24)
	$P_{loss}^{EZ} = \left\{R_{L1}(D i_{out})^2 + K(\Delta B_{L1}/2)^\beta f_{sw}^\alpha\right\}_{inductor1} + \left\{R_{L2} i_{out}^2 + K(\Delta B_{L2}/2)^\beta f_{sw}^\alpha\right\}_{inductor2}$ $+ \left\{D\left(R_{d1}(D i_{out})^2 + V_{f1} D i_{out}\right)\right\}_{diode1}$ $+ \left\{(1-D)\left(R_{d2}((1-D) i_{out})^2 + V_{f2}(1-D) i_{out}\right)\right\}_{diode2}$ $+ \left\{D\left(R_{d3} i_{out}^2 + V_{f3} i_{out}\right)\right\}_{diode3}$ $+ \left\{DR_d' i_{out}^2 + (0.5(1+D) v_{bus} i_{out} i_{sw} f_{sw})\right\}_{MOSFET} + P_{ctrl}$	(A25)
BESS converter (Bidirec-tional) [IGBT]	Boost mode. in = battery out = MG's bus	$D = \frac{v_{bus} - v_{in} \cdot \eta}{v_{bus}}$	(A26)
		$v_L = v_{in}$	(A27)
		$P_{loss}^{BESS} = \{R_L i_{in}^2 + K(\Delta B/2)^\beta f_{sw}^\alpha\}_{inductor}$ $+ \{D(R_{d1}' i_{in}^2 + V_{f1}' i_{in}) + (0.5 v_{bus} i_{in} i_{sw1} f_{sw})\}_{IGBT1}$ $+ \{(1-D)(R_{d2}' i_{in}^2 + V_{f2}' i_{in}) + (0.5 v_{bus} i_{in} i_{sw2} f_{sw})\}_{IGBT2}$ $+ P_{ctrl}$	(A28)
	Buck mode. in = MG's bus out = battery	$D = \frac{v_{out}}{v_{bus} \cdot \eta}$	(A29)
		$v_L = v_{bus} - v_{out}$	(A30)
		$P_{loss}^{BESS} = \left\{R_L i_{out}^2 + K(\Delta B/2)^\beta f_{sw}^\alpha\right\}_{inductor}$ $+ \left\{(1-D)\left(R_{d1}' i_{out}^2 + V_{f1}' i_{out}\right) + (0.5 v_{bus} i_{out} i_{sw1} f_{sw})\right\}_{IGBT1}$ $+ \left\{D\left(R_{d2}' i_{out}^2 + V_{f2}' i_{out}\right) + (0.5 v_{bus} i_{out} i_{sw2} f_{sw})\right\}_{IGBT2}$ $+ P_{ctrl}$	(A31)
INV converter (Bidirec-tional) [IGBT]	Boost mode. in = MG's bus out = inv DC	$D = \frac{v_{out} - v_{bus} \cdot \eta}{v_{out}}$	(A32)
		$v_L = v_{bus}$	(A33)
		$P_{loss}^{INV} = \{R_L i_{in}^2 + K(\Delta B/2)^\beta f_{sw}^\alpha\}_{inductor}$ $+ \{D(R_{d1}' i_{in}^2 + V_{f1}' i_{in}) + (0.5 v_{out} i_{in} i_{sw1} f_{sw})\}_{IGBT1}$ $+ \{(1-D)(R_{d2}' i_{in}^2 + V_{f2}' i_{in}) + (0.5 v_{out} i_{in} i_{sw2} f_{sw})\}_{IGBT2}$ $+ P_{ctrl}$	(A34)
	Buck mode. in = inv DC out = MG's bus	$D = \frac{v_{bus}}{v_{in} \cdot \eta}$	(A35)
		$v_L = v_{in} - v_{bus}$	(A36)
		$P_{loss}^{INV} = \{R_L i_{out}^2 + K(\Delta B/2)^\beta f_{sw}^\alpha\}_{inductor}$ $+ \{(1-D)(R_{d1}' i_{out}^2 + V_{f1}' i_{out}) + (0.5 v_{in} i_{out} i_{sw1} f_{sw})\}_{IGBT1}$ $+ \{D(R_{d2}' i_{out}^2 + V_{f2}' i_{out}) + (0.5 v_{in} i_{out} i_{sw2} f_{sw})\}_{IGBT2} + P_{ctrl}$	(A37)

Where V_C is the volume of the inductor's core.

References

1. Hirsch, A.; Parag, Y.; Guerrero, J. Microgrids: A review of technologies, key drivers, and outstanding issues. *Renew. Sustain. Energy Rev.* **2018**, *90*, 402–411. [CrossRef]
2. Dragicevic, T.; Lu, X.; Vasquez, J.C.; Guerrero, J.M. DC Microgrids–Part I: A Review of Control Strategies and Stabilization Techniques. *IEEE Trans. Power Electron.* **2016**, *31*, 4876–4891. [CrossRef]
3. Zhao, J.; Dörfler, F. Distributed control and optimization in DC microgrids. *Automatica* **2015**, *61*, 18–26. [CrossRef]
4. Guerrero, J.M.; Vasquez, J.C.; Matas, J.; De Vicuña, L.G.; Castilla, M. Hierarchical control of droop-controlled AC and DC microgrids—A general approach toward standardization. *IEEE Trans. Ind. Electron.* **2011**, *58*, 158–172. [CrossRef]
5. Nguyen, D.T.; Le, L.B. Optimal energy trading for building microgrid with electric vehicles and renewable energy resources. In Proceedings of the ISGT 2014, Washington, DC, USA, 19–22 February 2014; pp. 1–5.
6. Li, C.; de Bosio, F.; Chaudhary, S.K.; Graells, M.; Vasquez, J.C.; Guerrero, J.M. Operation cost minimization of droop-controlled DC microgrids based on real-time pricing and optimal power flow. In Proceedings of the IECON 2015–41st Annual Conference of the IEEE Industrial Electronics Society, Yokohama, Japan, 9–12 November 2015.
7. Lujano-Rojas, J.M.; Dufo-Lopez, R.; Bernal-Agustin, J.L.; Catalao, J.P.S. Optimizing Daily Operation of Battery Energy Storage Systems Under Real-Time Pricing Schemes. *IEEE Trans. Smart Grid* **2017**, *8*, 316–330. [CrossRef]
8. Wang, Y.; Huang, Y.; Wang, Y.; Li, F.; Zhang, Y.; Tian, C. Operation Optimization in a Smart Micro-Grid in the Presence of Distributed Generation and Demand Response. *Sustainability* **2018**, *10*, 847. [CrossRef]
9. Gonzalez-Garrido, A.; Saez-de-Ibarra, A.; Gaztanaga, H.; Milo, A.; Eguia, P. Annual Optimized Bidding and Operation Strategy in Energy and Secondary Reserve Markets for Solar Plants with Storage Systems. *IEEE Trans. Power Syst.* **2018**, *8950*, 1–10. [CrossRef]
10. Anand, S.; Fernandes, B.G. Optimal voltage level for DC microgrids. In Proceedings of the IECON 2010–36th Annual Conference on IEEE Industrial Electronics Society, Glendale, AZ, USA, 7–10 November 2010; pp. 3034–3039. [CrossRef]
11. Zhao, Z.; Hu, J.; Chen, H. Bus Voltage Control Strategy for Low Voltage DC Microgrid Based on AC Power Grid and Battery. In Proceedings of the 2017 IEEE International Conference on Energy Internet (ICEI), Beijing, China, 17–21 April 2017; pp. 349–354. [CrossRef]
12. Dahiya, R. Voltage regulation and enhance load sharing in DC microgrid based on Particle Swarm Optimization in marine applications. *Indian J. Geo-Mar. Sci.* **2017**, *46*, 2105–2113.
13. Liu, J.M.; Yu, C.J.; Kuo, Y.C.; Kuo, T.H. Optimizing the efficiency of DC-DC converters with an analog variable-frequency controller. In Proceedings of the APCCAS 2008–2008 IEEE Asia Pacific Conference on Circuits and Systems, Macao, China, 30 November–3 December 2008; pp. 910–913. [CrossRef]
14. Sizikov, G.; Kolodny, A.; Fridman, E.G.; Zelikson, M. Frequency dependent efficiency model of on-chip DC-DC buck converters. In Proceedings of the 2010 IEEE 26-th Convention of Electrical and Electronics Engineers in Israel, Eliat, Israel, 17–20 November 2010; pp. 651–654. [CrossRef]
15. Jauch, F.; Biela, J. Generalized modeling and optimization of a bidirectional dual active bridge DC-DC converter including frequency variation. In Proceedings of the 2014 International Power Electronics Conference (IPEC-Hiroshima 2014–ECCE ASIA), Hiroshima, Japan, 18–21 May 2014; pp. 1788–1795. [CrossRef]
16. Çelebi, M. Efficiency optimization of a conventional boost DC/DC converter. *Electr. Eng.* **2018**, *100*, 803–809. [CrossRef]
17. Al-Hoor, W.; Abu-Qahouq, J.A.; Huang, L.; Batarseh, I. Adaptive variable switching frequency digital controller algorithm to optimize efficiency. In Proceedings of the 2007 IEEE International Symposium on Circuits and Systems, New Orleans, LA, USA, 27–30 May 2007; pp. 781–784.
18. Liu, J.-M.; Wang, P.-Y.; Kuo, T.-H. A Current-mode DC–DC buck converter with efficiency-optimized frequency control and reconfigurable compensation. *IEEE Trans. Power Electron.* **2012**, *27*, 869–880. [CrossRef]
19. Zhao, L.; Li, H.; Liu, Y.; Li, Z. High efficiency variable-frequency full-bridge converter with a load adaptive control method based on the loss model. *Energies* **2015**, *8*, 2647–2673. [CrossRef]
20. *MATLAB R2017a and Simulink*; The MathWorks, Inc.: Natick, MA, USA, 2017.

Energies **2018**, *11*, 3090

21. Bosque-Moncusi, J.M. Ampliación, Mejora e Integración en la Red de un Sistema Fotovoltaico. Ph.D. Thesis, Universitat Rovira i Virgili, Tarragona, Spain, 2015.
22. Eichhorn, T. Boost Converter Efficiency through Accurate Calculations. *Power Electron. Technol. Mag. Online* **2008**, *9*, 30–35.
23. Stasi, F. *De Working with Boost Converters*; Texas Instruments Incorporated: Dallas, TX, USA, 2015; pp. 1–11.
24. Hurley, W.G.; Wölfle, W.H. *Transformers and Inductors for Power Electronics Transformers and Electronics*; Wiley: Hoboken, NJ, USA, 2013; Volume 8, ISBN 9780071594325.
25. Servei Meteorològic de Catalunya. Available online: http://www.meteo.cat/wpweb/climatologia/serveis-i-dades-climatiques/series-climatiques-historiques/ (accessed on 15 June 2018).

energies

MDPI

Article

Dynamic Analysis of a Permanent Magnet DC Motor Using a Buck Converter Controlled by ZAD-FPIC

Fredy E. Hoyos Velasco [1,*], John E. Candelo-Becerra [2] and Alejandro Rincón Santamaría [3]

[1] Escuela de Física, Facultad de Ciencias, Sede Medellín, Universidad Nacional de Colombia, Carrera 65 No. 59A, 110, Medellín 050034, Antioquia, Colombia

[2] Departamento de Energía Eléctrica y Automática, Facultad de Minas, Sede Medellín, Universidad Nacional de Colombia, Carrera 80 No. 65-223, Campus Robledo, Medellín 050041, Antioquia, Colombia; jecandelob@unal.edu.co

[3] Grupo de Investigación en Desarrollos Tecnológicos y Ambientales—GIDTA, Faculty of Engineering and Architecture, Universidad Católica de Manizales, Carrera 23 No. 60-63, Manizales 170002, Caldas, Colombia; arincons@ucm.edu.co

*   Correspondence: fehoyosve@unal.edu.co; Tel.: +57-4-4309000 (ext. 46532)

Received: 7 November 2018; Accepted: 30 November 2018; Published: 3 December 2018

Abstract: This paper presents the dynamic analysis of a permanent magnet DC motor using a buck converter controlled by zero average dynamics (ZADs) and fixed-point inducting control (FPIC). Initially, the steady-state behavior of the closed-loop system was observed and then transient behavior analyzed while maintaining a fixed ZAD control parameter and changing the FPIC parameter. Other behaviors were studied when the value of the ZAD control parameter changed and the FPIC parameter was maintained at the initial value. Besides, bifurcation diagrams were built with one and two delay periods by changing the control parameter of the FPIC and maintaining fixed ZAD parameters while some disturbances were carried out in the electric source. The results show that the ZAD-FPIC controller allowed good regulation of the speed for different reference values. The ZAD-FPIC control technique is effective for controlling the buck converter with the motor, even with two delay periods. The robustness of the system was checked by changing the voltage of the source. It was shown that the system used a fixed switching frequency because the duty cycle was not saturated for certain ranges of the control parameters shown in the research. This technique can be used for higher order systems with experimental phenomena such as quantization effects, time delays, and variations in the input signal.

Keywords: buck converter; DC motor; bifurcations in control parameter; sliding control; zero average dynamics; fixed-point inducting control

1. Introduction

Electric motors are designed to perform tasks with high accuracy when completing repetitive tasks [1]. The speed control helps maintain the frequency close to the reference value and allows the motor to offer continuous stability. Recent advances in materials for permanent magnets mean they are now lighter, less expensive, and easier to control at low speed, thus expanding their domestic and industrial applications [2]. Currently in the industry, digital signal processing (DSP) offers the following characteristics: greater versatility compared with analog designs [3]; ease of implementation for nonlinear controllers and advanced control techniques; low-power consumption; reduction of external passive components; low sensitivity to parameter variation; applications of high-frequency switching controllers; and others that have been described in References [4–8]. However, dynamic analyses must be performed to understand the optimal functioning of the controller with loads and the different actions to apply.

The control of motors have been studied previously using zero average dynamics (ZADs) and fixed-point induced control (FPIC) [8]. The FPIC control technique has been used to control chaotic systems applied to DC–DC and DC–AC converters. The time-delay autosynchronization (TDAS) control strategy has shown better convergence results and easy implementation of the digital modulation of centered pulse width (DPWMC). As an alternative, a sliding surface has been used to apply the ZAD control technique in a quasi-sliding manner [9]. Besides, ZAD was used to implement a buck inverter in a field-programmable gate array (FPGA), verifying that the ZAD technique with pulse next meets the requirements of a fixed switching frequency [10,11]. Other studies using the ZAD control strategy have considered the transition from periodic bands to chaotic bands in a buck DC–DC converter. This is useful to identify bifurcations by double period and by corner impact, known as "corner collision bifurcations", as well as chaotic phenomena, chaotic bands, and doubling of these bands [12–14].

The existence of bifurcations and chaos bands for a buck converter operated with DPWMC with and without a delay period controlled with ZAD was analyzed numerically [15]. The dynamic behavior of these systems has been also extensively studied through mathematical, numerical, and experimental analyses [16–18]. In Reference [19], the simulation of a buck converter configured as an inverter was carried out by using the Powersim (PSIM) simulation software professional version 6.1.3 and applying the ZAD technique with PWMC and FPIC. The reference signals to be followed were triangular and sinusoidal, and it was tested for various types of load (i.e., resistive, time-variable, nonlinear, and open-circuit operation).

In Reference [20], a quasi-sliding algorithm based on ZAD was proposed for the modular control of the DC–AC conversion system by connecting m single-phase inverters in parallel to feed the same load. In Reference [21], numerical and experimental results were obtained by applying digital control implemented in a DSP using the ZAD-FPIC control technique to a DC–DC and a DC–AC converter. It was shown that the bifurcation diagrams, calculated numerically in the design stage, agree quantitatively with those obtained in the experimental stage.

The integration of ZAD with an FPIC controller has been shown to work well for the buck converter, regulating resistive and motor loads [8]. However, the dynamics of the control connected to the motors must be studied. Therefore, this paper focuses on the dynamic analysis of a buck converter that uses the combined ZAD-FPIC control technique to control the speed of an electric motor. The system involves a buck power converter, a permanent magnet DC motor, and a dSPACE platform [22]. The load torque and the friction torque were considered as known. The ZAD-FPIC scheme was formulated and experimentally tested. Bifurcation diagrams were developed for the adaptive ZAD-FPIC control system for different values of the controller parameters. The tests performed in the research show how the numerical and experimental diagrams match for the initial conditions and the changes carried out in the study. The main differences of the present paper with respect to closely related papers on ZAD-FPIC [23,24] are the following: the ZAD-FPIC technique was proven for the first time in a higher order system (a permanent magnet DC motor). In previous works, the ZAD-FPIC technique was only applied to second-order systems and no analysis related to bifurcation for the FPIC technique has been performed [8]. In this work, some disturbance in the input voltage and the control with one and two delay periods have been applied, achieving effectiveness in speed control. It was proved that the FPIC technique is useful to control the chaos and to follow the speed to a reference given by the user.

This paper is organized as follows. Section 2 presents the materials and methods used in the research and the mathematics required to perform the simulation and experimental tests. Section 3 includes the results and analysis of the different simulation and experimental tests performed with the proposed changes in the control parameters and the reference source. Section 4 presents the conclusions of the work.

2. Materials and Methods

Figure 1 shows the block diagram of the system under study that considers a motor connected to a buck converter and controlled by the ZAD-FPIC technique, and referred to in this research as the "buck-motor system". Some sensors are used to measure the voltage, current, and speed of the buck-motor system. The controller considers these signals, and with the reference speed, takes actions on the buck converter to regulate the speed of the DC motor. In this research, the speed reference and the control parameters are varied in order to identify how the controller regulates the speed of the motor.

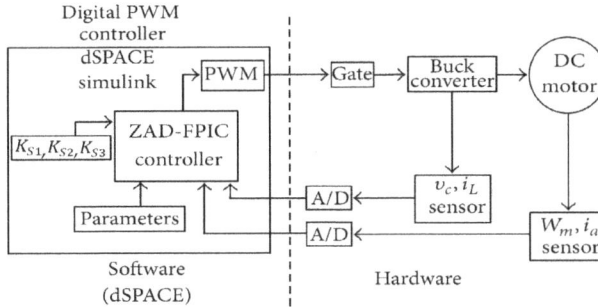

Figure 1. Buck converter connected to an electric DC motor and controlled by zero average dynamics-fixed-point induced control (ZAD-FPIC).

The system presented in Figure 1 considers a permanent magnet motor with the following characteristics as presented in Table 1.

Table 1. Rated values of the DC motor.

Parameter	Description	Value
P_r	Rated power	250 W
V_r	Rated voltage	42 VDC
I_r	Rated current	6 A
W_r	Rated speed	4000 RPM

The speed of the motor is measured by using an encoder of 1000 pulses per revolution. The state variables, established as the output voltage v_c, the inductor current i_L, and armature current i_a, are measured with an accurate resistance. The digital communication was performed by using a board with DS1104 of dSPACE, where the techniques of ZAD-FPIC are implemented. The output PWM is calculated by a DS1104 that sends the digital signal. The connection of the control system with the power system is performed with opto-couplers with fast response, such as HCPL-J312, which protect and isolate the digital circuit from possible currents and voltages generated in the power circuit. The state variables, v_c, i_L, and i_a, reach the controller based on the input ADC of 12 bits. The controlled variable W_m is measured by an encoder of 28 bits at sample frequencies of 6 kHz.

The parameters related to the DC electric motor parameters, buck converter, and law of ZAD-FPIC control are defined in the control blocks. The last parameters are related to the time constants and the dynamics of the error that is to be imposed in the control system, for example, K_{S1}, K_{S2}, and K_{S3}. At each sampling period, the microprocessor of the DS1104 calculates, with a resolution of 10 bits, the duty cycle d and its equivalent to the PWMC to control the solid-state switch S (metal-oxide-semiconductor field-effect transistor or MOSFET).

2.1. Model of the Buck Motor System

Equations (1) and (2) represent the DC motor and the mechanical load. This model considers a second order, where the state variables are the speed of the motor W_m (rad/s) and the armature current i_a (A). The term k_e represents the constant of the output voltage in the motor (V/rad/s), L_a is the armature inductance (mH), R_a is the armature resistance (Ω), $V_a = v_c$ is the voltage in the motor (V), B is the viscosity friction coefficient (N·m/rad/s), J_{eq} is the inertia moment (kg·m²), k_t is the torque constant of the motor (N·m/A), T_{fric} is the friction torque (N·m), T_L is the load torque (N·m), and J_L is the moment of inertia of the load (kg·m²):

$$\frac{dW_m(t)}{dt} = \frac{-BW_m(t)}{J_{eq}} + \frac{k_t i_a(t)}{J_{eq}} + \frac{-T_{fric}}{J_{eq}} + \frac{-T_L}{J_{eq}}, \tag{1}$$

$$\frac{di_a(t)}{dt} = \frac{-k_e}{L_a} W_m(t) + \frac{-R_a}{L_a} i_a(t) + \frac{v_c}{L_a}. \tag{2}$$

To model the buck-motor system and obtain a representation in state variables, the system of Figure 2 is analyzed. This figure shows that the equivalent diagram depends on the state of the switch S. For this system, the speed of the motor can be changed by manipulating the switch ON and OFF by changing the model as described next. The parameters of the buck-motor system are shown in Table 2.

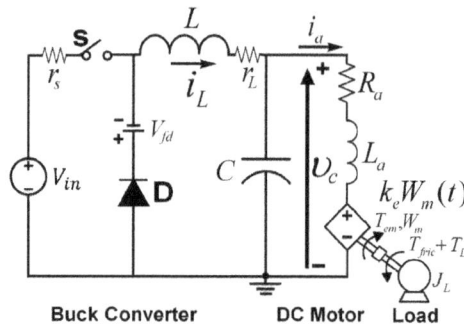

Figure 2. Diagram of the buck converter connected to a DC motor.

Table 2. Parameters of the buck-motor system.

Parameter	Description	Value
r_s	Internal resistance of the source	0.84 Ω
V_{in}	Input voltage	40.086 V
V_{fd}	Diode forward voltage	1.1 V
L	Inductance	2.473 mH
r_L	Internal resistance of the inductor	1.695 Ω
C	Capacitance	46.27 µF
R_a	Armature resistance	2.7289 Ω
L_a	Armature inductance	1.17 mH
B	Viscosity friction coefficient	0.000138 (N·m/rad/s)
J_{eq}	Inertia moment	0.000115 (kg·m²)
k_t	Motor torque constant	0.0663 (N·m/A)
k_e	Voltage constant	0.0663 (V/rad/s)
T_{fric}	Friction torque	0.0284 (N·m)
T_L	Load torque	Variable (N·m)
W_m	Speed of the motor [rad/s]	28 bits
i_a	Armature current [A]	12 bits
v_c	Voltage of the motor [V]	12 bits
i_L	Current in the inductor [A]	12 bits

Figure 3 illustrates the operation in continuous driving mode with the switch closed (S = ON) and the diode in cut (inactive). This circuit considers the main source E, the parasite resistance $r_s = r_{s1} + r_M$ as the sum of the internal resistance of the source r_{s1} and the parasite resistance of the MOSFET r_M, the inductance L, and the capacitance C. The load is a DC motor that considers the armature resistance R_a and armature inductance L_a.

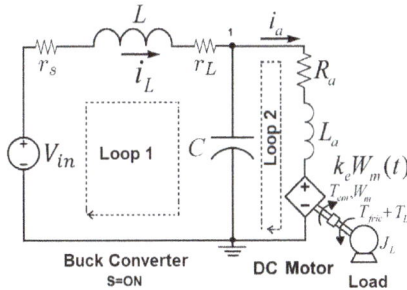

Figure 3. Electromechanical system with the switch ON.

Applying the Kirchhoff laws to the circuit in Figure 3, the mathematical model of this new equivalent circuit can be represented as in Equation (3). In a simple form, this expression can be presented as $\dot{x} = A_1 x + B_1$, where $x_1 = W_m$, $x_2 = i_a$, $x_3 = v_c$, and $x_4 = i_L$:

$$
\begin{bmatrix} \dot{W}_m \\ \dot{i}_a \\ \dot{v}_c \\ \dot{i}_L \end{bmatrix} = \begin{bmatrix} \frac{-B}{J_{eq}} & \frac{k_t}{J_{eq}} & 0 & 0 \\ \frac{-k_e}{L_a} & \frac{-R_a}{L_a} & \frac{1}{L_a} & 0 \\ 0 & \frac{-1}{C} & 0 & \frac{1}{C} \\ 0 & 0 & \frac{-1}{L} & \frac{-(r_L+r_s)}{L} \end{bmatrix} \begin{bmatrix} W_m \\ i_a \\ v_c \\ i_L \end{bmatrix} + \begin{bmatrix} \frac{-(T_{fric}+T_L)}{J_{eq}} \\ 0 \\ 0 \\ \frac{V_{in}}{L} \end{bmatrix}.
\tag{3}
$$

On the other hand, Figure 4 shows the diagram of the circuit with the diode in conduction mode and the switch open (S = OFF). Thus, the initial circuit is reduced to an equivalent circuit as that presented in Figure 4.

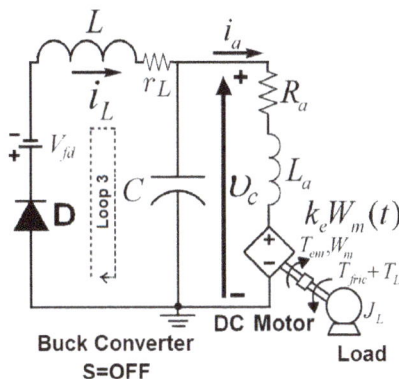

Figure 4. Electromechanical system with the switch OFF.

In this case, the mathematical model of this new equivalent circuit can be represented as Equation (4), what in a simple form can be presented as $\dot{x} = A_2 x + B_2$. The state variables are W_m, i_a, v_c, and i_L, and the parameters C and L are the capacitance and inductance of the converter, respectively. The parasite resistance $r_s = r_{s1} + r_M$ is equal to the sum of the internal resistance of the source and the parasite resistance of the MOSFET. The resistance $r_L = r_{L1} + r_{Med}$ is equal to the sum of the resistances of the winding and the current measurement, and V_{fd} is the voltage in the direct conduction diode. The source that feeds the buck-motor system is represented by the input voltage V_{in} and depends on the switch S controlled by the pulses of the PWMC; that is, the system feeds with V_{in} when S is active, or with $-V_{fd}$ when S is inactive:

$$
\begin{bmatrix} \dot{W}_m \\ \dot{i}_a \\ \dot{v}_c \\ \dot{i}_L \end{bmatrix} = \begin{bmatrix} \frac{-B}{J_{eq}} & \frac{k_t}{J_{eq}} & 0 & 0 \\ \frac{-k_e}{L_a} & \frac{-R_a}{L_a} & \frac{1}{L_a} & 0 \\ 0 & \frac{-1}{C} & 0 & \frac{1}{C} \\ 0 & 0 & \frac{-1}{L} & \frac{-(r_L)}{L} \end{bmatrix} \begin{bmatrix} W_m \\ i_a \\ v_c \\ i_L \end{bmatrix} + \begin{bmatrix} \frac{-(T_{fric}+T_L)}{J_{eq}} \\ 0 \\ 0 \\ \frac{-V_{fd}}{L} \end{bmatrix}. \tag{4}
$$

The buck-motor system can present a discontinuous conduction mode (DCM) when the switch is open and the currents in the inductor are zero. In this case, the diode stops conducting and the circuit can be represented as shown in Figure 5, with the differential equations shown in Equation (5). This last equation can be rewritten as $\dot{x} = A_{DCM}x + B_{DCM}$.

$$
\begin{bmatrix} \dot{W}_m \\ \dot{i}_a \\ \dot{v}_c \\ \dot{i}_L \end{bmatrix} = \begin{bmatrix} \frac{-B}{J_{eq}} & \frac{k_t}{J_{eq}} & 0 & 0 \\ \frac{-k_e}{L_a} & \frac{-R_a}{L_a} & \frac{1}{L_a} & 0 \\ 0 & \frac{-1}{C} & 0 & 0 \\ 0 & 0 & 0 & 0 \end{bmatrix} \begin{bmatrix} W_m \\ i_a \\ v_c \\ i_L \end{bmatrix} + \begin{bmatrix} \frac{-(T_{fric}+T_L)}{J_{eq}} \\ 0 \\ 0 \\ 0 \end{bmatrix}. \tag{5}
$$

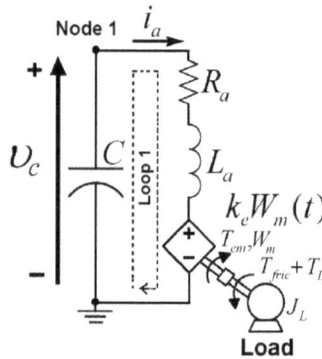

Figure 5. Buck-motor system working in discontinuous conduction mode (DCM).

Assuming that the PWM signal is configured with the pulse to the center and the buck-motor system works in continuous conduction mode (CCM), the dynamics of the system in a complete period are described by Equation (1) and can be written as:

$$
\dot{x} = \begin{cases} A_1 x + B_1 & \text{if} \quad kT \leq t \leq kT + dT/2 \\ A_2 x + B_2 & \text{if} \quad kT + dT/2 < t < kT + T - dT/2 \\ A_1 x + B_1 & \text{if} \quad kT + T - dT/2 < t < kT + T \end{cases} \tag{6}
$$

The term k represents the k-th iteration of the system with one sample period T or commutation period. The derivate of the state is defined as $\dot{x}$ and can be calculated as shown in Equation (7):

$$
\dot{x} = [\dot{x}_1, \dot{x}_2, \dot{x}_3, \dot{x}_4]^T \equiv \left[\frac{dx_1}{dt}, \frac{dx_2}{dt}, \frac{dx_3}{dt}, \frac{dx_4}{dt} \right]^T \equiv \left[\frac{dW_m}{dt}, \frac{di_a}{dt}, \frac{dv_c}{dt}, \frac{di_L}{dt} \right]^T. \tag{7}
$$

2.2. Analytical Solution of the Buck-Motor System

In this work, the system is assumed to operate in the CCM. The solution of the system equations presented in Equation (6) is obtained as shown in Equation (8):

$$x(t) = \begin{cases} e^{A_1 t} M_1 - V_1 & \text{if} \quad kT \le t \le (k+d/2)T \\ e^{A_2 t} M_2 - V_2 & \text{if} \quad (k+d/2)T < t < (k+1-d/2)T \\ e^{A_1 t} M_3 - V_1 & \text{if} \quad (k+1-d/2)T \le t \le (k+1)T \end{cases} \tag{8}$$

where

$$M_1 = x(0) + V_1$$
$$M_2 = Q_{12} M_1 - \Delta V e^{-A_2 T \frac{d}{2}}$$
$$M_3 = Q_{21} M_2 + \Delta V e^{-A_1 T (1 - \frac{d}{2})}$$
$$Q_{12} = e^{(A_1 - A_2)T(\frac{d}{2})}$$
$$Q_{21} = e^{(A_2 - A_1)T(1 - \frac{d}{2})}$$
$$V_1 = A_1^{-1} B_1$$
$$V_2 = A_2^{-1} B_2$$
$$\Delta V = V_1 - V_2$$

The solution when the system operates in DCM is given by Equation (9) and is presented when the current in the inductor is zero:

$$x(t) = e^{A_{DCM} t} \left[x(0) + A_{DCM}^{-1} B_{DCM} \right] - A_{DCM}^{-1} B_{DCM}. \tag{9}$$

Starting from the solution in Equation (8) and discretizing the output signals for each sampling period T, we have the expression in discrete time given by Equation (10), which is the solution in CCM for the buck-motor system:

$$x((k+1)T) = e^{A_1 T} Q x(kT) + e^{A_1 T} Q V_1 - Q_{12} e^{A_2 T(1 - \frac{d}{2})} \Delta V + e^{A_1 T \frac{d}{2}} \Delta V - V_1. \tag{10}$$

The term Q is obtained with Equation (11) and the expressions are defined in Equations (5) and (6):

$$Q = e^{(A_2 - A_1)T} e^{(A_1 - A_2)Td}. \tag{11}$$

The solution of the system operating in DCM is given in Equation (12):

$$x((k+1)T) = e^{A_{DCM} T} \left[x(kT) + A_{DCM}^{-1} B_{DCM} \right] - A_{DCM}^{-1} B_{DCM}. \tag{12}$$

2.3. Strategies to Control the Speed of the DC Motor

The speed W_m of the DC motor must follow a reference speed W_{mref}. Thus, for the sampling period kT, the tracking error is defined as presented in Equation (13):

$$e(kT) = W_m(kT) - W_{mref}(kT). \tag{13}$$

Besides, considering the system in fourth-order, the sliding surface is defined as $s(kT)$ [20], describing a third-order dynamic in the error variable $(e(kT))$, which is given by Equation (14):

$$s(kT) = e(kT) + k_{s1} \frac{de(kT)}{d(kT)} + k_{s2} \frac{d^2 e(kT)}{d(kT)^2} + k_{s3} \frac{d^3 e(kT)}{d(kT)^3}. \tag{14}$$

The constants $k_{s1} = K_{S1}\sqrt{LC}$, $k_{s2} = K_{S2}LC$, and $k_{s3} = K_{S3}LC\sqrt{LC}$ are parameterized in function of the constants applied (K_{S1}, K_{S2}, and K_{S3}). The constants K_{S1}, K_{S2}, and K_{S3} are the parameters of the ZAD and can be calibrated to impose a dynamic behavior in the closed-loop system. In addition, such parameters can be considered to construct dimensional bifurcation diagrams. When the reference signal W_{mref} is established as a constant value, Equation (14) can be written as in Equation (15) and the first derivate as in Equation (16):

$$s(kT) = W_m(kT) - W_{mref}(kT) + k_{s1}\frac{dW_m(kT)}{d(kT)} + k_{s2}\frac{d^2W_m(kT)}{d(kT)^2} + k_{s3}\frac{d^3W_m(kT)}{d(kT)^3}, \tag{15}$$

$$\dot{s}(kT) = \frac{dW_m(kT)}{d(kT)} + k_{s1}\frac{d^2W_m(kT)}{d(kT)^2} + k_{s2}\frac{d^3W_m(kT)}{d(kT)^3} + k_{s3}\frac{d^4W_m(kT)}{d(kT)^4}. \tag{16}$$

The duty cycle can be calculated as shown in Equation (17):

$$d_k(kT) = \frac{2s(kT) + T\dot{s}_-(kT)}{T(\dot{s}_-(kT) - \dot{s}_+(kT))}, \text{ is} \tag{17}$$

where $s(kT)$ is calculated as shown in Equation (15) at the beginning of each commutation period for the system in Equation (3). Thus, $s(kT) = s(kT)|_{S=ON}$; $\dot{s}_+(kT)$ is calculated as in Equation (16) for the system described in Equation (3) as $\dot{s}_+(kT) = \dot{s}(kT)|_{S=ON}$ [8].

Next, the necessary steps to calculate the variables are: the first derivatives taken from the system are obtained from Equation (3), which occurs when $S = ON$. To obtain $\dot{s}_-(kT)$, the first, second, third, and fourth derivates are calculated for the system and $\dot{s}_-(kT) = \dot{s}(kT)|_{S=OFF}$. With the delay period in the control action, the new duty cycle is calculated as in Equation (18):

$$d_k(kT) = \frac{2s((k-1)T) + T\dot{s}_-((k-1)T)}{T(\dot{s}_-((k-1)T) - \dot{s}_+((k-1)T))}. \tag{18}$$

With the ZAD-FPIC strategy, the new duty cycle is calculated to ensure that the load and the motor rotate at the desired speed W_{mref}, leading to the expression shown in Equation (19):

$$d_{ZAD-FPIC}(kT) = \frac{d_k(kT) + Nd^*}{N+1}. \tag{19}$$

Combining Equations (18) and (19), the control for the ZAD-FPIC is defined as in Equation (20):

$$d_{ZAD-FPIC}(kT) = \left(\frac{2s((k-1)T) + T\dot{s}_-((k-1)T)}{T(\dot{s}_-((k-1)T) - \dot{s}_+((k-1)T))} + Nd^* \right)(N+1)^{-1}. \tag{20}$$

where d^* is calculated at the beginning of the period with

$$d^* = d_k(kT)|_{stable\ state}. \tag{21}$$

Therefore, Equation (20) combines ZAD and FPIC techniques, and a saturation function must be applied to consider the actual physical limits of the duty cycle between 0 and 1. Such a saturation function is described in Equation (22):

$$d = \begin{cases} d_{ZAD-FPIC}(kT) \text{ if } 0 < d_{ZAD-FPIC}(kT) < 1 \\ 1 \text{ if } 1 \leq d_{ZAD-FPIC}(kT) \\ 0 \text{ if } d_{ZAD-FPIC}(kT) \leq 0 \end{cases}. \tag{22}$$

3. Results and Analysis

This section presents the results obtained from the dynamic analysis of a DC motor using a buck converter controlled by ZAD-FPIC. The quantization effects considered in the tests are: 28 bits of speed, 12 bits in the system variables (v_c, i_a, and i_L), and 10 bits for the duty cycle. MATLAB®/Simulink software was used for the simulation, and the simulation blocks that represent the system with ZAD-FPIC controller are shown in Figure 6. For the experimental test, the ZAD-FPIC technique is implemented in the rapid control prototyping card dSPACE DS1104; in this case, the card is programmed with MATLAB®/Simulink and then downloaded to the DSP. This platform has a graphical display interface called ControlDesk. In order to overlap the numerical with the experimental results, the experimental data were stored in matrices using the ControlDesk program and then read to perform calculations in MATLAB. Finally, the simulation is run and both the numerical and experimental results plotted.

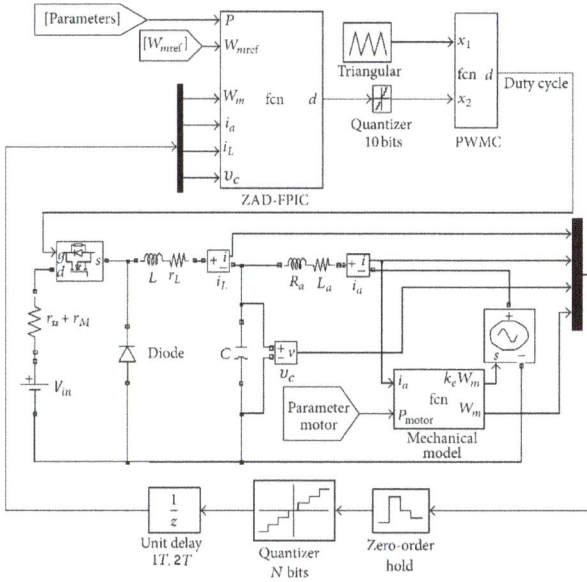

Figure 6. ZAD-FPIC controller simulation blocks.

3.1. Parameters of the Controller

Table 3 shows the parameters of the ZAD-FPIC controller used in the simulation and experimental tests.

Table 3. Parameters of the ZAD-FPIC controller.

Parameter	Description	Value
V_{in}	Input voltage	40.086 V
W_{mref}	Reference speed	Variable (rad/s)
N	Control parameter of FPIC	1
Fc	Commutation frequency	6 kHz
Fs	Sample frequency	6 kHz
$1T\_p$	1 delay period	166.6 µs
K_{S1}, K_{S2}, K_{S3}	Bifurcation parameter	Variables
d	Duty cycle	10 bits

3.2. Behavior of the Buck-Motor System in Closed Loops

Figure 7 shows the dynamic behavior of the mechanical speed and the error of the signals when the electric circuit works in closed loops. The speed is determined for both simulation and experimental tests following a change in the reference signal W_{mref}. Figure 7a,b show the mechanical speed and the error over time, respectively, when the circuit works in a closed loop. In this case, the system is working in a closed loop when the reference signal is W_{mref} = 400 rad/s at t = 1 s, K_{S1} = 2, K_{S2} = 2, K_{S3} = 30, and N = 1 with one delay period.

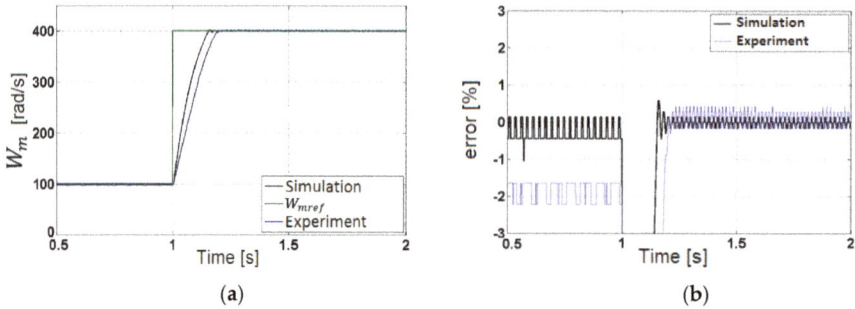

Figure 7. Numerical and experimental results for the system in closed loops: (**a**) mechanical speed W_m and W_{mref} in closed loop and (**b**) error of W_m in closed loop.

Figure 7a shows that the impulse is very small for the simulation and experimental tests; the settling time of W_m is t_s = 0.1473 s in the simulation test and t_s = 0.1859 s in the experimental test. Finally, Figure 7b shows that the steady-state error in the simulation test is −0.1645% and for the experimental test it is 0.4245%.

Table 4 summarizes the simulated and experimental results for the closed-loop systems shown in Figure 7.

Table 4. Transient responses of the buck-motor system in closed-loop with ZAD-FPIC.

Controller	M_p (%)	t_s (s)	Error (%)
Closed-loop system in the simulation test	0.5715	0.1473	−0.1645
Closed-loop system in the experimental test	Overdamped	0.1859	0.4245

Figure 8 shows the behavior of the system working closed loops for the simulation and experimental test when the reference signal W_{mref} changes from 100 to 400 rad/s in intervals of 4 s. In the experimental test, the closed-loop system did not present an impulse, whereas for the simulation test, some small impulses are presented for the closed-loop system. The signals (W_m) present a settling time of $t_s \cong$ 0.15 s. The steady-state error for the closed-loop system is less than 2%. There is a high coincidence between the numerical and experimental results. From the results shown in Figure 8a,b, the ZAD-FPIC controller is robust to the variations created in the reference signal.

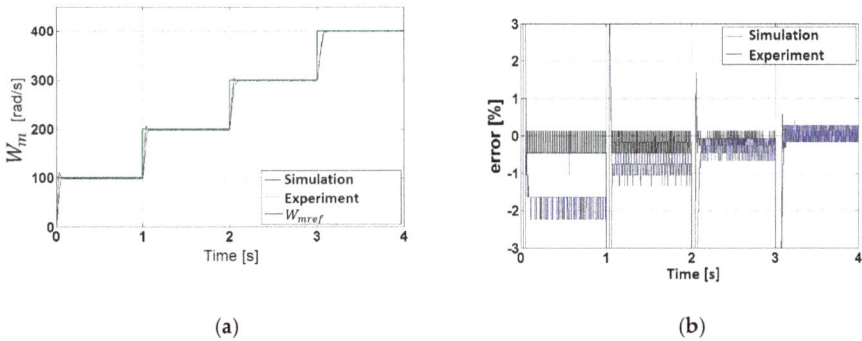

Figure 8. Numerical and experimental results in closed loops: (**a**) mechanical speed W_m and reference speed W_{mref} in a closed-loop system and (**b**) error of W_m in a closed-loop system.

3.3. Transient Response Changing the Control Parameter N with Parameter $K_{S3} = 35$

Figure 9 shows the behavior of the buck-motor system controlled by ZAD-FPIC. In this case, the control parameter is fixed as $K_{S3} = 35$, the reference signal is defined as $W_{mref} = 400$ rad/s, and some variations are made through the control parameter N in values 1, 3, 5, 7, and 9.

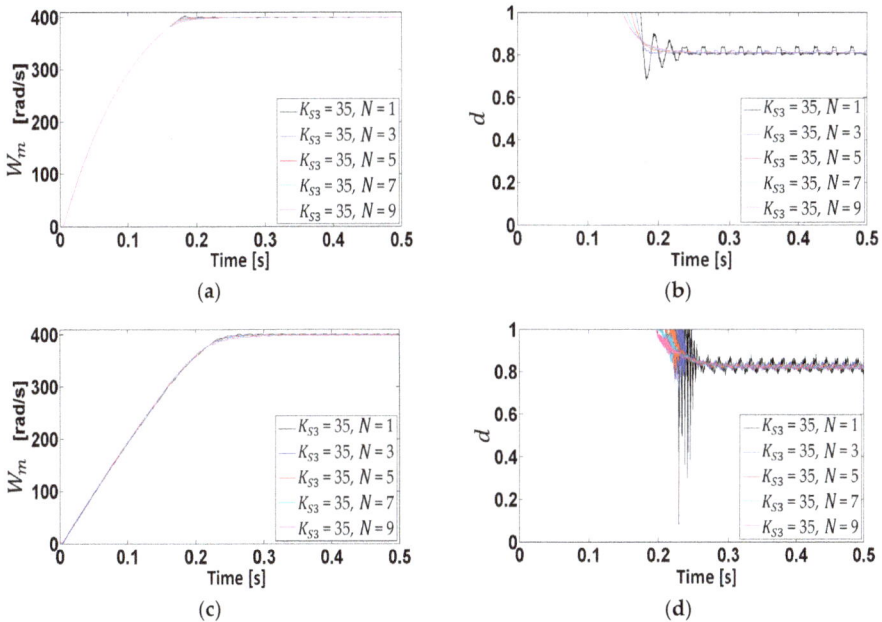

Figure 9. Behavior of the buck-motor system in closed-loop while changing N with $K_{S3} = 35$: (**a**) output W_m for the simulation; (**b**) d for the simulation; (**c**) output W_m for the experiment; and (**d**) d for the experiment.

Tables 5 and 6 summarize the results shown in Figure 9. Both the simulation and experimental test show that the steady-state error is less than 1% for the different values of the parameter N. If the value of N is increased or decreased, then it does not greatly affect the dynamics in the transient state. However, for small values of N, some small oscillations are presented and they will increase in amplitude if N is reduced below 1.

Table 5. Transient results obtained from the simulation in a closed loop with ZAD-FPIC.

Parameter	M_p (%)	t_s (s)	Error (%)
($N = 1$)	0.5717	0.1701	−0.0173
($N = 3$)	Overdamped	0.1701	−0.0173
($N = 5$)	Overdamped	0.1711	−0.0173
($N = 7$)	Overdamped	0.1737	−0.1473
($N = 9$)	Overdamped	0.1802	−0.0173

Table 6. Transient results for the experimental test when the system works in closed loop with ZAD-FPIC.

Parameter	M_p (%)	t_s (s)	Error (%)
($N = 1$)	Overdamped	0.2311	0.2775
($N = 3$)	Overdamped	0.2381	0.2772
($N = 5$)	Overdamped	0.2414	−0.3118
($N = 7$)	Overdamped	0.2455	−0.3118
($N = 9$)	Overdamped	0.2505	−0.4590

Because the model is not exact, for all values of N, the system responds faster in the simulation than in the experimental test. Regarding the duty cycle, Figure 9b,d show the saturation at the beginning; however, during the steady-state the duty cycle is not saturated for any value of N, which leads to a fixed commutation frequency that reduces the electrical and audible noises in the experimental test. Figure 9d shows electronic noise present in the measured signals. In general, for all values of N, the simulation and experimental tests show similar results.

3.4. Variation of the Parameter K_{S3}

Figures 10 and 11 show the bifurcation parameters of W_m and d for the experimental test when the parameter K_{S3} is changed. Figures 10 and 11–d show the behavior of W_m and d over time for the experimental test for $K_{S3} = 40$, $K_{S3} = 20$, and $K_{S3} = 5$, when the values of $K_{S1} = K_{S2} = 2$, and $N = 1$. The results show that for the value of $K_{S3} = 40$, some 4T periodic orbits are presented, which is the behavior clearly represented in Figure 10b. For the value of $K_{S3} = 20$, Figure 10a shows 6T periodic orbits, which can be represented in Figure 10c. For the value of $K_{S3} = 5$, Figure 10a shows 8T periodic orbits, which is easily represented as Figure 10d. For the duty cycle d in both diagrams, some chaotic and quasi-periodic behaviors are presented for different values of K_{S3} as shown in Figure 11.

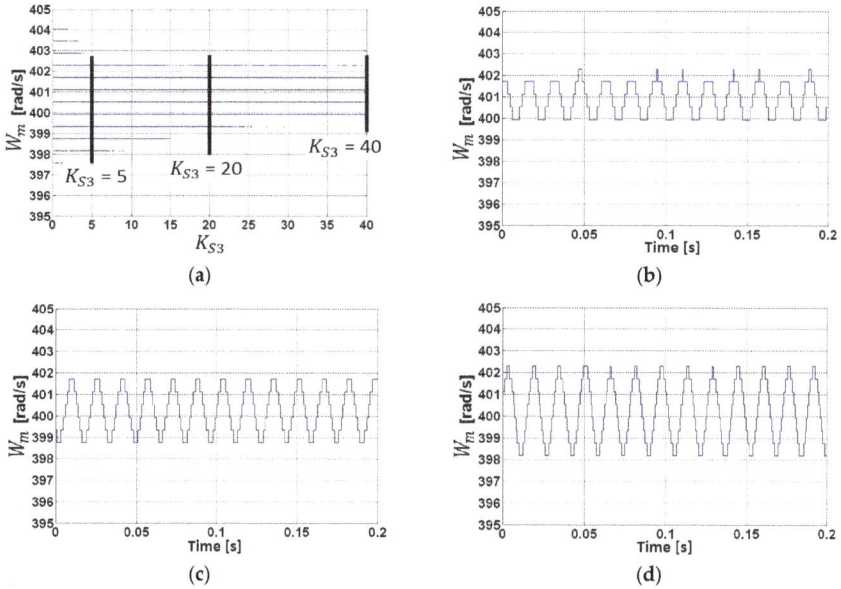

Figure 10. Bifurcation diagram of the variable W_m when the parameter K_{S3} is changed ($K_{S3} = 40$, $K_{S3} = 20$, and $K_{S3} = 5$) and the values of $K_{S1} = K_{S2} = 2$, and $N = 1$ are kept constant: (**a**) bifurcation diagram W_m vs. K_{S3} for the experimental test; (**b**) W_m in the experimental test when $K_{S3} = 40$; (**c**) W_m in the experimental test when $K_{S3} = 20$; and (**d**) W_m in the experimental test when $K_{S3} = 5$.

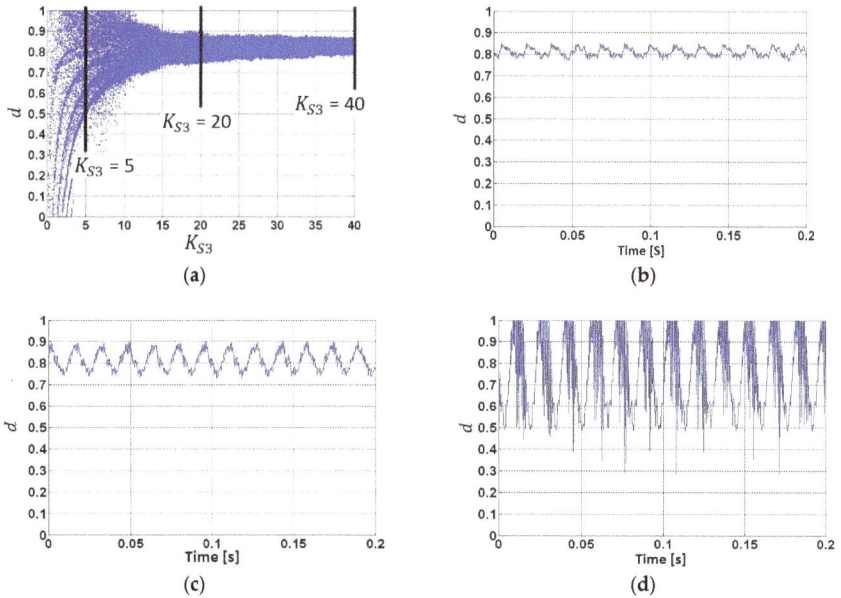

Figure 11. Bifurcation diagram of d vs. K_{S3} and behavior of $K_{S3} = 40$, $K_{S3} = 20$, and $K_{S3} = 5$ when the values of $K_{S1} = K_{S2} = 2$ and $N = 1$: (**a**) bifurcation diagram of d vs. K_{S3} in the experimental test; (**b**) behavior of d when $K_{S3} = 40$ in the experimental test; (**c**) behavior of d when $K_{S3} = 20$ in the experimental test; and (**d**) behavior of d when $K_{S3} = 5$ in the experimental test.

3.5. Behavior of the System When the Control Parameter N of the FPIC Is Changed

Figures 12 and 13 show the behavior of the buck-motor system controlled by ZAD when the parameter $K_{S3} = 35$ and the parameters K_{S1} and K_{S2} are equal to two. In this case, some bifurcations were created with the change of parameter N for the system with one delay period. The results show that the critical N in the simulation test was presented when $N \cong 0.7875$ and in the experimental test when $N \cong 0.5$. For values of N greater than the bifurcation point, the stability of the system was presented and the regulated variable (W_m) tends to the fixed point. Therefore, with values of $N = 1$ and $K_{S3} = 35$, there was good regulation as observed in Figure 12a,b. In the simulation test, with values of $N \leq 0.7875$, the system presents chaos and quasi-periodicity behaviors, whereas in the experimental test they occur for $N \leq 0.5$.

In the simulation test, stability was presented for values of $N \geq 0.7875$, whereas in the experimental test it was presented for values of $N \geq 0.5$. For these same values of N, a quasi-periodicity behavior occurs in the duty cycle, but a fixed switching frequency was achieved. Regarding the steady-state error, a value lower than 1% in the entire range of N was presented. In general, both the numerical and experimental diagrams were similar, thus validating the use of the model and the implemented circuit. Therefore, the ZAD-FPIC technique presented good performance to control the W_m and the FPIC technique demonstrated effectiveness to control chaos of the electric circuit.

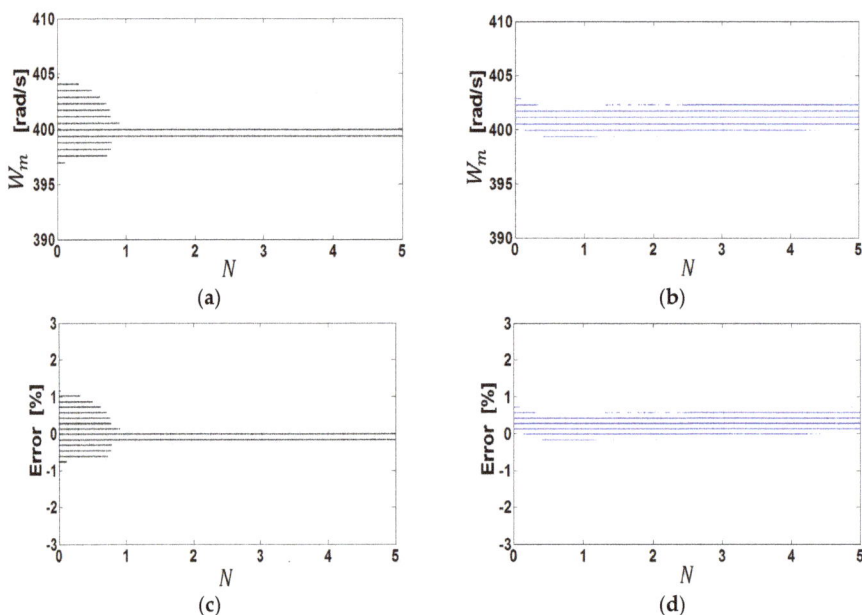

Figure 12. Bifurcation diagrams for the simulation and experimental test of the system with ZAD ($K_{S3} = 35$) and FPIC, changing the parameter N and maintaining fixed values of $K_{S1} = K_{S2} = 2$: (**a**) W_m vs. N for the simulation test; (**b**) W_m vs. N for the experimental test; (**c**) error vs. N for the simulation test; and (**d**) error vs. N for the experimental test.

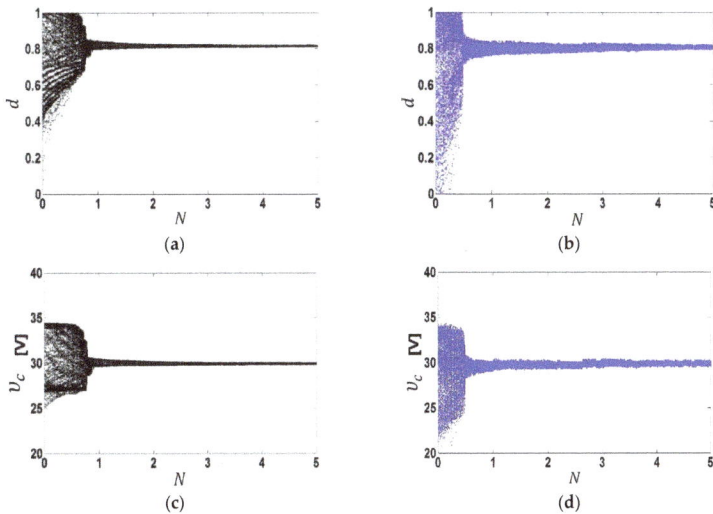

Figure 13. Bifurcation diagrams for the simulation and experimental test of the system with ZAD ($K_{S3} = 35$) and FPIC, changing the parameter N and maintaining fixed values of $K_{S1} = K_{S2} = 2$: (**a**) d vs. N for the simulation test; (**b**) d vs. N for the experimental test; (**c**) v_c vs. N for the simulation test; and (**d**) v_c vs. N for the experimental test.

Figures 14 and 15 show the behavior of the buck-motor system controlled with ZAD-FPIC with two delay periods. In this case, the equilibrium point has shifted slightly to the right when the critical N is approximately $N = 2.2$ in the experimental test and $N = 2.24$ in the simulation. Figures 14 and 15 show that the signals in both the numerical and experimental tests have similar behavior, validating that the circuits modeled and built for the tests are correct.

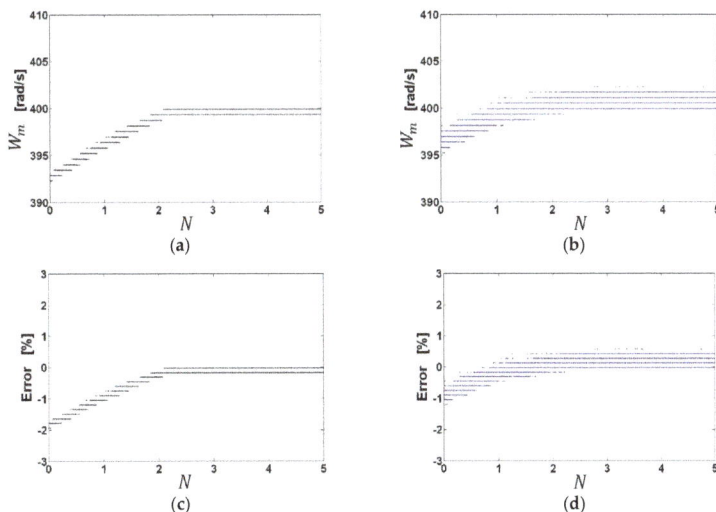

Figure 14. Bifurcation diagrams for the simulation and experimental test of the system with ZAD ($K_{S3} = 35$) and FPIC, changing the parameter N with $K_{S1} = K_{S2} = 2$ and a 2T delay period: (**a**) W_m vs. N for the simulation test; (**b**) W_m vs. N for the experimental test; (**c**) error vs. N for the simulation test; and (**d**) error vs. N for the experimental test.

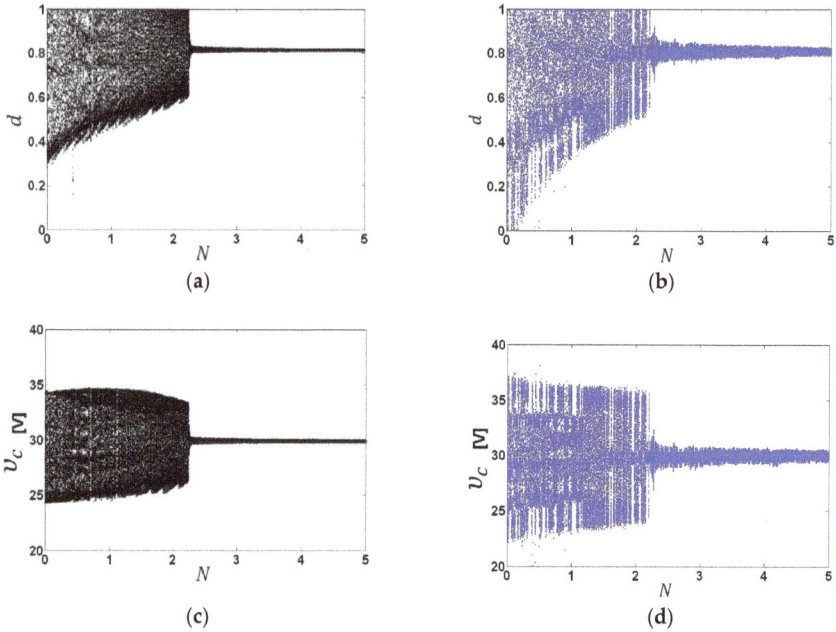

Figure 15. Bifurcation diagrams for the simulation and experimental test of the system with ZAD ($K_{S3} = 35$) and FPIC, changing the parameter N with $K_{S1} = K_{S2} = 2$ and a 2T delay period: (**a**) d vs. N for the simulation test; (**b**) d vs. N for the experimental test; (**c**) v_c vs. N for the simulation test; and (**d**) v_c vs. N for the experimental test.

3.6. Perturbation at the Input Voltage

This section shows the behavior of the system when V_{in} is changed. To carry out these experiments, the signal V_{in} was measured and registered with the DS1104MUXADC block, which was configured by the trigger signal at a frequency F_s. For these tests, the parameters of Table 3 were considered: $N = 1$, $K_{S1} = K_{S2} = 2$, and $K_{S3} = 35$.

Figure 16 shows the behavior of the buck-motor system described in Figure 2 and with the parameters of Table 3. Figure 16a shows the behavior of the system when instantaneous perturbations are created in V_{in} as performed for the experimental test. Figure 16b shows the effect in the regulated signal W_m by changing V_{in} and maintaining a value of $W_m = 198$ rad/s and reference signal $W_{mref} = 200$ rad/s. Figure 16c shows that the regulation error was maintained at the value of -1%. Figure 16d shows a phase diagram between the regulation error and V_{in} in which it is observed that the error was maintained at the value of -1%.

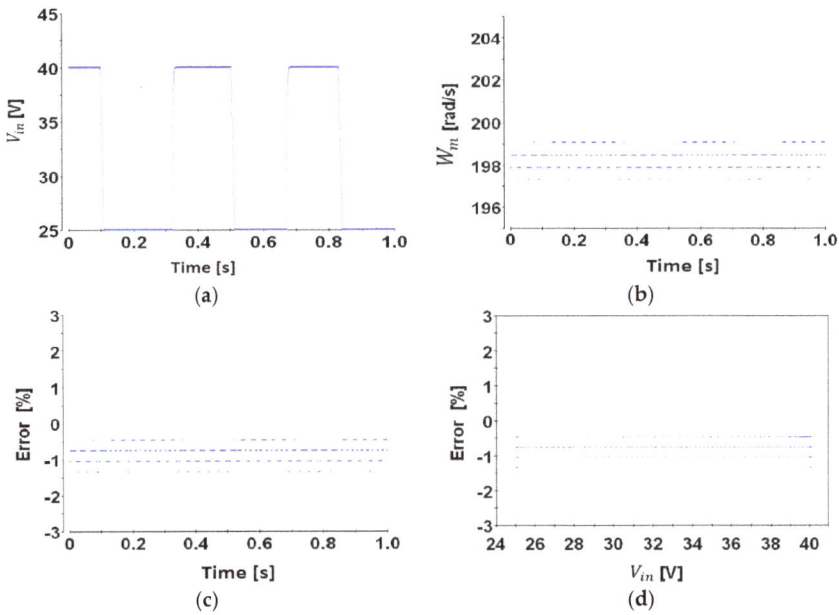

Figure 16. Behavior of the buck-motor system when perturbations in V_{in} are presented for the experimental test with $K_{S3} = 40$, $K_{S1} = K_{S2} = 2$, and $N = 1$: (**a**) variation of V_{in} over time; (**b**) regulated signal W_m over time; (**c**) error in the controlled variable; and (**d**) phase diagram of the error vs. V_{in}.

Figures 17 and 18 show the behavior of the regulation error in the experimental test when V_{in} was disturbed irregularly. In the first test, V_{in} was increased as shown in Figure 17a, and for this input, Figure 17b presents the results of the error vs. V_{in}, where the ZAD-FPIC controls the output signal with an error lower than -2%. For the second test, V_{in} was initially increased and then further decreased as shown in Figure 18a, and the error of the control variable vs. V_{in} does not exceed -2% as shown in Figure 18b.

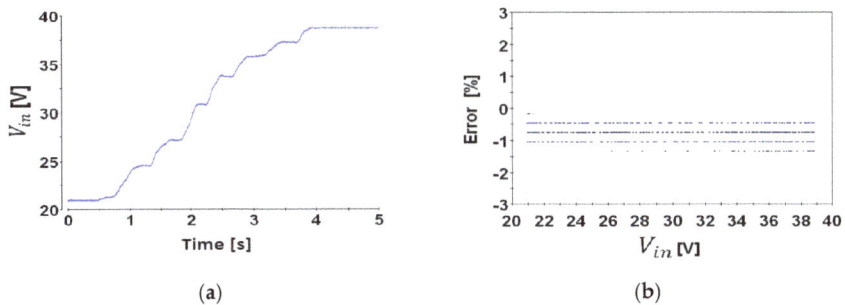

Figure 17. Experimental behavior of the buck-motor system after perturbations in V_{in} when $K_{S3} = 40$, $K_{S1} = K_{S2} = 2$, and $N = 1$: (**a**) variation in V_{in} over time and (**b**) phase diagram of error vs. V_{in}.

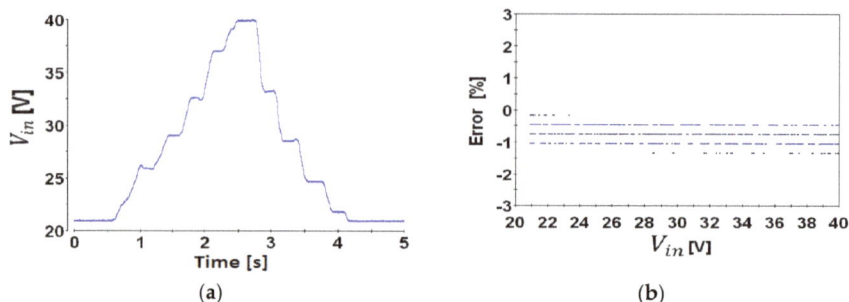

Figure 18. Experimental behavior of the buck-motor system after perturbations in V_{in} when $K_{S3} = 40$, $K_{S1} = K_{S2} = 2$, and $N = 1$: (**a**) variation of V_{in} over time and (**b**) phase diagram of the error vs. V_{in}.

4. Conclusions

This paper presented the dynamic behavior of a buck-motor system controlled with ZAD-FPIC. The results show that the control regulates very well the speed at the output W_m for different values of the reference signal (W_{mref}). When $K_{S3} = 35$ and increasing the value of $N > 1$, the dynamic response of the system was very similar for different values of N. The ZAD-FPIC control technique was effective to control the buck-motor system even for two delay periods and the robustness of the system was checked by making variations in V_{in}. When the control parameter of the ZAD was a fixed value and the FPIC control parameter changed, the transient behavior showed that neither the transient nor stationary regime changed with the change of N. Numerically and experimentally, orbits of periods $6T$, $8T$, and chaos were shown, which were plotted in the bifurcation diagrams against time. For the case where there were two periods of delay, the controller with ZAD-FPIC was able to regulate the output speed with low steady-state error.

Previous works have shown that the duty cycle is normally saturated and the electrical and audible noises increase. With the ZAD-FPIC control technique, greater stability and noise reduction were obtained in the controlled variable and in the voltage feeding the motor due to the fixed frequency switching implemented in the control technique. Some clear advantages of this controller were obtained when the control parameters had higher values because the system was more stable and did not present chaos; besides, the controller presented a low steady-state error in the controlled variable. However, when using the ZAD-FPIC control technique, the system became less robust to real-time variations of the system parameters because the duty cycle depended directly on these parameters. Furthermore, the controller depended on all the variables of the system, increasing the time for digital processing and real-time calculation of the duty cycle, requiring more powerful processors.

Author Contributions: F.E.H.V. conceived the theory, performed the experiments, and analyzed the data; J.E.C.-B. wrote the paper, reviewed the theory, performed simulations, and analyzed the data; A.R.S. helped with the design of the control and reviewed the theory.

Funding: This research received no external funding.

Acknowledgments: This work was supported by the Universidad Nacional de Colombia, Sede Manizales and Sede Medellín under the projects HERMES-34671 and HERMES-36911. The authors thank the School of Physics and the Department of Electrical Energy and Automation for their valuable support to conduct this research.

Conflicts of Interest: The authors declare no conflict of interest.

References

1. Cuong, N.D.; Van Lanh, N.; Dinh, G.T. An Adaptive LQG Combined With the MRAS—Based LFFC for Motion Control Systems. *J. Autom. Control Eng.* **2015**, *3*, 130–136. [CrossRef]
2. Sankardoss, V.; Geethanjali, P. PMDC Motor Parameter Estimation Using Bio-Inspired Optimization Algorithms. *IEEE Access* **2017**, *5*, 11244–11254. [CrossRef]

3. Hilairet, M.; Auger, F. Speed sensorless control of a DC-motor via adaptive filters. *IET Electr. Power Appl.* **2007**, *1*, 601. [CrossRef]

4. Hu, H.; Yousefzadeh, V.; Maksimovic, D. Nonuniform A/D Quantization for Improved Dynamic Responses of Digitally Controlled DC-DC Converters. *IEEE Trans. Power Electron.* **2008**, *23*, 1998–2005. [CrossRef]

5. Fung, C.W.; Liu, C.P.; Pong, M.H. A Diagrammatic Approach to Search for Minimum Sampling Frequency and Quantization Resolution for Digital Control of Power Converters. In Proceedings of the 2007 IEEE Power Electronics Specialists Conference, Orlando, FL, USA, 17–21 June 2007; pp. 826–832.

6. Taborda, J.A.; Angulo, F.; Olivar, G. Estimation of parameters in Buck converter with Digital-PWM control based on ZAD strategy. In Proceedings of the 2011 IEEE Second Latin American Symposium on Circuits and Systems (LASCAS), Bogota, Colombia, 23–25 February 2011; pp. 1–4.

7. Zane, M. Erickson Impact of digital control in power electronics. In Proceedings of the 16th International Symposium on Power Semiconductor Devices & IC's, Kawasaki, Japan, 24–27 May 2004; pp. 13–22.

8. Hoyos, F.E.; Rincón, A.; Taborda, J.A.; Toro, N.; Angulo, F. Adaptive Quasi-Sliding Mode Control for Permanent Magnet DC Motor. *Math. Probl. Eng.* **2013**, *2013*, 693685. [CrossRef]

9. Fossas, E.; Griñó, R.; Biel, D. Quasi-Sliding control based on pulse width modulation, zero averaged dynamics and the L2 norm. In *Advances in Variable Structure Systems, Proceedings of the 6th IEEE International Workshop on Variable Structure Systems, Coolangatta, Australia, 7–9 December 2000*; World Scientific: Singapore, 2000; pp. 335–344.

10. Biel, D.; Fossas, E.; Ramos, R.; Sudria, A. Programmable logic device applied to the quasi-sliding control implementation based on zero averaged dynamics. In Proceedings of the 40th IEEE Conference on Decision and Control (Cat. No.01CH37228), Orlando, FL, USA, 4–7 December 2001; Volume 2, pp. 1825–1830.

11. Ramos, R.R.; Biel, D.; Fossas, E.; Guinjoan, F. A fixed-frequency quasi-sliding control algorithm: Application to power inverters design by means of FPGA implementation. *IEEE Trans. Power Electron.* **2003**, *18*, 344–355. [CrossRef]

12. Angulo, F.; Fossas, E.; Olivar, G. Transition from Periodicity to Chaos in a PWM-Controlled Buck Converter with ZAD Strategy. *Int. J. Bifurc. Chaos* **2005**, *15*, 3245–3264. [CrossRef]

13. Angulo, F.; Olivar, G.; di Bernardo, M. Two-parameter discontinuity-induced bifurcation curves in a ZAD-strategy-controlled dc-dc buck converter. *IEEE Trans. Circuits Syst. I Regul. Pap.* **2008**, *55*, 2392–2401. [CrossRef]

14. Fossas, E.; Hogan, S.J.; Seara, T.M. Two-parameter bifurcation curves in power electronics converters. *Int. J. Bifurc. Chaos* **2009**, *19*, 349–357. [CrossRef]

15. Hoyos, F.E.; Candelo-Becerra, J.E.; Toro, N. Numerical and experimental validation with bifurcation diagrams for a controlled DC–DC converter with quasi-sliding control. *TecnoLógicas* **2018**, *21*, 147–167. [CrossRef]

16. Di Bernardo, M.; Budd, C.; Champneys, A. Grazing, skipping and sliding: Analysis of the non-smooth dynamics of the DC/DC buck converter. *Nonlinearity* **1998**, *11*, 859–890. [CrossRef]

17. Yuan, G.; Banerjee, S.; Ott, E.; Yorke, J.A. Border-collision bifurcations in the buck converter. *IEEE Trans. Circuits Syst. I Fundam. Theory Appl.* **1998**, *45*, 707–716. [CrossRef]

18. Hoyos, F.E.; Candelo, J.E.; Silva-Ortega, J.I. Performance evaluation of a DC-AC inverter controlled with ZAD-FPIC. *Inge CUC* **2018**, *14*, 9–18. [CrossRef]

19. Biel, D.; Cardoner, R.; Fossas, E. Tracking Signal in a Centered Pulse ZAD Power Inverter. In Proceedings of the International Workshop on Variable Structure Systems, Sardinia, Italy, 5–7 June 2006; pp. 104–109.

20. Biel, D.; Fossas, E.; Guinjoan, F.; Ramos, R. Interleaving quasi-sliding mode control of parallel-connected inverters. In Proceedings of the 2008 International Workshop on Variable Structure Systems, Antalya, Turkey, 8–10 June 2008; pp. 337–342.

21. Angulo, F.; Olivar, G.; Taborda, J.; Hoyos, F. Nonsmooth dynamics and FPIC chaos control in a DC-DC ZAD-strategy power converter. In Proceedings of the ENOC, EUROMECH Nonlinear Dynamics Conference, Saint Petersburg, Russia, 6–10 July 2008; pp. 1–6.

22. Rincón, A.; Hoyos, F.E.; Angulo, F. Controller Design for a Second-Order Plant with Uncertain Parameters and Disturbance: Application to a DC Motor. *Abstr. Appl. Anal.* **2013**, *2013*, 169519. [CrossRef]

23. Hoyos, F.E.; Candelo, J.E.; Taborda, J.A. Selection and validation of mathematical models of power converters using rapid modeling and control prototyping methods. *Int. J. Electr. Comput. Eng.* **2018**, *8*, 1551. [CrossRef]

24. Hoyos, F.E.; Burbano, D.; Angulo, F.; Olivar, G.; Toro, N.; Taborda, J.A. Effects of Quantization, Delay and Internal Resistances in Digitally ZAD-Controlled Buck Converter. *Int. J. Bifurc. Chaos* **2012**, *22*, 1250245. [CrossRef]

energies

MDPI

Article

A Novel Step-Up Converter with an Ultrahigh Voltage Conversion Ratio

Faqiang Wang [1], Herbert Ho-Ching Iu [2,*] and Jing Li [1,3]

[1] State Key Laboratory of Electrical Insulation and Power Equipment, School of Electrical Engineering,
    Xi'an Jiaotong University, Xi'an 710049, China; faqwang@mail.xjtu.edu.cn
[2] School of Electrical, Electronic and Computer Engineering, The University of Western Australia,
    Crawley, WA 6009, Australia
[3] Xi'an Institute of Space Radio Technology, Xi'an 710100, China; 1216threesun@163.com
* Correspondence: herbert.iu@uwa.edu.au; Tel.: +61-8-6488-7989

Received: 11 September 2018; Accepted: 8 October 2018; Published: 10 October 2018

Abstract: A new step-up converter with an ultrahigh voltage conversion ratio is proposed in this paper. Two power switches of such a converter, which conduct synchronically, and its output voltage, which has common ground and common polarity with its input voltage, lead to the simple control circuit. No abrupt changes in the capacitor voltage and the inductor current of the proposed step-up converter mean that it does not suffer from infinite capacitor current and inductor voltage. Two input inductors with different values can still allow the proposed step-up converter to work appropriately. An averaged model of the proposed step-up converter was built and one could see that it was still fourth-order even with its five storage elements. Some theoretical derivations, theoretical analysis, Saber simulations, and circuit experiments are provided to validate the effectiveness of the proposed step-up converter.

Keywords: new step-up converter; ultrahigh voltage conversion ratio; small-signal model; average-current mode control

1. Introduction

As part of a DC switching power supply, the step-up converter is important for transforming low input voltage into the desired high output voltage to satisfy the requirements of practical applications, such as photovoltaic (PV) systems, fuel-cell systems, etc. Step-up converters can be classified into two types: non-isolated and isolated. An isolated step-up converter is generally constructed by inserting a transformer into a non-isolated step-up converter to enlarge the voltage conversion ratio. However, switch voltage overshoot and EMI problems caused by the transformer make the whole system suffer from low efficiency and huge volume [1,2]. Therefore, the non-isolated step-up converter is the focus of many researchers and engineers. It is well known that the traditional boost converter with a voltage conversion ratio of $1/(1 - D)$, where D is the duty cycle, is a good topology to realize the boost ability because it has a simple structure [3]. Nevertheless, under certain input voltages, if an extremely high output voltage is required, the duty cycle must be close to 1.0, and this cannot generally be achieved because of the limitations of real semiconductors. Accordingly, in the last few decades, many researchers and engineers have made much effort to explore a novel step-up converter with a high voltage conversion ratio, and many effective topologies have been proposed. For example, for realizing a voltage conversion ratio of $(1 + D)/(1 - D)$, which is higher than that of the traditional boost converter, Yang et al. constructed a transformerless step-up converter [4], Gules et al. introduced a modified single-ended primary-inductor converter (Sepic) [5], and Mummadi proposed a fifth-order boost converter [6]. However, that voltage conversion ratio was limited to some extent.

To achieve a voltage conversion ratio which is higher than $(1 + D)/(1 - D)$ within a certain area of D, several step-up converters have been proposed. For example, for obtaining a voltage conversion ratio of $(2 - D)/(1 - D)$, the following converters have been proposed: KY boost converter constructed by combining a KY converter with a traditional synchronously rectified boost converter [7], a step-up converter constructed by combining KY and buck-boost converters [8], and an elementary positive output super-lift Luo converter [9]. For obtaining a voltage conversion ratio of $2/(1 - D)$, Hwu et al. combined the charge pump concept with the traditional boost converter to construct a fourth-order step-up converter [10], and Al-Saffar et al. integrated the traditional boost converter with a self-lift Sepic converter to introduce a sixth-order step-up converter [11]. Also, Hwu et al. proposed two voltage-boosting converters with a voltage conversion ratios of $(3 - D)/(1 - D)$ and $(3 + D)/(1 - D)$ by using bootstrap capacitors and boost inductors [12]. Chen et al. proposed an interleaved step-up converter with the voltage conversion ratio being $3/(1 - D)$ [13]. However, all of the above step-up converters possess an abrupt change in voltage across the capacitor, which limits them in practical applications to some extent. Moreover, like a boost converter, if an ultrahigh output voltage from those converters is required, the duty cycle D must be close to 1.0, and this also cannot generally be achieved because of the limitations of real semiconductors.

Therefore, for acquiring a higher output voltage with the same polarity as the input voltage with the duty cycle D being close to 0.5, which is very easy to implement in practical situations, some new DC-DC converters have been proposed. For example, in [14], based on a Sheppard-Taylor converter whose voltage conversion ratio of $-D/(1 - 2D)$ is negative, a modified Sheppard-Taylor converter with a voltage conversion ratio of $D/((1 - D)(1 - 2D))$ was proposed. Also, by removing some components of the Sheppard-Taylor converter, a simple modified Sheppard-Taylor converter with a voltage conversion ratio of $1/(1 - 2D)$ was proposed in [15]. However, its voltage conversion ratio of $1/(1 - 2D)$ was obtained under the unreasonable assumption that the voltages across its two capacitors were equal. In fact, its voltage conversion ratio was related to not only the duty cycle D, but also the load resistor and the switch frequency, so its load regulation is not good enough [16]. In addition, a fourth-order step-up converter with a voltage conversion ratio of $(1 - D)/(1 - 2D)$ was presented in [17] and a pulse-width modulation (PWM) Z-source DC-DC converter with the same voltage conversion ratio was investigated in [18]. However, their voltage conversion ratios were also limited to some extent. In particular, the output voltage of the PWM Z-source DC-DC converter was floating. Hence, exploring new step-up converters with good performance is very important and valuable. In this study, a new step-up converter with an ultrahigh voltage conversion ratio is proposed. In this converter, the output voltage is common-grounded with the input voltage, and its two power switches conduct synchronically. Even if the two input inductors have different values, the proposed step-up converter can still work appropriately. Additionally, there is no abruptly changing on the current through the inductors and the voltage across the capacitors.

This paper is organized as follows. In Section 2, the structure and basic principle of the proposed step-up converter in continuous conduction mode (CCM) is presented in detail. In Section 3, the averaged model and corresponding small-signal model are established and analyzed. Comparisons among existing step-up converters and the proposed step-up converter are presented in Section 4. Some Saber simulations and circuit experiments for confirmation are presented in Section 5. Finally, some concluding remarks and comments are given in Section 6.

2. Novel Topology's Structure and Its Basic Principle

A circuit schematic of the proposed step-up converter is shown in Figure 1. It consists of two power switches (Q_1 and Q_2), five diodes (D_1, D_2, D_3, D_4, and D_5), three inductors (L_1, L_2, and L_3), two capacitors (C_1 and C_2), and the resistive load R. Two power switches conduct synchronically and are driven by the same PWM signal v_d, with the period being T and duty cycle being d. The currents through the three inductors are denoted by i_{L1}, i_{L2} and i_{L3}. The voltages across the two capacitors are defined as v_{C1} and v_0. Notably, the proposed step-up converter operating in the continuous conduction mode (CCM) is the only concern here, and its possible stages are shown in Figure 2. Based on the relation between inductors L_1 and L_2, there are three cases for the proposed step-up converter, case 1: $L_1 = L_2$, case 2: $L_1 < L_2$ and case 3: $L_1 > L_2$. The principle of the proposed step-up converter under the three cases is discussed in detail in the following subsections.

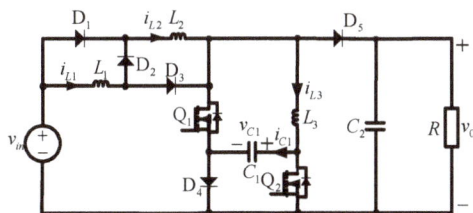

Figure 1. Circuit schematic of the proposed step-up converter.

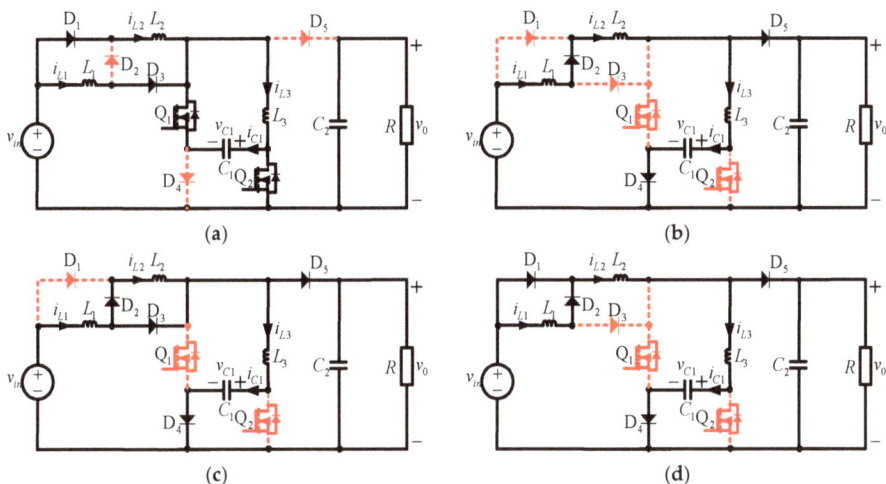

Figure 2. Equivalent circuits for possible stages of the proposed step-up converter in CCM. (a) Stage 1; (b) Stage 2; (c) Stage 3; (d) Stage 4.

2.1. Case 1: $L_1 = L_2$

For this case, the proposed step-up converter has only two operation stages; their equivalent circuits are shown in Figure 2a,b.

2.1.1. Stage 1

Figure 2a shows that two power switches (Q_1 and Q_2) are turned on for the high level of PWM signal v_d, and three of the diodes D_2, D_4 and D_5 do not conduct for the inverse biased voltage, whereas two of the diodes (D_1 and D_3) conduct for the forward biased voltage. That is, the state of power switches and diodes is: (Q_1, Q_2, D_1, D_2, D_3, D_4, $D_5 \equiv$ ON, ON, ON, OFF, ON, OFF, OFF) within $NT < t \le NT + dT$ where N is a natural number. Accordingly, the input voltage source supplies the energy to two input inductors (L_1 and L_2) so that both of them are magnetized and their currents increase. Because $L_1 = L_2$, the currents through these two inductors are equal, that is, $i_{L1} = i_{L2}$. Capacitor C_1 is in parallel with inductor L_3. Consequently, capacitor C_1 is discharged so that its voltage decreases and inductor L_3 is demagnetized so that its current also decreases. In addition, capacitor C_2 delivers energy to the resistive load R. The associated equations for stage 1 are:

$$\begin{cases} \frac{di_{L1}}{dt} = \frac{v_{in}+v_{C1}}{L_1} \\ \frac{di_{L2}}{dt} = \frac{v_{in}+v_{C1}}{L_2} \\ \frac{di_{L3}}{dt} = -\frac{v_{C1}}{L_3} \\ \frac{dv_{C1}}{dt} = \frac{i_{L3}-i_{L1}-i_{L2}}{C_1} \\ \frac{dv_0}{dt} = -\frac{v_0}{RC_2} \end{cases} \tag{1}$$

2.1.2. Stage 2

Figure 2b shows that two power switches (Q_1 and Q_2) are turned off for the low level of PWM signal v_d. Three diodes (D_2, D_4, and D_5) conduct, and the remaining diodes (D_1 and D_3) do not conduct. That is, the state of power switches and diodes is: (Q_1, Q_2, D_1, D_2, D_3, D_4, $D_5 \equiv$ OFF, OFF, OFF, ON, OFF, ON, ON) within $NT + dT < t \le NT + T$. Hence, inductor L_1 is in series with inductor L_2, leading to $i_{L1} = i_{L2}$, and together with the input voltage v_{in}, they supply the energy to inductor L_3 and capacitor C_1. The corresponding equations for stage 2 are:

$$\begin{cases} \frac{di_{L1}}{dt} = \frac{v_{in}-v_0}{L_1+L_2} \\ \frac{di_{L2}}{dt} = \frac{v_{in}-v_0}{L_1+L_2} \\ \frac{di_{L3}}{dt} = \frac{v_0-v_{C1}}{L_3} \\ \frac{dv_{C1}}{dt} = \frac{i_{L3}}{C_1} \\ \frac{dv_0}{dt} = \frac{i_{L1}}{C_2} - \frac{i_{L3}}{C_2} - \frac{v_0}{RC_2} \end{cases} \tag{2}$$

2.2. Case 2: $L_1 < L_2$

For case 2, besides stage 1 and stage 2, it has stage 3, as shown in Figure 2c, corresponding to (Q_1, Q_2, D_1, D_2, D_3, D_4, $D_5 \equiv$ OFF, OFF, OFF, ON, ON, ON, ON) within $NT + dT < t \le NT + dT + d_{11}T$ because diode D_3 is still conducting for $i_{L1} > i_{L2}$. Its mathematical model is:

$$\begin{cases} \frac{di_{L1}}{dt} = \frac{v_{in}-v_0}{L_1} \\ \frac{di_{L2}}{dt} = 0 \\ \frac{di_{L3}}{dt} = \frac{v_0-v_{C1}}{L_3} \\ \frac{dv_{C1}}{dt} = \frac{i_{L3}}{C_1} \\ \frac{dv_0}{dt} = \frac{i_{L1}}{C_2} - \frac{i_{L3}}{C_2} - \frac{v_0}{RC_2} \end{cases} \tag{3}$$

Notably, stage 3 will last until $i_{L1} = i_{L2}$, leading to the diode D_3 being off and the proposed step-up converter immediately operating in stage 2. Therefore, the sequence of operations of the proposed step-up converter in case 2 is: stage 1 (Figure 2a) during $NT < t \le NT + dT$, stage 3 (Figure 2c) during $NT + dT < t \le NT + dT + d_{11}T$, and stage 2 (Figure 2b) during $NT + dT + d_{11}T < t \le NT + T$.

2.3. Case 3: $L_1 > L_2$

For case 3, besides stage 1 and stage 2, it has stage 4, as shown in Figure 2d, corresponding to (Q_1, Q_2, D_1, D_2, D_3, D_4, $D_5 \equiv$ OFF, OFF, ON, ON, OFF, ON, ON) during $NT + dT < t \le NT + dT + d_{22}T$ because diode D_1 is still conducting for $i_{L1} < i_{L2}$. Its equations are:

$$\begin{cases} \frac{di_{L1}}{dt} = 0 \\ \frac{di_{L2}}{dt} = \frac{v_{in}-v_0}{L_2} \\ \frac{di_{L3}}{dt} = \frac{v_0-v_{C1}}{L_3} \\ \frac{dv_{C1}}{dt} = \frac{i_{L3}}{C_1} \\ \frac{dv_0}{dt} = \frac{i_{L2}}{C_2} - \frac{i_{L3}}{C_2} - \frac{v_0}{RC_2} \end{cases} \tag{4}$$

Please note that stage 4 will end if $i_{L1}=i_{L2}$, which prevents diode D_1 from conducting and the proposed step-up converter will immediately operate in stage 2. Therefore, the sequence of operations of the proposed step-up converter in case 3 is: stage 1 (Figure 2a) during $NT < t \leq NT + dT$, stage 4 (Figure 2d) during $NT + dT < t \leq NT + dT + d_{22}T$, and stage 2 (Figure 2b) during $NT + dT + d_{22}T < t \leq NT + T$.

3. Modeling and Theoretical Analysis

Based on the averaging method in [19], the averaged model for the proposed step-up converter under the three cases are established and analyzed. Firstly, some symbols are defined. x is defined as the variables of the proposed step-up converter, such as i_{L1}, i_{L2}, i_{L3}, v_{C1}, v_0, d, and v_{in}. $\langle x \rangle$, X and $\hat{x}$ are denoted by their averaged, DC and small AC values, respectively. Also, the following items are assumed:

$$\langle x \rangle = X + \hat{x} \text{ with } \hat{x} << X \tag{5}$$

3.1. Averaged Model

3.1.1. Case 1: $L_1 = L_2$

For this case, $L_1 = L_2$ so that $i_{L1} = i_{L2}$. From (1) and (2), and using the averaging method in [19], the averaged model of the proposed step-up converter in case 1 can be directly derived as follows:

$$\begin{cases} \frac{d\langle i_{L1}\rangle}{dt} = \frac{\langle v_{in}\rangle+\langle v_{C1}\rangle}{L_1}d + \frac{\langle v_{in}\rangle-\langle v_0\rangle}{2L_1}(1-d) \\ \frac{d\langle i_{L3}\rangle}{dt} = -\frac{\langle v_{C1}\rangle}{L_3} + \frac{\langle v_0\rangle(1-d)}{L_3} \\ \frac{d\langle v_{C1}\rangle}{dt} = -\frac{2\langle i_{L1}\rangle}{C_1}d + \frac{\langle i_{L3}\rangle}{C_1} \\ \frac{d\langle v_0\rangle}{dt} = \frac{\langle i_{L1}\rangle}{C_2}(1-d) - \frac{\langle i_{L3}\rangle}{C_2}(1-d) - \frac{\langle v_0\rangle}{RC_2} \end{cases} \tag{6}$$

3.1.2. Case 2: $L_1 < L_2$

As described in Section 2, there are three stages of the proposed step-up converter in case 2. The typical time-domain waveforms for the inductor currents i_{L1} and i_{L2} and the PWM signal v_d are plotted in Figure 3, where I_{LN}, I_{LM}, and I_{LN1} are the values of i_{L1} and i_{L2} at $t = NT$, $t = (N + d + d_{11})T$, and $t = (N + 1)T$, respectively, and I_{L1P} is the value of i_{L1} at $t = (N + d + d_{11})T$.

Based on (1), (2), (3), and Figure 3 and using the geometrical technique, the following equations can be derived:

$$\frac{\langle v_{in}\rangle + \langle v_{C1}\rangle}{L_1}dT + \frac{\langle v_{in}\rangle - \langle v_0\rangle}{L_1}d_{11}T = \frac{\langle v_{in}\rangle + \langle v_{C1}\rangle}{L_2}dT \tag{7}$$

$$\langle i_{L1}\rangle = \frac{I_{LN} + I_{L1P}}{2}d + \frac{I_{L1P} + I_{LM}}{2}d_{11} + \frac{I_{LM} + I_{LN1}}{2}(1 - d - d_{11}) \tag{8}$$

$$\langle i_{L2}\rangle = \frac{I_{LN} + I_{LM}}{2}d + I_{LM}d_{11} + \frac{I_{LM} + I_{LN1}}{2}(1 - d - d_{11}) \tag{9}$$

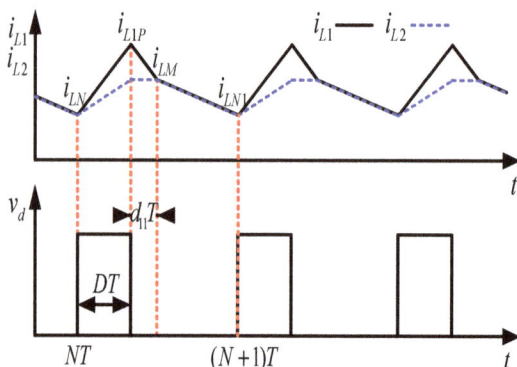

Figure 3. Typical time-domain waveforms about i_{L1}, i_{L2} and v_d for the proposed step-up converter under $L_1 < L_2$.

Hence, the expressions for d_{11} and $\langle i_{L2}\rangle$ can be derived as follows:

$$d_{11} = \frac{((\langle v_{in}\rangle + \langle v_{C1}\rangle)L_1 Kd}{\langle v_0\rangle - \langle v_{in}\rangle} \tag{10}$$

$$\langle i_{L2}\rangle = \langle i_{L1}\rangle - ((\langle v_{in}\rangle + \langle v_{C1}\rangle) + \frac{((\langle v_{in}\rangle + \langle v_{C1}\rangle)^2 L_1 K}{\langle v_0\rangle - \langle v_{in}\rangle})\frac{Kd^2}{2f} \tag{11}$$

where $K = 1/L_1 - 1/L_2$. Thereby, the completed averaged model of the proposed step-up converter in case 2 can be obtained by using the averaging method in (1)–(3), and then combining (10) and (11). The result is:

$$\begin{cases} \frac{d\langle i_{L1}\rangle}{dt} = \frac{\langle v_{in}\rangle + \langle v_{C1}\rangle}{L_1+L_2}2d + \frac{\langle v_{in}\rangle - \langle v_0\rangle}{L_1+L_2}(1-d) \\ \frac{d\langle i_{L3}\rangle}{dt} = -\frac{\langle v_{C1}\rangle}{L_3}d + \frac{\langle v_0\rangle - \langle v_{C1}\rangle}{L_3}(1-d) \\ \frac{d\langle v_{C1}\rangle}{dt} = \frac{\langle i_{L3}\rangle}{C_1} - 2\frac{\langle i_{L1}\rangle}{C_1}d + \frac{Kd^3}{2fC_1}(\langle v_{in}\rangle + \langle v_{C1}\rangle - \frac{((\langle v_{in}\rangle + \langle v_{C1}\rangle)^2 L_1 K}{\langle v_{in}\rangle - \langle v_0\rangle}) \\ \frac{d\langle v_0\rangle}{dt} = -\frac{\langle v_0\rangle}{RC_2} + (\frac{\langle i_{L1}\rangle}{C_2} - \frac{\langle i_{L3}\rangle}{C_2})(1-d) \end{cases} \tag{12}$$

3.1.3. Case 3: $L_1 > L_2$

As indicated in Section 2, for case 3, the proposed step-up converter also has three stages, and its typical domain waveforms are shown in Figure 4, where I_{LN}, I_{LM}, and I_{LN1} are the values of i_{L1} and i_{L2} at $t = NT$, $t = (N + d + d_{22})T$ and $t = (N + 1)T$, respectively, and I_{L2P} is the value of i_{L2} at $t = (N + d + d_{22})T$. Because case 3 is similar to case 2, the completed averaged model of the proposed step-up converter for case 3 is directly derived as follows:

$$\begin{cases} \frac{d\langle i_{L2}\rangle}{dt} = \frac{\langle v_{in}\rangle + \langle v_{C1}\rangle}{L_2+L_1}2d + \frac{\langle v_{in}\rangle - \langle v_0\rangle}{L_2+L_1}(1-d) \\ \frac{d\langle i_{L3}\rangle}{dt} = -\frac{\langle v_{C1}\rangle}{L_3}d + \frac{\langle v_0\rangle - \langle v_{C1}\rangle}{L_3}(1-d) \\ \frac{d\langle v_{C1}\rangle}{dt} = \frac{\langle i_{L3}\rangle}{C_1} - 2\frac{\langle i_{L2}\rangle}{C_1}d - \frac{Kd^3}{2fC_1}(\langle v_{in}\rangle + \langle v_{C1}\rangle + \frac{((\langle v_{in}\rangle + \langle v_{C1}\rangle)^2 L_2 K}{\langle v_{in}\rangle - \langle v_0\rangle}) \\ \frac{d\langle v_0\rangle}{dt} = -\frac{\langle v_0\rangle}{RC_2} + (\frac{\langle i_{L2}\rangle}{C_2} - \frac{\langle i_{L3}\rangle}{C_2})(1-d) \end{cases} \tag{13}$$

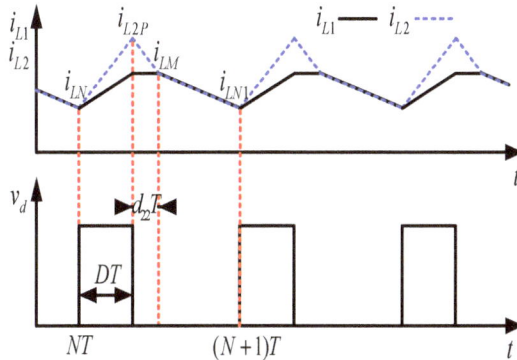

Figure 4. Typical time-domain waveforms about i_{L1}, i_{L2} and v_d for the proposed step-up converter under $L_1 > L_2$.

3.2. DC Equilibrium Point

By substituting (5) into (6), (12), and (13), and then separating DC items, the DC equilibrium points of the proposed step-up converter under three cases can be derived; they are shown in Table 1.

Table 1. DC equilibrium points of proposed step-up converter.

Items	Case 1: $L_1 = L_2$	Case 2: $L_1 < L_2$	Case 3: $L_1 > L_2$
I_{L1}	$\frac{M^2 V_{in}}{R(1+D)}$	$\frac{M^2 V_{in}}{R(1+D)} + K_1 I_{L0}$	$\frac{M^2 V_{in}}{R(1+D)} - K_4 I_{L0}$
I_{L2}	$\frac{M^2 V_{in}}{R(1+D)}$	$\frac{M^2 V_{in}}{R(1+D)} + K_2 I_{L0}$	$\frac{M^2 V_{in}}{R(1+D)} - K_3 I_{L0}$
I_{L3}	$\frac{2M^2 DV_{in}}{R(1+D)}$	$\frac{2M^2 DV_{in}}{R(1+D)} + K_1 I_{L0}$	$\frac{2M^2 DV_{in}}{R(1+D)} - K_3 I_{L0}$
V_{C1}	$\frac{1+D}{1-2D} V_{in}$	$\frac{1+D}{1-2D} V_{in}$	$\frac{1+D}{1-2D} V_{in}$
V_0	$M V_{in}$	$M V_{in}$	$M V_{in}$

where $I_{L0} = KD^2 V_{in}/(4f(1 - 2D)^2)$, $K_1 = (D - 2)((D - 1)L_1/L_2 + 1 + D)$, $K_2 = K_1 - (2D + L_1K(1 - D))(2 - D)$ $(1 - 2D)/D$, $K_3 = (D - 2) ((D - 1) L_2/L_1 + 1 + D)$, $K_4 = K_3 - (2D - L_2K(1 - D))(2 - D)(1 - 2D)/D$.

From Table 1, it can be seen that the expressions for the DC voltage V_0 under the three cases are equal. In other words, no matter what the relation between L_1 and L_2 is, the voltage conversion ratio M of the proposed step-up converter can be described as follows:

$$M = \frac{1+D}{(1-D)(1-2D)} \tag{14}$$

Also, the expressions for the DC voltage V_{C1} under three cases are also equal. In conclusion, the relation between L_1 and L_2 does not influence V_0 and V_{C1}.

3.3. Voltage Stress of Power Switches and Diodes under Three Cases

Based on the definition of the voltage stress on the power switch and diode, the corresponding results for the proposed step-up converter under the three cases can be derived; they are shown in Table 2. One can see that the voltage stresses on the power switches (Q_1 and Q_2) and diodes (D_2, D_4 and D_5) under the three cases are equal except for the diodes D_1 and D_3.

Table 2. Voltage stress on power switches and diodes.

Items	Case 1: $L_1 = L_2$	Case 2: $L_1 < L_2$	Case 3: $L_1 > L_2$
Q_1	MV_{in}	MV_{in}	MV_{in}
Q_2	$\frac{1+D}{1-2D}V_{in}$	$\frac{1+D}{1-2D}V_{in}$	$\frac{1+D}{1-2D}V_{in}$
D_1	$\frac{M-1}{2}V_{in}$	$(M-1)V_{in}$	$\frac{(M-1)L_1V_{in}}{L_1+L_2}$
D_2	$\frac{2-D}{1-2D}V_{in}$	$\frac{2-D}{1-2D}V_{in}$	$\frac{2-D}{1-2D}V_{in}$
D_3	$\frac{M-1}{2}V_{in}$	$\frac{(M-1)L_2V_{in}}{L_1+L_2}$	$(M-1)V_{in}$
D_4	$\frac{1+D}{1-2D}V_{in}$	$\frac{1+D}{1-2D}V_{in}$	$\frac{1+D}{1-2D}V_{in}$
D_5	$(2-D)MV_{in}$	$(2-D)MV_{in}$	$(2-D)MV_{in}$

3.4. Ripples for Inductor Currents and Capacitor Voltages

The ripples for the inductor currents and the capacitor voltages can be obtained by using (1) and Table 1. The results are shown in Table 3.

Table 3. Ripples for inductor currents and capacitor voltages.

Items	Case 1: $L_1 = L_2$	Case 2: $L_1 < L_2$	Case 3: $L_1 > L_2$
Δi_{L1}	$\frac{D(2-D)TV_{in}}{(1-2D)L_1}$	$\frac{D(2-D)TV_{in}}{(1-2D)L_1}$	$\frac{D(2-D)TV_{in}}{(1-2D)L_1}$
Δi_{L2}	$\frac{D(2-D)TV_{in}}{(1-2D)L_2}$	$\frac{D(2-D)TV_{in}}{(1-2D)L_2}$	$\frac{D(2-D)TV_{in}}{(1-2D)L_2}$
Δi_{L3}	$\frac{D(1+D)TV_{in}}{(1-2D)L_3}$	$\frac{D(1+D)TV_{in}}{(1-2D)L_3}$	$\frac{D(1+D)TV_{in}}{(1-2D)L_3}$
Δv_{C1}	$\frac{2(1-D)M^2V_{in}}{R(1+D)C_1}DT$	$(\frac{2(1-D)M^2V_{in}}{R(1+D)}+K_2I_{L0})\frac{DT}{C_1}$	$(\frac{2(1-D)M^2V_{in}}{R(1+D)}-K_4I_{L0})\frac{DT}{C_1}$
Δv_0	$\frac{MV_{in}DT}{RC_2}$	$\frac{MV_{in}DT}{RC_2}$	$\frac{MV_{in}DT}{RC_2}$

Hence, unlike the voltage ripple for capacitor C_1, the current ripples for inductors (L_1, L_2 and L_3) and the voltage ripples for capacitor C_2 under the three cases are equal. Generally, the ripple ratio, which is defined by the ripple over the corresponding DC value, can be used to select the values of the inductors and capacitors.

3.5. Transfer Functions

The transfer function is fundamental for the consequent controller design for DC-DC converters. By substituting (5) into (6), (12), and (13), and then separating AC items and ignoring the second- and higher-order AC terms because their values are very small, the corresponding transfer functions for the proposed step-up converter under the three cases can be derived by using their respective definitions.

The control-to-output voltage transfer function $G_{vd}(s)$, the input voltage-to-output voltage transfer function $G_{vv}(s)$, the control-to-inductor current i_{L1} transfer function $G_{i1d}(s)$, and the input voltage-to-inductor current i_{L1} transfer function $G_{i1v}(s)$ of the proposed step-up converter can be obtained as follows:

$$G_{vd}(s) = \left.\frac{\hat{v}_0(s)}{\hat{d}(s)}\right|_{\hat{v}_{in}(s)=0} = [0,\,0,\,0,\,1](s\mathbf{I}-\mathbf{A})^{-1}\mathbf{B_d} \tag{15}$$

$$G_{vv}(s) = \left.\frac{\hat{v}_0(s)}{\hat{v}_{in}(s)}\right|_{\hat{d}(s)=0} = [0,\,0,\,0,\,1](s\mathbf{I}-\mathbf{A})^{-1}\mathbf{B_v} \tag{16}$$

$$G_{i1d}(s) = \left.\frac{\hat{i}_{L1}(s)}{\hat{d}(s)}\right|_{\hat{v}_{in}(s)=0} = [1,\,0,\,0,\,0](s\mathbf{I}-\mathbf{A})^{-1}\mathbf{B_d} \tag{17}$$

$$G_{i1v}(s) = \left.\frac{\hat{i}_{L1}(s)}{\hat{v}_{in}(s)}\right|_{\hat{d}(s)=0} = [1,\,0,\,0,\,0](s\mathbf{I}-\mathbf{A})^{-1}\mathbf{B_v} \tag{18}$$

where the expressions for $\mathbf{A}$, $\mathbf{B_d}$ and $\mathbf{B_v}$ under the three cases are presented in Table 4.

Table 4. Expressions for **A**, $\mathbf{B}_d$ and $\mathbf{B}_v$ for the proposed step-up converter.

Case	A	$\mathbf{B}_d$	$\mathbf{B}_v$
Case 1: $L_1 = L_2 = L$	$\begin{bmatrix} 0 & 0 & \frac{D}{L} & \frac{D-1}{2L} \\ 0 & 0 & -\frac{1}{L_3} & \frac{1-D}{L_3} \\ -\frac{2D}{C_1} & \frac{1}{C_1} & 0 & 0 \\ \frac{1-D}{C_2} & \frac{D-1}{C_2} & 0 & -\frac{1}{RC_2} \end{bmatrix}$	$\begin{bmatrix} \frac{1+3M-2MD}{2L}V_{in} \\ -\frac{M}{L_3}V_{in} \\ \frac{2M^2V_{in}}{RC_1(1+D)} \\ \frac{(1-2D)M^2V_{in}}{RC_2(1+D)} \end{bmatrix}$	$\begin{bmatrix} \frac{1+D}{2L} \\ 0 \\ 0 \\ 0 \end{bmatrix}$
Case 2: $L_1 < L_2$	$\begin{bmatrix} 0 & 0 & \frac{2D}{L_1+L_2} & \frac{D-1}{L_1+L_2} \\ 0 & 0 & -\frac{1}{L_3} & \frac{1-D}{L_3} \\ -\frac{2D}{C_1} & \frac{1}{C_1} & \frac{\alpha_2}{C_1} & \frac{\alpha_2}{C_1} \\ \frac{1-D}{C_2} & \frac{D-1}{C_2} & 0 & -\frac{1}{RC_2} \end{bmatrix}$	$\begin{bmatrix} \frac{1+3M-2MD}{L_1+L_2}V_{in} \\ -\frac{M}{L_3}V_{in} \\ \frac{\beta_2}{C_1} \\ -\frac{(1-2D)M^2V_{in}}{RC_2(1+D)} \end{bmatrix}$	$\begin{bmatrix} \frac{1+D}{L_1+L_2} \\ 0 \\ \frac{KD^3(L_1K\beta_4-1)}{2fC_1} \\ 0 \end{bmatrix}$
Case 3: $L_1 > L_2$	$\begin{bmatrix} 0 & 0 & \frac{2D}{L_1+L_2} & \frac{D-1}{L_1+L_2} \\ 0 & 0 & -\frac{1}{L_3} & \frac{1-D}{L_3} \\ -\frac{2D}{C_1} & \frac{1}{C_1} & \frac{\alpha_3}{C_1} & \frac{\alpha_3}{C_1} \\ \frac{1-D}{C_2} & \frac{D-1}{C_2} & 0 & -\frac{1}{RC_2} \end{bmatrix}$	$\begin{bmatrix} \frac{1+3M-2MD}{L_1+L_2}V_{in} \\ -\frac{M}{L_3}V_{in} \\ \frac{\beta_3}{C_1} \\ -\frac{(1-2D)M^2V_{in}}{RC_2(1+D)} \end{bmatrix}$	$\begin{bmatrix} \frac{1+D}{L_1+L_2} \\ 0 \\ 0 \\ 0 \end{bmatrix}$

where $\alpha_2 = KD^3(1 - M - 2L_1K(1 + M - MD))/(2f(1 - M))$, $\alpha_3 = -KD^3(1 - M + 2L_2K(1 + M - MD))/(2f(1 - M))$, $\beta_2 = -2M^2V_{in}/(R(1 + D)) - 2K_1I_{L0} + 3D^2KV_{in}(1 + M(1 - D))(1 - L_1K(1 + M(1 - D))/(1 - M))/(2f)$, $\beta_3 = -2M^2V_{in}/(R(1 + D)) + 2K_3I_{L0} - 3D^2KV_{in}(1 + M(1 - D))(1 + L_2K(1 + M(1 - D))/(1 - M))/(2f)$, $\beta_4 = (2(1 + M(1 - D))(1 - M) - (1 + M)^2)/(1 - M)^2$.

4. Comparisons among Different Topologies

Table 5 shows the comparisons among the modified Sheppard-Taylor converter (MSTC) in [14], the PWM Z-source DC-DC converter (ZSC) in [18], the simple modified Sheppard-Taylor converter (SMSTC) in [15], and the proposed step-up converter (PSUC).

Table 5. Comparisons among the converters.

Topology	MSTC in [14]	ZSC in [18]	SMSTC in [15]	PSUC
M	$\frac{D}{(1-D)(1-2D)}$	$\frac{1-D}{1-2D}$	$\frac{1}{1-2D}$	$\frac{1+D}{(1-D)(1-2D)}$
Switches	2	1	2	2
Diodes	4	1	3	5
Inductors	2	3	1	3
Capacitors	2	3	2	2
Output floating	No	Yes	No	No

Although the proposed step-up converter has five diodes, while others have less, from Figure 5, which shows the comparisons of the voltage conversion ratio M among these converters under different duty cycle D, it can be seen that the proposed step-up converter possesses the highest voltage conversion ratio. For example, the voltage conversion ratio M of the proposed step-up converter is up to 46.224 at $D = 0.47$. Additionally, the ZSC's output voltage is floating, whereas those of the others are not.

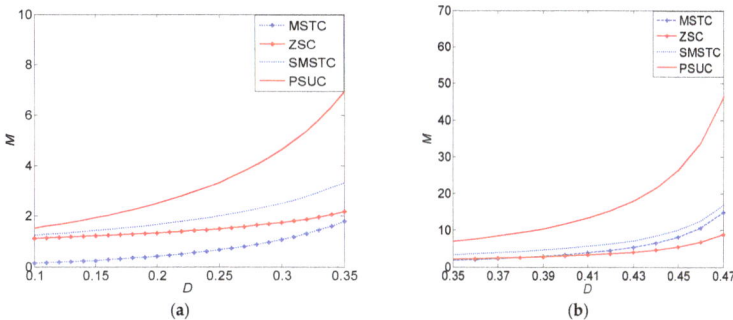

Figure 5. Comparisons about the voltage conversion ratio M among MSTC, ZSC, SMSTC and PSUC. (a) D is within 0.1–0.35; (b) D is within 0.35–0.47.

5. Saber Simulations and Circuit Experiments

For validation purposes, the circuit for the proposed step-up converter is designed. The given specifications are described as V_{in} = 12 V, V_0 = 90 V, f = 32 kHz, R = 300 Ω. Thus, from (14), the duty cycle D should be equal to 0.358742. Based on the voltage stresses of the power switches and diodes in Table 2, the HEXFET Power MOSFET IRFP4668 whose V_{DSS} = 200 V was selected for power switches Q_1 and Q_2, and the Switchmode Schotty Power Rectifier MBR40250 rated for 250 V was selected for diodes D_1, D_2, D_3, D_4, and D_5. Inductors (L_1, L_2 and L_3) can be designed by using their current ripple ratios $\varepsilon_L = \Delta i_L / I_L$ whose values should generally be less than 45%. Capacitors C_1 and C_2 can be designed by using their voltage ripple ratios $\varepsilon_C = \Delta v_C / V_C$ whose values should be less than 20% and 0.5%, respectively. Thereby, from Tables 1 and 3, the current ripple ratio for each inductor and the voltage ripple ratio for each capacitor under each case can be calculated, and accordingly the selected inductors and capacitors in each case should satisfy conditions: $L_1 > 1048$ μH, $L_2 > 1048$ μH, $L_3 > 1210$ μH, $C_1 > 2.06$ μF and $C_2 > 7.47$ μF. Here, C_1 = 4.7 μF with r_{C1} = 10 mΩ, C_2 = 40 μF with r_{C2} = 6 mΩ, and L_3 = 2.76 mH with r_{L3} = 164 mΩ were selected for the proposed step-up converter. Additionally, $L_1 = L_2 = 1.2$ mH with $r_{L1} = r_{L2} = 70$ mΩ were selected for case 1, L_1 = 1.2 mH with r_{L1} = 70 mΩ and L_2 = 2.27 mH with r_{L2} = 156 mΩ were selected for case 2, and L_1 = 2.27 mH with r_{L1} = 156 mΩ and L_2 = 1.2 mH with r_{L2} = 70 mΩ were selected for case 3.

From the above-designed circuit parameters, the simulated model in Saber software, which is widely used in the field of power electronics [20], for the proposed step-up converter is constructed, and some measured results on the output voltage V_0 from the saber simulations were presented in Table 6. One can see that the output voltages V_0 for the proposed step-up converter in the three cases were close, and their values were smaller than the required 90 V because the parasitic parameters were considered in the Saber simulations.

Table 6. Comparisons of the output voltage V_0 between the calculations and the saber simulations.

Cases	$L_1 = L_2$	$L_1 < L_2$	$L_1 > L_2$
Calculations for V_0	90.000 V	90.000 V	90.000 V
Simulations for V_0	84.217 V	83.362 V	83.363 V

Moreover, a hardware circuit for the proposed step-up converter with the same circuit parameters and selected power switches and diodes was also constructed. Notably, in the experiments, the photocoupler TLP250H was applied to drive the power switches. The averaged values of the input voltage V_{in}, the input current I_{in} and the output voltage V_0 with different duty cycle D for the proposed step-up converter under case 1 were measured. In addition, then, the voltage conversion ratio $M = V_0 / V_{in}$ and the efficiency $\eta = V_0^2 / (R V_{in} I_{in})$ with different duty cycle D for the proposed step-up converter in case 1 were calculated and plotted in Figure 6a,b, respectively. Simultaneously, the corresponding Saber simulations were also detected, calculated, and plotted in Figure 6a,b. It can be seen that the experimental results were in basic agreement with the Saber simulations.

As shown in Table 6, it is necessary to design an appropriate controller for this proposed step-up converter. Based on the circuit parameters, the zeros of $G_{vd}(s)$ (shown in (15)) can be calculated. The results showed that $G_{vd}(s)$ was a fourth-order and non-minimum phase since it had right-half side zeros, so that it was difficult to select only the single voltage loop to obtain good performance [21]. Alternatively, an average current mode controller shown in Figure 7 was selected and designed for the proposed step-up converter. This controller had an outer voltage loop and an inner current loop. For the outer voltage loop, it was necessary to detect the output voltage v_0 and design a voltage compensator. For the inner current loop, it was necessary to select one of the inductor currents in the proposed step-up converter and design a current compensator. Due to all the $G_{i1d}(s)$'s poles and zeros being in the left-half side of the s-plane, that is, $G_{i1d}(s)$ is stable and minimum phase, the inductor current i_{L1} was selected and measured for the average current mode controller. Notably, the inductor current i_{L1} here was transformed into a voltage with the same value through the current transducer LA55-A. The current compensator's output voltage is denoted by v_{vi}. AM1 and AM2 were realized by the operational amplifiers LF356 and COM was realized by the voltage comparator LM311. The corresponding parameters were: R_{vi} = 1000 kΩ, R_{vd} =20 kΩ, R_{vf} = 200 kΩ, C_{vf} = 10 nF, R_i = 20 kΩ, R_p = 180 kΩ, C_p = 10 nF, C_i = 100 pF, V_{ref} = 1.76 V. The PWM signal v_d was generated by comparing the voltage v_{vi} with the ramp signal V_{ramp}, whose expression was given as follows:

$$V_{ramp} = V_L + (V_U - V_L)\left(\frac{t}{T}\text{mod}\,1\right) \tag{19}$$

where V_L = 0 V, V_U = 10 V and T = $1/f$.

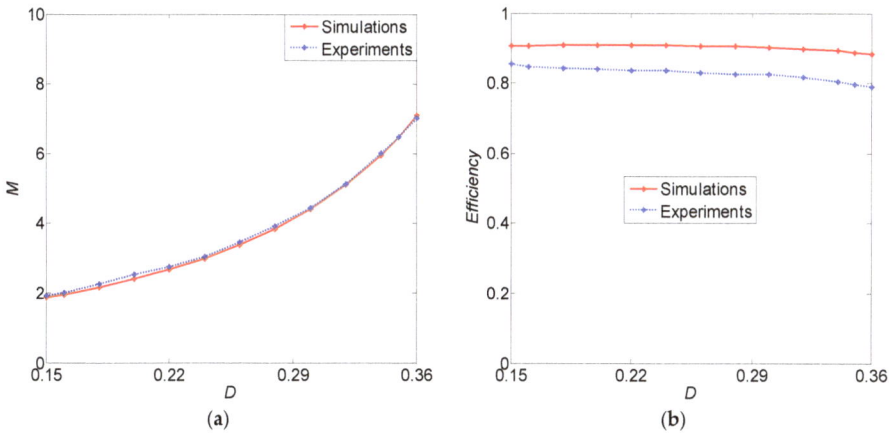

Figure 6. The Saber simulations and the experiments about the voltage conversion ratio M and the efficiency for the proposed step-up converter under case 1. (**a**) The voltage conversion ratio M; (**b**) The efficiency.

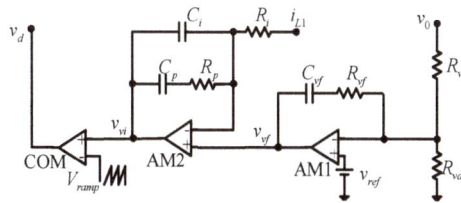

Figure 7. Circuit schematic for the average current mode controller.

The experimental results for the output voltage v_0, the inductor currents i_{L1} and i_{L2}, and the PWM signal v_d for the average-current mode controlled proposed step-up converter under the three cases are presented in Figure 8a–c, respectively. One can see that the output voltages v_0 for the systems under the three cases are really the same, despite different relations between the inductors L_1 and L_2. Moreover, the response of the output voltage v_0, the inductor current i_{L1}, and the PWM signal v_d for the average-current mode controlled proposed step-up converter with the step changing of the load R being 300 Ω–600 Ω–300 Ω is shown in Figure 9. One can see that the closed-loop controlled proposed step-up converter had good performance.

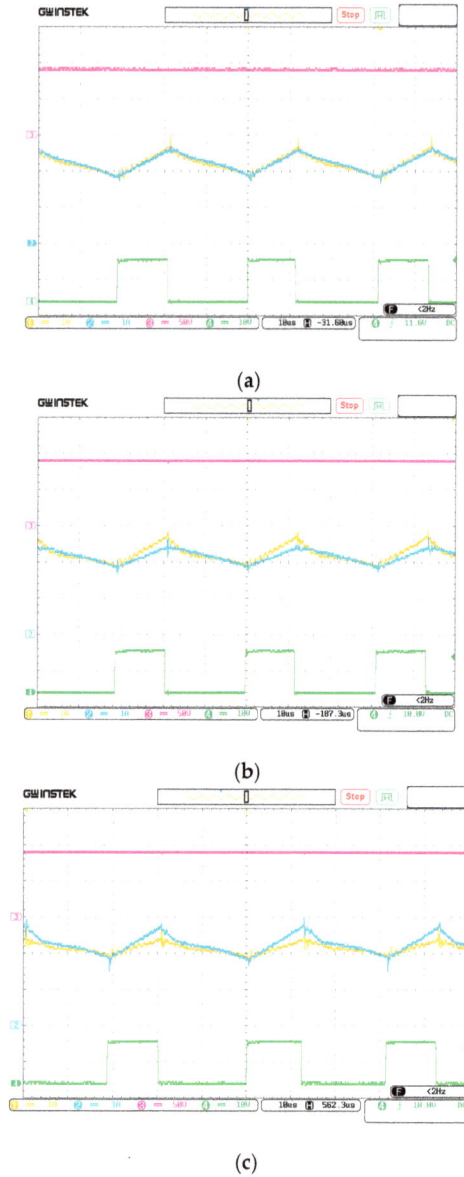

(a)

(b)

(c)

Figure 8. Experimental results for v_0 (Top: 50 V/div, Pink), i_{L1} (Middle: 1 A/div, Yellow), i_{L2} (Middle: 1 A/div, Blue) and v_d (Bottom: 10 V/div, Green) for the average current mode controlled proposed step-up converter under three cases. (**a**) Case 1: $L_1 = 1.2$ mH and $L_2 = 1.2$ mH; (**b**) Case 2: $L_1 = 1.2$ mH and $L_2 = 2.27$ mH; (**c**) Case 3: $L_1 = 2.27$ mH and $L_2 = 1.2$ mH.

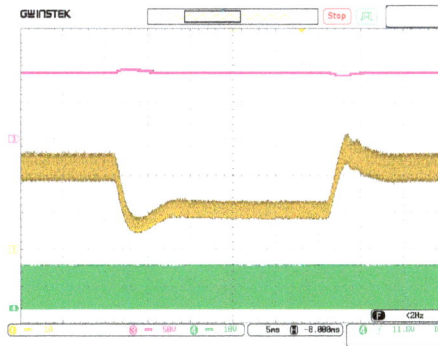

(a)

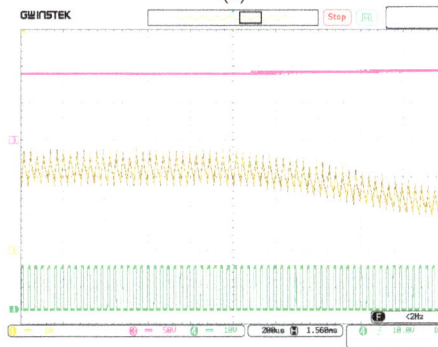

(b)

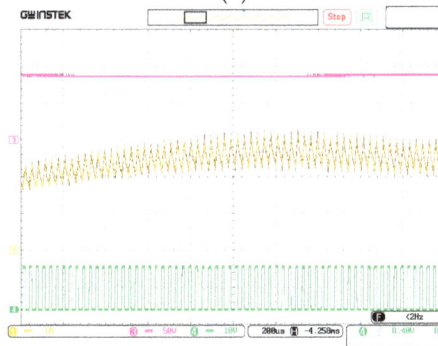

(c)

Figure 9. Response of v_0 (Top: 50 V/div, Pink), i_{L1} (Middle: 1 A/div, Yellow) and v_d (Bottom: 10 V/div, Green) with the step change of the resistive load R being 300 Ω–600 Ω–300 Ω. (**a**) Time: 5 ms/div; (**b**) Close-up view of response for the step changing of the resistive load R being 300 Ω–600 Ω, Time: 200 μs/div; (**c**) Close-up view of response for the step changing of the load R being 600 Ω–300 Ω, Time: 200 μs/div.

6. Conclusions

This paper introduces a new step-up converter. The results from theoretical analysis, the Saber simulations, and the circuit experiments show that, even if it has five diodes and five storage elements, it still has the following five good features:

(1) Although this new step-up converter has five storage elements, that is, three inductors and two capacitors, its averaged model is still fourth-order, because one of the input inductor currents can be expressed by another input inductor current.

(2) The relation between the inductor L_1 and L_2 has three cases: $L_1 = L_2$, $L_1 < L_2$, and $L_1 > L_2$. However, these relations do not influence its output voltage V_0, i.e., the output voltage V_0 are the same in the three cases.

(3) Compared to MSTC, ZSC, and SMSTC, the proposed step-up converter has ultrahigh voltage conversion ratio.

(4) The output voltage of the proposed step-up converter is common-ground and common-polarity with its input voltage, so its value is easy to detect.

(5) The proposed step-up converter has no abrupt changes in capacitor voltage and inductor current, so it does not suffer from infinite capacitor current and inductor voltage.

Thus, the proposed step-up converter with ultrahigh voltage conversion ratio is a good candidate topology for applications of photovoltaic systems and fuel cell systems, and these applications will be investigated in future work.

Author Contributions: F.W. conceived, validated and wrote the manuscript. H.H.-C.I. and J.L. revised the manuscript. All authors contributed to the manuscript.

Funding: This research was funded by the National Natural Science Foundation of China (grant nos. 51377124 and 51521065), the China Scholarship Council (grant no. 201706285022), a Foundation for the Author of National Excellent Doctoral Dissertation of PR China (grant no. 201337), and the New Star of Youth Science and Technology of Shaanxi Province (grant no. 2016KJXX-40).

Conflicts of Interest: The authors declare no conflict of interest.

Nomenclature of Main Symbols and Variables

Q_1, Q_2	Power mosfets
D_1, D_2, D_3, D_4, D_5	Power Diodes
L_1, L_2, L_3 (mH)	Inductors of power stage
C_1, C_2 (µF)	Capacitors of power stage
R (Ω)	Resistive load
V_{ramp}	Ramp signal
v_d (V)	PWM signal
f (kHz)	Switching frequency
$d, D, \hat{d}$	Instantaneous, DC and small signal of duty cycle
$v_{in}, V_{in}, \hat{v}_{in}$ (V)	Instantaneous, DC and small signal of input voltage
T (µs)	Switching period
i_{L1}, i_{L2}, i_{L3} (A)	Instantaneous values of inductor currents
I_{L1}, I_{L2}, I_{L3} (A)	DC values of inductor currents
$\langle i_{L1} \rangle, \langle i_{L2} \rangle, \langle i_{L3} \rangle$ (A)	Averaged values of inductor currents
$\hat{i}_{L1}, \hat{i}_{L2}, \hat{i}_{L3}$ (A)	Small signal of inductor currents
$\Delta i_{L1}, \Delta i_{L2}, \Delta i_{L3}$ (A)	Ripples of inductor currents
v_{C1}, v_0 (V)	Instantaneous values of capacitor voltage of C_1, C_2
$\langle v_{C1} \rangle, \langle v_0 \rangle$ (V)	Averaged values of capacitor voltage of C_1, C_2
V_{C1}, V_0 (V)	DC values of capacitor voltage of C_1, C_2
$\hat{v}_{C1}, \hat{v}_0$ (V)	Small signal of capacitor voltage of C_1, C_2
$\Delta v_{C1}, \Delta v_0$ (V)	Ripples of capacitor voltage of C_1, C_2

$G_{vd}(s)$	Control-to-output voltage transfer function
$G_{vv}(s)$	Input voltage-to-output voltage transfer function
$G_{i1d}(s)$	Control-to-inductor current i_{L1} transfer function
$G_{i1v}(s)$	Input voltage-to-inductor current i_{L1} transfer function
$R_{vi}, R_{vd}, R_{vf}, R_i, R_p$ (kΩ)	Resistors of average current mode controller
C_{vf}, C_p, C_i (μF)	Capacitors of average current mode controller
V_{ref} (V)	Reference voltage
V_L, V_U (V)	Lower and upper threshold of ramp signal

References

1. Hwu, K.I.; Jiang, W.Z.; Chien, J.Y. Isolated high voltage-boosting converter derived from forward converter. *Int. J. Circuit Theory Appl.* **2016**, *44*, 280–304. [CrossRef]
2. Tang, Y.; Wang, T.; He, Y.H. A switched-capacitor-based active-network converter with high voltage gain. *IEEE Trans. Power Electron.* **2014**, *29*, 2959–2968. [CrossRef]
3. Ben-Yaakov, S.; Zeltser, I. The dynamics of a PWM Boost converter with resistive input. *IEEE Trans. Ind. Eletron.* **1999**, *46*, 613–619. [CrossRef]
4. Yang, L.S.; Liang, T.J.; Chen, J.F. Transformerless DC-DC converters with high step-up voltage gain. *IEEE Trans. Ind. Eletron.* **2009**, *56*, 3144–3152. [CrossRef]
5. Gules, R.; Santos, W.M.D.; Reis, F.A.D.; Romaneli, E.F.R.; Badin, A.A. A modified SEPIC converter with high static gain for renewable applications. *IEEE Trans. Power Electron.* **2014**, *29*, 5860–5871. [CrossRef]
6. Mummadi, V.; Mohan, B.K. Robust digital voltage-mode controller for fifth-order Boost converter. *IEEE Trans. Ind. Eletron.* **2011**, *58*, 263–277. [CrossRef]
7. Hwu, K.I.; Yau, Y.T. A KY Boost converter. *IEEE Trans. Power Electron.* **2010**, *25*, 2699–2703. [CrossRef]
8. Hwu, K.I.; Huang, K.W.; Tu, W.C. Step-up converter combining KY and buck-boost converters. *Electron. Lett.* **2011**, *47*, 722–724. [CrossRef]
9. Luo, F.L.; Ye, H. Positive output super-lift converters. *IEEE Trans. Power Electron.* **2003**, *18*, 105–113.
10. Hwu, K.I.; Yau, Y.T. High step-up converter based on charge pump and Boost converter. *IEEE Trans. Power Electron.* **2012**, *27*, 2484–2494. [CrossRef]
11. Al-Saffar, M.A.; Ismail, E.H. A high voltage ratio and low stress DC-DC converter with reduced input current ripple for fuel cell source. *Renew. Energy* **2015**, *82*, 35–43. [CrossRef]
12. Hwu, K.I.; Chuang, C.F.; Tu, W.C. High voltage-boosting converters based on bootstrap capacitors and boost inductors. *IEEE Trans. Ind. Eletron.* **2013**, *60*, 2178–2193. [CrossRef]
13. Chen, Y.T.; Lin, W.C.; Liang, R.H. An interleaved high step-up DC-DC converter with double boost paths. *Int. J. Circuit Theory Appl.* **2015**, *43*, 967–983. [CrossRef]
14. Kanaan, H.Y.; Al-Haddad, K.; Hayek, A.; Mougharbel, I. Design, study, modelling and control of a new single-phase high power factor rectifier based on the single-ended primary inductance converter and the Sheppard-Taylor topology. *IET Power Electron.* **2009**, *2*, 163–177. [CrossRef]
15. Hua, C.C.; Chiang, H.C.; Chuang, C.W. New boost converter based on Sheppard-Taylor topology. *IET Power Electron.* **2014**, *7*, 167–176. [CrossRef]
16. Wang, F.Q. Improved transfer functions for modified sheppard-taylor converter that operates in CCM: Modeling and application. *J. Power Electron.* **2017**, *17*, 884–891.
17. Patidar, K.; Umarikar, A.C. A step-up PWM DC-DC converter for renewable energy applications. *Int. J. Circuit Theory Appl.* **2016**, *44*, 817–832. [CrossRef]
18. Galigekere, V.P.; Kazimierczuk, M.K. Analysis of PWM Z-Source DC-DC Converter in CCM for Steady State. *IEEE Trans. Circuits Syst. I Regul. Pap.* **2012**, *59*, 854–863. [CrossRef]
19. Middlebrook, R.D.; Cuk, S. A general unified approach to modelling switching-converter power stages. *Int. J. Electron.* **1977**, *42*, 521–550. [CrossRef]

20. Shi, J.J.; Liu, T.J.; Cheng, J.; He, X.N. Automatic current sharing of an input-parallel output-parallel (IPOP)-connected DC-DC converter system with chain-connected rectifiers. *IEEE Trans. Power Electron.* **2015**, *30*, 2997–3016. [CrossRef]

21. Morales-Saldana, J.A.; Galarza-Quirino, R.; Leyva-Ramos, J.; Carbajal-Gutierrez, E.E.; Ortiz-Lopez, M.G. Multiloop controller design for a quadratic boost converter. *IET Electr. Power Appl.* **2007**, *1*, 362–367. [CrossRef]

energies

MDPI

Article

Slope Compensation Design for a Peak Current-Mode Controlled Boost-Flyback Converter

Juan-Guillermo Muñoz [1], Guillermo Gallo [2], Fabiola Angulo [1,*] and Gustavo Osorio [1]

[1] Departamento de Ingeniería Eléctrica, Electrónica y Computación, Percepción y Control Inteligente, Facultad de Ingeniería y Arquitectura, Universidad Nacional de Colombia—Sede Manizales, Bloque Q, Campus La Nubia, Manizales 170003, Colombia; jgmunozc@unal.edu.co (J.-G.M.); gaosoriol@unal.edu.co (G.O.)
[2] Departamento de Ingeniería Electrónica y Telecomunicaciones, Automática, Electrónica y Ciencias Computacionales (AE&CC), Instituto Tecnológico Metropolitano, Medellín 050013, Colombia; ggalloh@unal.edu.co
* Correspondence: fangulog@unal.edu.co

Received: 27 September 2018; Accepted: 29 October 2018; Published: 1 November 2018

Abstract: Peak current-mode control is widely used in power converters and involves the use of an external compensation ramp to suppress undesired behaviors and to enhance the stability range of the Period-1 orbit. A boost converter uses an analytical expression to find a compensation ramp; however, other more complex converters do not use such an expression, and the corresponding compensation ramp must be computed using complex mechanisms. A boost-flyback converter is a power converter with coupled inductors. In addition to its high efficiency and high voltage gains, this converter reduces voltage stress acting on semiconductor devices and thus offers many benefits as a converter. This paper presents an analytical expression for computing the value of a compensation ramp for a peak current-mode controlled boost-flyback converter using its simplified model. Formula results are compared to analytical results based on a monodromy matrix with numerical results using bifurcations diagrams and with experimental results using a lab prototype of 100 W.

Keywords: slope compensation; monodromy matrix; current mode control; boost-flyback converter

---

1. Introduction

The main purpose of power converters is to change the level voltage. Currently, this task is achieved by controlling a converter through pulse width modulation (PWM) such that the system is described by a set of dynamic equations. Power converters can be modeled as a piecewise linear dynamic system [1–3], and all exhibit a plethora of nonlinear phenomena depending on the parameter values used. Such behaviors, which are currently being examined at length [4–7], include period-doubling bifurcations, subharmonics and chaos [2,8,9].

The main goal of a converter is usually to contribute a load with a desired voltage; in this sense, it is important to compute and analyze the stability of the Period-1 orbit and to study its complex dynamics (a complete revision of stability analysis methods applied to power converters can be found in [10]). The behaviors of a power converter are often determined by plotting bifurcation diagrams [4,5,11,12] using the Poincaré map [3]. In these diagrams, as a parameter value changes, the Poincaré map of the steady state is plotted. The stability of the Period-1 orbit is also analyzed by presenting the Poincaré map as a monodromy matrix such that by analyzing eigenvalues of the monodromy matrix (Floquet multipliers), it is possible to determine the stability of a Period-1 orbit [13]. Several studies also combine bifurcations and monodromy matrix analyses [6,14].

High step-up power converters are some of the main devices used in photovoltaic applications [15–19] due to the low output voltage of solar panels. With such applications, efficiency is a key issue, so single-stage converters are preferred over more complex converters [17,18]. Strong gains can be achieved through single-stage conversion by using coupled inductors where basic converters can be coupled, improving the advantages of every configuration to extend the voltage conversion ratio, suppress the switch voltage spike, recycle leakage energy and increase efficiency levels [17,18,20].

A converter that couples buck, boost and flyback topologies is presented in [17]. The converter consists of one MOSFET (metal-oxide-semiconductor field-effect transistor), four diodes, three inductors and three capacitors, rendering the system and controller difficult to model, analyze and design. This is the case due to the high-order dynamic equations used in the uncontrolled system (sixth order) and due to the number of electronic devices (five) used, which renders 32 topologies possible. A structure based on SEPIC (single-ended primary-inductor converter) and boost-flyback converters was proposed in [18]. This converter was composed of four semiconductors and eight energy storage elements. The system is difficult to use for analyses (eight differential equations and 16 topologies) and is less efficient than the converter presented in [17]. In a similar vein, a converter coupling several cells of flyback converters with switched capacitors was proposed in [16,21] Although this application considerably increases the voltage, the model is complex for the same reasons noted above, and its analysis and control design are difficult to use.

Because of the aforementioned drawbacks, researchers have returned to a more basic and efficient structure: the boost-flyback (BF) converter [22,23]. In a BF, boost and flyback converters are integrated via magnetic coupling between two inductors to form a BF converter to achieve a good trade-off between voltage gains, efficiency and complexity. Due to its high voltage gains, high efficiency and limited complexity relative to similar models, the boost-flyback converter is widely used in various applications such as in hybrid electric vehicles [24,25], for voltage balancing [26], in low-scale arrays of photovoltaic panels [27], in LED lighting [28,29] and for power factor correction [29]. A boost-flyback converter includes two capacitors, two coupled inductors, two diodes and one MOSFET such that the designer may use four differential equations and eight potential topologies. The efficiency and voltage gain were improved in [30,31], by adding other primary and secondary coils, leading to the same problems described above. In a similar way, the efficenciy was improved for gains greater than eight by adding switched coupled inductors, rendering the system more complex [32].

One of the most popular control techniques used in power converters is that of so-called peak current-mode control [33–36]. When the value of the slope is low, the system remains stable. As the duty cycle progresses, the system turns unstable, and when the slope value is very high, subharmonics are present [37,38], limiting the time response of the controlled system [39] and compromising performance [40]. In this way, it is necessary to find the correct compensation ramp value to avoid a fast scale related to the inner control loop [41,42] or a slow scale due to the outer control loop [43,44]. Both dynamic behaviors have been widely studied in reference to several converters [45–48]. Once slope compensation is designed, system behaviors can be improved by changing, retuning or controlling slope compensation. To improve the behavior of the controlled system, a polynomial curve slope compensation scheme was proposed [14]. This slope secures better results than a traditional linear ramp slope compensation scheme, though its practical implementation is less straightforward. The performance of current-mode control was optimized by means of an autotuning technique of the ramp slope, allowing for a broader control bandwidth than the traditional technique [49]. On the other hand, some peak current-mode control techniques avoid the use of an external signal generator alleviating the deviation of the inductor current peak value from its desired reference [10] and improving the range of the current reference [50].

In this paper, an analytical expression for determining the initial value of the slope compensation of a BF converter via peak current-mode control is determined. Unlike modern means of improving the range of stability of the compensation ramp [10,14,49,50], the proposed technique is less complex and therefore easier to implement and allows the precise calculation of the compensation ramp, which guarantees correct operation and prevents unwanted behaviors from manifesting. A complete analysis of stability and transitions to chaos for a peak-current mode-controlled BF converter is reported in [51], and the coexistence of Period-1, Period-2 and chaotic orbits with varying coupling coefficients has been proven via bifurcation analysis [52]. However, neither analytical expressions for computing the value of the compensation ramp of a peak-current controlled BF converter, nor experimental data that confirm the corresponding analytical expression have been even published.

An analytical expression is first obtained by simplifying the problem. Corresponding results are compared to those derived from three sources. (a) Results can first be derived from analytical expressions computed with a complete model and using the monodromy matrix [13]. In this case, the monodromy matrix is computed analytically, and its eigenvalues are calculated as the parameter varies. The largest absolute value of its eigenvalues is called the LAVE. (b) Numerical results can also be obtained from bifurcation diagrams computed by brute force and (c) from experiments carried out in a lab prototype of 100 W. All results show good agreement.

The rest of this paper is organized as follows. In Section 2, the operation mode of the boost-flyback converter is described, as well as that of peak-current mode control. In Section 3, computations for obtaining the mathematical expression for slope compensation are presented. In Section 4, numerical and experimental results are shown and compared. Numerical results are obtained using the derived formula for a particular converter using parameters similar to those used in the experimental setup, including the nonideal model (internal resistance for certain components). The experimental results are obtained from a 100-watt lab prototype and they are presented and compared to numerical ones. Finally, Section 5 concludes.

2. Boost-Flyback Converter: Modeling and Control

A peak-current controlled boost-flyback converter is depicted in Figure 1. The boost-flyback components are denoted with black lines while the controller is denoted with gray lines. The boost-flyback consists of two coupled inductors (L_p, L_s), two capacitors (C_1, C_2), one MOSFET (S) and two diodes (D_1, D_2). The MOSFET is controlled while the diodes commutate depending on their degree of polarization. As the name implies, the union of a boost and flyback converter achieves high gains and high levels of efficiency, while the stress voltage of semiconductor devices decreases relative to that of a standard flyback [23,53].

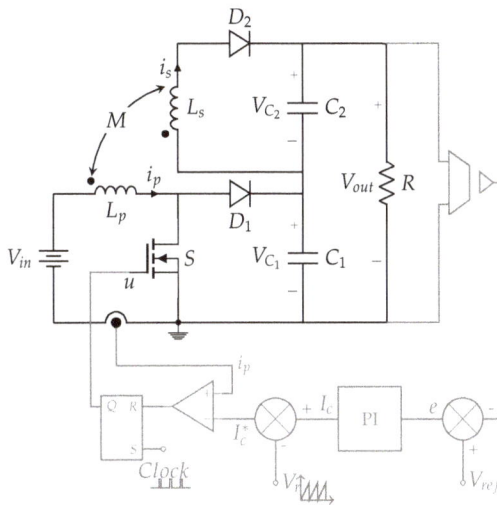

Figure 1. Boost-flyback converter with peak current-mode control.

As three semiconductor devices are used, there are eight possible switch configurations or states: E_1 ... E_8. However, it has been shown that only six states have physical meaning [54], and it has been also proven that the controlled system exhibits a Period-1 orbit switching between four states, as described in Table 1 [51]. A schematic diagram of the steady state current behavior of a Period-1 solution is presented in Figure 2. States E_1 and E_2 are present when the MOSFET is turned on, and states E_3 and E_4 are present when the MOSFET is turned off.

Table 1. States of the Period-1 orbit.

State	S	D_2	D_1
E_1	ON	ON	OFF
E_2	ON	OFF	OFF
E_3	OFF	ON	ON
E_4	OFF	ON	OFF

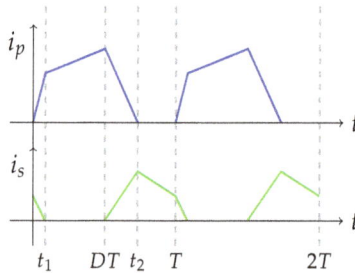

Figure 2. Typical behavior of the currents flowing by the coils in the steady state of a Period-1 orbit.

From E_1, the system evolves as follows: $E_1 \mapsto E_2 \mapsto E_3 \mapsto E_4$. A change from E_1 to E_2 occurs when $i_s = 0$ at $t = t_1$; the system switches from E_2 to E_3 when the switching condition is satisfied at $t = DT$, which is referred to as the duty cycle and which corresponds to the ratio between the time at which the MOSFET is turned on and the period T, i.e.,: $D = t_{u=1}/T$; E_3 changes to E_4 when $i_p = 0$ (at $t = t_2$), and finally, when $t = T$, the system returns to E_1. The set of differential equations describing the Period-1 orbit is shown below:

State 1: $E_1, t \in [kT \ \ kT + t_1]$:

$$
\begin{aligned}
\frac{di_p}{dt} &= \frac{(L_s V_{in} + M V_{C_2})}{n} \\
\frac{di_s}{dt} &= \frac{(-M V_{in} - L_p V_{C_2})}{n} \\
\frac{dV_{C_1}}{dt} &= -\frac{(V_{C_1} + V_{C_2})}{R C_1} \\
\frac{dV_{C_2}}{dt} &= \frac{i_s}{C_2} - \frac{(V_{C_1} + V_{C_2})}{R C_2}
\end{aligned}
\tag{1}
$$

State 2: $E_2, t \in (kT + t_1 \ kT + DT]$:

$$\frac{di_p}{dt} = \frac{V_{in}}{L_p}$$

$$\frac{di_s}{dt} = 0$$

$$\frac{dV_{C_1}}{dt} = -\frac{(V_{C_1} + V_{C_2})}{RC_1} \qquad (2)$$

$$\frac{dV_{C_2}}{dt} = -\frac{(V_{C_1} + V_{C_2})}{RC_2}$$

State 3: $E_3, t \in (kT + DT \ kT + t_2]$:

$$\frac{di_p}{dt} = \frac{(L_s(V_{in} - V_{C_1}) + MV_{C_2})}{n}$$

$$\frac{di_s}{dt} = \frac{(-M(V_{in} - V_{C_1}) - L_pV_{C_2})}{n} \qquad (3)$$

$$\frac{dV_{C_1}}{dt} = \frac{i_p}{C_1} - \frac{(V_{C_1} + V_{C_2})}{RC_1}$$

$$\frac{dV_{C_2}}{dt} = \frac{i_s}{C_2} - \frac{(V_{C_1} + V_{C_2})}{RC_2}$$

State 4: $E_4, t \in (kT + t_2 \ kT + T)$:

$$\frac{di_p}{dt} = 0$$

$$\frac{di_s}{dt} = -\frac{V_{C_2}}{L_s} \qquad (4)$$

$$\frac{dV_{C_1}}{dt} = -\frac{(V_{C_1} + V_{C_2})}{RC_1}$$

$$\frac{dV_{C_2}}{dt} = \frac{i_s}{C_2} - \frac{(V_{C_1} + V_{C_2})}{RC_2}$$

where V_{in} is the input voltage, i_p and i_s are the primary and secondary currents, respectively, V_{C_1} and V_{C_2} are the voltages across capacitors C_1 and C_2 and $M = \bar{k}\sqrt{L_pL_s}$ is the mutual inductance, with $\bar{k}$ as the coupling coefficient and $n = L_pL_s - M^2$. The output voltage is $V_{out} = V_{C_1} + V_{C_2}$.

Peak current-mode control is widely used for the control of power converters [41,47,51]. A general schematic diagram of the boost-flyback converter with the proposed controller is depicted in Figure 1. When peak current-mode control is used, a fixed switching frequency is obtained, and current behaviors are very similar to those shown in Figure 2. At the start of the period, the MOSFET is active, the current i_p increases and the current i_s declines to $i_s = 0$; at time t_1, dynamic equations describing the system change while the MOSFET continues on until i_p is equal to the reference current I_c^* at $t = DT$. At $t = DT$, switches stop until the next cycle begins. Signal I_c^* is composed of two parts: the first (denoted as I_c) is provided by a PI controller applied to the output voltage error $e = V_{ref} - V_{out}$. The second part corresponds to the signal supplied by the compensation ramp $V_r = \frac{A_r}{T}\text{mod}(t/T)$. Thus, the reference current can be expressed as:

$$I_c^* = k_pe + k_i \int e\, dt - \frac{A_r}{T}\text{mod}(t/T) \qquad (5)$$

where k_p and k_i [A/(V.t)] are parameters associated with the PI controller and A_r [A]corresponds to the amplitude of the compensation ramp. As a result of the controller, there is only one switching cycle per period. At the start of the period, the switch turns on, and it remains on until switching condition

$i_p = I_c^*$ is achieved (the corresponding duty cycle). When $i_p = I_c^*$, the switch opens and remains open until the next period starts. As sliding is not possible (i.e., there is only one round of commutation per cycle), the switching condition can be expressed as:

$$u = \begin{cases} 1 & \text{if } 0 \le t < DT, \\ 0 & \text{if } DT \le t < T. \end{cases} \tag{6}$$

where $D \in [0,1]$ is the duty cycle.

3. Slope Compensation Design

To our knowledge, the related literature has not reported on a means of determining the slope of the compensation ramp of a boost-flyback converter, which can be used to stabilize the Period-1 orbit. The objective of this section is to analyze the slopes of currents flowing through inductors to find an analytical expression to determine the slope of a compensation ramp and thus to guarantee the stability of the Period-1 orbit. Figure 3 presents the behavior of currents flowing through primary and secondary coils when the system operates within the Period-1 orbit described by states E_1, E_2, E_3 and E_4. Slopes are clearly marked in the figure.

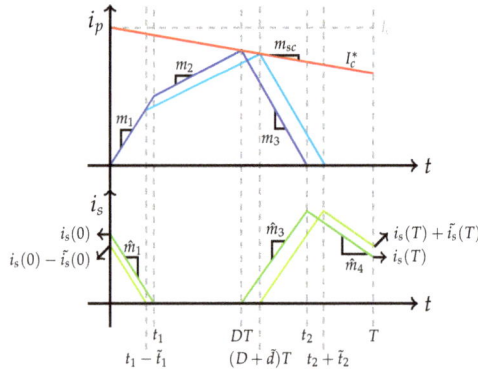

Figure 3. Primary- and secondary-coil currents for the Period-1 orbit and a perturbed solution.

3.1. Assumptions

In the analysis, the following approximations are considered. (i) For all elements and devices, the internal resistances are zero. (ii) The steady-state output of the PI-controller (I_c) is constant, and hence, its derivative is zero. However, as can be seen in the procedure, the constant value is not needed to compute the final expression. (iii) Voltages V_{C_1} and V_{C_2} are constant and can be computed as a function of the duty cycle D; V_{C_1} is the output of the boost component, and V_{C_2} is the output of the flyback component, taking into account the coupling factor $k < 1$ (see Appendix A for a complete derivation of the formula).

$$
\begin{aligned}
V_{C_1} &= \frac{1}{(1-D)} V_{in} \\[2mm]
V_{C_2} &= \frac{\left(1 - \frac{M}{Lp}\right)}{\left(\frac{M}{L_s} - 1\right)} \frac{D}{(1-D)} V_{in} \\[2mm]
V_{out} &= \frac{1 + \frac{\left(1 - \frac{M}{Lp}\right)}{\left(\frac{M}{L_s} - 1\right)} D}{1 - D} V_{in}.
\end{aligned}
\tag{7}
$$

(iv) Finally, all currents can be mathematically expressed as straight lines such that slopes associated with i_p include m_1, m_2 and m_3, while slopes associated with i_s include $\hat{m}_1$, $\hat{m}_3$ and $\hat{m}_4$ (see Figure 3). These slopes can be computed from Equations (2)–(5) as follows:

$$
\begin{aligned}
m_1 &= \frac{L_s V_{in} + M V_{C_2}}{n} \\[4pt]
\hat{m}_1 &= \frac{-M V_{in} - L_p V_{C_2}}{n} \\[4pt]
m_2 &= \frac{V_{in}}{L_p} \\[4pt]
m_3 &= \frac{L_s(V_{in} - V_{C_1}) + M V_{C_2}}{n} \\[4pt]
\hat{m}_3 &= \frac{-M(V_{in} - V_{C_1}) - L_p V_{C_2}}{n} \\[4pt]
\hat{m}_4 &= -\frac{V_{C_2}}{L_s}
\end{aligned}
\tag{8}
$$

In a similar way as the slope compensation in a boost power converter is designed considering the stability of the Period-1 orbit [33], in this paper, we present an analysis of the stability of the Period-1 orbit based on information on current slopes and on conditions that should be met to guarantee the stability of the controlled system. To analyze the stability of the Period-1 orbit, small perturbations are applied at the beginning of the cycle, and its corresponding value is measured at the end of period T. When the magnitude of perturbation increases, the Period-1 orbit is unstable; by contrast, when the magnitude of perturbation decreases, the orbit is stable.

3.2. Mathematical Procedure

3.2.1. Analysis of Currents in the Primary Coil

At switching time $t = DT$, a pair of equations is fulfilled in Figure 3: one to its left and the other to its right. When defining the slope of the compensation ramp as $m_{sc} = \frac{Ar}{T}$, at the switching time, the following equation is satisfied:

$$
I_c - m_{sc} DT = m_1 t_1 + m_2(DT - t_1)
\tag{9}
$$

Based on perturbation observed in the initial condition, the last equation can be expressed as follows:

$$
I_c - m_{sc}(D + \tilde{d})T = m_1(t_1 - \tilde{t}_1) + m_2((D + \tilde{d})T - (t_1 - \tilde{t}_1))
\tag{10}
$$

By subtracting Equation (10) from (9), we obtain the following:

$$
m_{sc}\tilde{d}T = m_1 \tilde{t}_1 - m_2(\tilde{d}T + \tilde{t}_1)
\tag{11}
$$

From (11),

$$
\tilde{t}_1 = \frac{(m_{sc} + m_2)}{(m_1 - m_2)} \tilde{d}T
\tag{12}
$$

In a similar way, the analysis illustrated to the right of the switching time leads to the following equation.

$$
I_c - m_{sc} DT - m_3(t_2 - DT) = 0
\tag{13}
$$

Taking into account the perturbation, this equation is given by:

$$
I_c - m_{sc}(D + \tilde{d})T - m_3((t_2 + \tilde{t}_2) - (D + \tilde{d})T) = 0
\tag{14}
$$

Subtracting (14) from (13),

$$m_{sc}\check{d}T + m_3(\tilde{t}_2 - \check{d}T) = 0 \tag{15}$$

From (15),

$$\tilde{t}_2 = \frac{(m_3 - m_{sc})}{m_3}\check{d}T \tag{16}$$

3.2.2. Analysis of Currents in the Secondary Coil

Now, the expressions for the current i_s and its perturbation $\tilde{i}_s(0)$ are computed. At $t = t_1$, they are as follows:

$$i_s(0) - \hat{m}_1 t_1 = 0 \tag{17}$$

and:

$$i_s(0) - \tilde{i}_s(0) - \hat{m}_1(t_1 - \tilde{t}_1) = 0 \tag{18}$$

Subtracting (18) from (17), it is obtained:

$$\tilde{i}_s(0) = \hat{m}_1 \tilde{t}_1 \tag{19}$$

Replacing (12) in (19), we have:

$$\tilde{i}_s(0) = \hat{m}_1 \frac{(m_{sc} + m_2)}{(m_1 - m_2)}\check{d}T \tag{20}$$

From this equation, $\check{d}T$ can be expressed as:

$$\check{d}T = \frac{\tilde{i}_s(0)}{\hat{m}_1 \frac{(m_{sc} + m_2)}{(m_1 - m_2)}} \tag{21}$$

Now, at $t = t_2$, the following equation is fulfilled,

$$\hat{m}_3(t_2 - DT) - \hat{m}_4(T - t_2) = i_s(T) \tag{22}$$

At the same time $t = t_2$, the perturbed equation is:

$$\hat{m}_3((t_2 + \tilde{t}_2) - (D + \check{d})T) - \hat{m}_4(T - (t_2 + \tilde{t}_2)) = i_s(T) + \tilde{i}_s(T) \tag{23}$$

Now, subtracting (22) from (23), we have:

$$\tilde{i}_s(T) = (\hat{m}_3 + \hat{m}_4)\tilde{t}_2 - \hat{m}_3\check{d}T \tag{24}$$

Replacing (16) in (24), we obtain:

$$\tilde{i}_s(T) = \left(\hat{m}_4 - m_{sc}\frac{(\hat{m}_3 + \hat{m}_4)}{m_3}\right)\check{d}T \tag{25}$$

Finally, by replacing Equation (21) in (25), we find an expression that relates to the secondary coil current at the beginning of the cycle with its value shown at the end. This expression is given by:

$$\tilde{i}_s(T) = \alpha\tilde{i}_s(0) \tag{26}$$

where:

$$\alpha = \left[\frac{\left(\hat{m}_4 - m_{sc}\frac{(\hat{m}_3 + \hat{m}_4)}{m_3}\right)}{\hat{m}_1\frac{(m_{sc} + m_2)}{(m_1 - m_2)}}\right] \tag{27}$$

3.2.3. Stability Condition

Then, the stability of the Period-1 orbit is given by the absolute value of α. When $|\alpha| > 1$, the periodic orbit is unstable; when $|\alpha| < 1$, it is asymptotically stable; and when $|\alpha| = 1$, it corresponds to the limit of stability. To guarantee that the system operates within a Period-1 orbit, the slope of the compensation ramp must satisfy the following expression:

$$m_{sc} = \frac{A_r}{T} > \frac{m_3(\hat{m}_4(m_1 - m_2) - \hat{m}_1 m_2)}{\hat{m}_1 m_3 + (\hat{m}_3 + \hat{m}_4)(m_1 - m_2)} \tag{28}$$

4. Results

Parameters associated with the converter and experiment are presented in Table 2. The first two columns are needed to simulate the system, and the other parameters describe the electronic circuit.

Table 2. Converter and experimental setup parameters.

Converter's Parameters			Experiment's Parameters		
Element	Value	Element	Value	Electronic Device	Reference
V_{in}	18 V	R_a	1 MΩ	IC_1	INA128p
L_p	129.2 μH	R_b	20 kΩ	IC_2	TL084
L_s	484.9 μH	R_1	100 kΩ	IC_3	LM311
r_p	0.0268 Ω	R_2	5.7 kΩ	IC_4	555
r_s	0.1307 Ω	R_3	10 kΩ	IC_5	74XX02
k	0.995	R_4	200 kΩ	IC_6	IRF2110
C_1	220 μF	R_t	2.2 kΩ	IC_7	74XX08
C_2	220 μF	C_3	0.1 μF	Q_T	2N3906
R	200 Ω	C_4	10 nF	D	1N4148
k_p	2 A/V	r_{shunt}	0.01 Ω		
k_i	350 A/(V.s)				
T	50 μs				

4.1. Numerical Results

To compare the results obtained using Equation (28) with the analytical results, we determined the stability of the Period-1 orbit from the saltation matrix associated with switching times [13,55,56]. A complete analysis of the stability and computation of the saltation matrix for this system can be found in [51]. Parameter values used for the simulations and experiments are given in Table 2. Voltages V_{C_1} and V_{C_2} are computed from Equation (7); the slopes of straight lines are calculated from Equation (8); and the output voltage V_{out} corresponds to the desired output voltage V_{ref} and $|\alpha| = 1$. With these data, the desired output voltage varies, and the limit value of slope compensation m_{sc} is obtained.

Figure 4a shows results obtained from the proposed approach (see (28)) and $V_{ref} \in (90, 130)$ V. Figure 4b presents the exact computation using the saltation matrix. Values of A_r exceeding the stability limit guarantee the stability of a Period-1 orbit. In addition, for $V_{ref} = 100$ V, the limit value of the compensation ramp is close to $A_r = 1.94$ A, and for $V_{ref} = 120$ V, it is close to $A_r = 3.25$ A (see Figure 4). Figure 4c compares the analytical approach proposed in this paper with the exact value obtained from the saltation matrix; the result is expressed as a percentage. As is shown, the lower the reference voltage, the higher the error value is. In fact, for gain factors greater than six, the approach generates better results. This is due to the assumptions in (7): as the gain factor decreases, the gains of boost and flyback parts cannot be separated. Even more, for gains close to two, only the boost converter works, and the flyback part is voided.

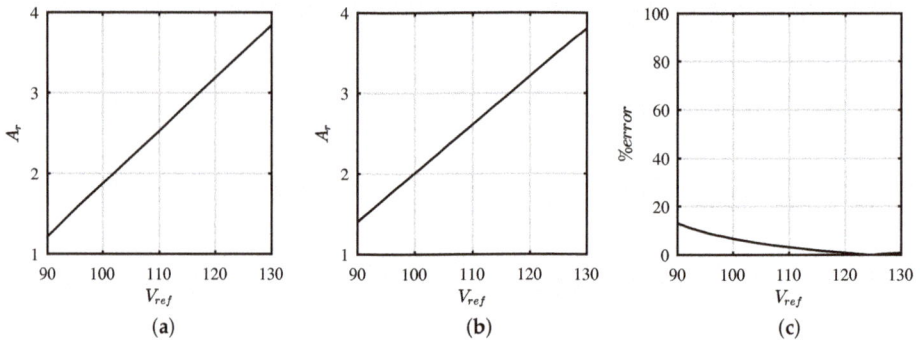

Figure 4. Value of the slope compensation. (a) Approach proposed in this paper. (b) Exact value obtained with the saltation matrix and (c) percentage error.

To verify these results, we consider a more complete model of the converter that uses internal resistances of the primary inductor, secondary inductor and MOSFET, as well as the shunt resistance to measure the current (see Table 2). In Figure 5a,c, bifurcation diagrams varying the slope of the ramp are computed while the desired output voltage remains fixed (see Figure 5a for $V_{ref} = 100$ V and Figure 5c for $V_{ref} = 120$ V). In Figure 5b,d, the behavior of the largest absolute value of eigenvalues (LAVE) is shown for the same reference voltage values. In these cases, limit values of the compensation ramp are $A_r = 2.035$ A and $A_r = 3.21$ A. These results complement those computed from Equation (27) and Figure 4. Slight displacement is observed between numerical values obtained from equations and from the LAVE due to the assumptions applied.

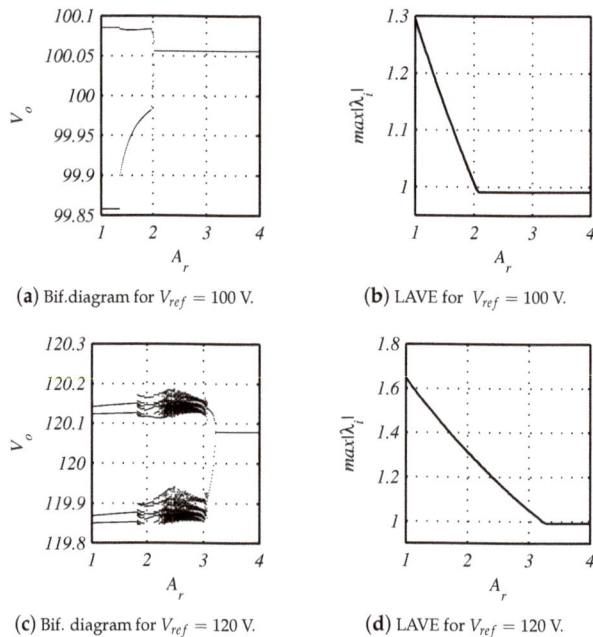

(a) Bif.diagram for $V_{ref} = 100$ V.

(b) LAVE for $V_{ref} = 100$ V.

(c) Bif. diagram for $V_{ref} = 120$ V.

(d) LAVE for $V_{ref} = 120$ V.

Figure 5. Bifurcation diagrams and largest absolute value of eigenvalues (LAVE) evolution for A_r variations.

Using the previous limit values, we can find the ramp slopes to stabilize the Period-1 orbit. To prove the robustness of the controller and to compare the system's behaviors for two values of A_r, changes in the load are induced first by applying $V_{ref} = 100$ V and $A_r = 1.8$ A. When the load is changed, it can be observed that for the full range of load resistance values, the Period-1 orbit is unstable (see Figure 6a). However, when we fix the slope of the ramp at $A_r = 2.2$ A (the limit of stability is close to two), the Period-1 orbit is stable for the full range of load resistance values (see Figure 6b). In the second case, we establish $V_{ref} = 120$ V and $A_r = 3$ A. When the load is changed, it can be observed that for the full range of load resistance values, the Period-1 orbit is unstable (see Figure 6c). However, when we fix the slope of the ramp at $A_r = 3.4$ A (the limit of stability is close to 3.2), the Period-1 orbit is stable for the full load range, as is shown in Figure 6d.

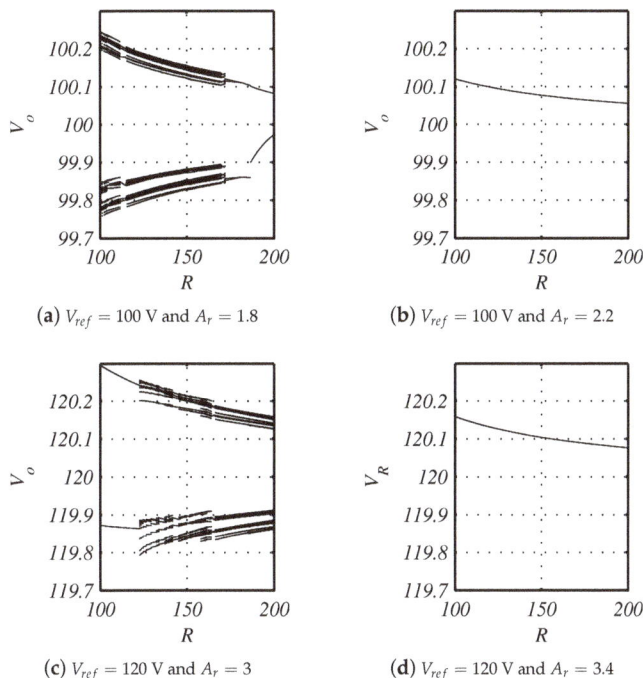

(**a**) $V_{ref} = 100$ V and $A_r = 1.8$

(**b**) $V_{ref} = 100$ V and $A_r = 2.2$

(**c**) $V_{ref} = 120$ V and $A_r = 3$

(**d**) $V_{ref} = 120$ V and $A_r = 3.4$

Figure 6. Bifurcation diagrams varying the load resistance R.

4.2. Experimental Results

To validate the numerical results, an experimental lab prototype that can deliver 100 watts to the load was designed and implemented. The complete design of the circuit is shown in Figure 7. A ferrite core type E is used to design coupled inductors, and the number of turns is calculated using the approach proposed in [57]. Values of the different circuit elements are given in Table 2. The current in the primary coil is measured with non-inductive shunt-resistance r_{shunt} (LTO050FR0100FTE3) and then with an instrumentation amplifier IC_1; the output voltage is measured through a voltage divider consisting of R_a and R_b. The signal generated from the voltage divider feeds amplifier IC_1. The MOSFET is an $IRFP260N$ with low internal resistance. Finally, two ultrafast diodes $RHRP30120$ (D_1 and D_2) are used.

The controller is applied using operational amplifiers (IC_2). The compensation ramp and clock signals are generated using an $LM555(IC_4)$. The amplitude of the compensation ramp is adjusted with a span resistor R_{span}, and V_B compensates for the offset. Constants k_p and k_i are associated with the PI controller and are obtained from R_2, R_3, R_4 and C_3. The measured signals are scaled to 0.196 from voltage gains (A_{g_1} and A_{g_2}). Constant G_v is given by voltage divider $R_b/(R_a + R_b)$.

Four experiments employed to validate the results shown in the previous section were carried out. All figures of the experimental results show reference current I_c^*, primary coil current i_p, secondary coil current i_s and output voltage V_{out}. Therefore, the output voltage and current in the secondary coil are scaled by a factor of 10. The reference current and the current in the primary coil are scaled by a factor of 0.196, as mentioned above.

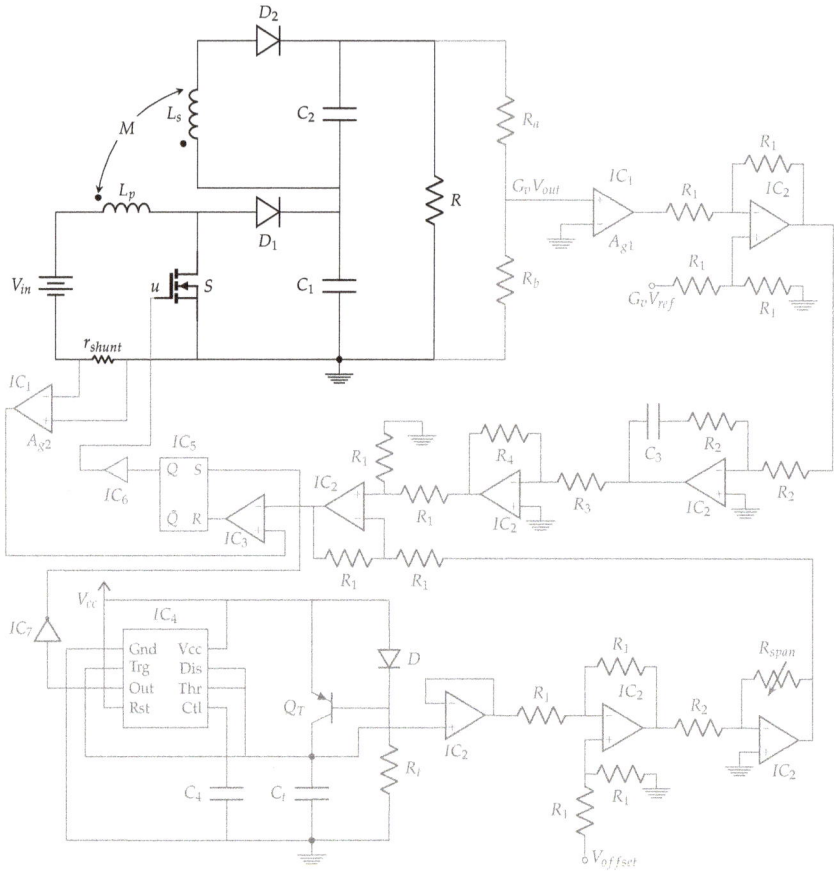

Figure 7. Experimental circuit.

For $V_{ref} = 100$ V (the load resistance is fixed at $R = 200\ \Omega$; see Table 2), two values of slope compensation are tuned: $A_r = 1.8$ A and $A_r = 2.2$ A. When $A_r = 1.8$ A, the limit set is a Period-2 orbit, as is shown in Figure 8a, but when the ramp compensation increases to $A_r = 2.2$ A, it converts to a Period-1 orbit (Figure 8b).

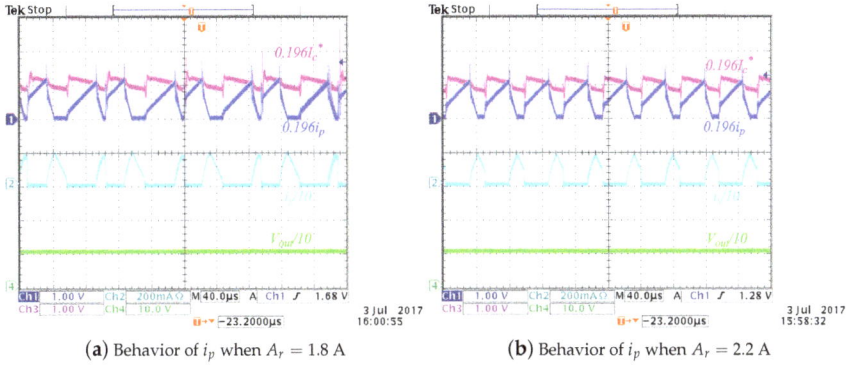

(**a**) Behavior of i_p when $A_r = 1.8$ A

(**b**) Behavior of i_p when $A_r = 2.2$ A

Figure 8. Experimental results with two different values of the compensation ramp for $V_{ref} = 100$ V.

For the second experiment, $V_{ref} = 120$ V. In a similar way, two values of slope compensation are tuned: $A_r = 3$ A and $A_r = 3.4$ A. The behaviors of I_c^*, i_p, i_s and V_{out} are shown in Figure 9a,b. For $A_r = 3$ A, a high-period orbit appears, and for $A_r = 3.4$ A, the Period-1 orbit is stable. These results complement information provided by Equation (28), and the formula can be used to tune the slope of the compensation ramp.

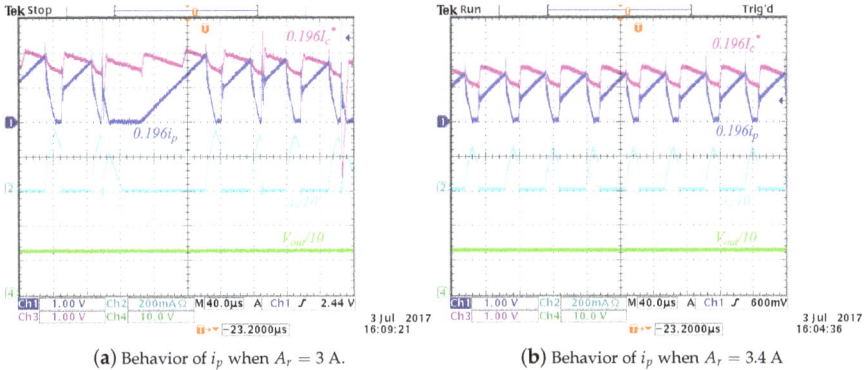

(**a**) Behavior of i_p when $A_r = 3$ A.

(**b**) Behavior of i_p when $A_r = 3.4$ A

Figure 9. Experimental results with two different values of the compensation ramp for $V_{ref} = 120$ V.

5. Conclusions

This paper enhances the knowledge of the controller design for a boost-flyback converter, which is currently a topic of study.

To achieve high gains with a stable Period-1 orbit when a boost-flyback converter is used, it is necessary to add a compensation ramp to the design. In this work, an analytical expression for computing the value of the compensation ramp slope is presented and mathematically proven. For gains of greater than six, the approach given in this paper has an error of less than 5%.

In general, the results of the equation derived from our computations agree with those of experiments, with minor discrepancies in exact solutions observed for gains of less than six, mainly because certain assumptions are too strong to apply to the real system, and these are not included in the model for the sake of simplicity. This difference is negligible for high step-up gains, for which our approach offers the major benefit of using a formula that guarantees stability while preventing overcompensation and the use of very complex computations.

Author Contributions: Conceptualization, F.A. Formal analysis, J.-G.M. and F.A. Funding acquisition, F.A. and G.O. Investigation, J.-G.M. Project administration, F.A. Software, J.-G.M., G.G. and G.O. Supervision, F.A. and G.O. Validation, J.-G.M. Writing, original draft, J.-G.M. and G.G. Writing, review and editing, F.A. and G.O.

Funding: This work was supported by Universidad Nacional de Colombia, Manizales, Project 31492 from Vicerrectoría de Investigación, DIMA, and COLCIENCIASunder Contract FP44842-052-2016 and program Doctorados Nacionales 6172-2013.

Acknowledgments: The authors would like to thank Ángel Cid Pastor and Abdelali el Aroudi from GAEIResearch Center, Universitat Rovira i Virgili, Spain, for their assistance in obtaining experimental results.

Conflicts of Interest: The authors declare no conflict of interest. The founding sponsors had no role in the design of the study; in the collection, analyses or interpretation of data; in the writing of the manuscript; nor in the decision to publish the results.

Appendix A

In this Appendix, the procedure to find the ratio between input and output voltages for a flyback converter when coupling factor $\bar{k}$ is different from zero is presented:

$$V_{C_2} = \frac{n_2}{n_1} \frac{D}{1 - D} \tag{A1}$$

The flyback converter operates in two topologies named State 1 and State 2, which are depicted in Figure A1. Voltage equations in primary and secondary coils are given in the general form as:

$$v_{L_p} = L_p \frac{di_p}{dt} + M \frac{di_s}{dt}$$
$$v_{L_s} = L_s \frac{di_s}{dt} + M \frac{di_p}{dt}. \tag{A2}$$

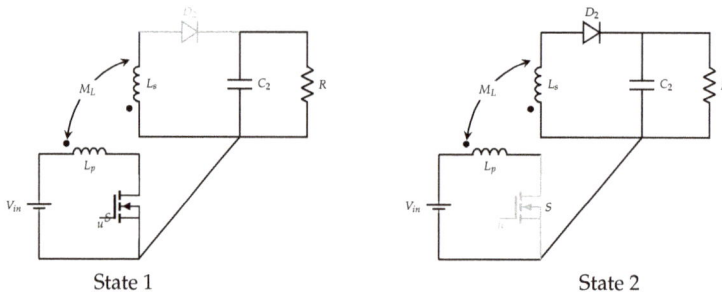

State 1   State 2

Figure A1. Flyback converter topologies.

Depending on the state, voltages and currents can be approximated as:
State 1:

$$v_{L_{p1}} \approx V_{in}$$
$$v_{L_{s1}} \approx \frac{M}{L_p} V_{in}$$
$$i_{C1} \approx -V_{C_2}/R \tag{A3}$$

State 2:

$$v_{L_{p2}} \approx -\frac{M}{L_s}V_{C_2}$$
$$v_{L_{s2}} \approx -V_{C_2}$$
$$i_{C2} \approx i_{L_s} - V_{C_2}/R, \tag{A4}$$

such that the average values can be calculated as:

$$<v_{L_p}> = DV_{in} - (1-D)\frac{M}{L_s}V_{C_2} = 0$$
$$<v_{L_s}> = D\frac{M}{L_p}V_{in} - (1-D)V_{C_2} = 0$$
$$<i_C> = -DV_{C_2}/R + (1-D)(i_{L_s} - V_{C_2}/R) = 0. \tag{A5}$$

Taking into account $k < 1$, i.e., $\frac{M}{L_p} \neq \frac{L_s}{M}$, we have:

$$DV_{in} - (1-D)\frac{M}{L_s}V_{C_2} = D\frac{M}{L_p}V_{in} - (1-D)V_{C_2}, \tag{A6}$$

to finally find

$$\frac{V_{C_2}}{V_{in}} = \frac{(1-\frac{M}{L_p})}{(\frac{M}{L_s}-1)}\frac{D}{(1-D)} \tag{A7}$$

Doing $k = 1$, it is easy to prove that this ratio is the same as that reported for a non-magnetically coupled flyback converter.

References

1. Acary, V.; Bonnefon, O.; Brogliato, B. *Nonsmooth Modeling and Simulation for Switched Circuits*; Springer: Berlin, Gremany, 2011.
2. Banerjee, S.; Verghese, G. *Nonlinear Phenomena in Power Electronics: Bifurcations, Chaos, Control, and Applications*; Wiley-IEEE Press: Hoboken, NJ, USA, 2001.
3. Di Bernardo, M.; Budd, C.; Champneys, A.; Kowalczyk, P. *Piecewise-Smooth Dynamical Systems, Theory and Applications*; Springer: Berlin, Gremany, 2008.
4. Zhang, X.; Bao, B.; Bao, H.; Wu, Z.; Hu, Y. Bi-Stability Phenomenon in Constant On-Time Controlled Buck Converter with Small Output Capacitor ESR. *IEEE Access* **2018**, *6*, 46227–46232. [CrossRef]
5. Bandyopadhyay, A.; Parui, S. Bifurcation behavior of photovoltaic panel fed Cuk converter connected to different types of loads. In Proceedings of the 2018 International Symposium on Devices, Circuits and Systems (ISDCS), Howrah, India, 29–31 March 2018; pp. 1–5. [CrossRef]
6. Zhang, R.; Hu, M.; Zhang, Y.; Ma, W.; Zhao, J. Control of bifurcation in the buck converter using novel resonant parametric perturbation. In Proceedings of the IECON 2017—43rd Annual Conference of the IEEE Industrial Electronics Society, Beijing, China, 29 October–1 November 2017; pp. 982–986. [CrossRef]
7. Samanta, B.; Ghosh, S.; Panda, G.K.; Saha, P.K. Chaos control on current-mode controlled DC-DC boost converter using TDFC. In Proceedings of the 2017 IEEE Calcutta Conference (CALCON), Kolkata, India, 2–3 December 2017; pp. 159–163. [CrossRef]
8. Deane, J.H.B.; Hamill, D.C. Chaotic behaviour in current-mode controlled DC-DC convertor. *Electron. Lett.* **1991**, *27*, 1172–1173. [CrossRef]
9. di Bernardo, M.; Garefalo, F.; Glielmo, L.; Vasca, F. Analysis of chaotic buck, boost and buck-boost converters through switching maps. In Proceedings of the PESC97, Record 28th Annual IEEE Power Electronics Specialists Conference. Formerly Power Conditioning Specialists Conference 1970–71. Power Processing and Electronic Specialists Conference 1972, Saint Louis, MO, USA, 27 June 1997; Volume 1, pp. 754–760. [CrossRef]

10. Aroudi, A.E.; Mandal, K.; Giaouris, D.; Banerjee, S. Self-compensation of DC–DC converters under peak current mode control. *Electron. Lett.* **2017**, *53*, 345–347. [CrossRef]

11. Fossas, E.; Olivar, G. Study of chaos in the buck converter. *IEEE Trans. Circuits Syst. I Fundam. Theory Appl.* **1996**, *43*, 13–25. [CrossRef]

12. Zhusubaliyev, Z.T.; Mosekilde, E. Torus birth bifurcations in a DC/DC converter. *IEEE Trans. Circuits Syst. I Regul. Pap.* **2006**, *53*, 1839–1850. [CrossRef]

13. Leine, R.I.; Nijmeijer, H. *Dynamics and Bifurcations of Non-Smooth Mechanical Systems*; Springer: Berlin, Germnay, 2004.

14. Wu, H.; Pickert, V.; Deng, X.; Giaouris, D.; Li, W.; He, X. Polynomial Curve Slope Compensation for Peak-Current-Mode-Controlled Power Converters. *IEEE Trans. Ind. Electron.* **2019**, *66*, 470–481. [CrossRef]

15. Wu, Y.E.; Chiu, P.N. A High-Efficiency Isolated-Type Three-Port Bidirectional DC/DC Converter for Photovoltaic Systems. *Energies* **2017**, *10*, 434. [CrossRef]

16. Choudhury, T.R.; Dhara, S.; Nayak, B.; Santra, S.B. Modelling of a high step up DC-DC converter based on Boost-flyback-switched capacitor. In Proceedings of the 2017 IEEE Calcutta Conference (CALCON), Kolkata, India, 2–3 December 2017; pp. 248–252. [CrossRef]

17. Shen, C.L.; Chiu, P.C. Buck-boost-flyback integrated converter with single switch to achieve high voltage gain for PV or fuel-cell applications. *IET Power Electron.* **2016**, *9*, 1228–1237. [CrossRef]

18. Lodh, T.; Majumder, T. Highly efficient and compact Sepic-Boost-Flyback integrated converter with multiple outputs. In Proceedings of the 2016 International Conference on Signal Processing, Communication, Power and Embedded System (SCOPES), Paralakhemundi, India, 3–5 October 2016; pp. 6–11. [CrossRef]

19. Arango, E.; Ramos-Paja, C.A.; Calvente, J.; Giral, R.; Serna, S. Asymmetrical Interleaved DC/DC Switching Converters for Photovoltaic and Fuel Cell Applications-Part 1: Circuit Generation, Analysis and Design. *Energies* **2012**, *5*, 4590–4623. [CrossRef]

20. Liu, H.; Hu, H.; Wu, H.; Xing, Y.; Batarseh, I. Overview of High-Step-Up Coupled-Inductor Boost Converters. *IEEE J. Emerg. Sel. Top. Power Electron.* **2016**, *4*, 689–704. [CrossRef]

21. Wang, Y.F.; Yang, L.; Wang, C.S.; Li, W.; Qie, W.; Tu, S.J. High Step-Up 3-Phase Rectifier with Fly-Back Cells and Switched Capacitors for Small-Scaled Wind Generation Systems. *Energies* **2015**, *8*, 2742–2768. [CrossRef]

22. Zhao, Q.; Lee, F.C. High performance coupled-inductor DC-DC converters. In Proceedings of the Eighteenth Annual IEEE Applied Power Electronics Conference and Exposition (APEC '03), Miami Beach, FL, USA, 9–13 February 2003; Volume 1, pp. 109–113.

23. Tseng, K.; Liang, T. Novel high-efficiency step-up converter. *IEE Proc. Electr. Power Appl.* **2004**, *151*, 182–190. [CrossRef]

24. Lai, C.M.; Yang, M.J. A High-Gain Three-Port Power Converter with Fuel Cell, Battery Sources and Stacked Output for Hybrid Electric Vehicles and DC-Microgrids. *Energies* **2016**, *9*, 180. [CrossRef]

25. Tseng, K.C.; Lin, J.T.; Cheng, C.A. An Integrated Derived Boost-Flyback Converter for fuel cell hybrid electric vehicles. In Proceedings of the 2013 1st International Future Energy Electronics Conference (IFEEC), Tainan, Taiwan, 3–6 November 2013; pp. 283–287.

26. Park, J.H.; Kim, K.T. Multi-output differential power processing system using boost-flyback converter for voltage balancing. In Proceedings of the 2017 International Conference on Recent Advances in Signal Processing, Telecommunications Computing (SigTelCom), Da Nang, Vietnam, 9–11 January 2017; pp. 139–142. [CrossRef]

27. Chen, S.M.; Wang, C.Y.; Liang, T.J. A novel sinusoidal boost-flyback CCM/DCM DC-DC converter. In Proceedings of the 2014 IEEE Applied Power Electronics Conference and Exposition—APEC 2014, Fort Worth, TX, USA, 16–20 March 2014; pp. 3512–3516.

28. Lee, S.W.; Do, H.L. A Single-Switch AC-DC LED Driver Based on a Boost-Flyback PFC Converter with Lossless Snubber. *IEEE Trans. Power Electron.* **2017**, *32*, 1375–1384. [CrossRef]

29. Divya, K.M.; Parackal, R. High power factor integrated buck-boost flyback converter driving multiple outputs. In Proceedings of the 2015 Online International Conference on Green Engineering and Technologies (IC-GET), Coimbatore, India, 27 November 2015; pp. 1–5. [CrossRef]

30. Xu, D.; Cai, Y.; Chen, Z.; Zhong, S. A novel two winding coupled-inductor step-up voltage gain boost-flyback converter. In Proceedings of the 2014 International Power Electronics and Application Conference and Exposition, Shanghai, China, 5–8 November 2014; pp. 1–5.

31. Zhang, J.; Wu, H.; Xing, Y.; Sun, K.; Ma, X. A variable frequency soft switching boost-flyback converter for high step-up applications. In Proceedings of the 2011 IEEE Energy Conversion Congress and Exposition, Phoenix, AZ, USA, 17–22 September 2011; pp. 3968–3973.

32. Ding, X.; Yu, D.; Song, Y.; Xue, B. Integrated switched coupled-inductor boost-flyback converter. In Proceedings of the 2017 IEEE Energy Conversion Congress and Exposition (ECCE), Cincinnati, OH, USA, 1–5 October 2017; pp. 211–216. [CrossRef]

33. Erickson, R.W.; Maksimovic, D. *Fundamentals of Power Electronics*; Springer: Berlin, Germnay, 2001.

34. Wang, Y.; Xu, J.; Zhou, S.; Zhao, T.; Liao, K. Current-mode controlled single-inductor dual-output buck converter with ramp compensation. In Proceedings of the 2017 IEEE Energy Conversion Congress and Exposition (ECCE), Cincinnati, OH, USA, 1–5 October 2017; pp. 4996–5000. [CrossRef]

35. Abdelhamid, E.; Bonanno, G.; Corradini, L.; Mattavelli, P.; Agostinelli, M. Stability Properties of the 3-Level Flying Capacitor Buck Converter Under Peak or Valley Current-Programmed-Control. In Proceedings of the 2018 IEEE 19th Workshop on Control and Modeling for Power Electronics (COMPEL), Padua, Italy, 25–28 June 2018; pp. 1–8. [CrossRef]

36. Zhou, S.; Zhou, G.; Zeng, S.; Zhao, H.; Xu, S. Unified modelling and dynamical analysis of current-mode controlled single-inductor dual-output switching converter with ramp compensation. *IET Power Electron.* **2018**, *11*, 1297–1305. [CrossRef]

37. Ridley, R.B. A new continuous-time model for current-mode control with constant frequency, constant on-time, and constant off-time, in CCM and DCM. In Proceedings of the 21st Annual IEEE Conference on Power Electronics Specialists, San Antonio, TX, USA, 11–14 June 1990; pp. 382–389. [CrossRef]

38. Unitrode-Corporation. Modeling, Analysis and Compensation of the Current-Mode Converter. 1999. Available online: http://www.ti.com/general/docs/litabsmultiplefilelist.tsp?literatureNumber=slua101 (accessed on 7 October 2018).

39. Yang, Y.Z.; Xie, G.J. Research and design of a self-adaptable slope compensation circuit with simple structure. In Proceedings of the 2010 IEEE International Conference on Intelligent Computing and Intelligent Systems, Xiamen, China, 29–31 October 2010; Volume 2, pp. 333–335. [CrossRef]

40. Yang, C.; Wang, C.; Kuo, T. Current-Mode Converters with Adjustable-Slope Compensating Ramp. In Proceedings of the APCCAS 2006—2006 IEEE Asia Pacific Conference on Circuits and Systems, Singapore, 4–7 December 2006; pp. 654–657. [CrossRef]

41. Jiuming, Z.; Shulin, L. Design of slope compensation circuit in peak-current controlled mode converters. In Proceedings of the 2011 International Conference on Electric Information and Control Engineering (ICEICE), Wuhan, China, 15–17 April 2011; pp. 1310–1313.

42. Grote, T.; Schafmeister, F.; Figge, H.; Frohleke, N.; Ide, P.; Bocker, J. Adaptive digital slope compensation for peak current mode control. In Proceedings of the Energy Conversion Congress and Exposition (ECCE 2009), San Jose, CA, USA, 20–24 September 2009; pp. 3523–3529.

43. Chen, Y.; Tse, C.K.; Wong, S.C.; Qiu, S.S. Interaction of fast-scale and slow-scale bifurcations in current-mode controlled DC/DC converters. *Int. J. Bifurc. Chaos* **2007**, *17*, 1609–1622. [CrossRef]

44. Chen, Y.; Tse, C.K.; Qiu, S.S.; Lindenmuller, L.; Schwarz, W. Coexisting Fast-Scale and Slow-Scale Instability in Current-Mode Controlled DC/DC Converters: Analysis, Simulation and Experimental Results. *IEEE Trans. Circuits Syst. I Regul. Pap.* **2008**, *55*, 3335–3348. [CrossRef]

45. El Aroudi, A. A New Approach for Accurate Prediction of Subharmonic Oscillation in Switching Regulators Part I: Mathematical Derivations. *IEEE Trans. Power Electron.* **2017**, *32*, 5651–5665. [CrossRef]

46. El Aroudi, A. A New Approach for Accurate Prediction of Subharmonic Oscillation in Switching Regulators Part II: Case Studies. *IEEE Trans. Power Electron.* **2017**, *32*, 5835–5849. [CrossRef]

47. Fang, C.C.; Redl, R. Subharmonic Instability Limits for the Peak-Current-Controlled Buck Converter with Closed Voltage Feedback Loop. *IEEE Trans. Power Electron.* **2015**, *30*, 1085–1092. [CrossRef]

48. Fang, C.C.; Redl, R. Subharmonic Instability Limits for the Peak-Current-Controlled Boost, Buck-Boost, Flyback, and SEPIC Converters With Closed Voltage Feedback Loop. *IEEE Trans. Power Electron.* **2017**, *32*, 4048–4055. [CrossRef]

49. Liu, P.; Yan, Y.; Lee, F.C.; Mattavelli, P. Universal Compensation Ramp Auto-Tuning Technique for Current Mode Controls of Switching Converters. *IEEE Trans. Power Electron.* **2018**, *33*, 970–974. [CrossRef]

50. Lu, W.G.; Lang, S.; Li, A.; Iu, H.H.C. Limit-cycle stable control of current-mode dc-dc converter with zero-perturbation dynamical compensation. *Int. J. Circuit Theory Appl.* **2015**, *43*, 318–328. [CrossRef]

51. Munoz, J.G.; Gallo, G.; Osorio, G.; Angulo, F. Performance Analysis of a Peak-Current Mode Control with Compensation Ramp for a Boost-Flyback Power Converter. *J. Control Sci. Eng.* **2016**, *2016*, 7354791. [CrossRef]
52. Muñoz, J.G.; Gallo, G.; Angulo, F.; Osorio, G. Coexistence of solutions in a boost-flyback converter with current mode control. In Proceedings of the 2017 IEEE 8th Latin American Symposium on Circuits Systems (LASCAS), Bariloche, Argentina, 20–23 February 2017; pp. 1–4. [CrossRef]
53. Liang, T.; Tseng, K. Analysis of integrated boost-flyback step-up converter. *IEE Proc. Electr. Power Appl.* **2005**, *152*, 217–225. [CrossRef]
54. Carrero Candelas, N.A. Modelado, Simulación y Control de un Convertidor Boost Acoplado Magnéticamente. Ph.D. Thesis, Universidad Politécnica de Catalunya, Barcelona, Spain, 2014.
55. Elbkosh, A.; Giaouris, D.; Pickert, V.; Zahawi, B.; Banerjee, S. Stability analysis and control of bifurcations of parallel connected DC/DC converters using the monodromy matrix. In Proceedings of the IEEE International Symposium on Circuits and Systems (ISCAS 2008), Seattle, WA, USA, 18–21 May 2008; pp. 556–559.
56. Morcillo Bastidas, J.; Burbano Lombana, D.; Angulo, F. Adaptive Ramp Technique for Controlling Chaos and Sub-harmonic Oscillations in DC-DC Power Converters. *IEEE Trans. Power Electron.* **2016**, *31*, 5330–5343. [CrossRef]
57. Dixon, L. Coupled Inductor Design. 1993. Available online: http://www.ti.com/lit/ml/slup105/slup105.pdf (accessed on 4 August 2018).

![energies logo] *energies*

MDPI

Article

Passivity-Based Robust Output Voltage Tracking Control of DC/DC Boost Converter for Wind Power Systems

Seok-Kyoon Kim

Department of Creative Convergence Engineering, Hanbat National University, Daejeon 341-58, Korea; lotus45kr@gmail.com; Tel.: +82-042-828-8801

Received: 8 May 2018; Accepted: 4 June 2018; Published: 6 June 2018

Abstract: This paper exhibits a passivity-based robust output voltage controller for DC/DC boost converters for wind power system applications. The proposed technique has two features. The first one is to introduce a nonlinear disturbance observer for estimating the disturbances arising from the load and parameter variations. The second one is to derive a proportional-type passivity-based output voltage tracking controller incorporating the disturbance observer output, which simplifies the control algorithm by removing the use of tracking error integrators and an anti-windup algorithm. These two features constitute the useful closed-loop properties called the performance recovery and offset-free properties. Numerical simulation results confirm the efficacy of the proposed scheme, where a wind power system including the proposed controller is emulated using the PowerSIM software.

Keywords: power conversion; model–plant mismatches; disturbance observer; performance recovery; offset-free

---

1. Introduction

A DC/DC boost converter driven by pulse-width modulation (PWM) provides an acceptable output voltage and current regulation performance with a power factor correction property. Because of these two beneficial properties, the DC/DC boost converter has wide industrial applications, including variable home appliances, electrical vehicles, and solar/wind power systems [1–6].

Conventionally, the cascade-type output voltage regulator has primarily been adopted for DC/DC boost converter control systems where the outer-loop voltage control output is used as the reference signal for the inner-loop current controller [7]. These inner- and outer-loops can be implemented using a simple proportional-integral (PI) controller with two degree of freedom for each loop, and the resulting closed-loop performance can be adjusted through the frequency domain using the Bode and Nyquist techniques [7,8]. The feedback-linearization technique was applied by combining the classical PI scheme and the converter parameter dependent feed-forward compensation terms, in which the PI gains are tuned for the cut-off frequency of the closed-loop transfer function using the converter parameters [7]. Thus, the parameter identification accuracy critically affects the closed-loop performance. The parameter dependency can be reduced by also incorporating the gain-scheduling techniques in the control algorithms, as in [9,10].

It was reported that closed-loop performance improvement could be achieved by applying several advanced techniques, such as deadbeat [11], predictive [12], sliding mode [13], adaptive [14,15], model predictive [16–19], and robust controllers [20]. However, these advanced techniques still have the parameter dependency problem, and, even in the case of an adaptive controller, knowledge of the true inductance value is required. Recently, a sliding mode technique [21] was devised through a multi-variable approach and efficiently alleviated the chattering effect. The upper and lower bounds of the disturbances should be found using a trial-and-error procedure, which determines the feed-forward compensation terms dominating the disturbances coming from the load and parameter variations.

This paper presents a robust output voltage tracking controller for DC/DC boost converters, which considers the nonlinear dynamic behavior and model–plant mismatches arising from load and parameter variations. The proposed technique is devised through a multi-variable approach in the port-controlled Hamiltonian (PCH) framework introduced in [22]. This study made three contributions. First, a nonlinear disturbance observer (DOB) was constructed to exponentially estimate the disturbances given in the perturbed converter dynamical equations. Second, a proportional-type output voltage controller was devised by solving a partial differential equation (PDE) for the desired closed-loop energy function, including the DOB state variables. Third, it is rigorously proven that the closed-loop system driven by the proposed technique ensures two beneficial properties called the performance recovery and offset-free properties without the use of the tracking error integrators. These three contributions could simplify the control algorithms by: (1) reducing the dependency on converter information, such as the parameters and load current; and (2) removing the tracking error integral actions with anti-windup algorithms, which is a stark contrast to previous studies. Realistic simulations verify the effectiveness of the proposed technique by implementing a wind power system comprised of a wind turbine, permanent magnet synchronous generator (PMSG), and three-phase diode rectifier.

2. DC/DC Boost Converter Nonlinear Dynamics

The application of the averaging technique to the DC/DC boost converter depicted in Figure 1 leads to the nonlinear differential equations as [7]

$$L\dot{i}_L(t) = -(1-u(t))v_{dc}(t) + v_{in}(t), \tag{1}$$

$$C\dot{v}_{dc}(t) = (1-u(t))i_L(t) - i_{Load}(t), \forall t \geq 0, \tag{2}$$

where the averaged inductor current of $i_L(t)$ and output voltage of $v_{dc}(t)$ are treated as the state variables, and the duty ratio of $u(t)$ acts as the control input constrained in the closed-interval of $[0,1]$. The inductance and capacitance values are denoted as L and C, respectively. The input DC source voltage of $v_{in}(t)$ comes from a wind power system comprised of a wind turbine, PMSG, and rectifier, and the load current of $i_L(t)$ acts as the external disturbance.

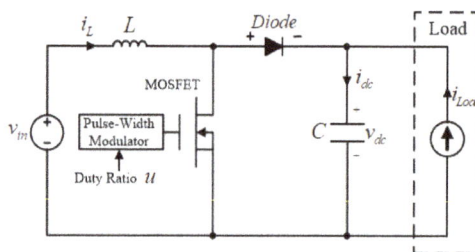

Figure 1. DC/DC boost converter topology.

Considering the wind power system implementations, the following assumptions are made:

- The true inductance and capacitance values are unknown but their nominal values, denoted as L_0 and C_0, are known.
- The input DC source voltage of $v_{in}(t)$ is time-varying but unknown except for its initial value, i.e., $v_{in,0} = v_{in}(0)$ is known.
- The load current of $i_{Load}(t)$ is unknown and time-varying.

- The inductor current of $i_L(t)$ and output voltage of $v_{dc}(t)$ are available for feedback.

The converter dynamics of Equations (1) and (2) show that the output voltage tracking control problem is not trivial because of the nonlinear terms presented in the inductor and output voltage dynamics and the unstable zero-dynamics. This paper tackles this difficulty by combining the passivity approach introduced in [22] and DOB techniques. For details, see the following section.

3. Output Voltage Tracking Controller Design

This section develops an output voltage tracking algorithm that allows the closed-loop output voltage behavior to be convergent to the low-pass filter (LPF) dynamics as

$$\frac{V_{dc}(s)}{V_{dc,ref}(s)} = \frac{\omega_{vc}}{s + \omega_{vc}}, \ \forall s \in \mathbb{C}, \tag{3}$$

with a desired cut-off frequency of $\omega_{vc} > 0$, where $V_{dc}(s)$ and $V_{dc,ref}(s)$ stand for the Laplace transforms of $v_{dc}(t)$ and $v_{dc,ref}(t)$, respectively. For this purpose, Section 3.1 analyzes the open-loop stability using a positive-definite energy function, which is used to constitute a PDE. Section 3.2 designs the PDE using the open-loop and closed-loop energy functions and proposes an output voltage tracking controller for solving the resulting PDE. Finally, Section 3.3 presents two useful closed-loop properties, called the performance recovery and offset-free properties, by analyzing the closed-loop dynamics.

3.1. Open-Loop Stability Analysis

First, to remove the true parameter dependency, rewrite the nonlinear dynamics of Equations (1) and (2) using the nominal converter parameters of L_0 and C_0 with the initial input DC source voltage of $v_{in,0}$ as:

$$L_0 \dot{i}_L(t) = -(1 - u(t))v_{dc}(t) + v_{in,0} + d_{L,o}(t), \tag{4}$$
$$C_0 \dot{v}_{dc}(t) = (1 - u(t))i_L(t) + d_{v,o}(t), \ \forall t \geq 0, \tag{5}$$

with $d_{L,o}(t)$ and $d_{v,o}(t)$ being unknown lumped disturbances caused by the model–plant mismatches, which can be written in a vector form:

$$\mathbf{M}\dot{\mathbf{x}}(t) = \mathbf{J}(u(t))\mathbf{M}^{-1}\nabla H(\mathbf{x}(t)) + \mathbf{g} + \mathbf{d}_o(t), \ \forall t \geq 0, \tag{6}$$

where the state vector of $\mathbf{x}(t)$ is defined as $\mathbf{x}(t) := \begin{bmatrix} i_L(t) & v_{dc}(t) \end{bmatrix}^T$, $\nabla H(\mathbf{x}(t))$ denotes the gradient of the open-loop energy function of $H(\mathbf{x}(t))$ given by $H(\mathbf{x}(t)) := \frac{1}{2}\mathbf{x}^T(t)\mathbf{M}\mathbf{x}(t), \forall t \geq 0$ with the positive definite matrix of $\mathbf{M} := \text{diag}\{L_0, C_0\}$, and the rest of the system matrices are defined as

$$\mathbf{J}(u(t)) := \begin{bmatrix} 0 & -(1 - u(t)) \\ (1 - u(t)) & 0 \end{bmatrix}, \ \mathbf{g} := \begin{bmatrix} v_{in,0} \\ 0 \end{bmatrix}, \ \mathbf{d}_o(t) := \begin{bmatrix} d_{L,o}(t) \\ d_{v,o}(t) \end{bmatrix}, \ \forall t \geq 0.$$

The open-loop stability can easily be seen using the open-loop energy function of $H(\mathbf{x}(t))$ and the state equation of Equation (6) as

$$\begin{aligned} \dot{H}(\mathbf{x}(t)) &= \nabla H^T(\mathbf{x}(t))\dot{\mathbf{x}}(t) = \mathbf{x}^T(t)\mathbf{M}\left(\mathbf{M}^{-1}\left[\mathbf{J}(u(t))\mathbf{M}^{-1}\nabla H(\mathbf{x}(t)) + \mathbf{g} + \mathbf{d}_o(t)\right]\right) \\ &= \mathbf{x}^T(t)\left(\mathbf{J}(u(t))\mathbf{x}(t) + \mathbf{g} + \mathbf{d}_o(t)\right) = \mathbf{x}^T(t)(\mathbf{g} + \mathbf{d}_o(t)), \ \forall t \geq 0, \end{aligned}$$

which shows the passivity for the input-output mapping of $(\mathbf{g} + \mathbf{d}_o(t)) \mapsto \mathbf{x}(t)$.

3.2. Controller Design

The control problem for the target dynamics of Equation (3) can be solved by deriving a control law that forces the closed-loop output voltage trajectory of $v_{dc}(t)$ to exponentially converge to the desired trajectory of $v_{dc}^*(t)$ governed by

$$\dot{v}_{dc}^*(t) = \omega_{vc}\left(v_{dc,ref}(t) - v_{dc}^*(t)\right), \ \forall t \geq 0, \tag{7}$$

because the dynamical Equation (7) is the inverse Laplace transform of Equation (3). To this end, define the tracking error vector as $\tilde{\mathbf{x}}(t) := \mathbf{x}_{ref}(t) - \mathbf{x}(t) = \begin{bmatrix} \tilde{i}_L(t) & \tilde{v}_{dc}^*(t) \end{bmatrix}^T$ with the reference vector of $\mathbf{x}_{ref}(t) := \begin{bmatrix} i_{L,ref}(t) & v_{dc}^*(t) \end{bmatrix}^T$, where $i_{L,ref}(t)$ refers to the inductor current reference determined later. Then, the tracking error dynamics are obtained as

$$
\begin{aligned}
\mathbf{M}\dot{\tilde{\mathbf{x}}}(t) &= \mathbf{M}(\dot{\mathbf{x}}_{ref}(t) - \dot{\mathbf{x}}(t)) \\
&= -\mathbf{J}(u(t))\mathbf{M}^{-1}\nabla H(\mathbf{x}(t)) - \mathbf{g} + \mathbf{d}(t), \ \forall t \geq 0,
\end{aligned}
\tag{8}
$$

where the disturbance vector of $\mathbf{d}(t)$ is defined as $\mathbf{d}(t) := \mathbf{M}\dot{\mathbf{x}}_{ref}(t) - \mathbf{d}_o(t), \ \forall t \geq 0$. Now, consider the closed-loop positive definite energy function given by

$$H_{cl}(\tilde{\mathbf{x}}(t)) := \frac{1}{2}\tilde{\mathbf{x}}^T(t)\mathbf{M}\tilde{\mathbf{x}}(t), \ \forall t \geq 0, \tag{9}$$

which gives its time-derivative along the trajectory of Equation (8):

$$
\begin{aligned}
\dot{H}_{cl}(\tilde{\mathbf{x}}(t)) &= \nabla H_{cl}^T(\tilde{\mathbf{x}}(t))\dot{\tilde{\mathbf{x}}}(t) \\
&= \nabla H_{cl}^T(\tilde{\mathbf{x}}(t))\mathbf{M}^{-1}\left(-\mathbf{J}(u(t))\mathbf{M}^{-1}\nabla H(\mathbf{x}(t)) - \mathbf{g} + \mathbf{d}(t)\right), \ \forall t \geq 0.
\end{aligned}
\tag{10}
$$

Through a further analysis using Lemma 1, it can be proven that a useful inequality of

$$\dot{H}_{cl}(\tilde{\mathbf{x}}(t)) \leq -\alpha_{cl}H_{cl}(\tilde{\mathbf{x}}(t)) + \tilde{\mathbf{d}}^T(t)\tilde{\mathbf{x}}(t), \ \forall t \geq 0, \tag{11}$$

holds for some $\alpha_{cl} > 0$ where $\tilde{\mathbf{d}}(t) := \mathbf{d}(t) - \hat{\mathbf{d}}(t)$ with $\hat{\mathbf{d}}(t) = \begin{bmatrix} \hat{d}_L(t) & \hat{d}_v(t) \end{bmatrix}^T$ being the estimated disturbance vector, $\forall t \geq 0$, if the PDE of

$$-\mathbf{J}(u(t))\mathbf{M}^{-1}\nabla H(\mathbf{x}(t)) - \mathbf{g} = \left(\mathbf{J}_{cl}(u(t)) - \mathbf{R}_{cl}\right)\mathbf{M}^{-1}\nabla H_{cl}(\tilde{\mathbf{x}}(t)) - \hat{\mathbf{d}}(t), \ \forall t \geq 0, \tag{12}$$

is solvable for some skew-symmetric matrix of $\mathbf{J}_{cl}(u(t))$ and positive definite matrix of $\mathbf{R}_{cl}$.

The PDE of Equation (12) can be solved using the proposed control law $u(t)$, with the inductor current reference $i_{L,ref}(t)$ given by

$$u(t) = \frac{1}{v_{dc}^*(t)}\left(L_0 k_{cc}\tilde{i}_L(t) + v_{dc}^*(t) - v_{in,0} + \hat{d}_L(t)\right), \tag{13}$$

$$i_{L,ref}(t) = \frac{1}{1-u(t)}\left(C_0 k_{vc}\tilde{v}_{dc}^*(t) + \hat{d}_v(t)\right), \ \forall t \geq 0, \tag{14}$$

with $\mathbf{J}_{cl}(u(t)) = \mathbf{J}(u(t))$ and $\mathbf{R}_{cl} = \mathrm{diag}\left\{L_0 k_{cc}, C_0 k_{vc}\right\}$ for any given tuning parameters of $k_{cc} > 0$ and $k_{vc} > 0$. Meanwhile, the estimated disturbance vector of $\hat{\mathbf{d}}(t)$ is given by

$$\hat{\mathbf{d}}(t) = \mathbf{z}(t) + \mathbf{L}\mathbf{M}\tilde{\mathbf{x}}(t), \ \forall t \geq 0, \tag{15}$$

with the diagonal DOB gain matrix of $\mathbf{L} = \mathrm{diag}\{l_{cc}, l_{vc}\} > 0$. The DOB state vector of $\mathbf{z}(t)$ is updated as

$$\dot{\mathbf{z}}(t) = -\mathbf{L}\mathbf{z}(t) - \mathbf{L}^2\mathbf{M}\tilde{\mathbf{x}}(t) + \mathbf{L}\Big(\mathbf{J}(u(t))\mathbf{x}(t) + \mathbf{g}\Big), \ \forall t \geq 0, \tag{16}$$

Lemma 1 presents a beneficial inequality of Equation (11) to derive a closed-loop property through investigating the closed-loop energy function behavior driven by the proposed controller of Equation (13) with the inductor current reference of Equation (14). The proof is given in the Appendix A.

Lemma 1. *For any given $k_x > 0$, $x = cc$, vc, the proposed controller of Equation (13) with the inductor current reference of Equation (14) solves the PDE of Equation (12) such that the inequality of Equation (11) holds true.*

3.3. Closed-Loop Property Analysis

This subsection rigorously analyzes the closed-loop properties. First, Theorem 1 derives a closed-loop property, called the performance recovery property, which includes the DOB output of Equation (15) and the DOB state equation of Equation (16) based on the inequality of Equation (11) derived by Lemma 1. The Appendix A presents the proof of Theorem 1.

Theorem 1. *Under the assumption of Lemma 1, for any $k_x > 0$ and l_x, $x = cc$, vc, the proposed controller of Equation (13) with the inductor current reference of Equation (14) and DOB of Equations (15) and (16) ensures strict passivity for the input–output mapping:*

$$\mathbf{w}(t) \mapsto \mathbf{y}(t), \tag{17}$$

where $\mathbf{w}(t) := \begin{bmatrix} 0 & 0 & (\Gamma\dot{\mathbf{d}}(t))^T \end{bmatrix}^T$ and $\mathbf{y}(t) := \begin{bmatrix} \tilde{\mathbf{x}}^T(t) & \tilde{\mathbf{d}}^T(t) \end{bmatrix}^T$ for some $\Gamma = \Gamma^T > 0$, $\forall t \geq 0$.

The proof of Theorem 1 can be accomplished by showing that the composite-type positive-definite function of $V(t)$ defined as

$$V(\tilde{\mathbf{x}}(t), \tilde{\mathbf{d}}(t)) := H_{cl}(\tilde{\mathbf{x}}(t)) + \frac{1}{2}\tilde{\mathbf{d}}^T(t)\Gamma\tilde{\mathbf{d}}(t), \ \forall t \geq 0, \tag{18}$$

with a positive definite weighting matrix of Γ gives

$$\dot{V}(\tilde{\mathbf{x}}(t), \tilde{\mathbf{d}}(t)) \leq -\beta V(\tilde{\mathbf{x}}(t), \tilde{\mathbf{d}}(t)) + \mathbf{w}^T(t)\mathbf{y}(t), \ \forall t \geq 0, \tag{19}$$

for some $\beta > 0$. For details, see the Appendix A. The resulting inequality of Equation (19) implies that the closed-loop system driven by the proposed control algorithm exponentially recovers the target output voltage tracking performance of Equation (7) as the disturbance vector of $\mathbf{d}(t)$ exponentially reaches its steady state, i.e, $v_{dc}(t) \to v_{dc}^*(t)$ as $\dot{\mathbf{d}}(t) \to 0$, exponentially.

Theorem 2 shows that the closed-loop system does not suffer from an offset error despite the absence of the tracking error integrators in the proposed controller, thanks to the DOB dynamics of Equations (15) and (16). This property is called the offset-free property to simplify the controller by removing the additional anti-windup algorithms. The proof is given in the Appendix A.

Theorem 2. *The closed-loop system controlled by the proposed control algorithm of Equations (13)–(16) always removes the output voltage steady state error, i.e, $v_{dc}(\infty) = v_{dc,ref}(\infty)$ where $v_{dc}(\infty)$ and $v_{dc,ref}(\infty)$ denote the steady states of $v_{dc}(t)$ and $v_{dc,ref}(t)$, respectively.*

4. Simulation Results

This section describes the simulation results of a wind power system that includes the DC/DC boost converter to numerically demonstrate the effectiveness of the proposed technique. Section 4.1 gives the simulation setup. The numerical verification results are presented in Section 4.2. Section 4.3 concludes this section by discussing the numerical verification results.

4.1. Simulation Setup

The wind power system was emulated by the powerSIM (PSIM) software using its wind turbine and permanent magnet synchronous machine (PMSM) model. The following values were selected for the nominal output power, inertia, base wind and rotational speed, and initial rotational speed of the wind turbine: 10-kW, 0.1 kg·m², 20 m/s, 50 rpm, and 10 rpm, respectively. The PMSM parameters were chosen as $R_s = 0.099 \ \Omega$ (stator resistance), $L_d = L_q = 4.07$ mH (d-q inductances), $\lambda_{PM} = 0.3166$ Wb (flux), $P = 40$ (number of poles), $J = 0.12$ kg·m² (inertia), $B = 0.000425$ Nm/rad/s (viscous damping). The Weibull distribution-based wind model [23] was used to randomly determine the wind speed for the wind turbine, which is shown in Figure 2. As components of the DC/DC boost converter, the inductance of L and the capacitance of C were selected as

$$L = 460 \ \mu H, \ C = 470 \ \mu F, \tag{20}$$

and their nominal values were determined to be

$$L_0 = 0.5L, \ C_0 = 1.5C, \tag{21}$$

to take the model–plant mismatches into account, where the input DC source voltage was supplied by the PMSM with a three-phase diode rectifier. Figure 3 shows the emulated wind power system configuration, whose output DC-link voltage of $v_{dc}(t)$ was controlled by the DC/DC boost converter. The control algorithms were implemented using the dynamic link library (DLL) block written in the C language, where the pulse-width modulation (PWM) and control periods were chosen to be synchronized to 0.1 ms.

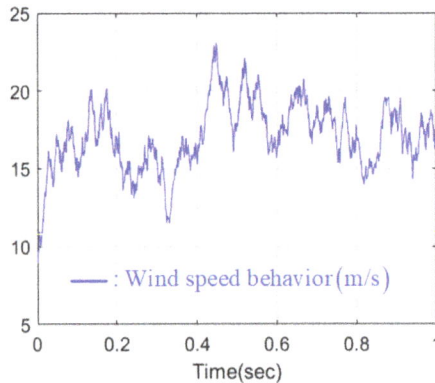

Figure 2. Wind speed pattern from Weibull distribution-based wind model.

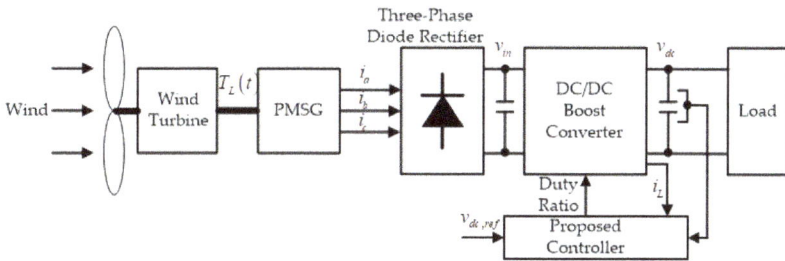

Figure 3. Wind power system configuration.

For comparison, the feedback linearization (FL) technique was considered [7], which is given as

$$u(t) = 2L_0\omega_{cc}\tilde{i}_L(t) + L_0\omega_{cc}^2 \int_0^t \tilde{i}_L(\tau)d\tau, \ \forall t \geq 0, \tag{22}$$

with the inductor current tracking error of $\tilde{i}_L(t) := i_{L,ref}(t) - i_L(t), \forall t \geq 0$, where the inductor current reference of $i_{L,ref}(t)$ is updated by the outer-loop voltage regulator as

$$i_{L,ref}(t) = 2C_0\omega_{vc}\tilde{v}_{dc}(t) + C_0\omega_{vc}^2 \int_0^t \tilde{v}_{dc}(\tau)d\tau, \ \forall t \geq 0,, \tag{23}$$

with the output voltage tracking error of $\tilde{v}_{dc}(t) := v_{dc,ref}(t) - v_{dc}(t), \forall t \geq 0$. The resulting closed-loop system controlled by the FL technique gives the closed-loop transfer function

$$\frac{I_L(s)}{I_{L,ref}(s)} \approx \frac{\omega_{cc}}{s + \omega_{cc}}, \ \frac{V_{dc}(s)}{V_{dc,ref}(s)} \approx \frac{\omega_{vc}}{s + \omega_{vc}}, \ \forall s \in \mathbb{C}, \tag{24}$$

as long as the nominal parameters of L_0 and C_0 exactly match their true values of L and C for all operating points, where $I_L(s)$, $I_{L,ref}(s)$, $V_{dc}(s)$, and $V_{dc,ref}(s)$ represent the Laplace transforms of $i_L(t)$, $i_{L,ref}(t)$, $v_{dc}(t)$, and $v_{dc,ref}(t)$, respectively. It is easy to see that the control objective of the FL technique is the same as that of the proposed technique. The FL controller of Equations (22) and (23) was implemented using the nominal parameters of L_0 and C_0 with the cut-off frequencies of $\omega_{cc} = 2\pi f_{cc} = 600\pi$ rad/s and $\omega_{vc} = 2\pi f_{vc} = 2\pi$ rad/s, i.e., $f_{cc} = 300$ Hz and $f_{vc} = 4$ Hz.

The proposed algorithm was also constructed using the nominal parameters of L_0 and C_0, where the cut-off frequency of ω_{vc} was set the same as the FL controller. The design parameters of k_{cc} and k_{vc} and DOB gains of l_{cc} and l_{vc} were tuned as $k_{cc} = \omega_{cc}$, $k_{vc} = 95$, $l_{cc} = 62.8$, and $l_{vc} = 62.8$, respectively, which are summarized in Figure 4.

	Control Gains			DOB Gains	
	ω_{vc}	k_{cc}	k_{vc}	l_{cc}	l_{vc}
Values	6.28	1884	95	62.8	62.8

Figure 4. Simulation parameter summary table.

4.2. Simulation Results

The first simulation evaluates the output voltage tracking performance for a time-varying output voltage reference that was increased from $v_{dc,ref}(t) = 250$ V to $v_{dc,ref}(t) = 350$ V and afterwards was restored to $v_{dc,ref}(t) = 250$ V. This simulation was performed for three-types of resistive loads ($R_L = 30, 60, 100 \, \Omega$) to evaluate the closed-loop robustness against load variations. The resulting closed-loop output voltage behaviors are shown in Figure 5, and the trajectories of the estimated disturbances from the DOBs are depicted in Figure 6. Figure 7 shows the corresponding input DC voltage and PMSM speed responses that originated from the wind turbine and velocity. These results indicate that the proposed technique precisely assigned the desired output voltage tracking performance to the closed-loop system in the presence of model–plant mismatches and load variations, thanks to the DOBs.

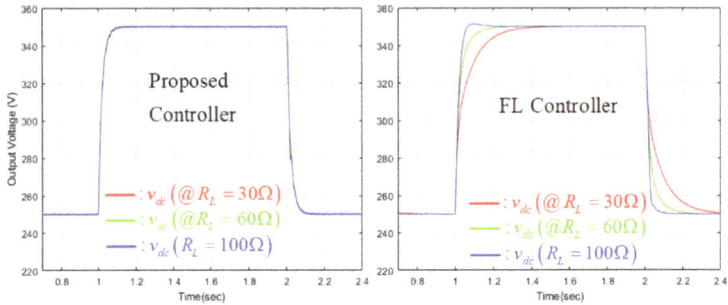

Figure 5. Output voltage tracking performance change behaviors for $R_L = 30, 60, 100 \, \Omega$.

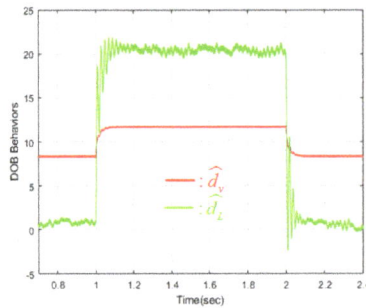

Figure 6. Estimated disturbance behaviors.

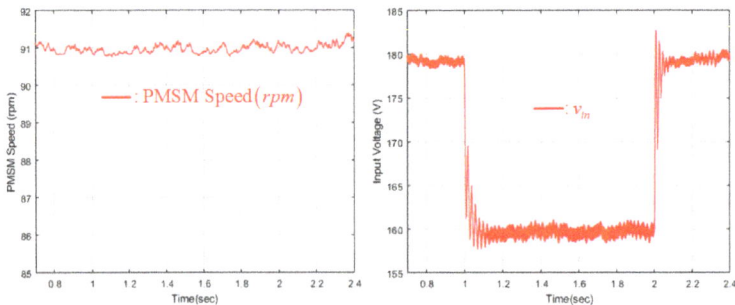

Figure 7. Permanent magnet synchronous machine (PMSM) speed and input voltage behaviors.

The second simulation investigates the output voltage regulation performance under a sudden load variation, where the resistive load of R_L was changed from $R_L = 60\ \Omega$ to $R_L = 30\ \Omega$, and it was restored to $R_L = 60\ \Omega$ in a sequential manner. The resulting closed-loop responses are depicted in Figure 8, which observes that the proposed technique considerably enhanced the output voltage regulation performance by speeding up the output voltage restoring rate with a rapid inductor current response. This feature was also obtained because the DOB exponentially estimated the disturbances coming from load current variations and model–plant mismatches.

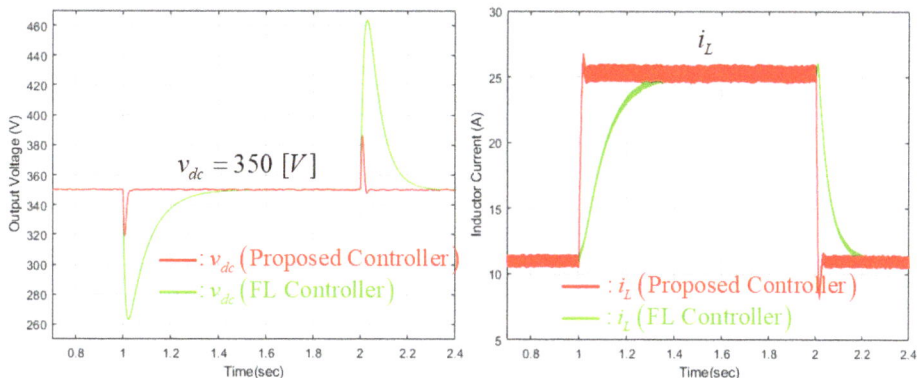

Figure 8. Output voltage regulation performance comparison for sudden load change from $R_L = 60\ \Omega$ to $R_L = 30\ \Omega$.

The last simulation examines the output voltage tracking performance change behaviors as increasing the cut-off frequency of f_{vc} to $f_{vc} = 0.7, 2, 4$ Hz. In this simulation, the output voltage reference was increased from $v_{dc,ref}(t) = 250$ V to $v_{dc,ref}(t) = 350$ V under a resistive load of $R_L = 30\ \Omega$. The comparison results are presented in Figure 9, which indicates that the closed-loop output voltage tracking performance was precisely adjusted by the proposed technique using a fixed control parameter set in the presence of model–plant mismatches.

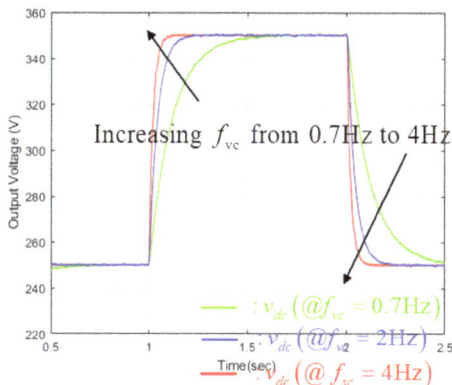

Figure 9. Output voltage tracking performances for several cut-off frequencies, $f_{vc} = 0.7, 2, 4$ Hz.

4.3. Discussion

These numerical verifications confirmed the beneficial closed-loop properties proven in Theorems 1 and 2, and this section clearly shows that a considerable output voltage tracking and regulation performance improvement can be obtained using the proposed technique in the presence of input voltage variations caused by wind speed changes. Therefore, it can be concluded that the proposed technique is qualified as a promising solution for wind power system applications.

5. Conclusions

For DC/DC boost converter applications, this paper suggests a passivity-based proportional-type output voltage tracking control algorithm incorporating the DOB under the PCH framework, which results in a classical cascade structure. The proposed control algorithm was derived by solving a PDE so that the closed-loop system had the desired positive-definite energy function. Moreover, it is shown that the proposed controller guarantees the performance recovery property by rendering the closed-loop energy function to be decreased exponentially, and it also ensures the offset-free property in the absence of tracking error integrators by analyzing the closed-loop steady-state equations. The beneficial closed-loop properties were numerically verified by emulating a wind power system equipped with a wind turbine, PMSM, three-phase diode rectifier, and DC/DC boost converter. In this study, the closed-loop cut-off frequency was manually found through a trial and error process, and it was fixed for all time. An auto-tuning mechanism for the closed-loop cut-off frequency will be devised and experimentally verified in a future study.

Acknowledgments: This research was supported by Basic Science Research Program through the National Research Foundation of Korea(NRF) funded by the Ministry of Education(2018R1A6A1A03026005).

Conflicts of Interest: The authors declare no conflict of interest.

Appendix A

Proofs for Lemma 1, Theorem 1, and Theorem 2 are presented in this section, sequentially. First, Lemma 1 is proven as:

Proof. Letting $\mathbf{J}_{cl} = \begin{bmatrix} 0 & J_{cl,1} \\ J_{cl,2} & 0 \end{bmatrix}$ and $\mathbf{R}_{cl} = \text{diag}\{R_{cl,1}, R_{cl,2}\} > 0$, the PDE of (12) can be written as

$$\begin{bmatrix} 0 & (1-u) \\ -(1-u) & 0 \end{bmatrix} \begin{bmatrix} i_L \\ v_{dc} \end{bmatrix} - \begin{bmatrix} v_{in,0} \\ 0 \end{bmatrix}$$

$$= \left(\begin{bmatrix} 0 & J_{cl,1} \\ J_{cl,2} & 0 \end{bmatrix} - \begin{bmatrix} R_{cl,1} & 0 \\ 0 & R_{cl,2} \end{bmatrix} \right) \begin{bmatrix} \tilde{i}_L \\ \tilde{v}_{dc}^* \end{bmatrix} - \begin{bmatrix} \hat{d}_L(t) \\ \hat{d}_v(t) \end{bmatrix},$$

$\forall t \geq 0$, which gives the two equations of

$$(1-u)v_{dc} - v_{in,0} = J_{cl,1}\tilde{v}_{dc}^* - R_{cl,1}\tilde{i}_L - \hat{d}_L, \tag{A1}$$

$$-(1-u)i_L = J_{cl,2}\tilde{i}_L - R_{cl,2}\tilde{v}_{dc}^* - \hat{d}_v, \ \forall t \geq 0. \tag{A2}$$

Then, together with $J_{cl,2} := (1-u)$ and $R_{cl,2} := C_0 k_{vc}$, Equation (A2) yields the inductor current reference of Equation (14), and the control law of Equation (13) is obtained by setting $J_{cl,1}$ and $R_{cl,1}$ as $J_{cl,1} := -(1-u)$ and $R_{cl,1} := L_0 k_{cc}$ for Equation (A1). Therefore, the proposed controller of Equation (13) with the inductor current reference of Equation (14) is a solution to the

PDE of Equation (12). It is easy to see that the time-derivative of H_{cl} can be obtained by combining Equations (10) and (12) as

$$
\begin{aligned}
\dot{H}_{cl} &= \nabla H_{cl}^T \mathbf{M}^{-1}\left(\mathbf{J}_{cl} - \mathbf{R}_{cl}\right)\mathbf{M}^{-1}\nabla H_{cl} + \tilde{\mathbf{d}}\right) \\
&= -\tilde{\mathbf{x}}^T \mathbf{R}_{cl}\tilde{\mathbf{x}} + \tilde{\mathbf{d}}^T \tilde{\mathbf{x}} \\
&\leq -\alpha_{cl}H_{cl} + \tilde{\mathbf{d}}^T \tilde{\mathbf{x}}, \ \forall t \geq 0,
\end{aligned}
\tag{A3}
$$

with $\alpha_{cl} := \frac{2\lambda_{min}(\mathbf{R}_{cl})}{\lambda_{max}(\mathbf{M})}$ where $\lambda_{min}((\cdot))$ and $\lambda_{max}((\cdot))$ represent the minimum and maximum eigenvalues of the square matrix of $(\cdot)$ which satisfies that $\lambda_{min}((\cdot))\|\mathbf{x}\|^2 \leq \mathbf{x}^T(\cdot)\mathbf{x} \leq \lambda_{max}((\cdot))\|\mathbf{x}\|^2$ for any vector of $\mathbf{x} \in \mathbb{R}^n$. Therefore, the inequality of Equation (11) holds true. □

Second, Theorem 1 is proven as:

Proof. The substitution of the DOB output in Equation (15) to the DOB state equation of Equation (16) yields

$$
\dot{\hat{\mathbf{d}}} - \mathbf{LM}\dot{\tilde{\mathbf{x}}} = -\mathbf{L}(\hat{\mathbf{d}} - \mathbf{LM}\tilde{\mathbf{x}}) - \mathbf{L}^2\mathbf{M}\tilde{\mathbf{x}} + \mathbf{L}(\mathbf{Jx} + \mathbf{g}), \ \forall t \geq 0,
$$

which is equivalent to

$$
\dot{\tilde{\mathbf{d}}} = -\mathbf{L}\tilde{\mathbf{d}} + \dot{\mathbf{d}}, \ \forall t \geq 0,
\tag{A4}
$$

since it holds that $\mathbf{d} = \mathbf{M}\dot{\tilde{\mathbf{x}}} + \mathbf{J}\mathbf{M}^{-1}\nabla H + \mathbf{g} = \mathbf{M}\dot{\tilde{\mathbf{x}}} + \mathbf{Jx} + \mathbf{g}$ (See Equation (8)). Consider the positive definite function of Equation (A5) as

$$
V = H_{cl} + \frac{1}{2}\tilde{\mathbf{d}}^T\Gamma\tilde{\mathbf{d}}, \ \forall t \geq 0,
\tag{A5}
$$

with a positive definite weighting matrix of $\Gamma := \mathrm{diag}\{\gamma_{cc}, \gamma_{vc}\} > 0$ determined later, which gives

$$
\begin{aligned}
\dot{V} &= \dot{H}_{cl} + \tilde{\mathbf{d}}^T\Gamma\dot{\tilde{\mathbf{d}}} \\
&= -\alpha_{cl}H_{cl} + \tilde{\mathbf{d}}^T\tilde{\mathbf{x}} - \tilde{\mathbf{d}}^T\Gamma\mathbf{L}\tilde{\mathbf{d}} + \tilde{\mathbf{d}}^T\Gamma\dot{\mathbf{d}} \\
&\leq -\frac{\alpha_{cl}}{2}H_{cl} - \tilde{\mathbf{d}}^T(\Gamma\mathbf{L} - \frac{1}{2\alpha_{cl}\lambda_{min}(\mathbf{M})})\tilde{\mathbf{d}} + \dot{\mathbf{d}}^T\Gamma\dot{\mathbf{d}}, \ \forall t \geq 0,
\end{aligned}
\tag{A6}
$$

where the inequality of Equation (A3) and the DOB dynamics of Equation (A4) are used for the second equality, and the last inequality is obtained by the Young's inequality of $\mathbf{x}^T\mathbf{y} \leq \frac{\epsilon}{2}\|\mathbf{x}\|^2 + \frac{1}{2\epsilon}\|\mathbf{y}\|^2$, $\forall \mathbf{x}, \mathbf{y} \in \mathbb{R}^n, \forall \epsilon > 0$. Then, the weighting matrix of $\Gamma := \mathbf{L}^{-1}(\frac{1}{2} + \frac{1}{2\alpha_{cl}\lambda_{min}(\mathbf{M})})$ renders for $\dot{V}$ to be

$$
\begin{aligned}
\dot{V} &\leq -\frac{\alpha_{cl}}{2}H_{cl} - \frac{1}{2}\|\tilde{\mathbf{d}}\|^2 + \dot{\mathbf{d}}^T\Gamma\dot{\mathbf{d}} \\
&\leq -\beta V + \mathbf{w}^T\mathbf{y}, \ \forall t \geq 0,
\end{aligned}
\tag{A7}
$$

which indicates the strict passivity of the input-output mapping of Equation (17), where $\beta := \min\{\frac{\alpha_{cl}}{2}, \frac{1}{\lambda_{max}(\Gamma)}\}$. □

Finally, Theorem 2 is proven as:

Proof. The closed-loop tracking error dynamics can be obtained by combining the tracking error dynamics of Equation (8) and the PDE of Equation (12) as

$$
\mathbf{M}\dot{\tilde{\mathbf{x}}} = (\mathbf{J}_{cl} - \mathbf{R}_{cl})\tilde{\mathbf{x}} + \tilde{\mathbf{d}}, \ \forall t \geq 0,
\tag{A8}
$$

which gives the simplified steady-state equation

$$0 = (\mathbf{J}_{cl} - \mathbf{R}_{cl})\tilde{\mathbf{x}}(\infty) \tag{A9}$$

since it always holds that $\tilde{\mathbf{d}}(\infty) = \mathbf{0}$ in the steady-state (see Equation (A4)), where $\tilde{\mathbf{x}}(\infty)$ and $\tilde{\mathbf{d}}(\infty)$ denote the steady states of $\tilde{\mathbf{x}}(t)$ and $\tilde{\mathbf{d}}(t)$, respectively. Furthermore, it follows from the skew-symmetricity of the matrix $\mathbf{J}_{cl}$, i.e., $\mathbf{J}_{cl} = -\mathbf{J}_{cl}^T$, that

$$\begin{aligned}0 &= \tilde{\mathbf{x}}^T(\infty)(\mathbf{J}_{cl} - \mathbf{R}_{cl})\tilde{\mathbf{x}}(\infty)\\ &= -\tilde{\mathbf{x}}^T(\infty)\mathbf{R}_{cl}\tilde{\mathbf{x}}(\infty),\end{aligned}$$

which shows that $\tilde{\mathbf{x}}(\infty) = \mathbf{0}$ because the matrix $\mathbf{R}_{cl}$ is positive definite. Therefore, it concludes that $\tilde{v}_{dc}(\infty) = v_{dc}^*(\infty) = v_{dc,ref}(\infty)$. $\square$

References

1. Zhai, L.; Zhang, T.; Cao, Y.; Yang, S.; Kavuma, S.; Feng, H. Conducted EMI Prediction and Mitigation Strategy Based on Transfer Function for a High-Low Voltage DC-DC Converter in Electric Vehicle. *Energies* **2018**, *11*, 1028. [CrossRef]
2. Tran, V.T.; Nguyen, M.K.; Choi, Y.O.; Cho, G.B. Switched-Capacitor-Based High Boost DC-DC Converter. *Energies* **2018**, *11*, 987. [CrossRef]
3. Padmanaban, S.; Bhaskar, M.S.; Maroti, P.K.; Blaabjerg, F.; Fedak, V. An Original Transformer and Switched-Capacitor (T & SC)-Based Extension for DC-DC Boost Converter for High-Voltage/Low-Current Renewable Energy Applications: Hardware Implementation of a New T & SC Boost Converter. *Energies* **2018**, *11*, 783. [CrossRef]
4. Park, Y.J.; Khan, Z.H.N.; Oh, S.J.; Jang, B.G.; Ahmad, N.; Khan, D.; Abbasizadeh, H.; Shah, S.A.A.; Pu, Y.G.; Hwang, K.C.; et al. Single Inductor-Multiple Output DPWM DC-DC Boost Converter with a High Efficiency and Small Area. *Energies* **2018**, *11*, 725. [CrossRef]
5. Bi, H.; Wang, P.; Wang, Z. Common Grounded H-Type Bidirectional DC-DC Converter with a Wide Voltage Conversion Ratio for a Hybrid Energy Storage System. *Energies* **2018**, *11*, 349. [CrossRef]
6. Zhang, S.; Wang, Y.; Chen, B.; Han, F.; Wang, Q. Studies on a Hybrid Full-Bridge/Half-Bridge Bidirectional CLTC Multi-Resonant DC-DC Converter with a Digital Synchronous Rectification Strategy. *Energies* **2018**, *11*, 227. [CrossRef]
7. Erickson, R.W.; Maksimovic, D. *Fundamentals of Power Electronics*, 2nd ed.; Springer: New York, NY, USA, 2001.
8. Alexander, G.P.; Feng, G.; Yan-Fei, L.; Paresh, C.S. A Design Method for PI-like Fuzzy Logic Controllers for DC/DC Converter. *IEEE Trans. Ind. Electron.* **2007**, *54*, 2688–2696.
9. Olalla, C.; Leyva, R.; Queinnec, I.; Maksimovic, D. Robust Gain-Scheduled Control of Switched-Mode DC-DC Converters. *IEEE Trans. Power Electron.* **2012**, *27*, 3006–3019. [CrossRef]
10. Su, J.T.; Liu, C.W. Gain scheduling control scheme for improved transient response of DC-DC converters. *IET Power Electron.* **2012**, *5*, 678–692. [CrossRef]
11. Bibian, S.; Jin, H. High Performance Predictive Dead-Beat Digital Controller for DC Power Supplies. *IEEE Trans. Power Electron.* **2002**, *17*, 420–427. [CrossRef]
12. Zhang, Q.; Min, R.; Tong, Q.; Zou, X.; Liu, Z.; Shen, A. Sensorless Predictive Current Controlled DC-DC Converter With a Self-Correction Differential Current Observer. *IEEE. Trans. Ind. Electron.* **2014**, *61*, 6747–6757. [CrossRef]
13. Salimi, M. Sliding Mode Control of the DC-DC Fly back Converter with Zero Steady-State Error. *J. Basic Appl. Sci. Res.* **2012**, *2*, 10693–10705.
14. Oucheriah, S.; Guo, L. PWM-Based Adaptive Sliding-Mode Control for Boost DC/DC Converters. *IEEE Trans. Ind. Electron.* **2013**, *60*, 3291–3294. [CrossRef]
15. Linares-Flores, J.; Mendez, A.H.; G.-Rodriguez, C.; S.-Ramirez, H. Robust Nonlinear Adaptive Control of a Boost Converter via Algebraic Parameter Identification. *IEEE Trans. Ind. Electron.* **2014**, *61*, 4105–4114. [CrossRef]

16. Kim, S.K.; Park, C.R.; Lee, Y.I. A Stabilizing Model Predictive Controller for Voltage Regulation of a DC/DC Boost Converter. *IEEE Trans. Control Syst. Technol.* **2014**, *41*, 2107–2114. [CrossRef]
17. Kim, S.K.; Kim, J.S.; Park, C.R.; Lee, Y.I. Output-feedback model predictive controller for voltage regulation of a DC/DC converter. *IET Control Theory Appl.* **2013**, *7*, 1959–1968. [CrossRef]
18. Beccuti, A.G.; Mariethoz, S.; Cliquennois, S.; Wang, S.; Morari, M. Explicit Model Predictive Control of DC/DC Switched-Mode Power Supplies With Extended Kalman Filtering. *IEEE Trans. Ind. Electron.* **2009**, *56*, 1864–1874. [CrossRef]
19. Karamanakos, P.; Geyer, T.; Manias, S. Direct Voltage Control of DC-DC Boost Converters Using Enumeration-Based Model Predictive Control. *IEEE Trans. Power Electron.* **2014**, *29*, 968–978. [CrossRef]
20. Wang, Y.X.; Yu, D.H.; Kim, Y.B. Robust Time-Delay Control for the DC/DC Boost Converter. *IEEE Trans. Ind. Electron.* **2014**, *61*, 4829–4837. [CrossRef]
21. Wai, R.J.; Shih, L.C. Design of Voltage Tracking Control for DC-DC Boost Converter Via Total Sliding-Mode Technique. *IEEE Trans. Ind. Electron.* **2011**, *58*, 2502–2511. [CrossRef]
22. Ortega, R.; van der Schaft, A.; Maschke, B.; Escobar, G. Interconnection and damping assignment passivity-based control of port-controlled Hamiltonian systems. *Automatica* **2002**, *38*, 585–596. [CrossRef]
23. Mathew, S. *Wind Energy: Fundamentals, Resource Analysis and Economics*; Springer: New York, NY, USA, 2006.

energies

MDPI

Article

Improvement of the Response Speed for Switched Reluctance Generation System Based on Modified PT Control

Xiaoshu Zan [1], Mingliang Cui [1], Dongsheng Yu [1,*], Ruidong Xu [1] and Kai Ni [2]

[1] Jiangsu Province Laboratory of Mining Electric and Automation,
China University of Mining and Technology, Xuzhou 221116, China;
zanxiaoshu@126.com (X.Z.); Cuiml1993@163.com (M.C.); ruidongxu@163.com (R.X.)

[2] Department of Electrical engineering & Electronic, University of Liverpool, Liverpool L69 3BX, UK;
k.ni@student.liverpool.ac.uk

* Correspondence: dongsiee@163.com

Received: 1 July 2018; Accepted: 19 July 2018; Published: 7 August 2018

Abstract: The Switched Reluctance Generator (SRG) is suitable for wind power generation due to its good reliability and robustness. However, The SRG system adopting the conventional control algorithm with Pulse Width Modulation (PWM) method has a drawback, low response speed. The pulse train (PT) control has been widely used in dc/dc power converters operating in the discontinuous conduction mode due to its advantages of simple implementation and fast response. In this paper, for the first time, the PT control method is modified and adopted for controlling the output voltage of SRG system in order to achieve fast response. The capacitor current on the output side is sampled and combined with the output voltage to select the pulse trains and the low frequency oscillation cased by PT can be suppressed by tuning the feedback coefficient of the capacitor current. Also, good performance can be guaranteed with a wide range of voltage regulations, fast response, and no overshoot. The experimental platform of an 8/6 SRG system is built, and the experimental results show that the PT control can be used for SRG system with good practicability.

Keywords: switched reluctance generator; capacitance current pulse train control; voltage ripple; capacitance current; feedback coefficient

1. Introduction

With the rapid development of the economy energy shortages have become an inevitable problem. As a clean and abundant energy source, wind energy has been widely used. Wind power has the good features of wide distribution, large reserves, cleanness, security, etc. Therefore, wind power generation has attracted general attention from countries all over the world. Following this trend, there is a rapid development in wind power generation technologies and the wind power industry [1–3].

Switched reluctance generators (SRG) have advantages such as simple structure, high fault-tolerant ability, high operating efficiency, and hard mechanical properties, which make them widely used in aerospace, mining, textile, papermaking, and other industries [4–6]. The rotor of the SRG has no winding, no brush, and no permanent magnet so that the manufacturing cost is low and no copper loss is encountered. Since the SRG power generation system can be operated synchronously with a wind turbine, it is not restricted by stability issues when the frequency is low. Therefore, even if the wind speed is low, it can also ensure that the system runs with a high efficiency of power generation through reasonable design and control. At the same time, the application of current direct drives has become the developing trend of wind power generation systems, the advantages of SRG are attracting more and more attention than those of other generators. In particular, the magnetic and electric circuits are independent of each other, and hence when a fault occurs in a phase, the system can still function by isolating the faulty phase. Therefore, it has high operational reliability and fault tolerance [7].

The traditional PID control strategy is the classical method of power generation control for switched reluctance generators (SRGs) [8]. However, under PID control, the output overshoot of SRG could be large under the speed regulation, and the transient response could be quite slow due to the error compensation network [9]. In addition, PID controller parameters are normally preset and fixed, so the control performance will deteriorate when the working conditions are changed. It has been pointed out that the Fuzzy control strategy is suitable for SRG systems [10]. The fuzzy control theory refers to a bionic controller based on fuzzy knowledge and rule inference [11]. According to certain rules of inference, it can achieve the control goal in a simple way [12]. However, the independent use of a fuzzy controller may lead to an unavoidable overshoot and larger steady-state error. The sliding-mode variable structure control could also be a competitor for the control of SRGs. As compared with other control strategies, sliding-mode variable structure control has the invariance of perturbation and external disturbances to the system, making it more widely used in cases with high requirements of reliability and robustness [13]. However, the phenomenon of inherent buffeting and the inevitable steady-state error result in poor steady-state performance.

The large fluctuation of electricity loads and the regular capacitor charging and discharging processes lead to high difficulty of control for the irregular excitation current. In addition, an output voltage fluctuation can occur, and the original characteristics of power generation are affected. Besides, more harmonic contents could be injected into the grid by the grid-connected inverter. Until now, there has been some research progress of reducing the output voltage ripple and enhancing the voltage stability of SRG power systems. Aiming at reducing the output voltage ripple of SRG, a capacitor filter, as designed in a previous paper [14], is used to effectively reduce the voltage harmonics, but the parameters of the filter are difficult to determine. The switched reluctance generation system can achieve better control of the output voltage by introducing the double-closed-loop PID control with variable parameters [15]. However, the controller only considers the output voltage and phase current as the control targets, and hence ignores the impacts of other states of the system, which compromises the dynamic performance of SRG [16]. In a previous paper [17], a sliding mode controller of variable structure based on a genetic algorithm was designed to eliminate the voltage ripple caused by factors such as speed and load which change during the operation. However, the introduced genetic algorithm optimization module makes the system complicated.

Recently, pulse train (PT) control has been proposed for switching dc–dc converters [18,19] operated in the discontinuous conduction mode (DCM). The PT control method can adjust the system output voltage by using two or more sets of combination of control pulses. Without the requirement of an error amplifier and corresponding compensation network, the PT controller has excellent control performance of a fast response and simple circuit structure [20–22]. PT control methods are also attempted to be used in converters operated under the continuous conduction mode (CCM). In a previous paper [23], a buck converter operated in CCM is carried out with PT control for the first time, the research results point out that when the equivalent series resistance (ESR) of output capacitance is small, the circuit gives the phenomenon of low frequency oscillation. Though increasing the output capacitance ESR can suppress the low frequency oscillation of the converter, the ripple of output voltage is increased. To broaden the application scope of PT control, a peak capacitance current pulse train (PCC-PT) control method is described [24]. For the application of PCC-PT in converters, both excellent steady-state and transient response characteristics are revealed.

As a new contribution, a modified CC-PT control method is put forward based on the analysis of the conventional pulse train (PT) control, which is then applied in an SRG power generator system for the first time. The proposed CC-PT controlled method could adjust the output voltage by combining two or more groups of preset control pulses without the need of the error compensation network, which has the advantage of quick response compared with the PWM method.

2. SRG Power Generation System

2.1. Principle of SRG

The SRG is a doubly salient varied reluctant motor, and the salient poles of stators and rotors are superposed by the common silicon steel sheet [25]. Figure 1a shows the structure of a SRG, of which phase A is taken as an example to illustrate the principle of its power generation. The SRG rotates clockwise to the position as shown in Figure 1a under the external force (wind power), and the switches S_1, S_2 are closed, then excite phase A and form the circuit $S_1 \rightarrow A \rightarrow A' \rightarrow S_2$. Figure 1b shows the excitation and freewheeling phases: First, the winding of phase A absorbs electrical energy from the source. Then, S_1 and S_2 are switched off and the winding of phase A is subjected to freewheeling through the freewheeling diodes, thus feeding back the energy stored by the winding to the DC power supply.

Because both the excitation and armature windings of the SRG are stator windings, the process of excitation and power generation must be controlled periodically. Its excitation process is controllable, while the generation process is not controllable, so the power generation is commonly controlled by adjusting the excitation current.

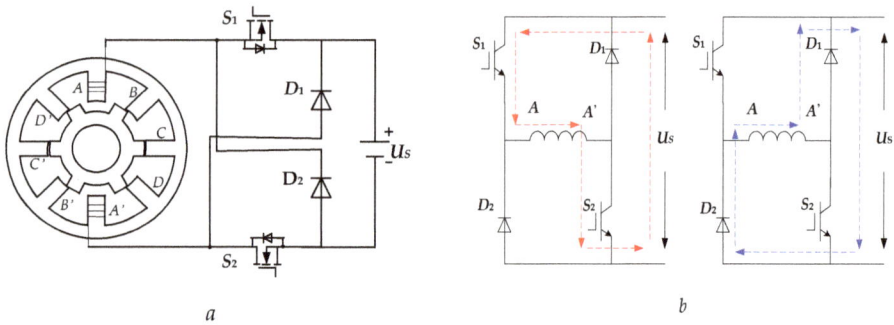

Figure 1. Principles of SRG. (**a**) Profile of 8/6 SRG; (**b**) two phases of SRG.

2.2. Control Methods of SRG

A SRG power generation system has two main excitation modes, which includes self-excitation and separate excitation. This paper chooses the separate excitation mode shown in Figure 2, which can separate the power generated by the winding from the excitation power completely and ensure high stability [26].

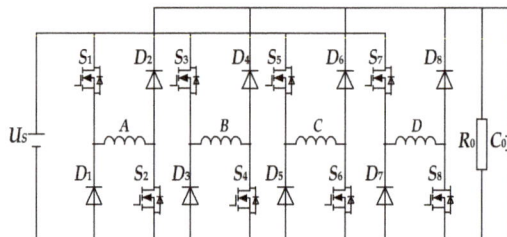

Figure 2. The main circuit structure diagram of separate excitation.

The control methods have been proposed for regulating the output power of SRG, such as angle position control (APC), current chopping control (CCC), and pulse width modulation (PWM).

The APC control mode regulates the current waveforms by changing the turn-on angle (θ_{on}) and turn off angle (θ_{off}). The output current can be changed by tuning the turn-on angle (θ_{on}) and turn off angle (θ_{off}), as shown in Figure 3a. It can be seen that, higher conduction angles could lead to a larger output current. APC control can also be divided into the modes of changing θ_{on}, θ_{off} and both θ_{on} and θ_{off} together [27].

CCC is a method that directly chops the phase currents during the specific position of each phase. In CCC, the current i is compared with the chopper current (i_{chop}). When the rotor position angle is between θ_{on} and θ_{off}, where $\theta_{on} < \theta < \theta_{off}$, if $I \leq i_{chop}$, the switches are turned on, then the current i rises and gradually reaches the chopper; or else, the switches are turned off and the current i declines, as shown in Figure 3b. As compared with the APC control, where the current is not controllable, CCC directly controls the phase current and can get more accurate control performance [28].

The PWM control applies the PWM modulation signals on the main switches and adjusts the excitation voltage by changing the duty cycle to realize the control of excitation current. The PWM control of SRG can also be divided into two methods: one uses the PWM cut double tubes, and the other uses the PWM cut single tube. These two methods adjust the size of the excitation current and ultimately realize the control of output voltage. The waveforms of phase current by using PWM control is shown in Figure 3.

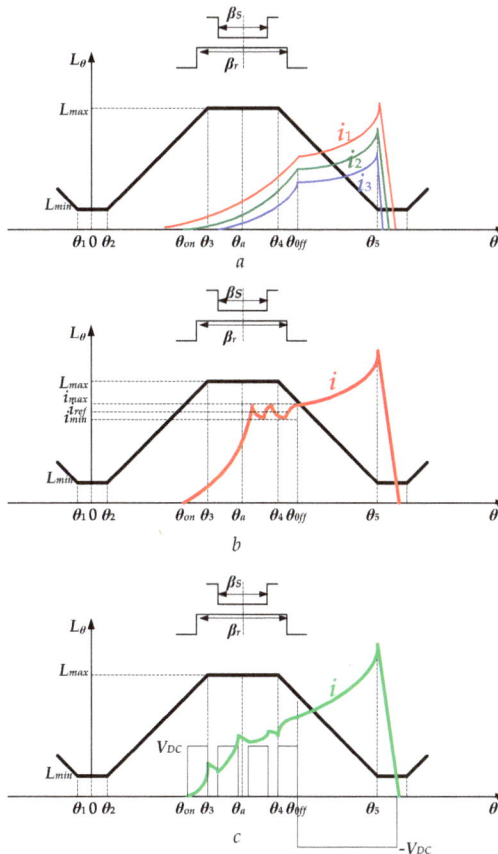

Figure 3. The waveforms of the phase current using three control methods. (**a**) APC control method; (**b**) CCC control method; (**c**) PWM control method.

2.3. Analysis of Output Voltage Ripple of SRG

At the stage of building up the voltage field in the self-excited mode of SRG, the excited voltage source U_S provides an initial excitation voltage to the system. The self-excited mode generator system has heavy weight and high efficiency. However, when the loads fluctuate seriously and the capacitor charges or discharges, control difficulty of the irregular excitation current is encountered. The irregular excitation will finally result in the output voltage ripple, having an impact on the power generation performance and causing certain damage to the SRG's body that shortens the lifetime of SRG.

The voltage balancing equation of the kth phase winding is given by

$$U_k = R_k i_k + \frac{d\psi_k}{dt} \tag{1}$$

The phase inductance $L_k(i_k, \theta)$ is a function of the phase current i_k and the rotor position angle θ. Therefore, (1) can be converted to;

$$U_k = R_k i_k + L_k \frac{di_k}{dt} + i_k \omega \frac{dL}{d\theta} \tag{2}$$

where U_k is kth phase output voltage, L_k is the phase inductance, ω is the generator angular speed, R_k is the resistance of the phase winding, and i_k is the phase current.

In the nonlinear model, the voltage at generator windings in excitation and power generation are given by;

$$U_k = \begin{cases} i_k R_k + \frac{d\psi_k(i_k,\theta)}{dt} \\ -i_k R_k - \frac{d\psi_k(i_k,\theta)}{dt} \end{cases} \tag{3}$$

It is supposed that the load in the power generation state is R, and the voltage on capacitor is U_c. In the process of excitation, the capacitor supplies power to both the winding excitation and the load, which is expressed by the following equation;

$$C\left(\frac{dU_c}{dt}\right)\omega = -i_R - i_c \tag{4}$$

During the freewheeling process, the capacitor is charged by the winding and the capacitor supplies power to the load, which is expressed as;

$$C\left(\frac{dU_c}{dt}\right)\omega = -i_R - i_z \tag{5}$$

with i_z is the armature current.

From (3), the capacitance voltage U_c can be obtained, then the variation of the capacitance voltage $\Delta U = \Delta U_c$ can be derived. Afterwards, the variation of the output voltage in the nonlinear model can be expressed by;

$$\Delta U = \Delta U_c(\theta_{on})\{\exp(-\theta_{off}/(\tau\omega)) - \exp(-\theta_{on}/(\tau\omega))\} - Q/(C\omega) \tag{6}$$

where ω is the angular velocity of SRG, U_c is the capacitor voltage, R is the load resistance, and Q is the energy storage during the excitation phase, $\tau = RC$.

From (6), it can be seen that the parameters that affect the output voltage ripple include the capacitor voltage U_c, rpm of SRG n, load R, and capacitance C.

3. Capacitance Current Pulse Train Control

3.1. The Principle of Pulse Train Control

The pulse train (PT) controlled method is put forward based on the linear control theory of the pulse width modulation (PWM) method [29]. The PT controlled method can adjust the output voltage by using two or more sets of a combination of preset control pulses, which has the advantages of simple circuit structure, no compensation network, and fast response. There is a wide application of the PT controlled method in systems with a switching power supply that need high performance [30,31]. Figure 4a shows the schematic diagram of the PT controlled buck converter.

A PT controller consists of a voltage comparator, clock signal, flip-flop delay, and logic gate circuit. The control circuit is simple and easy to implement [32]. The main work process is as follows. In the case that the nth clock trigger signal comes, the comparator transfers the comparison V_0 of the output voltage value and the reference value V_{ref} to delay the flip-flop. When $V_0(n) < V_{ref}$, the Q of delay flip-flop outputs 1, $\overline{Q}$ outputs 0, and the pulse control chooses P_H as the effective control signal, which causes the converter to absorb more energy from the input U_S, forcing the output voltage to rise; similarly, when the $(n + 2)^{th}$ clock trigger signal comes, the sampling of the output voltage value is $V_0(n + 2)$. Because $V_0(n + 2) > V_{ref}$, the Q of delay flip-flop outputs 0, $\overline{Q}$ outputs 1, and the pulse control chooses P_L as the effective control signal, where the converter absorbs less energy from the input U_S, forcing the output voltage to decline. The waveforms of the output voltage, inductance current, and control pulse are as shown in Figure 4b.

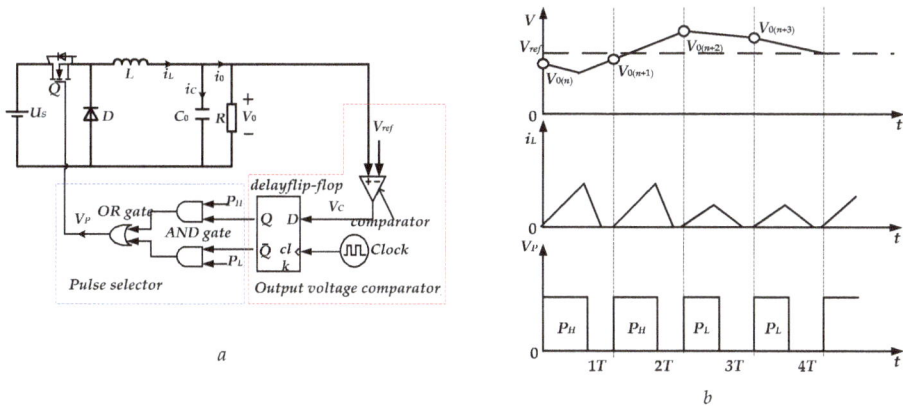

Figure 4. The principle of PT and CC-PT control. (**a**)VM-PT controlled Buck converter schematic; (**b**) schematic of working principle.

3.2. The Principle of Capacitor Current Train Pulse

As the full flow of current ripple flows into the output filter capacitance, the phase position of the inductance current ripple can be reflected through the capacitive current. By sampling the capacitive current of converter load side and adding the capacitive current to the output voltage as the condition of pulse choice, one can adjust the hysteresis of output voltage and realize the suppression of low frequency oscillation behavior [33,34].

The accurate circuit model of SRG does not calculate the electromagnetic torque, therefore the model is worthless. Here, the linear model of SRG is applied, since it excludes the influence of magnetic saturation, and the inductance of phase winding is irrelevant to the current. Therefore, the output voltage ripple can be decreased by controlling CC-PT.

In a duty cycle, the inductance of phase is a function of rotor angle. The corresponding phase inductance of SRG in the linear model is expressed as:

$$L(\theta) = \begin{cases} L_{min} & \theta_1 \leq \theta < \theta_2 \\ \delta(\theta - \theta_2) + L_{min} & \theta_2 \leq \theta < \theta_3 \\ L_{max} & \theta_3 \leq \theta < \theta_4 \\ L_{max} - \delta(\theta - \theta_4) & \theta_4 \leq \theta < \theta_5 \end{cases} \tag{7}$$

where $\delta = (L_{max} - L_{min})/\beta_S$, β_S is the arc angle of the stator.

Under the control of CC-PT, the structure of the phase of a SRG power generation system is shown in Figure 5.

The difference between the CC-PT controlled method and the traditional PT controlled method is in the matter of the sampling information. When the clock signal comes, the sampling circuit samples the output voltage V_0 and the capacitor current i_c of the circuit, and the sampling results are compared with the reference voltage V_{ref}. When $V_0 + \alpha i_c \leq V_{ref}$, the pulse control chooses D_H as the effective control signal, which causes the winding of phase A to absorb more energy from the input U_S, forcing the output voltage to rise; when $V_0 + \alpha i_c > V_{ref}$, the pulse control chooses D_L as the effective control signal, where the converter absorbs less energy from the input U_S, forcing the output voltage to decline. The selection of pulse control is expressed as:

$$D = \begin{cases} D_H, & V_0 + \alpha i_c \leq V_{ref} \\ D_L, & V_0 + \alpha i_c > V_{ref} \end{cases} \tag{8}$$

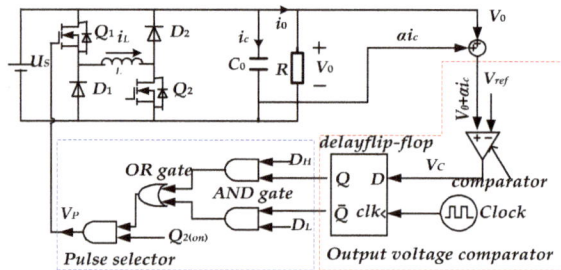

Figure 5. The structure of SRG power system under the control of CC-PT.

In the circuit shown in Figure 6, the conduction of switches is controlled by cutting the single pipe. Figure 6 shows the waveforms of winding current and output voltage when switch Q_2 is conducted ($\theta_2 \leq \theta < \theta_5$). When switch Q_2 is conducted, the inductance of winding changes with θ, and the winding current is nonlinearly increased; when switch Q_2 is turned off, the winding current is nonlinearly declined. When Q_1 is on and $\theta_2 \leq \theta < \theta_4$, the winding of phase A absorbs energy and is excited by U_S. When $\theta_4 \leq \theta < \theta_5$, the winding of phase A converts the mechanical energy of the rotor to magnetic energy. When Q_1 and Q_2 are off, the magnetic energy stored in the magnetic field is released from phase A. Thus, the conversion between mechanical energy and electrical energy is completed in the form of magnetic energy.

The winding current $i_L(\theta)$ satisfies the relationship with the current of the load i_0, which is given by;

$$i_{L(\theta)} = \begin{cases} \int_{t_{on}}^{t_{off}} \frac{U_S}{L(\theta)} dt & t_{on} \leq t < t_{off} \\ i_0 + C \frac{dU_c}{dt} & others \end{cases} \tag{9}$$

where U_c is the instantaneous value of the output filter capacitor voltage.

The change of the output voltage (capacitance voltage) in a switching cycle can be determined by:

$$\Delta v_0(nT) = \frac{1}{C} \int_{nT}^{(n+1)} (i_L - i_0) dt \tag{10}$$

According to Figure 6, the value of the integral item in the right-hand side of (10) equals the difference between the area of trapezium *ABCD* and that of the rectangle *ADFE*. The area of rectangle *ADFE* is given as:

$$S_{ADFE} = |\Delta i_L(nT)| T \tag{11}$$

The area of trapezium *ABCD* equals the sum of the areas of A_1, A_2, and A_3. The area of A_3 can be regarded as *L*, which equals the constant value L_0. The area of trapezium *ABCD* is expressed as

$$
\begin{aligned}
S_{ABCD} &= S_{ABG} + S_{BCDG} + A_1 + A_2 \\
&= \frac{1}{2} DT \frac{U_S - V_0}{L_0} DT + \frac{1}{2} \left(\frac{U_S - V_0}{L_0} DT + \frac{U_S D - V_0}{L_0} T \right)(1-D)T + A_1 + A_2 \\
&= \frac{T^2}{2L_0} [(U_S D - V_0) + U_S D(1-D)] + A_1 + A_2
\end{aligned} \tag{12}
$$

From (10) to (12), the change in the output voltage can be derived as

$$
\begin{aligned}
\Delta v_0(nT) &= \frac{T^2}{2L_0 C} [(U_S D - V_0) + U_S D(1-D)] + \frac{\Delta i_L(nT)T}{C} + Z_1 \\
&= Z_1 + Z_2 + \frac{i_c(nT)T}{C}
\end{aligned} \tag{13}
$$

with $Z_2 = \frac{T^2}{2L_0 C} [(U_S D - V_0) + U_S D(1-D)]$, $Z_1 = \frac{S_1 + S_2}{C}$.

Since $T^2 \ll LC$, from (13), the change in the output voltage can be approximately expressed as

$$\Delta v_0(nT) \approx \frac{i_c(nT)T}{C} \tag{14}$$

The PT controlled method adjusts the output voltage through adjusting the current of winding indirectly. When the output voltage is larger than the reference voltage, the PT controller chooses D_H as the control signal, and the output voltage may vary. Similarly, when the output voltage is lower than the reference voltage, the PT controller selects D_L as the control signal, and the output voltage may vary as well. The phenomenon of voltage hysteresis leads to the occurrence of low frequency fluctuations in PT control. It can be seen from (14) that the change of output voltage in a switching cycle is mainly determined by $i_c(nT)$. The output voltage increases when $i_c(nT) > 0$, otherwise the output voltage declines.

The above analysis shows that the phenomenon of low frequency fluctuations is caused by not adjusting the output voltage in a timely fashion. If the output voltage rises when the system chooses D_H, or it declines when the system chooses D_L, the low frequency fluctuations can be suppressed. As the full flow of the current ripple of the winding flows into the output filter capacitance, the phase position of the inductance current ripple can be reflected in the capacitive current. In addition, as can be seen from (14), the value of output filter capacitance only affects the output voltage.

From (14), it can be seen that;

$$\Delta v_0(nT) = \frac{T^2}{2L_0 C} [(U_S D - V_0) + U_S D(1-D)] + \frac{\Delta i_L(nT)T}{C} \tag{15}$$

where $i_L(n)$ is the inductance current at the beginning of the nth switching cycle.

From (15) and the working principle of PT controlled method, the synchronous switching mapping equation of the output voltage and inductance current under the control of CC-PT is shown as:

$$Y_{n+1} = \begin{cases} AY_n + B^H, CY_n \leq V_{REF} \\ AY_n + B^L, CY_n > V_{REF} \end{cases} \tag{16}$$

where:

$$A = \begin{bmatrix} 1 - \frac{T^2}{2LC} - \frac{T}{RC} & \frac{T}{C} \\ -\frac{T}{L} & 1 \end{bmatrix}, Y_{n+1} = \begin{bmatrix} v_0(n+1) \\ i_L(n+1) \end{bmatrix}, Y_n = \begin{bmatrix} v_0(n) \\ i_L(n) \end{bmatrix},$$

$$B^H = \begin{bmatrix} \frac{U_S D_H T^2(2-D_H)}{2LC} \\ \frac{U_S D_H T}{L} \end{bmatrix}, B^L = \begin{bmatrix} \frac{U_S D_L T^2(2-D_L)}{2LC} \\ \frac{U_S D_L T}{L} \end{bmatrix}, C = \begin{bmatrix} 1 - \frac{\alpha}{R} & \alpha \end{bmatrix}$$

According to (16), when the system parameters are determined, the combination of high and low pulse signals in a pulse cycle is determined as well. The inductance currents of windings and the output voltage change as the feedback coefficient of the capacitor and the value of load resistance change in each switching cycle, which results in the change of the combination of high and low pulses in the pulse cycle. The Lyapunov exponent function of the CC-PT control system is obtained as:

$$\lambda(\alpha, R) = \ln\left|1 - \frac{T^2}{2LC} - \frac{T}{RC}\right| \tag{17}$$

λ is always less than 0, which indicates that the CC-PT controlled system is in the steady working state when the capacitance current regulation coefficient α and load resistance R change. The output voltage of the system is oscillating at the base level and is measured by defining the standard deviation, which is expressed as:

$$\sigma(\alpha, R) = \sqrt{\frac{1}{N}\sum_{i=1}^{N}\left[v_{oi}(\alpha, R) - V_{0ref}\right]} \tag{18}$$

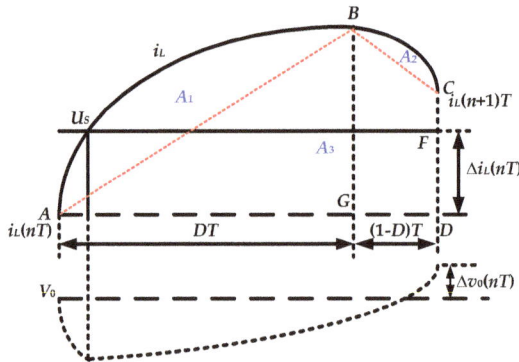

Figure 6. The waveforms of winding current and output voltage during a switching cycle.

4. Optimization of Power Generation Efficiency of SRG

As shown in Figure 7, the ratio of the total instantaneous current value during the phase of electric power generation and excitation is calculated to measure the efficiency of power generation, which is named as ε, and it is obtained by;

$$\varepsilon = \frac{B_1}{B_2} = \frac{\int_{\theta_{on}}^{\theta_{end}} i_k d\theta}{\int_{\theta_{on}}^{\theta_{off}} i_k d\theta} \tag{19}$$

where i_k is the current of phase k.

Therefore, ε can be obtained simply by calculating the integration of the phase generation current and the excitation current. In practical applications, θ_{on}, θ_{off}, and θ_{end} are needed, and the total instantaneous current value during the phase of electric power generation and excitation is calculated, which can be used to determine the efficiency of power generation.

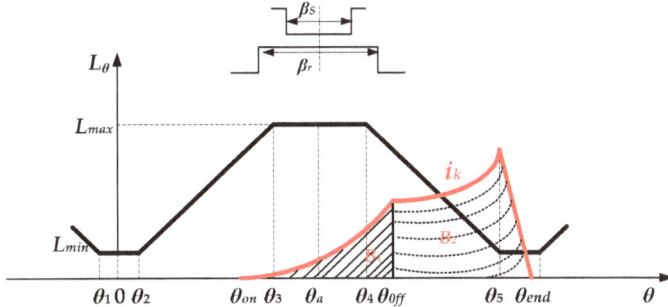

Figure 7. The schematic diagram of ε.

5. Experiment Model and Verification

In order to verify the validity of theoretical analysis, this paper has constructed a platform of 8/6 SRG power generation shown in Figure 8a,b which is based on the analysis of SRG and CC-PT control.

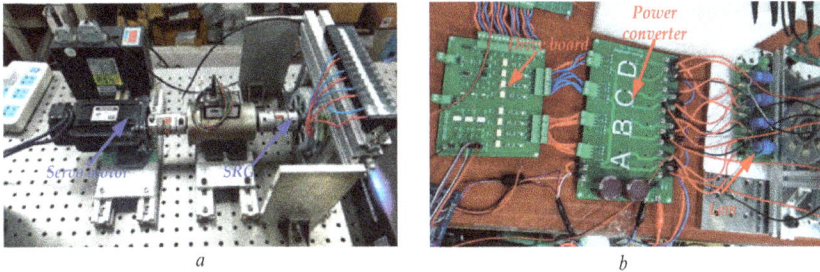

Figure 8. The photograph of the experimental platform of SRG power generation. (**a**) Experiment platform of SRG; (**b**) control circuits and the power converter.

Figure 9 shows the control scheme of the 8/6 SRG system. It consists of a SRG, a servo motor, a drive board, controller (*STM32*), and so on. The main parts of the circuit are the encoder and sensors, which detect the location of the rotors, the values of current i_A, i_B, i_C, i_D, V_0, and i_c, where i_c is the current though the filter capacitor.

The detail control of PI and CC-PT are shown in Figure 10. It can be easily seen when Q_2 is open, which is controlled by the APC controller, and determined by the value of ε (the efficiency of power generation). The PI control uses the difference of V_0^* and $V_0(t)$ to form a voltage loop and the PT control leads i_c into the control. By limiting the overshoot of i_c, the ripple of U_{C0} is reduced. Figure 11 shows the working principle flowchart of the *STM32* controller.

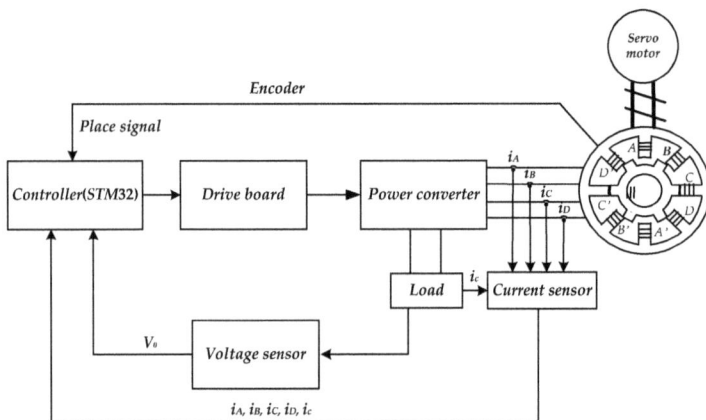

Figure 9. The control scheme of the 8/6 SRG system.

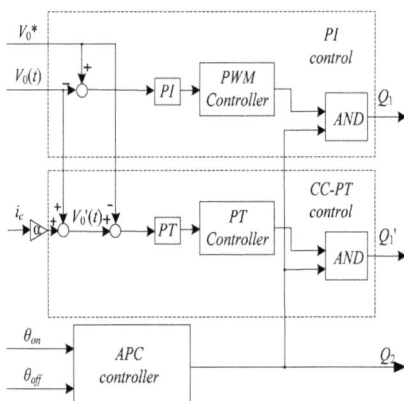

Figure 10. SRG controller.

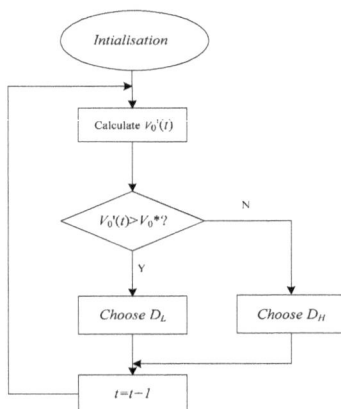

Figure 11. Flowchart of the *STM*32 controller.

Figure 12a shows the trend of generation power at different speeds and different θ_{off} when $\theta_{on} = 20°$. From Figure 12a, it can be seen that at a certain speed, the generation power of SRG gradually increases with the increase of θ_{off}.

Figure 12b shows the trend of ε with the change of θ_{off} and speed. It can be seen that ε increases as θ_{off} increases at different speeds. However, when θ_{off} reaches a certain value, the value of ε decreases, according to which, the optimal θ_{off} can be determined, as well as the optimal net generation power. The optimal θ_{off} at different speeds with the condition of constant θ_{on} are shown in Table 1. In the same way, the optimal θ_{on} can be obtained at different speeds with the condition of constant θ_{off}, which is also shown in Table 1.

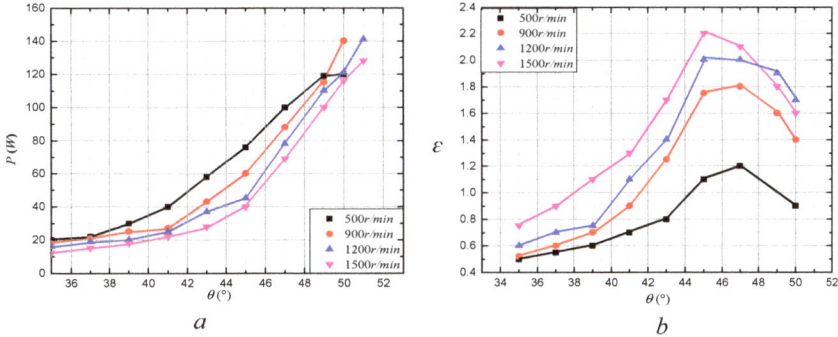

Figure 12. Optimize the θ_{off}. (**a**) The trend of generation power; (**b**) the trend of ε.

Table 1. The optimal θ_{on} and θ_{off}.

n, r/min	θ_{on}, deg	θ_{off}, deg
500	25	47
900	23	47
1200	25	45
1500	25	45

From (18), the feedback coefficient of capacitor α with i ranging from 0 to 1, and the value of the load resistance R with i ranging from 0 to 10, are combined as parameters. With the change of feedback coefficient, the output voltage error approximation is small when the feedback coefficient is larger than 0.2, and the low frequency oscillation can be suppressed by the CC-PT controlled method. In this paper, the PI parameters of traditional PWM control is determined by the critical ratio method ($P = 0.5$, $I = 0.032$). In order to compare the ripple of the output voltage with that under PI control, a small capacitor is used in the experiment, the capacitance of which is 680 uF.

In summary, the parameters used in the experiment are shown in Table 2.

Table 2. Parameters of experiment model.

Parameters	Value	Parameters	Value
m	4	R, Ω	8
Ns	8	C, uF	680
N_r	6	P	0.5
Us, V	80	I	0.032
n,r/min	1200	D_H	0.6
θ_{on}, deg	25	D_L	0.4
θ_{off}, deg	45	f, Hz	6000

When the desired output voltage $V_0 = 15$ V and the load $R = 8\ \Omega$, the experiment waveforms of V_0 under PWM control are shown in Figure 13a. It can be seen from Figure 13a, when the SRG runs at a stable state with a constant speed $n = 1200$ r/min, the ripple of V_0 is large, which is up to 5 V.

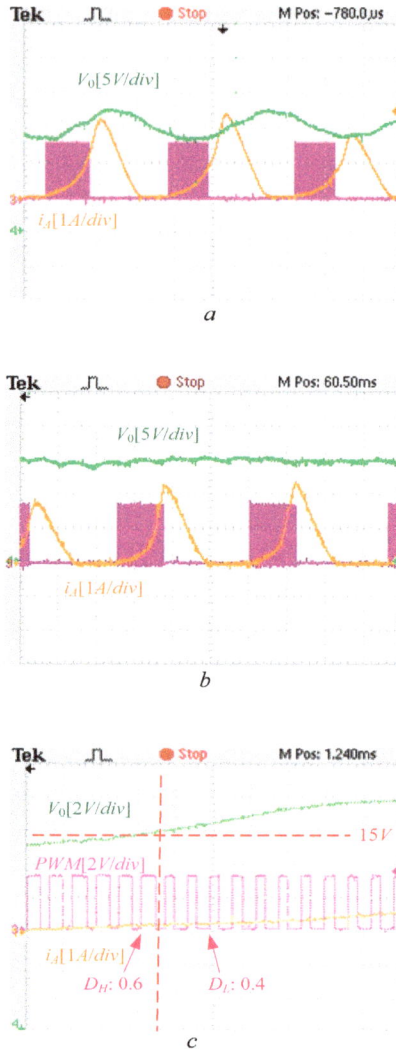

Figure 13. The experiment waveforms of SRG. (**a**) Experiment waveforms of output voltage V_0 under PWM control; (**b**) experiment waveforms of output voltage V_0 under PT control; (**c**) the details of experiment waveforms under PT control.

The parameters of the SRG power generator system under the PT controlled method are: $D_H = 0.6$, $DL = 0.4$, and $f = 6$ kHz. Figure 13b shows the experiment waveforms of V_0 under the PT controlled method. The ripple of V_0 is smaller than that obtained by applying the PI controlled method, but the ripple is still up to 2 V.

Figure 13c shows the details of the PT controlled method. It can be seen that when V_0 is smaller than the target value of 15 V, the controller chooses the PWM scheme with $D_H = 0.6$, and the output voltage increases. On the other hand, when the value of V_0 is larger than 15 V, the controller chooses the PWM scheme with $D_L = 0.4$ to decrease the output voltage value.

Figure 14 shows the experiment waveforms of V_0 under the CC-PT controlled method. The coefficient of ratio α is 1. From (14), the ripple of output voltage $\Delta v_0(nT)$ can be calculated. The largest $\Delta v_0(nT)$ is about ±0.49 V. It can be seen from Figure 14 that the ripple of V_0 is the smallest among these three controlled methods. The value of the output voltage ripple is about 1 V, which is consistent with the calculation results of (14).

In order to investigate the response of the SRG system starting process, a larger filter capacitance is used in the experiment and the value of it is 0.27 F. Figure 15 shows the starting processes under the control of PI and CC-PT.

It can be easily seen that the output voltage derived by applying the PI controlled method takes about 1.4 s to achieve a stable state. However, the output voltage obtained by using the CC-PT controlled method only takes about 0.7 s to achieve a stable state. Therefore, the system under the CC-PT controlled method responds more quickly than that under the PI controlled method. Meanwhile, from Figure 15, it also can be seen that V_0 has a bigger overshoot under the control of PI control, but there is nearly no overshoot of V_0 under the control of the CC-PT control. The ripple of V_0 derived by the CC-PT controlled method is smaller than that obtained by the PI controlled method.

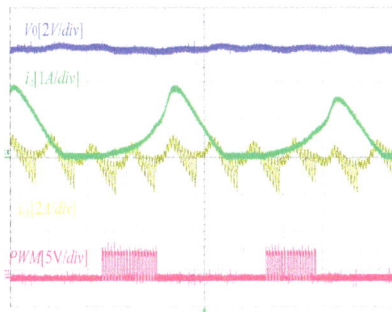

Figure 14. The experiment waveforms of output voltage under CC-PT control.

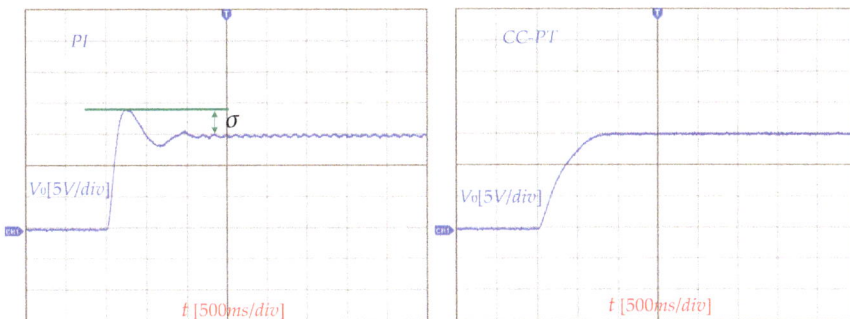

Figure 15. The starting process under the control of PI and CC-PT.

6. Conclusions

Based on the traditional PT control, a CC-PT controlled method is proposed in this paper to improve the performance of an SRG power system. Through experiment verification, as compared with the traditional SRG power generation system under PWM control, the CC-PT controlled SRG has the following advantages:

(a)   The CC-PT controlled method can adjust the output voltage by using two or more sets of combinations of preset control pulses, which has the advantages of simple circuit structure and no compensation network.

(b)   The SRG power generation system under the CC-PT controlled method can realize a fast response during start-up. By reasonably configuring the value of D_H, the output voltage overshoot can be eliminated.

(c)   The start-up phase current is reduced, which makes the system more secure and cost saving. The output voltage ripple is smaller than 5% of the rated value.

Author Contributions: X.Z. and D.Y. suggested the method of CC-PT control used in SRG generation system; M.C. and R.X. analyzed the data. M.C. and X.Z. wrote the paper. K.N. reviewed and edited.

Funding: This research received no external funding.

Acknowledgments: The authors acknowledge the financial support from the Key Laboratory of Control of Power Transmission and Conversion (SJTUM), Ministry of Education (2016AC06).

Conflicts of Interest: The authors declare no conflicts of interest.

References

1.   Zhao, Z.Y.; Hu, J.; Zuo, J. Performance of wind power industry development in China: A diamondmodel study. *Renew. Energy* **2009**, *34*, 2883–2891. [CrossRef]

2.   Liu, Y.; Ren, L.; Li, Y.; Zhao, X.G. The industrial performance of wind power industry in China. *Renew. Sustain. Energy Rev.* **2015**, *43*, 644–655. [CrossRef]

3.   Zhao, S.; Nair, N.K.C. Assessment of wind farm models from a transmission system operator perspective using field measurements. *IET Renew. Power Gener.* **2011**, *5*, 455–464. [CrossRef]

4.   Cheng, H.; Chen, H.; Wang, Q.; Xu, S.; Yang, S. Design and control of switched reluctance motor drive for electric vehicles. In Proceedings of the 2016 14th International Conference on Control, Automation, Robotics and Vision (ICARCV), Phuket, Thailand, 13–15 November 2016; pp. 1–6.

5.   Dang, J.; Harley, R.G. Sensorless control scheme for ultra high speed switched reluctance machine. In Proceedings of the 2013 IEEE Energy Conversion Congress and Exposition, Denver, CO, USA, 15–19 September 2013; pp. 3830–3836.

6.   Shen, L.; Wu, J.; Yang, S.; Huang, X. Fast flux linkage measurement for switched reluctance motors excluding rotor clamping devices and position sensors. *IEEE Trans. Instrum. Meas.* **2012**, *62*, 185–191. [CrossRef]

7.   Chen, H.; Gu, J.J. Switched reluctance motor drive with external rotor for fan in air conditioner. *IEEE/ASME Trans. Mechatron.* **2013**, *18*, 1448–1458. [CrossRef]

8.   Song, S.; Zhang, M.; Ge, L. A new fast method for obtaining flux-linkage characteristics of SRM. *IEEE Trans. Ind. Electron.* **2015**, *62*, 4105–4117. [CrossRef]

9.   Morimoto, S.; Nakayama, H.; Sanada, M.; Takeda, Y. Sensorless output maximization control for variable-speed wind generation system using ipmsg. *IEEE Trans. Ind. Appl.* **2015**, *41*, 60–67. [CrossRef]

10.   Liu, J.; Qiao, F.; Liu, M.; Wu, C. Neural fuzzy control to minimise torque ripple of SRM. *Int. J. Model. Identif. Control* **2010**, *10*, 132–137. [CrossRef]

11.   Muyeen, S.M.; Al-Durra, A.; Hasanien, H.M. Application of an adaptive neuro-fuzzy controller for speed control of switched reluctance generator driven by variable speed wind turbine. In Proceedings of the 2015 Modern Electric Power Systems (MEPS), Wroclaw, Poland, 6–9 July 2015; pp. 1–6.

12.   Oliveira, E.S.L.; Aguiar, M.L.; Silva, I.N.D. Strategy to control the terminal voltage of a SRG based on the excitation voltage. *IEEE Lat. Am. Trans.* **2015**, *13*, 975–981. [CrossRef]

13.   Liu, Y.Z.; Zhou, Z.; Song, J.L.; Fan, B.J. Based on sliding mode variable structure of studying control for status switching of switched reluctance starter/generator. In Proceedings of the 2015 Chinese Automation Congress, Wuhan, China, 27–29 November 2015; pp. 934–939.

14.   Le-Huy, H.; Chakir, M. Optimizing the performance of a switched reluctance generator by simulation. In Proceedings of the XIX International Conference on Electrical Machines—ICEM 2010, Rome, Italy, 6–8 September 2010; pp. 1–6.

15. Chang, Y.C.; Liaw, C.M. On the design of power circuit and control scheme for switched reluctance generator. *IEEE Trans. Power Electron.* **2008**, *23*, 445–454. [CrossRef]

16. Liptak, M.; Hrabovcova, V.; Rafajdus, P. Equivalent circuit of switched reluctance generator based on dc series generator. *J. Electr. Eng.* **2008**, *59*, 23–28.

17. Liu, Y.Z.; Zheng, Z.; Sheng, Z.J.; Fan, B.J.; Song, J.L. Study of control strategy for status switching of switched reluctance starter/generator. *Electr. Mach. Control* **2015**.

18. Ferdowsi, M.; Emadi, A. Pulse regulation control technique for integrated high-quality rectifier-regulators. *IEEE Trans. Ind. Electr.* **2005**, *52*, 116–124. [CrossRef]

19. Ferdowsi, M.; Emadi, A.; Telefus, M.; Davis, C. Pulse regulation control technique for flyback converter. *IEEE Trans. Power Electr.* **2004**, *20*, 798–805. [CrossRef]

20. Khaligh, A.; Rahimi, A.M.; Emadi, A. Modified pulse-adjustment technique to control dc/dc converters driving variable constant-power loads. *IEEE Trans. Ind. Electr.* **2008**, *55*, 1133–1146. [CrossRef]

21. Luo, F.; Ma, D. An integrated switching DC–DC converter with dual-mode pulse-train/pwm control. *IEEE Trans. Circuits Syst. II Express Briefs* **2015**, *56*, 152–156.

22. Wang, J.; Xu, J.; Zhou, G.; Bao, B. Pulse-train-controlled CCM buck converter with small ESR output-capacitor. *IEEE Trans. Ind. Electr.* **2013**, *60*, 5875–5881. [CrossRef]

23. Xu, J.-P.; Mu, Q.-B.; Wang, J.-P.; Qin, M. Output voltage ripple of pulse train controlled DCM buck converter. *Electr. Mach. Control* **2010**, *14*, 1–6.

24. Sha, J.; Xu, D.; Chen, Y.; Xu, J.; Williams, B.W. A peak-capacitor-current pulse-train-controlled buck converter with fast transient response and a wide load range. *IEEE Trans. Ind. Electr.* **2016**, *63*, 1528–1538. [CrossRef]

25. Songyan, K.; Li, K.; Fengping, H.; Zhao, S. Position sensorless technology of switched reluctance motor considering mutual inductances. *Trans. China Electrotech. Soc.* **2017**, *11*, 1085–1094.

26. Gameiro, N.S.; Cardoso, A.J.M. A new method for power converter fault diagnosis in SRM drives. *IEEE Trans. Ind. Appl.* **2012**, *48*, 653–662. [CrossRef]

27. Cai, J.; Deng, Z. A position sensorless control of switched reluctance motors based on phase inductance slope. *J. Power Electr.* **2013**, *13*, 264–274. [CrossRef]

28. Zhang, Z.; Zhaoyu, P.; Feng, G. Research on new control model for switched reluctance motor. In Proceedings of the International Conference on Computer Application and System Modeling, Taiyuan, China, 22–24 October 2010; pp. 198–202.

29. Sha, J.; Xu, J.; Zhong, S.; Liu, S. Control pulse combination-based analysis of pulse train controlled DCM switching DC–DC converters. *IEEE Trans. Ind. Electr.* **2015**, *62*, 246–255. [CrossRef]

30. Ferdowsi, M.; Emadi, A.; Telefus, M.; Shteynberg, A. Suitability of pulse train control technique for bifred converter. *IEEE Trans. Aerosp. Electr. Syst.* **2015**, *41*, 181–189. [CrossRef]

31. Kapat, S. Analysis and synthesis of reconfigurable digital pulse train control in a DCM buck converter. In Proceedings of the IECON 2014—40th Annual Conference of the IEEE Industrial Electronics Society, Dallas, TX, USA, 29 October–1 November 2014; pp. 1254–1260.

32. Kapat, S. Voltage-mode digital pulse train control for light load DC-DC converters with spread spectrum. In Proceedings of the 2015 IEEE Applied Power Electronics Conference and Exposition (APEC), Charlotte, NC, USA, 15–19 March 2015; pp. 966–971.

33. Yang, P.; Bao, B.-C.; Sha, J.; Xu, J.-P. Dynamical mechanism of ramp compensation for switching converter. *Acta Phys. Sin.* **2013**, *62*, 709–712.

34. Sha, J.; Bao, B.-C.; Xu, J.-P.; Gao, Y. Dynamical modeling and border collision bifurcation in pulse train controlled discontinuous conduction mode buck converter. *Acta Phys. Sin.* **2012**, *61*, 855–865.

energies

MDPI

Article

Sliding-Mode Control of Distributed Maximum Power Point Tracking Converters Featuring Overvoltage Protection

Carlos Andres Ramos-Paja [1], Daniel Gonzalez Montoya [2,*] and Juan David Bastidas-Rodriguez [3]

[1] Departamento de Energia Eléctrica y Automática, Universidad Nacional de Colombia, Carrera 80 No 65-223—Facultad de Minas, Medellín 050041, Colombia; caramosp@unal.edu.co
[2] Departamento de Electrónica y Telecomunicaciones, Instituto Tecnológico Metropolitano, Carrera 31 No 54-10, Medellín 050013, Colombia
[3] Escuela de Ingenierías, Eléctrica, Electrónica y de Telecomunicaciones, Universidad Industrial de Santander, Bucaramanga 68002, Colombia; jdbastir@uis.edu.co
*   Correspondence: danielgonzalez@itm.edu.co; Tel.: +57-4-4600727

Received: 10 July 2018; Accepted: 15 August 2018; Published: 24 August 2018

Abstract: In Photovoltaic (PV) systems with Distributed Maximum Power Point Tracking (DMPPT) architecture each panel is connected to a DC/DC converter, whose outputs are connected in series to feed a grid-connected inverter. The series-connection forces the output voltage of those converters to be proportional to the converter' output power; therefore, under mismatched conditions, the output voltage of a highly-irradiated converter may exceed the rating (safe) value, causing an overvoltage condition that could damage the converter. This paper proposes a sliding-mode controller (SMC) acting on each converter to regulate both the input and output voltages, hence avoiding the overvoltage condition under partial shading. The proposed control strategy has two operation modes: maximum power point tracking (MPPT) and Protection. In MPPT mode the SMC imposes to the PV panel the voltage reference defined by an MPPT technique. The Protection mode is activated when the output voltage reaches the safety limit, and the SMC regulates the converter' output voltage to avoid overvoltage condition. The SMC has a bilinear sliding surface designed to provide a soft transition between both MPPT and Protection modes. The SMC analysis, parameters design and implementation are presented in detail. Moreover, simulation and experimental results illustrate the performance and applicability of the proposed solution.

Keywords: distributed architecture; maximum power point tracking; sliding mode control; overvoltage

1. Introduction

The continuous growing of Photovoltaic (PV) systems in the last years has consolidated PV technology as one of the most important renewable energy sources. Only in 2017 approximately 96 GW were installed, i.e., 29% more with respect to 2016, reaching a global installed PV capacity of 402.5 GW, approximately [1].

Most of the PV installed capacity corresponds to grid-connected PV systems (GCPVS) aimed at supplying electricity demand in different applications. In general, a GCPVS is composed by a PV generator, one or more DC/DC power converters, an inverter and a control system [2]. The PV generator transforms the sunlight into electric power, which depend on the environmental conditions (irradiance and temperature) and the operation point. The DC/DC converters allows the modification of the PV generator operation point and the DC/AC converter delivers the electrical power to the grid. The control system can be divided into two main parts: maximum power point tracking (MPPT) and inverter control. On the one hand, the MPPT uses the DC/DC power converters to find and track the PV generator operation point where it delivers the maximum power (MPP). On the other hand, the inverter control has two main tasks, the first one is to synchronize the AC voltage with the grid, and the second one is to inject the AC current to the grid, which is proportional to the power delivered by the PV generator and the DC/DC converters [3,4].

The inverter control is particularly important in a GCPVS because the stability and the power quality injected to the grid depend on it [2–4]. For this controller, the PV generator, the DC/DC converters and the MPPT are represented by a voltage [2] or a current [3,4] source, which feeds a link capacitor to form a DC bus. The DC voltage is converted to AC with a set of switches, and a filter eliminates the high frequency components [2]. The voltage-source two-level inverters with L, LC or LCL filters are widely used in commercially available inverters [2,5] and the inverter controller is usually a cascaded control where inner loop regulates voltage and the outer loop controls the current injected to the grid and keeps the DC bus voltage around its reference value [2]. Nonetheless, other authors propose cascaded controller where the inner loop regulates the current injected to the grid [3,4,6,7] and the current references are generated from a Droop controller [7], active and reactive power references [6] or form the maximum power provided by the PV source and the reactive power demanded by the load [3,4]. Moreover, some papers propose linear current controllers [6], while other papers combine linear regulators with state feedback [3,7] or Lyapunov-based [4] controllers to regulate the current injected to the grid.

Notwithstanding the important role of the inverter controller in a GCPVS, the maximum power delivered by the PV generator does not depend on this controller, since the MPPT is in charge of finding and tracking the MPP of the PV generator for different irradiance and temperature conditions. When all the the PV panels in a generator are operating under the same irradiance and temperature conditions (i.e., homogeneous conditions), there is a single MPP in the power vs. voltage (P-V) curve of the generator. However, GCPVS in urban environments (i.e., homes, buildings, companies, etc.) are surrounded by different objects, which may produce partial shadings over the PV array, which forces the PV panels of the array to operate under different (mismatched) irradiance and temperature conditions. Moreover, mismatching conditions may also be produced by the aging, soiling, early degradation and manufacturing tolerances in the PV panels [8].

When a PV generator is operating under mismatching conditions, the power produced is significantly reduced [9,10]; therefore, it is important to mitigate their effects. In general, it is possible to find three different architectures to mitigate the adverse effects of the mismatching conditions in PV installations: centralized systems (CMPPT), distributed systems (DMPPT), and reconfiguration systems [11]. However, CMPPT and DMPPT architectures are the most widely used architectures in urban applications; hence they are briefly discussed below.

In CMPPT systems, depicted in Figure 1, the complete PV array is connected to a single DC/DC power converter, whose output is connected to the grid through an inverter. The DC/DC converter modifies the operation voltage of the PV array, in order to track the MPP through the MPPT. Under mismatching conditions, the maximum current (i.e., the short-circuit current) produced by a shaded PV panel is less than the short-circuit current of the unshaded panels; hence, when the array current is greater than the short-circuit current of the shaded panel, the excess of current flows through the bypass diode (BD) connected in antiparallel to the panel (see Figure 1). As consequence, for a particular shading profile over the PV panels and a particular array current, some BDs are active and the rest are inactive. This activation and deactivation of the bypass diode produce multiple MPPs in the array P-V curve, which means that there are local MPPs and one global MPP (GMPP) [12].

In general, MPPT techniques for CMPPT architectures are complex [11,12] because they should be able to track the global MPP of the PV array in any condition. Moreover, mismatching conditions continuously change along the day and year due to the sun trajectories in the sky, and also due to the changes in the surrounding objects. As consequence, the number of MPPs and the location of the global MPP continuously change in the P-V curve of a PV array. CMPPT techniques can be classified into three main groups [13]: conventional techniques, soft computing techniques and other techniques.

The first group includes techniques based on Perturb & Observe (P&O), incremental conductance and hill climbing, as well as other GMPP search techniques and adaptive MPPTs. Soft computing techniques uses artificial intelligence methods to find the GMPP, like evolutionary algorithm, genetic programming, fuzzy system, among others. The last group includes methods like Fibonacci search, direct search, segmentation search, and others to locate the GMPP.

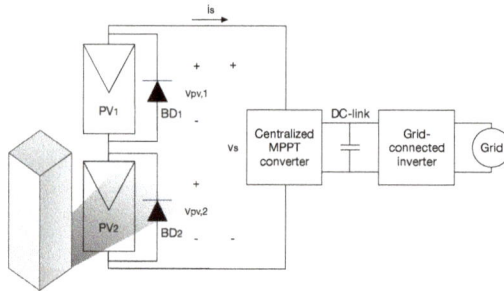

Figure 1. Centralized MPPT system based on a double-stage structure.

In DMPPT architectures the PV array is divided into smaller arrays, or sub-arrays, to reduce the number of MPPs in each sub-array. Then, each sub-array is connected to a DC/DC power converter, which has an MPPT technique much more simple than the ones used in CMPPT systems [11,14,15]. The double stage DMPPT system, presented in Figure 2, is one of the most widely adopted architectures in literature [11,14,15], where each panel is connected to a DC/DC converter to form a DMPPT unit (DMPPT-U) and all DMPPT-Us are connected in series to feed an inverter. Boost converters are widely used as DC/DC converter in double stage DMPPT system, while other approaches uses buck, buck-boost or more complex converters to improve the voltage gain or the efficiency [13,15].

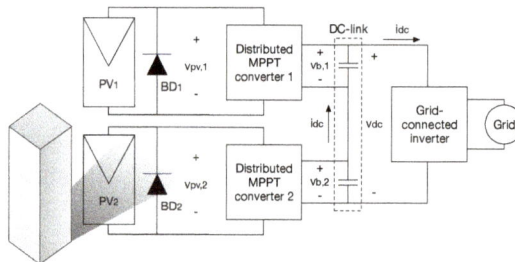

Figure 2. Distributed MPPT system based on a double-stage structure.

The main advantage of the double stage DMPPT systems is that each PV panel can operate at its MPP even under mismatching conditions [15]. Moreover, no communication is required among the DMPPT-Us or with the inverter, and the the dynamics of the DMPPT-Us are decoupled from the dynamics of the GCPVS inverter, due to the high capacitance in the DC link that forms the DC bus [15]. However, one of the main limitations of double stage DMPPT systems is that the output voltage of each DMPPT-U is proportional to its output power; therefore, under mismatching conditions, the output voltage of a DMPPT-U with a highly irradiated PV panel may exceed the maximum voltage of the DMPPT-U output capacitor and the maximum open-circuit voltage of the switching devices. Such a condition is denominated overvoltage and must be avoided to protect the DC/DC converter [16–18]. Although overvoltage condition is important to assure a secure operation of the DMPPT-Us, it is not discussed in some papers devoted to analyzing double stage DMPPT systems, like [19,20], nor in review papers about MPPTs for PV generators under mismatching conditions [13,15,21,22].

In general, overvoltage can be faced by two main approaches. The first one is to design the DMPPT-U with an output capacitor and switching devices able to endure voltages that may be close to the DC bus voltage in the link with the inverter [18]. Nevertheless, this solution increases the size and cost of each DMPPT-U, hence, this effect is multiplied by the number of DMPPT-Us in the PV system. The second approach is to monitor the DMPPT-U output voltage and if it is greater than a reference value, the control objective must be changed to regulate the DMPPT-U output voltage under its maximum value. This operating mode is denominated Protection mode.

Therefore, the DMPPT-U control strategy must consider two basic operation modes: MPPT and Protection. In MPPT mode the control objective is to extract the maximum power from the PV generator, while monitors the output voltage of the DMPPT-U. If such a voltage surpasses a reference value, then MPPT mode is disabled and Protection mode is activated to keep the DMPPT-U output voltage below its maximum value. Although in literature there is a significant number of control systems for double stage DMPPTs, as shown in different review papers [13–15], after an exhaustive review the authors have found just a few control systems that consider the overvoltage problem and implement MPPT and Protection modes [16,23–30]. That is why, the literature review in this paper is focused on these references.

In [23–25] the authors propose centralized strategies to perform the MPPT and to avoid the condition $v_b > V_{max}$ on DMPPT-Us implemented with Boost converters, where v_b and V_{max} are the output voltage of the DC/DC converter and its maximum value, respectively. In [23,24] the authors propose to monitor v_b of each DMPPT-U, if there is at least one DMPPT-U with $v_b > V_{max}$, then the input voltage of the inverter (v_{dc}) is reduced. Moreover, when v_{dc} is reduced below 80% of its nominal value, the DMPPT-Us with $v_b > V_{max}$ change their operating mode from MPPT to v_b regulation. Nevertheless, the authors use linear controllers for v_{pv} and v_b, which no not guarantee the DMPPT-U stability in the full operation range. Additionally, the paper does not provide information about the implementation of the v_b controller and it does not discuss how to perform the transition between MPPT and Protection mode (and viceversa) or the stability issues of those transitions. Finally, the paper does not provide guidelines or a design procedure of the proposed control system.

Another centralized control strategy for a DMPPT system, based on Particle-Swarm Optimization (PSO), is proposed in [25]. The objective of the control strategy is to find the values of v_{pv} of each DMPPT-U that maximizes the output power of the whole system. However, the constraints of the PSO algorithm include the condition $v_b < V_{max}$ for each DMPPT-U. Therefore, the proposed control system is able to track the MPPT in each DMPPT-U avoiding the overvoltage condition. Although the authors provide some considerations to set the PSO parameters, they not explain how regulate the PV panel voltage with the power converters and they do not analyze the stability of the DMPPT-Us. Moreover, the authors do not provide information for the implementation of the proposed control system because they implemented it on a dSpace control board. It is worth noting that the centralized strategies proposed in [23–25] require additional hardware to implement the centralized controllers and monitoring systems, hence these solutions require high calculation burden compared with other DMPPT-U control approaches like [16,26–29].

The authors in [16,26] consider DMPPT-Us implemented with Boost converters and propose to limit the duty-cycle (*d*) of each DMPPT-U to avoid the condition $v_b > V_{max}$. The limit of *d* is defined as $d < 1 - v_{pv}/v_b$, where, v_{pv} is the PV panel voltage [16,26]. Nevertheless, the DMPPT-U control operates in open-loop during the saturation of *d*, which may lead to the instability of the DMPPT-U controller. Additionally, the papers do not provide a clear explanation about how to define the duty cycle limit, since the voltage v_{pv} of a DMPPT-U varies with the irradiance and temperature conditions as well as the mismatching profile over the PV panels. Finally, the authors in [16] focus on the analysis of double stage DMPPT systems implemented with boost converters, but they do not provide a design procedure for the DMPPT-U control in MPPT mode.

In [27–29] the authors propose two different control strategies for each DMPPT-U, one for MPPT mode and another for Protection mode, where the trigger for the Protection mode is the condition $v_b > V_{max}$. On the one hand, the strategy presented in [27] for Protection mode is to adopt a P&O strategy, i.e., perturb v_{pv} and observe v_b in order to fulfill the condition $v_b < V_{max}$. On the other hand, in [28,29] two PI-type regulators are proposed for each DMPPT-U: one for v_{pv} in MPPT mode and another for v_b in Protection mode. The reference of v_b and v_{pv} regulators are V_{max} and the MPPT reference, respectively. The voltage regulators presented in [27–29] are linear-based, with fixed parameters, and designed with a linearized model in a single operation point of the DMPP-U; therefore, they cannot guarantee a consistent dynamic performance and stability of the DMPPT-U in the entire operation range. Moreover, the authors in [27–29] do not provide a design procedure of the proposed controllers and only [28] provide relevant information for the controller implementation.

A Sliding-Mode Controller (SMC) designed to regulate v_{pv} and v_b on a Boost-based DMPPT-U is proposed in [30]. The sliding surface (Ψ) has three terms: $\Psi = i_L - k_{pv} \cdot (v_{pv} - v_{mppt}) \cdot (1 - OV) - k_b (v_b - V_{max}) \cdot (OV)$, where i_L is the inductor current of the Boost converter, v_{mppt} is the v_{pv} reference provided by the MPPT algorithm, OV is a binary value assuming $OV = 1$ when $v_b > V_{max}$ and $OV = 0$ when $v_b < V_{max}$, and the constants k_{pv} and k_b are SMC parameters. During MPPT mode the first and second terms of Ψ are active to regulate v_{pv} according to the MPPT algorithm; while during Protection mode the first and third terms of Ψ are active to regulate $v_b = V_{max}$. The main advantage of the SMC proposed in [30] is the capability to guarantee the global stability of the DMPPT-U in the entire operation range. Nonetheless, that paper does not analyze the dynamic restrictions of the SMC reference in MPPT to guarantee the DMPPT-U stability; additionally, the paper does not provide a design procedure for the SMC parameters (k_{pv} and k_b) and the sliding surface does not include integral terms, which introduces steady-state error in the regulation of v_{pv} and v_b. Finally, the authors do not provide information for the real implementation and the proposed control system is validated by simulation results only.

This paper introduces a control strategy with MPPT and Protection modes for DMPPT-Us implemented with boos converters, where the regulation of v_{pv}, in MPPT mode, and v_b, in Protection mode, is performed by a single SMC. In MPPT mode v_{pv} reference is provided by a P&O algorithm and v_b is monitored to verify if its value is less than a safe limit named V_{max}. If $v_b \geq V_{max}$, Protection mode is activated and v_b is regulated to V_{max} by the SMC. During Protection mode v_b is monitored to verify the condition $v_b < V_{max}$, if so, MPPT mode is activated. The proposed SMC has the same structure of the SMC introduced in [30] to adapt the SMC switching function with the operation mode. However, the proposed SMC introduces two integral terms to guarantee null steady-state error in the regulation of v_{pv} and v_b; moreover, the paper analyzes the dynamic restrictions in the P&O references to ensure the stability of the DMPPT-U in the entire operation range. The design procedure of the proposed SMC parameters is analyzed in detail as well as its implementation using embedded systems and analog circuits.

There are three main contributions of this paper. The first one is a single SMC that guarantees global stability and null steady-state error of the DMPPT-Us in the entire operation range of MPPT and Protection modes. The second contribution is a detailed design procedure of the proposed SMC parameters and the definition of the dynamic limits of P&O references that guarantee the global stability. Finally, the last contribution is the detailed description of the SMC implementation that helps the reader to reproduce the results.

The rest of the paper is organized as follows: Section 2 explains the effects of the mismatching conditions on a DMPPT system; Section 3 introduces the model of the DMPPT-U and the structure of the proposed SMC. Sections 4 and 5 provide the analysis and parameters design of the proposed SMC in both MPPT and Protection modes. Then, Section 6 describes the implementation of the proposed SMC, and Sections 7 and 8 present both the simulation and experimental results, respectively. Finally, the conclusions given in Section 9 close the paper.

2. Mismatched Conditions and DMPPT

In a CMPPT system operating under mismatching conditions, there are some panels subjected to a reduced irradiance due to, for example, the shadows produced by surrounding objects (see Figure 1); hence, the maximum current (short-circuit current) of those panels is lower than the short-circuit current of the non-shaded panels. Moreover, when the string current is lower than the short-circuit current of the shaded PV panel, the protection diode connected in antiparallel, i.e., bypass diode (BD), is reverse biased (inactive) and both panels contribute to the string voltage. However, when the string current is higher than the short-circuit current of the shaded panel, the BD of such panel is forward biased (active) to allow the flow of the difference between the string current and the short-circuit current of the shaded panel.

The Current vs. Voltage (I-V) and P-V curves of a PV array, composed by two PV panels, is simulated to illustrate the mismatching effects on the CMPPT system presented in Figure 1. For the simulation, the non-shaded (PV$_1$) and shaded (PV$_2$) panels irradiances are $S_1 = 1000$ W/m^2 and $S_2 = 500$ W/m^2, respectively. The panels are represented by using the single-diode model expression given in Equation (1) [31], where v_{pv} and i_{pv} are the current and voltage of the panel, i_{ph} is the photovoltaic current, A is the inverse saturation current, R_s is the series resistance and R_h is the parallel resistance. B is defined as $B = N_s \cdot \eta \cdot k \cdot T/q$ where N_s is the number of cells in the panel, η is the ideality factor, k is the Boltzmann constant, q is the electron charge, and T is panel temperature in K. The parameters used for the simulations are calculated using the equations presented in [32]: $A = 154.15\ \mu A$, $B = 1.1088$ V^{-1}, $R_s = 0.0045\ \Omega$ and $R_h = 109.405\ \Omega$.

$$i_{pv} = i_{ph} - A \cdot \left[\exp \left(\frac{v_{pv} + R_s \cdot i_{pv}}{B} \right) - 1 \right] - \frac{v_{pv} + R_s \cdot i_{pv}}{R_h} \tag{1}$$

The BD activation of the shaded PV panel in Figure 1 produces an inflection point in the I-V curve, which in turns produces two MPPs in the P-V curve as it is shown in Figure 3. Therefore, the maximum power produced by the CMPPT system (86.52 W) is less than 123.28 W, which is the sum of the maximum power that can be produced by both PV$_1$ (84.25 W) and PV$_2$ (39.03 W).

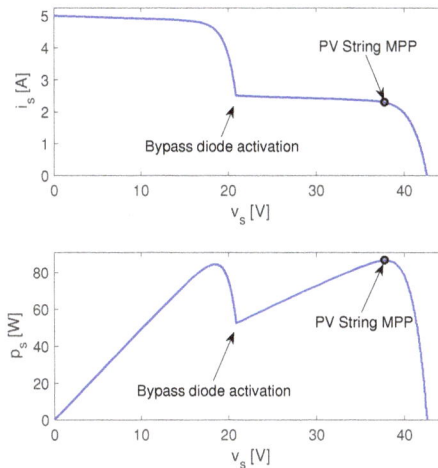

Figure 3. Electrical behavior of a mismatched PV string with two PV panels.

In a double stage DMPPT system, each panel is connected to a DC/DC converter to form a DMPPT unit (DMPPT-U), and the converters' outputs are connected in series to obtain the input voltage of an inverter, as reported in Figure 2. The boost converter is a widely used topology to implement the DMMPT-Us [11,14,16,23–27,29,30], since it is necessary to step-up the PV panel voltage to match the inverter input voltage. Moreover, the boost structure is simple and the stress voltages of both output capacitor and switch are smaller in comparison with other step-up topologies [33]. Furthermore, the series connection of the DMPPT-U outputs impose low boosting factors to the boost converters, which enables those topologies to operate in a high efficiency condition.

To illustrate the theoretical power extraction provided by a double stage DMPPT solution, the system of Figure 2 is simulated considering the same mismatching conditions adopted for the CMPPT solution: $S_1 = 1000$ W/m^2 and $S_2 = 500$ W/m^2. The simulation results are presented in Figure 4. In this case, both PV panels are able to operate at any voltage, hence the maximum power achievable in each panel is extracted. Therefore, the theoretical optimal operation conditions $(v_{pv,1} = 18.43$ V, $v_{pv,2} = 17.64$ V$)$ correspond to the MPP conditions in each panel as reported in Figure 4, in which PV$_1$ has a maximum power of 84.25 W and PV$_2$ has a maximum power of 39.03 W, hence the maximum power provided by the DMMPT system is 123.28 W; this is considering loss-less converters.

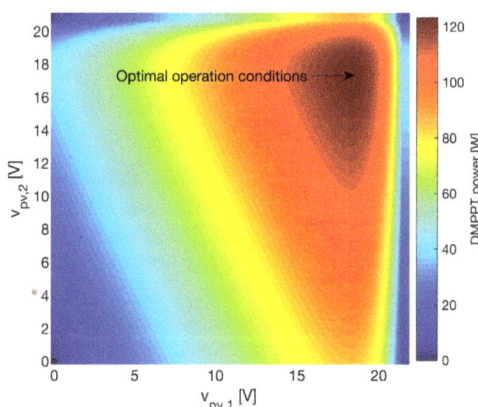

Figure 4. Theoretical power production of the DMPPT system in Figure 2.

However, from Figure 2 it is observed that the DC-link is formed by the output capacitors of the DMPPT converters, which are connected in series. Therefore, the DC-link voltage v_{dc} is equal to the sum of the output capacitors voltages $v_{b,1}$ and $v_{b,2}$. For a general system, with N DMPPT converters associated to N PV panels, such a voltage condition is expressed in Equation (2):

$$v_{b,1} + v_{b,2} + \cdots + v_{b,i} + \cdots + v_{b,N} = \sum_{j=1}^{N} v_{b,j} = v_{dc} \qquad (2)$$

Moreover, the series connection of the output capacitors imposes the same current at the output of the DMPPT converters. Therefore, the power delivered to the DC-link p_{dc}, which is transferred to the grid-connected inverter, is equal to the sum of the power delivered by each converter, $p_{b,1}$ and $p_{b,2}$. In the general system formed by N converters, the following expression holds:

$$p_{b,1} + p_{b,2} + \cdots + p_{b,i} + \cdots + p_{b,N} = \sum_{j=1}^{N} p_{b,j} = p_{dc} \qquad (3)$$

Finally, the voltage imposed to the *i*-th output capacitor is obtained from Equations (2) and (3) as follows:

$$v_{b,i} = v_{dc} \cdot \frac{p_{b,i}}{\sum\limits_{j=1}^{N} p_{b,j}} \qquad (4)$$

That expression put into evidence that the voltage imposed to any of the output capacitors depends on the power delivered by all the DMPPT converters. Moreover, grid-connected inverters, like the one described in Figure 2, regulate the DC-link voltage at its input terminals to ensure a correct and safe operation [34]. In light of the previous operation conditions, Equations (2) and (4) reveal that the DC-link voltage v_{dc}, imposed by the inverter, is distributed into the output capacitor voltages $v_{b,i}$ proportionally to the power delivered by the associated PV panel $p_{pv,i}$ with respect to the total power delivered by all the PV sources. Hence, the converter providing the higher power will exhibit the higher output voltage, which could lead to over-voltage conditions.

Considering the DMPPT system of Figure 2 with a DC-link voltage imposed by the inverter equal to $v_{dc} = 80$ V, and output capacitors with maximum voltage rating equal to $V_{max} = 50$ V, the DMPPT system operates safely if both PV panels produce the same power since $v_{b,1} = v_{b,2} = 40$ V. However, in the mismatched conditions considered ($S_1 = 1000$ W/m^2 and $S_2 = 500$ W/m^2), the DMPPT system is subjected to overvoltage conditions as it is reported in Figure 5: at the theoretical optimal operation conditions $(v_{pv,1} = 18.43$ V, $v_{pv,2} = 17.64$ V$)$ the output voltage of the first converter is 54.67 V, which is higher than the rating voltage V_{max} producing an overvoltage condition that could damage the converter. Figure 5 shows the conditions for safe operation, overvoltage in the first converter ($v_{b,1} > V_{max}$) and overvoltage in the second converter ($v_{b,2} > V_{max}$).

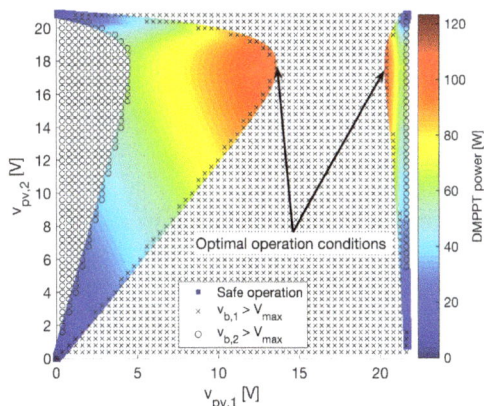

Figure 5. Safe power production of the DMPPT system in Figure 2.

The simulation puts into evidence that new optimal operation condition appear due to the overvoltage conditions. In this example, the first optimal operation points of the PV panels are $v_{pv,2} = 17.64$ V (MPP voltage) and $v_{pv,1} = 13.33$ V (no MPP voltage), while the second optimal operation point is $v_{pv,2} = 17.64$ V (MPP voltage) and $v_{pv,1} = 20.39$ V (no MPP voltage). This result is analyzed as follows: the first PV panel must be driven far enough from the MPP condition so that the power provided to the DC-link by the associated converter is, at most, 62.5% of the total power.

That percentage is calculated from Equation (4) replacing the output voltage by the rating voltage V_{max} and using the values of the DC-link voltage v_{dc} and the total power delivered to the DC-link as follows:

$$\max\left(p_{b,i}\right) = V_{max} \cdot \frac{\sum_{j=1}^{N} p_{b,j}}{v_{dc}} \tag{5}$$

Equation (5) shows that, in the cases when the theoretical optimal operation conditions are out of the safe voltages, the new optimal operation voltages are located at the frontier of the safe conditions, which ensures the maximum power extraction from the PV panel associated to the converter near the overvoltage condition. This analysis is confirmed by the simulation results presented in Figure 5.

Therefore, to ensure the maximum power extraction for any irradiance and mismatching profile, the DMPPT converters must be operated in two different modes:

- *MPPT mode:* when the output capacitor voltage $v_{b,i}$ is under the safe (rating) limit, the converter must be controlled to track the MPP condition.
- *Protection mode:* when the output capacitor voltage $v_{b,i}$ reaches the safe limit, the converter must be controlled to set $v_{b,i}$ at the maximum safe value V_{max}.

The following sections propose a control system, based on the sliding-mode theory, to impose the previous behavior to the DMPPT converters.

3. Converter Model and Structure of the Control System

As discussed before, boost converters are widely used in DMPPT systems; hence, this paper considers a DMPPT-U system implemented with a boost converter. The electrical model of the adopted DMPPT converter is presented in Figure 6, which includes the MPPT algorithm that provides the reference for the SMC. Moreover, a current source is used to model the current i_{dc} imposed by the inverter to regulate the DC-link voltage.

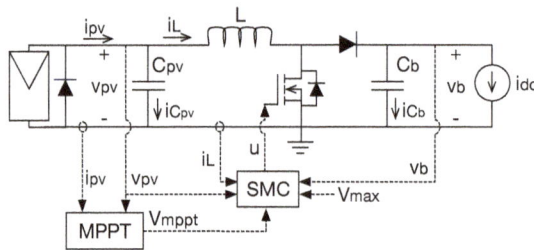

Figure 6. Electrical model of a DMPPT boost converter.

The differential equations describing the dynamic behavior of the DMMPT converter are given in Equations (6)–(8), in which u represents the binary signal that defines the MOSFET and diode states: $u = 1$ for MOSFET on and diode off; $u = 0$ for MOSFET off and diode on.

$$\frac{dv_{pv}}{dt} = \frac{i_{pv} - i_L}{C_{pv}} \tag{6}$$

$$\frac{dv_b}{dt} = \frac{i_L \cdot (1-u) - i_{dc}}{C_b} \tag{7}$$

$$\frac{di_L}{dt} = \frac{v_{pv} - v_b \cdot (1-u)}{L} \tag{8}$$

Sliding-mode controllers are widely used to regulate DC/DC converters because they provide stability and satisfactory dynamic performance in the entire current and voltage operation ranges [35,36]. Furthermore, SMCs also provide robustness against parametric and non-parametric uncertainties [37]. In particular, in PV systems implemented with boost converters, SMCs have been adopted to improve the dynamic performance of the DC/DC converter in CMPPT systems [37,38] and to regulate the input and output voltages of a DMPPT-U operating in both MPPT and Protection modes [30]. Therefore, this paper adopts that type of controllers.

The proposed control system uses one switching function for each operation mode: Ψ_{pv} for MPPT mode and Ψ_b for protection mode, which leads to the unified sliding surface (Φ) given in Equation (9). Therefore, the system operating at $\Psi = 0$ is in the sliding-mode with null error, while $\Psi \neq 0$ corresponds to a system operating far from the reference, hence with an error. The surface includes a binary parameter P_r to switch between the two operation modes, depending on the voltage value v_b exhibited by the output capacitor, as it is reported in expression (10).

$$\Phi = \left\{ \Psi_{pv} \cdot (1 - P_r) + \Psi_b \cdot (P_r) = 0 \right\} \tag{9}$$

$$\begin{aligned} \text{if} \quad v_b < V_{max} \quad &\rightarrow \quad P_r = 0 \,, \; \Phi = \{\Psi_{pv} = 0\} \\ \text{if} \quad v_b \geq V_{max} \quad &\rightarrow \quad P_r = 1 \,, \; \Phi = \{\Psi_b = 0\} \end{aligned} \tag{10}$$

The switching functions Ψ_{pv} and Ψ_b, designed for each mode, are given in Equations (11) and (12), respectively, in which k_{pv}, λ_{pv}, k_b and λ_b are parameters, i_L corresponds to the inductor current of the boost converter, v_{pv} corresponds to the voltage at the PV panel terminals, v_{mppt} corresponds to the reference provided by the MPPT algorithm, v_b corresponds to the output voltage of the DMPPT converter and V_{max} is the maximum safe voltage at the converter output terminals.

$$\Psi_{pv} = i_L - k_{pv} \cdot \left(v_{pv} - v_{mppt}\right) - \lambda_{pv} \cdot \int \left(v_{pv} - v_{mppt}\right) \, dt \tag{11}$$

$$\Psi_b = i_L + k_b \cdot \left(v_b - V_{max}\right) + \lambda_b \cdot \int \left(v_b - V_{max}\right) \, dt \tag{12}$$

Both switching functions were designed to share the inductor current, so that the transition between such sliding-mode controllers is not abrupt since the inductor current keeps the same value when Pr changes the active sliding function. Figure 7 illustrates the concept of the two operation modes in the proposed control system.

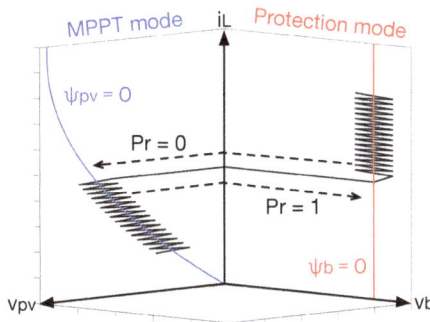

Figure 7. Concept of the proposed operation modes and sliding surfaces.

The following section analyzes the stability conditions of the proposed SMC, the equivalent dynamics of the closed loop system, the SMC parameters design, and the implementation of the proposed control system, in both MPPT ($\Phi = \{\Psi_{pv} = 0\}$) and Protection ($\Phi = \{\Psi_b = 0\}$) modes.

4. Analysis of the Proposed SMC

The design process of the sliding-mode control is performed by means of equivalent control method [35]. This technique was used to develop a method for testing convergence, global stability and performance of sliding-mode controllers acting on DC/DC converters, which is based on three considerations: transversality, reachability and equivalent control. Nevertheless, the authors in [35] demonstrated that sliding-mode controllers for DC/DC converters fulfilling the reachability conditions also fulfill the equivalent control condition. Transversality and reachability conditions of the proposed SMC in MPPT and Protection modes are analyzed in Sections 4.1 and 4.2, respectively. Moreover, the equivalent dynamic model of the DC/DC converter with the SMC is analyzed in Section 4.3 for MPPT and Protection modes.

4.1. Transversality Condition

The transversality condition analyses the ability of the controller to modify the sliding function trajectory. This condition is formalized in Equation (13), which evaluates that the MOSFET control signal u is present into the sliding function derivative [35,36] for MPPT and Protection modes. If the transversality conditions, given in Equation (13), are not fulfilled, the SMC has no effect on the sliding function trajectory and the system is not controllable. The left and right parts of the transversality condition must be fulfilled in MPPT and Protection modes, respectively; therefore, the following subsections analyze the transversality condition in each operation mode.

$$\frac{d}{du}\left(\frac{d\Psi_{pv}}{dt}\right)\bigg|_{P_r=0} \neq 0 \wedge \frac{d}{du}\left(\frac{d\Psi_b}{dt}\right)\bigg|_{P_r=1} \neq 0 \tag{13}$$

4.1.1. Transversality Condition in MPPT Mode

In this mode the SMC follows a voltage reference v_{mppt} provided by an external MPPT algorithm as depicted in Figure 6. In this work it is considered a Perturb and Observe (P&O) MPPT algorithm due to its positive compromise between efficiency and simplicity [39]. In MPPT mode the derivative of the switching function is obtained from Equation (11) as:

$$\frac{d\Psi_{pv}}{dt} = \frac{di_L}{dt} - k_{pv}\cdot\left(\frac{dv_{pv}}{dt} - \frac{dv_{mppt}}{dt}\right) - \lambda_{pv}\cdot(v_{pv} - v_{mppt}) \tag{14}$$

Replacing the PV voltage and inductor current derivatives, given in Equations (6)–(8), into Equation (14):

$$\frac{d\Psi_{pv}}{dt} = \frac{v_{pv} - v_b\cdot(1-u)}{L} - k_{pv}\cdot\left(\frac{i_{pv} - i_L}{C_{pv}}\right) + k_{pv}\frac{dv_{mppt}}{dt} - \lambda_{pv}\cdot(v_{pv} - v_{mppt}) \tag{15}$$

Finally, the transversality condition is evaluated by replacing Equation (15) into Equation (13), which leads to Equation (16).

$$\frac{d}{du}\left(\frac{d\Psi_{pv}}{dt}\right) = \frac{v_b}{L} > 0 \tag{16}$$

Since the output voltage is always positive, the transversality value (16) is also positive, which ensures that the transversality condition (13) is fulfilled in any operation condition of MPPT mode. Therefore, the switching function Ψ_{pv} designed for the MPPT mode is suitable to implement a SMC.

Moreover, the positive sign of the transversality value provides information concerning the behavior of a SMC implemented with Ψ_{pv}: $\frac{d}{du}\left(\frac{d\Psi_{pv}}{dt}\right) > 0$ implies that a positive values of u ($u = 1$) causes a positive change in $\frac{d\Psi_{pv}}{dt}$ [35]. In contrast, negative values of u ($u = 0$) causes a negative change in $\frac{d\Psi_{pv}}{dt}$; those considerations are used in Section 4.2 to analyze the reachability conditions.

4.1.2. Transversality Condition in Protection Mode

In this mode, the SMC limits the output voltage v_b to the maximum acceptable voltage V_{max} using the switching function Ψ_b introduced in Equation (12). In this mode the derivative of the switching function is obtained from Equation (12) as:

$$\frac{d\Psi_b}{dt} = \frac{di_L}{dt} + k_b \cdot \left(\frac{dv_b}{dt} - \frac{dV_{max}}{dt}\right) + \lambda_b \cdot (v_b - V_{max}) \tag{17}$$

In this mode, the reference V_{max} is constant, hence, $\frac{dV_{max}}{dt} = 0$. Replacing that value and the output voltage and inductor current derivatives, in Equations (7) and (8), into Equation (17):

$$\frac{d\Psi_b}{dt} = \frac{v_{pv} - v_b \cdot (1-u)}{L} + k_b \cdot \left(\frac{i_L \cdot (1-u) - i_{dc}}{C_b}\right) + \lambda_b \cdot (v_b - V_{max}) \tag{18}$$

The transversality condition is evaluated by replacing Equation (18) into Equation (13), which leads to expression (19).

$$\frac{d}{du}\left(\frac{d\Psi_b}{dt}\right) = \frac{v_b}{L} - k_b \cdot \frac{i_L}{C_b} > 0 \tag{19}$$

In expression (19), the transversality condition is defined positive to simplify the circuital implementation of the proposed SMC, as will be shown in Section 6.1. Therefore, the following restriction must be fulfilled by k_b:

$$k_b < \frac{v_b \cdot C_b}{L \cdot i_L} \tag{20}$$

Since the design of k_b, presented afterwards, takes into account the restriction imposed by expression (20), the transversality condition in expression (13) is fulfilled in any operation condition of the Protection mode. Therefore, the switching function Ψ_b designed for the Protection mode is suitable to implement a SMC.

Similar to the MPPT mode, the positive sign of the transversality value in Equation (19) imposes the switching conditions for Ψ_b, which are used in the next subsection to analyze the reachability conditions.

4.2. Reachability Conditions and Equivalent Control

The reachability conditions enables the analysis of the conditions in which the SMC successfully tracks the desired surface $\Phi = \{\Psi_{pv} = 0\}$ in MPPT mode and $\Phi = \{\Psi_b = 0\}$ in Protection mode. Considering that the transversality condition is positive for MPPT and Protection modes, the reachability analysis is based on the following conditions [35]: when the switching function of the system is under the surface, the derivative of the switching function must be positive to reach the surface; on the contrary, when the switching function is above the surface, the derivative of the switching function must be negative. Those conditions are formalized in Equations (21) and (22) for MPPT and Protection modes, respectively, which take into account the effect of the transversality value on the switching function derivative explained at the end of Section 4.1.2.

$$\lim_{\Psi_{pv}\to 0^-} \frac{d\Psi_{pv}}{dt}\bigg|_{u=1} > 0 \quad \wedge \quad \lim_{\Psi_{pv}\to 0^+} \frac{d\Psi_{pv}}{dt}\bigg|_{u=0} < 0 \tag{21}$$

$$\lim_{\Psi_b\to 0^-} \frac{d\Psi_b}{dt}\bigg|_{u=1} > 0 \quad \wedge \quad \lim_{\Psi_b\to 0^+} \frac{d\Psi_b}{dt}\bigg|_{u=0} < 0 \tag{22}$$

It is worth noting that the equivalent control condition is not included in the stability analysis of the proposed SMC, because Sira-Ramirez demonstrated in [35] that sliding-mode controllers for DC/DC converters fulfilling the reachability conditions also fulfill the equivalent control condition.

4.2.1. Reachability in MPPT Mode

Replacing the explicit expression of the switching function derivative, shown in Equation (15), into expression (21) becomes:

$$k_{pv} \cdot \frac{dv_{mppt}}{dt} > -\frac{v_{pv}}{L} + k_{pv} \cdot \left(\frac{i_{pv} - i_L}{C_{pv}}\right) + \lambda_{pv} \cdot \left(v_{pv} - v_{mppt}\right) \tag{23}$$

$$k_{pv} \cdot \frac{dv_{mppt}}{dt} < -\frac{v_{pv} - v_b}{L} + k_{pv} \cdot \left(\frac{i_{pv} - i_L}{C_{pv}}\right) + \lambda_{pv} \cdot \left(v_{pv} - v_{mppt}\right) \tag{24}$$

From the electrical model in Figure 6 it can be observed that the current of the input capacitor can be defined as $i_{Cpv} = (i_{pv} - i_L)$. According to the charge balance principle [40], $\langle i_{Cpv} \rangle = 0$ A, which implies that i_L and PV current i_{pv} exhibit the same average value, i.e., $\langle i_{pv} \rangle = \langle i_L \rangle$, otherwise the PV voltage will not be stable. Hence, the only difference between i_L and i_{pv} is the high-frequency current ripple present in the inductor, which produces ripples around zero in $(i_{pv} - i_L)$. Therefore, assuming that both inductor and PV currents are approximately equal $(i_{pv} \approx i_L)$ does not introduce a significant error in the analysis of expressions (23) and (24). This assumption will be validated in simulation results shown in Section 7, where the switching function remains within the hysteresis band in MPPT mode for different operation conditions.

Moreover, the maximum and minimum values of the term $(v_{pv} - v_{mppt})$, assuming a correct operation of the SMC, are Δv_{mppt} and $-\Delta v_{mppt}$, respectively, where Δv_{mppt} is the size of the voltage perturbation introduced by the P&O algorithm, i.e., $\max(v_{pv} - v_{mppt}) = \Delta v_{mppt}$ and $\min(v_{pv} - v_{mppt}) = -\Delta v_{mppt}$. Finally, the most restrictive case for expression (23) occurs at the minimum values of v_{pv} and $(v_{pv} - v_{mppt})$, while the most restrictive case for expression (24) occurs for the maximum values of v_{pv} and $(v_{pv} - v_{mppt})$, and the minimum value of v_b, in which $v_{pv} < v_b$ is ensured by boost topology.

In light of the previous considerations, expressions (23) and (24) are rewritten as follows:

$$\frac{dv_{mppt}}{dt} > -\frac{1}{k_{pv}} \left[\frac{\min(v_{pv})}{L} - \lambda_{pv} \cdot \Delta v_{mppt} \right] \tag{25}$$

$$\frac{dv_{mppt}}{dt} < -\frac{1}{k_{pv}} \left[\frac{\max(v_{pv}) - \min(v_b)}{L} + \lambda_{pv} \cdot \Delta v_{mppt} \right] \tag{26}$$

Inequalities (25) and (26) impose a dynamic restriction to the reference provided by the MPPT algorithm to guarantee the reachability of the sliding-surface. The main effect of these restrictions is that changes in vmppt cannot be performed in steps, but in ramps that fulfill expressions (25) and (26) [38]. Therefore, if the output of the P&O algorithm fulfills those restrictions the SMC will be able to track the reference in any operation condition. However, those limits depend on the SMC parameters, hence, inequalities (25) and (26) must be evaluated after the design of k_{pv} and λ_{pv}. It is important to note that k_{pv} and λ_{pv} need to be designed in order to provide the highest possible values of dv_{mppt}/dt limit, in this way, the dynamic restriction of the MPPT algorithm will be reduced. Section 5.3 shows an analysis of dv_{mppt}/dt limits as well as a numerical example, which illustrates that dv_{mppt}/dt limit may be in the order of tens of $mV/\mu s$ (kV/s); hence, the voltage variations can be performed in a small time compared with the perturbation period of the P&O algorithm, which means that restrictions imposed by expressions (25) and (26) do not affect considerably the dynamic performance of the DMPPTU.

In conclusion, the SMC in MPPT mode, i.e., operating with Ψ_{pv} given in Equation (11), is stable if restrictions (25) and (26) are fulfilled.

4.2.2. Reachability in Protection Mode

Replacing the explicit expression of the switching function derivative, Equation (18), into the inequalities introduced in expression (22) leads to:

$$\frac{v_{pv}}{L} - k_b \cdot \frac{i_{dc}}{C_b} + \lambda_b \cdot (v_b - V_{max}) > 0 \tag{27}$$

$$\frac{v_{pv} - v_b}{L} + k_b \cdot \frac{(i_L - i_{dc})}{C_b} + \lambda_b \cdot (v_b - V_{max}) < 0 \tag{28}$$

From the electrical model reported in Figure 6, and the power balance principle [40], the loss-less relation between input and output currents and voltages gives $i_{dc} \approx i_L \cdot v_{pv}/v_b$, which is used to simplify the reachability analysis. Moreover, fulfilling the reachability conditions ensures a correct operation of the SMC, hence inside the sliding-mode $v_b = V_{max}$. Finally, reorganizing expressions (27) and (28) it can be demonstrated that the most restrictive case occurs at the maximum value of i_L, which corresponds to the maximum PV current max $(i_L) = i_{ph}$ due to the charge balance condition. The values of v_{pv} and v_b are not considered in the worst case, since v_b is constant $(v_b = V_{max})$ and v_{pv} do not influence in the inequalities that define the worst case.

Under the light of the previous considerations, expressions (27) and (28) lead to the same restriction for k_b given in expression (20). Therefore, the SMC in Protection mode, i.e., operating with Ψ_b given in Equation (12), is stable if the inequality (20) is fulfilled.

4.3. Equivalent Dynamics

The equivalent dynamics correspond to the closed-loop behavior of the system under the action of the SMC. In this case, the equivalent dynamics are calculated by replacing the open-loop differential equation describing the inductor current, Equation (8), with the sliding-surface imposed by the SMC: $\{\Psi_{pv} = 0\}$ in MPPT mode (i.e., Equation (11)), and the sliding-surface imposed by $\{\Psi_{pv} = 0\}$ in Protection mode (i.e., Equation (12)).

4.3.1. Equivalent Dynamics in MPPT Mode

Expressions given in (29) describe the dynamic behavior of the system in MPPT mode, which are obtained replacing Equation (11) in Equation (8). In expression (29), the differential equation describing v_{pv} (Equation (6)) is the same, but the differential equation describing v_b (Equation (7)) has been modified to depend on the converter duty cycle d. The converter duty cycle d is defined as the the average value of the signal u within the switching period T_{sw}, as shown in Equation (30). This modification is performed because the MOSFET signal u is imposed by the SMC; hence, the equivalent dynamics disregards the switching ripple in u and it only depends on the average value of the control signal u within the switching period T_{sw} (i.e., d).

$$\begin{cases} \dfrac{dv_{pv}}{dt} = \dfrac{i_{pv} - i_L}{C_{pv}} \\ \dfrac{dv_b}{dt} = \dfrac{i_L \cdot (1 - d) - i_{dc}}{C_b} \\ i_L = k_{pv} \cdot (v_{pv} - v_{mppt}) + \lambda_{pv} \cdot \displaystyle\int (v_{pv} - v_{mppt}) \, dt \end{cases} \tag{29}$$

$$d = \frac{1}{T_{sw}} \cdot \int_0^{T_{sw}} u \, dt \tag{30}$$

The main challenge to analyze the dynamic behavior of Equation (29) corresponds to the non-linear relation between i_{pv} and v_{pv} shown in Equation (1). To overcome this problem, it is necessary to linearize the relation between i_{pv} and v_{pv} around a given operation point. Then, it is possible to obtain the transfer function between the PV panel voltage (controlled variable) and the voltage reference provided by the P&O algorithm. However, it is worth noting that the locations of the poles and zeros of the transfer function vary depending on the operation point where the relation between i_{pv} and v_{pv} is linearized. Hence, the transfer function must be analyzed in different operation points to analyze the dynamic behavior of the system under the action of the proposed SMC.

The small signal relationship between i_{pv} and v_{pv} in a given operation point is reported in Equation (31), where i_{PV} and v_{PV} (uppercase subscripts) are the panel small signal current and voltage, respectively, and Y_{pv} is the PV panel admittance evaluated in a given operation point. Y_{pv} is defined in Equation (32), which is obtained by deriving i_{pv} in Equation (1) with respect to v_{pv}.

$$i_{PV} = Y_{pv} \cdot v_{PV} \tag{31}$$

$$Y_{pv} = \frac{\partial i_{pv}}{\partial v_{pv}} = -\frac{\frac{A}{B} \exp\left(\frac{v_{pv} + R_s \cdot i_{pv}}{B}\right) + \frac{1}{R_h}}{1 + \frac{A \cdot R_s}{B} \exp\left(\frac{v_{pv} + R_s \cdot i_{pv}}{B}\right) + \frac{R_s}{R_h}} \tag{32}$$

Replacing Equation (31) into Equation (29), and applying the Laplace transformation, leads to the transfer function between the PV voltage and the reference voltage provided by the P&O algorithm shown in Equation (33), in which $V_{pv}(s)$ and $V_{mppt}(s)$ are the Laplace transformations of v_{PV} and v_{mppt}, respectively.

$$\frac{V_{pv}(s)}{V_{mppt}(s)} = \frac{\frac{k_{pv}}{C_{pv}} \cdot s + \frac{\lambda_{pv}}{C_{pv}}}{s^2 + \frac{k_{pv} - Y_{pv}}{C_{pv}} \cdot s + \frac{\lambda_{pv}}{C_{pv}}} \tag{33}$$

Equations (29) and (33) put into evidence that v_{pv} is decoupled from v_b due to the action of the SMC, hence, the variations in v_b caused by mismatched conditions will not disturb the MPPT action.

However, the transfer function in Equation (33) depends on Y_{pv}, which in turn depends on the operation point of the PV panel; therefore, the variation range of Y_{pv} must be analyzed to perform a correct design of the SMC parameters k_{pv} and λ_{pv}. Considering the same BP585 PV panel used in the previous examples, the current and power curves of such a PV panel are given in Figure 8 for multiple photo-induced currents (i.e., different irradiance conditions) at the expected PV panel temperature (298 K). The data show that the MPP conditions are constrained within the voltage range 16 V $< v_{pv} <$ 19 V, and taking into account that the SMC reference is provided by an MPPT algorithm, then the analysis of Y_{pv} must be performed within the same voltage range.

Figure 9 shows the admittance of the BP585 PV panel, calculated using Equation (32), for the interest voltage range 16 V $< v_{pv} <$ 19 V. The figure put into evidence that Y_{pv} is almost independent from the photo-induced current, which is also observed in Equation (32) because $\partial i_{pv}/\partial v_{pv}$ does not depend directly on i_{ph}. Finally, the admittance range used to design the parameters k_{pv} and λ_{pv} is $-0.40 \ \Omega^{-1} \le Y_{pv} \le -0.03 \ \Omega^{-1}$. It must be noted that the analysis of Y_{pv} must be performed for the particular PV panels to be used in the photovoltaic installation.

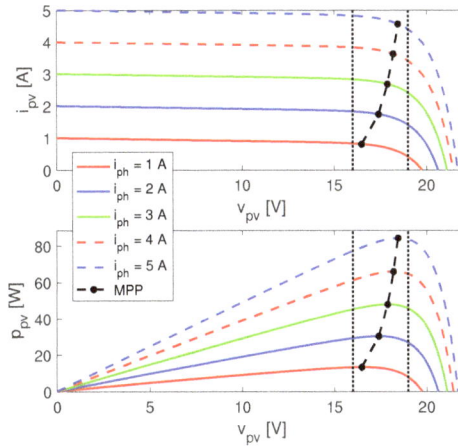

Figure 8. Current and power curves of the BP585 PV panel for $1\,A < i_{ph} < 5\,A$.

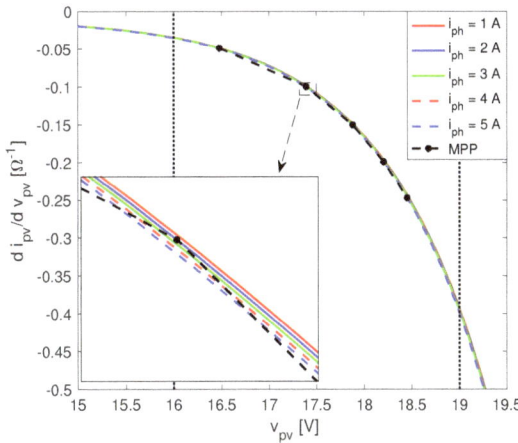

Figure 9. Admittance of the BP585 PV panel for $16\,V < v_{pv} < 19\,V$ and $1\,A < i_{ph} < 5\,A$.

4.3.2. Equivalent Dynamics in Protection Mode

As in the previous case, the equivalent dynamics are calculated by replacing the open-loop differential equation describing the inductor current, Equation (8), with the sliding-surface imposed by $\{\Psi_b = 0\}$ in Equation (12). This procedure is reported in expression (34), in which the differential equation describing v_b, Equation (7), has been modified to depend on the converter duty cycle d. Hence the equivalent dynamics disregard the switching ripple.

$$
\begin{cases}
\dfrac{dv_{pv}}{dt} &= \dfrac{i_{pv} - i_L}{C_{pv}} \\[2mm]
\dfrac{dv_b}{dt} &= \dfrac{i_L \cdot (1 - d) - i_{dc}}{C_b} \\[2mm]
i_L &= -k_b \cdot (v_b - V_{max}) - \lambda_b \cdot \displaystyle\int (v_b - V_{max})\, dt
\end{cases}
\tag{34}
$$

The dynamic system in Equation (34) is used to analyze the deviation of v_b from V_{max} caused by perturbations in the other DMPPT converters connected in series. Applying the Laplace transformation to the previous expression leads to the following transfer function between the output $V_b(s)$ and reference $V_{max}(s)$ voltages:

$$\frac{V_b(s)}{V_{max}(s)} = \frac{\frac{(1-d)\cdot k_b}{C_b}\cdot s + \frac{(1-d)\cdot\lambda_b}{C_b}}{s^2 + \frac{(1-d)\cdot k_b}{C_b}\cdot s + \frac{(1-d)\cdot\lambda_b}{C_b}} \tag{35}$$

The dynamic behavior of Equation (35) changes with the duty cycle d, which must be analyzed to design k_b and λ_b. For the example developed in this paper, the range of the PV voltage is 16 V $< v_{pv} <$ 19 V, which leads to $0.37 < d < 0.68$ because $v_b = V_{max} = 50$ V. As in the MPPT mode, this analysis of d must be performed for the particular PV panel and dc-link voltage to be used in the photovoltaic installation.

5. Parameters Design of the Proposed SMC

Equivalent dynamic models, introduced in Section 4.3, are used in Sections 5.1 and 5.2 for the design of the surface parameters k_{pv}, λ_{pv}, k_b and λ_b to impose a desired close loop dynamics of v_{pv} and v_b. Such dynamic behavior is defined as a maximum settling time and a maximum overshoot for all the operation conditions of the DMPPT-U. Furthermore, dynamic restrictions of the proposed SMC are discussed in Section 5.3, while the switching frequencies and hysteresis bands are analyzed in Section 5.4.

5.1. Parameters Design in MPPT Mode

The proposed procedure starts by defining a maximum settling-time (t_s^*) and a maximum overshoot (MO^*) for v_{pv} considering the restrictions imposed by the MPPT technique. The next step is to identify the feasible couples of parameters (k_{pv}, λ_{pv}). A couple (k_{pv}, λ_{pv}) is feasible if the small signal voltage (v_{PV}) settling time (t_s) and maximum overshoot (MO) fulfill $t_s \leq t_s^*$ and $MO \leq MO^*$, for all the possible operation points defined by min $(Y_{pv}) \leq Y_{pv} \leq$ max (Y_{pv}); where t_s and MO, for a given value of Y_{pv}, are calculated from Equation (33). Finally, the feasible couples (k_{pv}, λ_{pv}) are compared using a proposed indicator and the exact values are selected close to the indicator's maximum value.

The definition of t_s^* and MO^* is based on the MPPT parameters and power efficiency to provide the following time response criteria:

- Settling-time t_s, which must be shorter than the perturbation period T_a of the P&O algorithm to ensure the MPPT stability [41].
- Maximum Overshoot MO, which must be limited to avoid large deviations from the MPP voltage that produces high power losses.

The evaluation of those criteria requires the calculation of the time response of the PV voltage. Taking into account that the P&O produce step perturbations of Δv_{mppt} volts each T_a seconds, the PV voltage response is given by Equation (36).

$$V_{pv}(s) = \frac{\frac{k_{pv}}{C_{pv}}\cdot s + \frac{\lambda_{pv}}{C_{pv}}}{s^2 + \frac{k_{pv}-Y_{pv}}{C_{pv}}\cdot s + \frac{\lambda_{pv}}{C_{pv}}} \cdot \frac{\Delta v_{mppt}}{s} \tag{36}$$

The time-domain expression of the small signal PV voltage for a particular Y_{pv} ($v_{PV}(t)$) is calculated by applying the inverse Laplace transformation to Equation (36), i.e., $v_{PV}(t) = \mathcal{L}^{-1}\{V_{pv}(s)\}$, which corresponds to the step-response of a second-order system with a real zero.

Appendix A reports the time-domain expressions for the step-response of a canonical second-order system with a real zero for the three possible types of poles: real and different, real and equal, and conjugated complex values. Therefore, $V_{pv}(s)$ is rewritten as given in Equation (37) to take profit of the Appendix A expressions.

$$V_{pv}(s) = \frac{a \cdot s + b}{(s+p) \cdot (s+q)} \cdot \frac{1}{s} \quad \text{with} \quad \begin{cases} a = \frac{k_{pv}}{C_{pv}} \cdot \Delta v_{mppt} \\ b = \frac{\lambda_{pv}}{C_{pv}} \cdot \Delta v_{mppt} \\ p + q = \frac{k_{pv} - Y_{pv}}{C_{pv}} \\ p \cdot q = \frac{\lambda_{pv}}{C_{pv}} \end{cases} \tag{37}$$

Appendix A also reports the expressions for the voltage derivative $\frac{dv_{PV}(t)}{dt}$ and for the time t_{MO} at which the maximum overshoot MO occurs, i.e., the earliest time for $\frac{dv_{PV}(t)}{dt} = 0$.

Then, the maximum overshoot MO is calculated as shown in Equation (38).

$$MO = v_{PV}(t_{MO}) \tag{38}$$

Similarly, the settling time t_s is calculated from Equation (39), which corresponds to the instant in which $v_{PV}(t)$ enters into a band of $\pm\epsilon$ % around the final value Δv_{mppt} and keeps trapped inside. Commonly accepted values for the band are $\epsilon = 2\%$, $\epsilon = 5\%$ and $\epsilon = 10\%$ [42].

$$\begin{cases} \left| \frac{v_{PV}(t_s)}{\Delta v_{mppt}} - 1 \right| = \epsilon \\ \left| \frac{v_{PV}(t)}{\Delta v_{mppt}} - 1 \right| < \epsilon \quad \forall\, t > t_s \end{cases} \tag{39}$$

Equations (38) and (39) can be solved using different tools: processing the time-domain expressions for $v_{pv}(t)$, given in Appendix A, to calculate both MO and t_s as reported in [43]; transforming Equation (36) into differential equations, which must be simulated using numerical methods [44] to find the solutions of Equations (38) and (39); or using specialized functions like *stepinfo()* from the Control systems toolbox of Matlab [45], which calculates both MO and t_s values.

To ensure a correct behavior of the PV voltage, within the MPP range defined in Section 4.3.1, by using the small signal approximation, the SMC parameters k_{pv} and λ_{pv} must ensure that the small signal PV voltage exhibits settling times and maximum overshoots lower than the desired limits t_s^* and MO^*, respectively, for all the admittance values within the interesting range:

$$\begin{cases} MO \leq MO^* \\ t_s \leq t_s^* \end{cases} \quad \forall\ \min(Y_{pv}) \leq Y_{pv} \leq \max(Y_{pv}) \tag{40}$$

Therefore, a feasible couple (k_{pv}, λ_{pv}) must fulfill expression (40), where t_s and MO are evaluated by using expressions (38) and (39) for each value of Y_{pv}. In this paper, feasible couples (k_{pv}, λ_{pv}) are identified by using a Monte Carlo analysis [46] to evaluate a wide range of k_{pv} and λ_{pv} values. An example of the obtained results, for $\{t_s^* = 0.5$ ms, $\epsilon = 5\%$, $MO^* = 10\%$, $C_{pv} = 22\ \mu F$, $\Delta v_{mppt} = 0.5$ V$\}$, is shown in Figure 10, which reports the valid k_{pv} and λ_{pv} values that fulfill restrictions (40) at the minimum and maximum values of Y_{pv} considering a DMPPT-U formed by a BP585 PV panel, a boost converter constructed with an input capacitor $C_{pv} = 22\ \mu F$, and governed by a P&O algorithm with a perturbation magnitude $\Delta v_{mppt} = 0.5$ V.

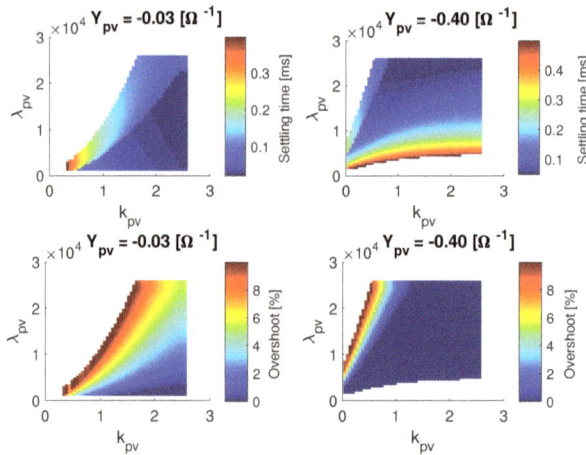

Figure 10. k_{pv} and λ_{pv} values that fulfill restrictions in expression (40) for $\min(Y_{pv})$ and $\max(Y_{pv})$ conditions.

The results reported in Figure 10 are useful to analyze the influence of k_{pv} and λ_{pv} into the performance criteria: increasing k_{pv} and λ_{pv} reduce both the settling time and maximum overshoot. However, increasing the values of k_{pv} and λ_{pv} also increases the magnitude of the switching noise transferred into the control system [47]. Therefore, this paper proposes to select k_{pv} and λ_{pv} near the lowest values fulfilling restrictions (40). The selection is performed using the Balance Ratio BR defined in Equation (41), which enables to compare the k_{pv} and λ_{pv} values fulfilling (40) in the entire interest range of the PV admittance.

$$BR = \max\left(\left[\frac{1}{2}\cdot\frac{t_s}{t_s^*} + \frac{1}{2}\cdot\frac{MO}{MO^*}\right]\Big|_{Y_{pv}\in[\min(Y_{pv}),\max(Y_{pv})]}\right) \forall \; \{t_s \le t_s^* \wedge MO \le MO^*\} \tag{41}$$

The Balance Ratio for a couple (k_{pv}, λ_{pv}) is not valid if $MO > MO^*$ or $t_s > t_s^*$ in at least one admittance condition. Moreover, the Balance Ratio is equal to one if $MO = MO^*$ and $t_s = t_s^*$ in at least one admittance condition. Hence, k_{pv} and λ_{pv} must be selected near the highest Balance Ratio calculated for the DMPPT converter, since a low value of BR implies an increment in k_{pv} and λ_{pv} and, as consequence, an unnecessary increment in both the control effort and the switching noise transferred to the control system [47]. Figure 11 shows the BR values for the example developed in this subsection, where it is observed that the higher values of BR are obtained for the lower feasible values of k_{pv} and λ_{pv}. These results help to select the values $k_{pv} = 0.6878$ and $\lambda_{pv} = 4347$. Those values provide a $BR = 0.8678$, which is close to the maximum condition $\max(BR) = 0.9618$, but it is not at the validity frontier. This selection provides a safety margin against tolerances in the elements of the PV system and small differences between the PV voltage and its small signal approximation used to calculate t_s and MO for $\min(Y_{pv}) \le Y_{pv} \le \max(Y_{pv})$.

Figure 12 shows the simulation of equivalent dynamics in the PV voltage, given in Equation (36), considering the designed k_{pv} and λ_{pv} values for the admittance values obtained in the previous subsection. The simulation confirms that both settling time and maximum overshoot of the PV voltage are below the imposed limits in all the admittance conditions. Therefore, the SMC based on Ψ_{pv} (11), and implemented with the selected designed parameters, always fulfills the performance criteria imposed by the expression (40).

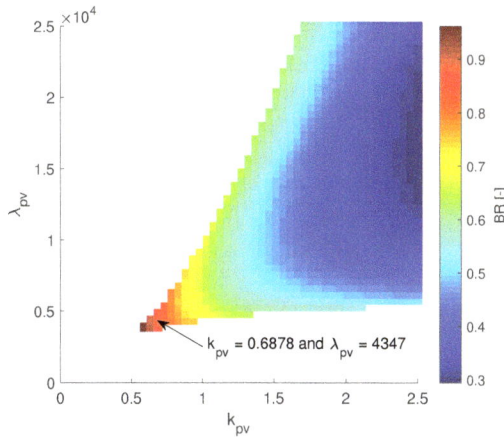

Figure 11. Balance Ratio for k_{pv} and λ_{pv} values that fulfill restrictions in expression (40).

In conclusion, this section presented a design process to calculate the parameters of Ψ_{pv} to fulfill both $t_s \leq t_s^*$ and $MO \leq MO^*$, which ensures a correct operation of the associated P&O algorithm and avoids excessive power losses due to transient voltage deviations from the MPP.

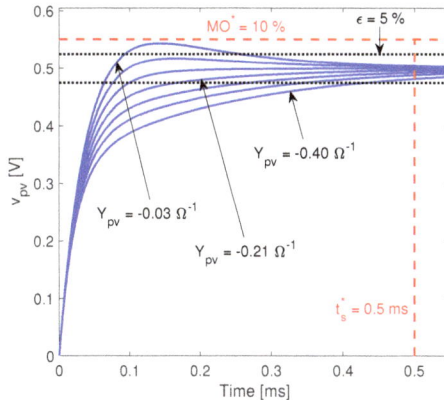

Figure 12. Simulation of the equivalent dynamics in the PV voltage for $k_{pv} = 0.6878$ and $\lambda_{pv} = 4347$.

5.2. Parameters Design in Protection Mode

The proposed procedure is similar to the one in MPPT mode. The first step is to define the maximum settling-time (t_s^*) and the maximum overshoot (MO^*) for v_b. The second step is to identify the feasible couples of parameters (k_b, λ_b). Finally, the feasible couples (k_b, λ_b) are compared by using the Balance Ratio (BR) and the exact values are defined close to the maximum value of BR. A couple (k_b, λ_b) is feasible if the output voltage (v_b) t_s and MO fulfill $t_s \leq t_s^*$ and $MO \leq MO^*$, for all the possible operation points defined by min (d) $\leq d \leq$ max (d); where t_s and MO, for a given value of d, are calculated from Equation (35).

The evaluation of t_s and MO requires to calculate the time response of the output voltage for a perturbation. In this case, it is considered the fastest perturbation possible, which corresponds to a deviation step of magnitude ΔV_{max} in v_b.

The time-domain expression of the output voltage $v_b(t)$, in response to the step perturbation ΔV_{max}, corresponds to the step-response of a second-order system with a real zero. Appendix A reports the time-domain expressions for this type of system in canonical form. The Laplace representation of $V_b(s)$ is rewritten as given in (42) to take profit from the Appendix A expressions.

$$V_b(s) = \frac{a \cdot s + b}{(s+p) \cdot (s+q)} \cdot \frac{1}{s} \quad \text{with} \quad \begin{cases} a = \frac{(1-d) \cdot k_b}{C_b} \cdot \Delta V_{max} \\ b = \frac{(1-d) \cdot \lambda_b}{C_b} \cdot \Delta V_{max} \\ p+q = \frac{(1-d) \cdot k_b}{C_b} \\ p \cdot q = \frac{(1-d) \cdot \lambda_b}{C_b} \end{cases} \tag{42}$$

From the expressions of the voltage derivate and the time t_{MO}, at which the maximum overshoot MO occurs, the following conditions are formulated:

$$MO = v_b(t_{MO}) \tag{43}$$

$$\begin{cases} \left| \frac{v_b(t_s)}{\Delta V_b} - 1 \right| = \epsilon \\ \left| \frac{v_b(t)}{\Delta V_b} - 1 \right| < \epsilon \quad \forall\, t > t_s \end{cases} \tag{44}$$

As discussed in the MPPT mode, Equations (43) and (44) can be solved using different tools. To ensure a correct behavior of the output voltage within the range defined in the previous subsection, the SMC parameters k_b and λ_b must be analyzed in all the operation range of the Protection mode $(\min(d) \le d \le \max(d))$ as given in expression (45), in which t_s^* and MO^* are the desired maximum settling time and maximum overshoot, respectively.

$$\begin{cases} MO \le MO^* \\ t_s \le t_s^* \end{cases} \quad \forall\, \min(d) \le d \le \max(d) \tag{45}$$

A feasible couple (k_b, λ_b) fulfills expression (45), where MO and t_s are calculated by using expressions (43) and (44) for each value of d. As in MPPT mode, feasible (k_b, λ_b) are identified using a Monte Carlo analysis for different values of k_b and λ_b. An example of the obtained results for $\max(d)$ and $\min(d)$ is shown in Figure 13 for $\{t_s^* = 0.5 \text{ ms}, \epsilon = 2\%, MO^* = 5\%, C_b = 44\ \mu F\}$. Such a figure reports the valid k_b and λ_b values at the minimum and maximum values of d.

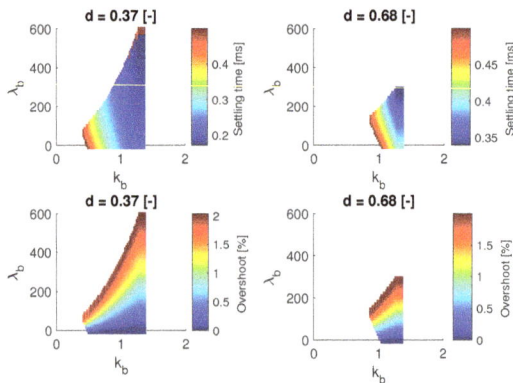

Figure 13. k_b and λ_b values that fulfill restrictions in expression (45) for $\min(d)$ and $\max(d)$ conditions.

The results reported in Figure 13 help to analyze the influence of the parameters into the performance criteria. Moreover, Figure 14 shows the Balance Ratio (BR) values for the example developed in this paper, which helps to select the values $k_b = 1.303$ and $\lambda_b = 221$. Those values provide a tradeoff between settling-time and overshoot; furthermore, the selected $BR = 0.76$ provides a safety margin between t_s and MO of v_b and the limits t_s^* and MO^* for the different operating conditions of the Protection mode.

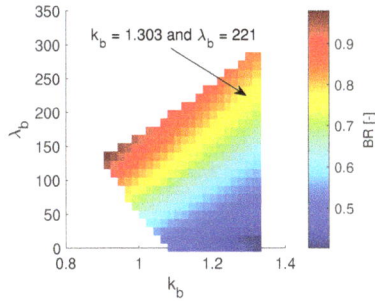

Figure 14. Balance Ratio for k_b and λ_b values that fulfill restrictions in expression (45).

5.3. Dynamic Restrictions

Dynamic restrictions are only present in MPPT mode, since in Protection mode the reference V_{max} is a constant value. Moreover, the reachability analysis in Protection mode (Section 4.2.2) showed that the proposed SMC is stable if expression (20) is fulfilled. Nevertheless, in the analysis of the reachability conditions in MPPT mode (Section 4.2.1) it was demonstrated that dynamic restrictions reported in expressions (25) and (26) must be fulfilled to ensure a stable operation. Those restrictions impose limits to the slew-rate of the voltage reference v_{mppt} provided by the P&O algorithm.

The example developed up to now is used to illustrate the evaluation of expressions (25) and (26), adopting an inductor $L = 330 \ \mu H$ for the construction of the DMPPT converter. Moreover, the same DC-link voltage levels analyzed in Section 2 are considered, i.e., $v_{dc} = 80$ V and $V_{max} = 50$ V, and the interesting range of the PV voltage defined in Section 4.3 is also needed, i.e., 16 V $< v_{pv} <$ 19 V. From that information the voltage parameters needed to compute expressions (25) and (26) are calculated: min $(v_{pv}) = 16$ V, max $(v_{pv}) = 19$ V, min $(v_b) = v_{dc} - V_{max} = 30$ V.

Figure 15 reports the limit values for $\frac{dv_{mppt}}{dt}$ to fulfill the dynamic restrictions imposed by expressions (25) and (26). Moreover, the figure also puts into evidence that high values of k_{pv} and λ_{pv} reduce significantly the maximum slew-rate allowed for v_{mppt}, which could constraint the speed of the P&O algorithm. Therefore, as proposed in the previous subsection, k_{pv} and λ_{pv} must be selected near to the smallest valid values. For example, the adopted values $k_{pv} = 0.6878$ and $\lambda_{pv} = 4347$ impose a $\frac{dv_{mppt}}{dt} = 0.0453 \ V/\mu s$, which is near to the maximum limit $0.0532 \ V/\mu s$ achieved at the left-side frontier in Figure 15. In contrast, the highest values for k_{pv} and λ_{pv} reported in Figure 15 will impose a maximum slew-rate equal to $0.0083 \ V/\mu s$, which is 5.5 times smaller than the adopted one, hence slowing-down the system response.

This slew-rate limitation for the P&O output signal could be done inside the micro-processor running the MPPT algorithm or using an analog circuit based on operational amplifiers.

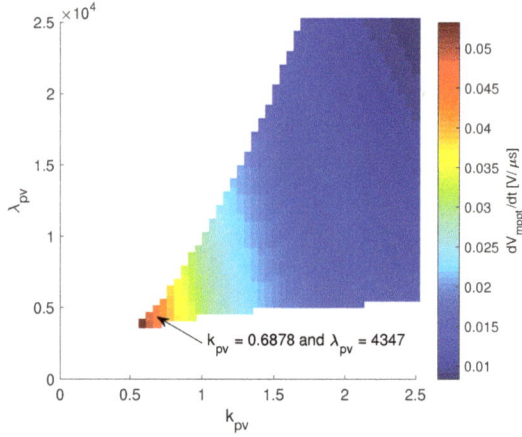

Figure 15. Limit values for $\frac{dv_{mppt}}{dt}$ to fulfill the dynamic restrictions imposed by expressions (25) and (26).

5.4. Switching Frequency and Hysteresis Band

Practical implementations of sliding-mode controllers require to add an hysteresis band H around the sliding-surface to constrain the switching frequency to the limits supported by commercial MOSFETs [38]. This section shows the procedure to define H in MPPT and Protection modes to warrant a switching frequency less than a maximum value.

5.4.1. Switching Frequency and Hysteresis Band in MPPT Mode

The practical implementation of Ψ_{pv} requires the transformation of the sliding-surface from $\{\Psi_{pv} = 0\}$ to:

$$\left|\Psi_{pv}(t)\right| < \frac{H}{2} \tag{46}$$

Due to the SMC operation, in steady state $v_{pv} = v_{mppt}$, which imposes an almost constant PV voltage, hence the integral term of $(v_{pv} - v_{mppt})$ in Ψ_{pv} is constant in steady-state. In addition, due to the flux balance principle [40], the steady-state inductor current is formed by two components, a constant average value I_L and a triangular current ripple $\delta i_L(t)$ with peak amplitude Δi_L. Those conditions impose the following steady-state behavior:

$$\text{In steady} - \text{state}: \begin{cases} k_{pv} \cdot (v_{pv} - v_{mppt}) \approx 0 \\ \lambda_{pv} \cdot \int (v_{pv} - v_{mppt}) \ dt \approx I_L \end{cases} \tag{47}$$

Therefore, the steady-state expression for the modified switching function $\Psi_{pv}(t)$ is equal to the waveform of the inductor current ripple:

$$-\frac{H}{2} < \Psi_{pv}(t) = \delta i_L(t) < \frac{H}{2} \tag{48}$$

The peak value Δi_L of $\delta i_L(t)$ is calculated from the differential equation of the inductor current, Equation (8), as given in Equation (49), in which F_{sw} represents the switching frequency and $d = \frac{v_b - v_{pv}}{v_b}$.

is the duty cycle. Since $\Psi_{pv}(t)$ has peak values $\pm\frac{H}{2}$ imposed by the hysteresis band in Equation (48), the value of H that ensures the desired steady-state switching frequency is given by Equation (50).

$$\Delta i_L = \frac{v_{pv} \cdot d}{2 \cdot L \cdot F_{sw}} \tag{49}$$

$$H = \frac{v_{pv} \cdot (v_b - v_{pv})}{v_b \cdot L \cdot F_{sw}} \tag{50}$$

The value of H must be designed for the worst-case scenario of Equation (50) to limit the switching frequency to the MOSFET's admissible conditions; such a worst-case scenario is obtained by analyzing the minimum values of H. The worst-case value of v_b is analyzed using the partial derivative of H, with respect to v_b, given in Equation (51): increments in v_b produce reductions in $\partial H / \partial v_b$, hence the worst-case corresponds to the maximum value of v_b, i.e., V_{max}.

$$\frac{\partial H}{\partial v_b} = \frac{v_{pv}^2}{L \cdot F_{sw} \cdot v_b^2} > 0 \tag{51}$$

Similarly, the worst-case value of v_{pv} is analyzed using the partial derivative given in Equation (52): if $v_b > 2 \cdot v_{pv}$ then the worst-case corresponds to the maximum value of v_{pv}; if $v_b < 2 \cdot v_{pv}$ then the worst-case corresponds to the minimum value of v_{pv}.

$$\frac{\partial H}{\partial v_{pv}} = \frac{v_b - 2 \cdot v_{pv}}{L \cdot F_{sw} \cdot v_b} \Rightarrow \begin{cases} \frac{\partial H}{\partial v_{pv}} > 0 & \text{if} \quad v_b > 2 \cdot v_{pv} \\[2mm] \frac{\partial H}{\partial v_{pv}} < 0 & \text{if} \quad v_b < 2 \cdot v_{pv} \end{cases} \tag{52}$$

The conditions of the example developed up to now impose $v_b > 2 \cdot v_{pv}$: $V_{max} = 50$ V, $\min (v_{pv}) = 16$ V and $\max (v_{pv}) = 19$ V. Hence, to ensure a maximum switching frequency $F_{sw} = 40$ kHz, the hysteresis band must be set to $H = 0.8924$ A.

5.4.2. Switching Frequency and Hysteresis Band in Protection Mode

The practical implementation of Ψ_b requires to transform the sliding-surface from $\{\Psi_b = 0\}$ to:

$$|\Psi_b(t)| < \frac{H}{2} \tag{53}$$

For the steady-state operation of the SMC it is possible to assume $v_b = V_{max}$, hence the integral of $(v_b - V_{max})$ in Ψ_b is constant. Moreover, as in the MPPT mode, the steady-state inductor current is formed by a constant average value I_L and a triangular current ripple $\delta i_L(t)$ with peak amplitude Δi_L. Those conditions impose the following steady-state behavior:

$$\text{In steady} - \text{state} : \begin{cases} k_b \cdot (v_b - V_{max}) \approx 0 \\ \lambda_b \cdot \int (v_b - V_{max}) \, dt \approx -I_L \end{cases} \tag{54}$$

Therefore, the steady-state expression of $\Psi_b(t)$ is given in Equation (55). This expression is analogous to the modified switching function $\Psi_{pv}(t)$ of the MPPT mode given in Equation (48), hence the value of H that ensures the desired steady-state switching frequency is the same one obtained for the MPPT mode, i.e., expression (50).

$$-\frac{H}{2} < \Psi_b(t) = \delta i_L(t) < \frac{H}{2} \tag{55}$$

Moreover, the analysis of H developed for the MPPT mode also holds for the Protection mode. Therefore, $H = 0.8924$ A is calculated for the example developed in the paper, which is the same value obtained for the MPPT mode.

6. Implementation of the Proposed SMC

The explanation of the proposed SMC implementation is divided into two main parts. The first one is introduced in Section 6.1, which focuses on the explanation of the block diagrams to calculate the switching functions Ψ_{pv} and Ψ_b, as well as the block diagram of the switching circuit to generate u from Ψ_{pv}, Ψ_b and H. The second part is presented in Section 6.2 and it explains the proposed approach to implement the SMC by using a combination of a microprocessor and an analog circuit.

6.1. Implementation Structure

The implementation of the proposed SMC based on Ψ_{pv} and Ψ_b includes three main block diagrams: the synthesis of the sliding function Ψ_{pv}, synthesis of the sliding function Ψ_b and the switching circuit. It is worth noting that only one switching circuit is required because the transversality sign and value of H are the same in MPPT and Protection modes.

6.1.1. Implementation Structure in MPPT Mode

The on-line calculation of Ψ_{pv}, shown in Equation (11), requires the measurement of the inductor current i_L, the PV voltage v_{pv} and the reference provided by the MPPT algorithm with the dynamic restriction analyzed in Section 5.3. Figure 16 presents the block diagram proposed to synthesize Ψ_{pv}, which could be implemented using analog circuits, i.e., Operational Amplifiers (OPAM), or using a digital microprocessor with Analog-to-Digital Converters (ADC) and Digital-to-Analog Converters (DAC). In the digital case, both the calculation of Ψ_{pv} and the limitation of the slew-rate of v_{mppt} can be implemented in the same microprocessor in charge of processing the P&O algorithm to reduce the complexity, size and cost of the implementation.

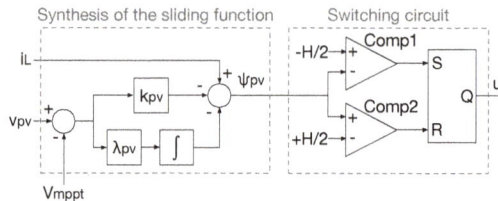

Figure 16. Block diagram implementation of the SMC for the MPPT mode.

The switching law, shown in expression (21), is in charge of producing the MOSFET control signal u. However, due to the introduction of the hysteresis band (see Equation (48)), the modified switching law that must be implemented is introduced in Equation (56).

$$\begin{cases} \Psi_{pv} \leq -\frac{H}{2} \rightarrow \text{set } u = 1 \text{ (MOSFET on)} \\ \\ \Psi_{pv} \geq +\frac{H}{2} \rightarrow \text{set } u = 0 \text{ (MOSFET off)} \end{cases} \tag{56}$$

The switching circuit implementing this law is constructed using two analog comparators and a Flip-Flop S-R. The comparators, Comp1 and Comp2, detect the switching conditions to trigger the change of u in the Flip-Flop. Figure 16 presents the proposed switching circuit.

6.1.2. Implementation Structure in Protection Mode

The implementation of the SMC based on Ψ_b requires the synthesis of the sliding function and the switching circuit. However, since the transversality sign and value of H are the same ones required to implement Ψ_{pv}, the switching circuit is the same one described in Figure 16.

The block diagram to calculate on-line Ψ_b, Equation (12), is presented in Figure 17. This circuit measures the inductor current i_L, the output voltage v_b and the reference. As in the MPPT mode, the proposed structure could be implemented using analog circuits or a digital microprocessor. The advantage of using a digital implementation concerns the integration of the sliding function calculation for both MPPT and Protection modes into a single device.

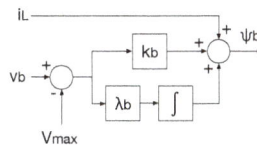

Figure 17. Block diagram to synthesize Ψ_b in Protection mode.

6.2. Control System Implementation and Modes Transition

Both SMC components based on Ψ_{pv} and Ψ_b are implemented into a single circuit to provide a simple and low cost solution. This is possible, in part, due to the fact that both SMC components exhibit a positive transversality value, Equations (16) and (19), which enable the adoption of the same switching circuit for both modes. Moreover, the structure of the switching functions makes it simple to unify the online calculation of Ψ into a single device. For that purpose, this paper proposes to divide the calculation process of Ψ into two steps: a digital step to process the calculations based on the voltages, and an analog step to process the calculations based on the faster changes present in the inductor current. This approach has been successfully used to implement other SMC for PV systems [38] and to implement SMC with variable switching functions [48], which is the type of solution proposed in this paper.

Then, the calculation of Ψ_{pv} and Ψ_b is performed as follows:

- **Digital step:** it is executed inside a microprocessor, which measures v_{pv}, i_{pv} and v_b, to calculate the intermediate variables $i^*_{L,pv}$ and $i^*_{L,b}$ reported in Equations (57)–(60), respectively, where the terms int_{pv} and int_b are the discrete integral terms of Ψ_{pv} and Ψ_b, respectively, processed with the forward Euler method, while δt corresponds to the time between two measurements performed by the ADC of the microprocessor. Intermediate variables $i^*_{L,pv}$ and $i^*_{L,b}$ correspond to the algebraic sum of second and third terms of Ψ_{pv} (Equation (11)) and Ψ_b (Equation (12)), respectively.

$$i^*_{L,pv} = -k_{pv} \cdot (v_{pv} - v_{mppt}) - \lambda_{pv} \cdot \text{int}_{pv} \tag{57}$$

$$\text{int}_{pv} = \text{int}_{pv} + \delta t \cdot (v_{pv} - v_{mppt}) \tag{58}$$

$$i^*_{L,b} = k_b \cdot (v_b - V_{max}) + \lambda_b \cdot \text{int}_b \tag{59}$$

$$\text{int}_b = \text{int}_b + \delta t \cdot (v_b - V_{max}) \tag{60}$$

- **Analog step:** it adds the measurement of i_L with $i^*_{L,pv}$ or $i^*_{L,b}$, provided by DAC of the microprocessor, to complete the calculation of $\Psi = \Psi_{pv}$ or $\Psi = \Psi_b$ depending on the active mode as reported in Equation (61), where Pr was already defined in Equation (10). This process is performed using OPAMs to provide a negligible delay between i_L and Ψ; this is needed to detect the instants in which u must be changed following Equation (48).

$$\Psi = i_L + i_L^* \text{ with } i_L^* = \begin{cases} i_{L,pv}^* & \text{for } Pr = 0 \\ \\ i_{L,b}^* & \text{for } Pr = 1 \end{cases} \tag{61}$$

Finally, the analog value of Ψ is delivered to the switching circuit, which produces the control signal u driving the MOSFET of the DMPPT converter in both modes.

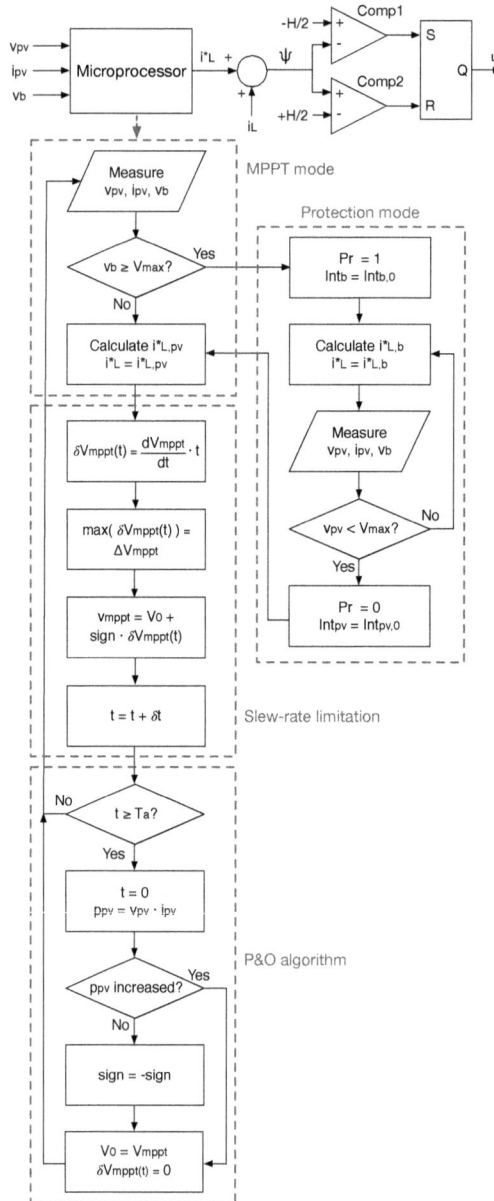

Figure 18. Implementation of the control system.

Figure 18 summarizes the hybrid analog-digital implementation of the proposed solution. The figure also shows the digital implementation of the P&O algorithm and the slew-rate limitation, both operating only in MPPT mode. This avoids the divergence of the P&O algorithm from the MPP zone when the DMPPT converter is operating in Protection mode. In the slew-rate limitation block the term $\delta V_{mppt}(t)$ describes the variation of the MPPT reference (v_{mppt}) for each time step (δt), which must fulfill the dynamic restriction in $\frac{dv_{mppt}}{dt}$ imposed by expressions (25) and (26).

Figure 18 also describes the transitions between MPPT and Protection modes:

- **MPPT to Protection:** this transition occurs when the output voltage v_b reaches the maximum safe value V_{max}, which activates the routine for Protection mode (Pr is set to 1). This routine starts by initializing the integral term of Ψ_b as given in Equation (62), which forces the inductor current to be close to the value previously defined by the MPPT mode; without this initialization the inductor current will be reset to zero, or to any other value far from its previous condition, which could produce strong perturbations on the DMPPT voltages.

$$\text{int}_b = \text{int}_{b,0} = \frac{i_L^*}{\lambda_b} \tag{62}$$

- **Protection to MPPT:** this transition occurs when the PV voltage v_{pv} enters the MPPT range, which in the example is 16.5 V $< v_{pv} <$ 18.5 V for the adopted BP585 panel. In such a condition the algorithm sets the variable Pr equal to 0, which activates the MPPT mode routine. This routine starts by initializing the integral term of Ψ_{pv} as given in Equation (63) to ensure a stable inductor current in the transition. Entering the MPPT mode enables the operation of the P&O algorithm, which delivers the reference value calculated at the end of the last MPPT mode activation.

$$\text{int}_{pv} = \text{int}_{pv,0} = \frac{i_L^*}{\lambda_{pv}} \tag{63}$$

7. Simulation Results

The DMPPT system formed by two DMPPT converters connected in series, previously presented in Figure 2, was implemented in the power electronics simulator PSIM to validate the previous analyses. Each DMPPT converter drives a BP585 PV panel with the same circuital implementation presented in Figure 6. The SMC in each DMPPT converter corresponds to the hybrid analog-digital implementation described in Figure 18. Finally, the BP585, boost converter and controller parameters were the same ones defined in the previous sections of the paper: $C_{pv} = 22\ \mu F$, $L = 330\ \mu H$ and $C_b = 44\ \mu F$ for the boost converters, $A = 154.15\ \mu A$, $B = 1.1088\ V^{-1}$, $R_s = 0.0045\ \Omega$ and $R_h = 109.405\ \Omega$ for the BP585 panels and $k_{pv} = 0.6878$, $\lambda_{pv} = 4347$, $k_b = 1.303$, $\lambda_b = 221$, $\Delta v_{mppt} = 0.5$ V, $T_a = 1$ ms, $H = 0.8924$, $V_{max} = 50$ V, and $\frac{dv_{mppt}}{dt} = 0.0453\ V/\mu s$ for the controller.

The simulation starts considering the two PV panels operating at 1000 W/m², i.e., in uniform conditions, which forces the output voltages of the DMPPT converters to be equal to 40 V. Figure 19 presents the simulation results, which depicts the operation in MPPT mode of both converters. Then, at 10 ms the irradiance of the second panel drops to 500 W/m², producing a mismatched condition that forces the output voltage of the first DMPPT converter to grow. After 1.1 ms $v_{b,1}$ reaches the maximum safe voltage $V_{max} = 50$ V, which triggers the Protection mode. From that moment the PV voltage $v_{pv,1}$ of the first panel diverges from the MPP value to reduce the power production, so that the output voltage is limited.

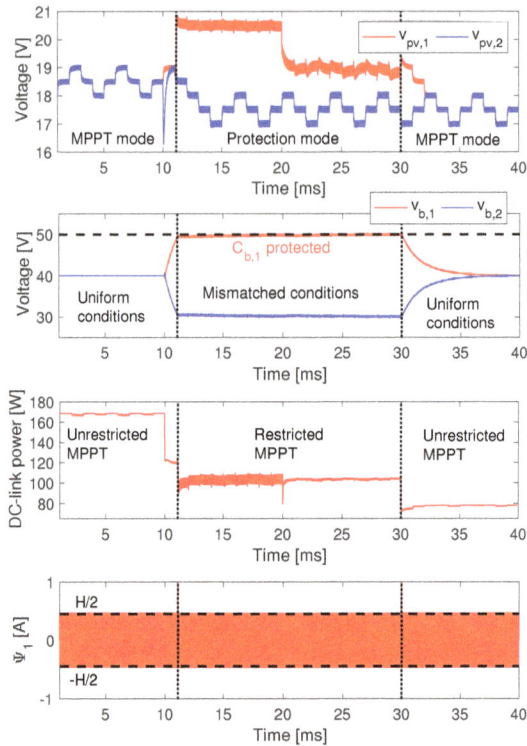

Figure 19. Simulation of the DMPPT system with the proposed control structure.

At 20 ms the irradiance of the first PV panel drops to 800 W/m², which requires the system to remain in Protection mode to avoid an overvoltage in $C_{b,1}$. Finally, at 30 ms the irradiance of the first panel drops to 500 W/m², leaving both panels in uniform conditions. Hence, 2.5 μs latter, the system enters in MPPT mode to start again the tracking of the MPP under safe conditions. The simulation also puts into evidence that the SMC is always stable: the switching function Ψ_1, corresponding to the DMPPT converter entering in both MPPT and Protection modes, operates inside the hysteresis band $-\frac{H}{2} < \Psi_1 < \frac{H}{2}$ in both modes under the presence of perturbations in the irradiance, output voltage and P&O reference. However, at the instants in which the modes transition occur (11.1060 ms and 30.0025 ms) the SMC leaves the hysteresis band, but the fulfillment of the reachability conditions forces the SMC to enter again in the band.

Figure 20 shows a zoom of the circuital simulation to verify the design requirements. The figure at the top shows the waveforms of the PV voltage and P&O reference for the first DMPPT-U operating in MPPT mode, which occurs between 6 ms and 8 ms. During that time the PV panel of the first DMPPT-U operates under an irradiance equal to 1000 W/m², and the SMC successfully fulfills the desired settling time $t_s \leq 0.5$ ms. Similarly, the overshoot is under the 10%. The figure at the middle also shows the waveforms of the PV voltage and P&O reference for the first DMPPT-U operating in MPPT mode, but this time under at irradiance equal to 500 W/m², which occurs between 36 ms and 38 ms. Again, the SMC successfully fulfills the desired settling time $t_s \leq 0.5$ ms and overshoot ($MO \leq 10\%$). The waveforms described in both MPPT conditions are in agreement with the equivalent dynamics analyses: at 1000 W/m² the MPP voltage is near 19 V, which corresponds to a PV module admittance near -0.4 Ω^{-1} according to the data reported in Figure 9. Then, from Figure 12 it is noted that such an admittance describes a PV voltage waveform without any overshoot and with a settling time equal to 0.5 ms, which corresponds to the waveform described by $v_{pv,1}$ at the top of Figure 20. Similarly, at 500 W/m^2 the MPP voltage is near 18 V, which corresponds to a PV module admittance near -0.16 Ω^{-1}; and Figure 12 reports that such an admittance describes a PV voltage waveform without any overshoot and with a settling time much shorter than 0.5 ms, which is equal to the waveform described by $v_{pv,1}$ in the middle of Figure 20.

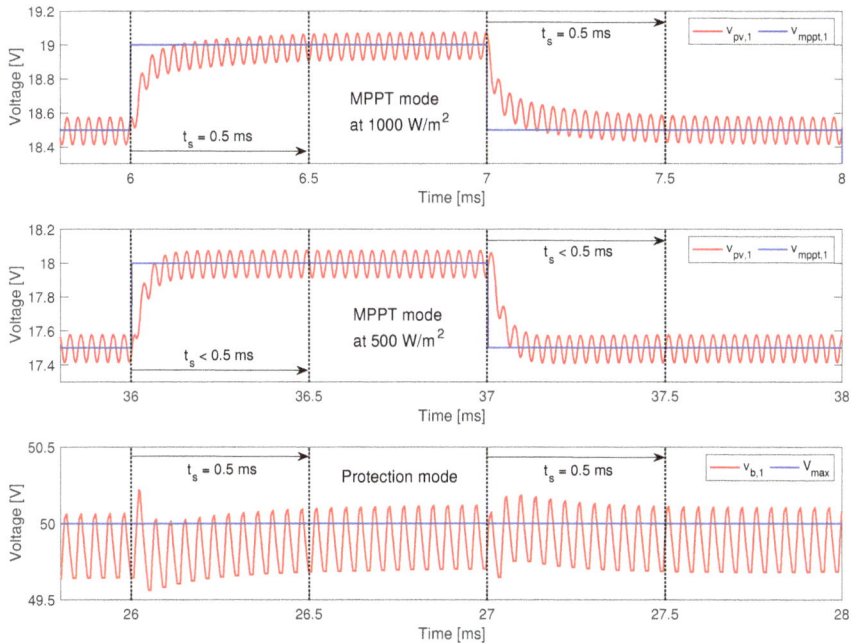

Figure 20. Zoom of the simulations in both MPPT and Protection modes.

Finally, the bottom of Figure 20 shows the waveform of the output voltage $v_{b,1}$ for the first DMPPT-U operating in Protection mode, which occurs between 26 ms and 28 ms. During that time the SMC regulates $v_{b,1}$ to avoid an overvoltage condition. The perturbations in $v_{b,1}$ are caused by the MPPT action of the second DMPPT-U, which perturbs the overall output power, thus changing the relation between the output voltages of both DMPPT-Us. For example, at 25.9 ms the first DMPPT converter provides 65 W while the second one provides 39 W, which imposes $v_{b,1} = 50$ V and $v_{b,2} = 30$ V; at 26 ms the SMC of the second DMPPT converter receives a perturbation command from the P&O algorithm, forcing that converter to provide 38.64 W, which in turns changes the output voltages to $v_{b,1} = 49.72$ V and $v_{b,2} = 30.28$ V. However, the simulation confirms that the SMC imposes the desired settling time $t_s = 0.5$ ms to the first DMPPT-U in the regulation of the output voltage $v_{b,1}$ under Protection mode. In this case no overshoot is observed.

In contrast, Figure 21 shows the simulation of the DMPPT system without activating both the Protection mode and slew-rate limitation. This simulation shows the overvoltage condition that occurs due to the operation in MPPT mode under the mismatching condition, which could destroy $C_{b,1}$ and subsequently the DMPPT converters. Moreover, the SMC exhibits loss of the sliding-mode since the switching function Ψ_1 operates outside the hysteresis band due to the lack of dynamic constraints in the P&O reference.

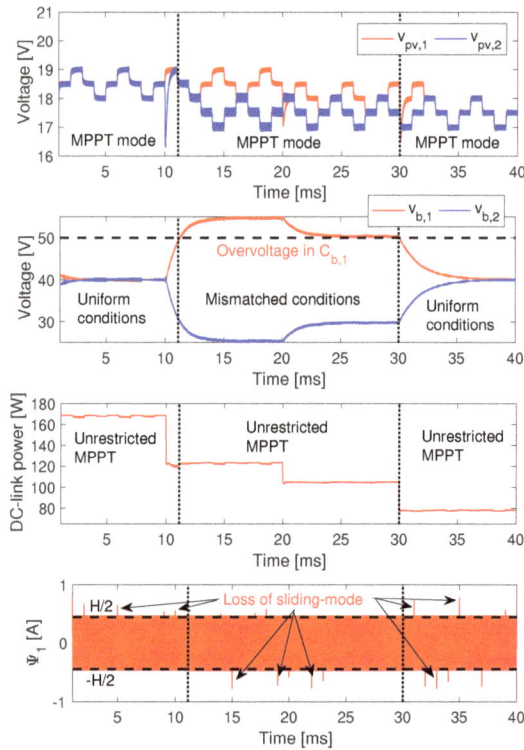

Figure 21. Simulation of the DMPPT system without activating both the Protection mode and slew-rate limitation.

Three DMPPT solutions were implemented to compare their performance with the proposed control strategy, where two of them are some of the most cited papers in double stage DMPPT systems, [16,23,24], and the other is based on SMC [30]. Simulation results introduced in Figure 22 show the comparison of the proposed control strategy with the solutions proposed in [16,23,24,30] respectively.

In [16] the authors use P&O in MPPT mode and fix the duty cycle to keep v_b below its maximum value in Protection mode. Results in Figure 22a shows an overshoot in $v_{b,1}$ in the transition of the DMPPT-U from MPPT to Protection mode. Such an overshoot surpasses V_{max}, which may damage the output capacitor or the switching devices of the DMPPT-U. Moreover, the solution proposed in [16] operates in open-loop during Protection mode and it cannot guarantee the regulation of v_b if there are perturbations like variations in the operation points of the other DMPPT-Us or oscillations introduced by the inverter. It is worth noting that the oscillations of v_{pv} obtained with linear regulator are greater than the ones of the proposed SMC. Those oscillations are smaller for high values of v_{pv} and larger for low values of v_{pv}. Additionally, the amplitude of the oscillations increments when one DMPPT-U is in Protection mode.

Solution proposed in [23,24] uses a proportional controller to regulate v_b when the DMPPT-U operates in Protection mode. The effect of the proportional controller produces an overshoot in $v_{b,1}$ (see Figure 22b) that may damage the output capacitor and switching devices of the boost converter. Additionally, the proportional controller may introduce steady-state errors in and it is not able to reject perturbations produced by the inverter or changes in the operation condition. Even though solution in [23,24] uses Extremum Seeking Control in MPPT mode, the same P&O used in the other solutions were implemented in order to perform a fair comparison in the performance of the DMPPT-U during the transition and regulation in Protection mode.

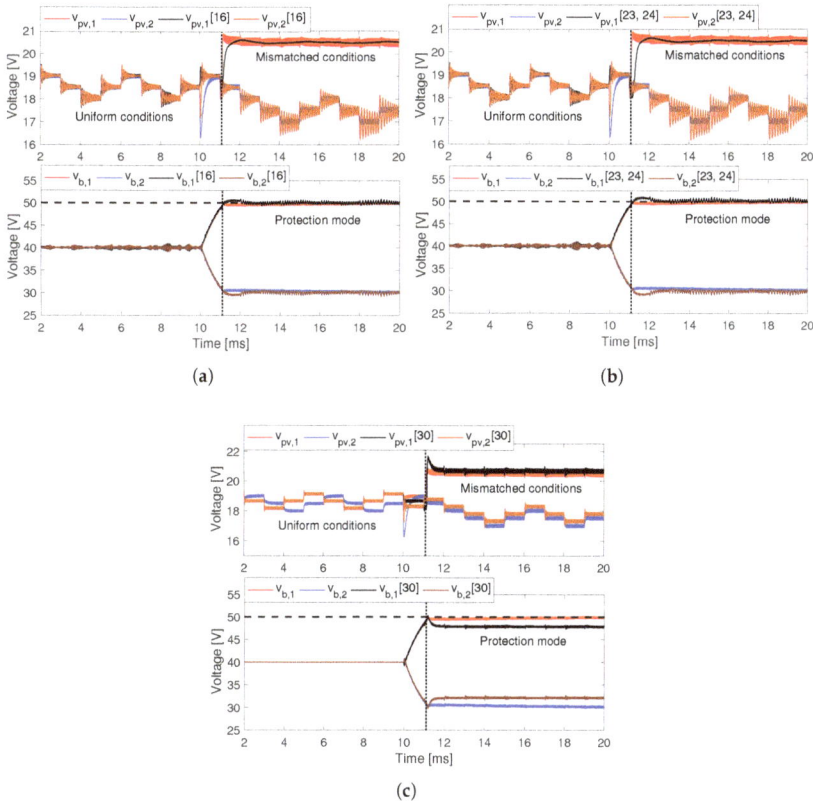

Figure 22. Comparison of the proposed solution with other DMPPT control strategies with Protection mode. (**a**) Comparison with DMPPT control proposed in [16]. (**b**) Comparison with DMPPT control proposed in [23,24]. (**c**) Comparison with DMPPT control proposed in [30].

In [30] the DMPPT-U control is implemented with a SMC in MPPT and Protection modes, however, the SMC does not include integral terms to regulate v_{pv} and v_b in the proposed switching function. There is no overshoot in the transition between MPPT and Protection modes (see Figure 22c). Nevertheless, there is a steady-state error in $v_{b,1}$, which forces the DMPPT system to operate in a non-optimal condition, because the optimal condition of a DMPPT in Protection mode is $v_b = V_{max}$, as demonstrated in Section 2 and Figure 5. Moreover, the steady-state error in v_b is proportional to the current of the DMPPT-U to the DC link, therefore, it is difficult to predict. Solution introduced in [30] also exhibits a steady-state error in v_{pv} and small overshoots, with respect to the proposed solution. That steady-state error is partially compensated by the P&O v_{pv} but deviates the MPPT technique from the MPP.

In conclusion, the simulation results put into evidence the correctness of the design equations and considerations developed in this paper. Moreover, the proposed solution guarantees zero steady-state error in MPPT and Protection modes, no overshoots in v_b, and predictable dynamic behaviors in v_{pv} and v_b in the entire operation range of the DMPPT-Us.

8. Experimental Implementation and Validation

An experimental prototype was developed to validate the proposed solution. The prototype follows the structure adopted in the simulations: it is formed by two DMPPT converters connected in series, each one of them interacting with a BP585 PV panel. The circuital scheme of the prototype is depicted in Figure 23, which reports the implementation of the proposed SMC. The digital steps of the SMC are processed using a DSP F28335 controlCARDs, which have ADC to acquire the current and voltage measurements needed. Both PV and inductor currents are measured using AD8210 circuits and shunt resistors to provide a high-bandwidth, and a MCP4822 DAC (labeled DAC in Figure 23) was used to produce the signals $i_{L,1}^*$ and $i_{L,2}^*$ needed to generate the switching functions Ψ. The DSP executes the designed sliding function presented in the structure defined in Figure 18, the result of this operation is converted to an analog value and injected to a circuit based on operational amplifiers, which performs the control action u by means of the TS555 device, based on the implementation presented in [49]. This implementation gives the advantage of computing the high frequency signal (i_L) by means of analog circuits and the low frequency signal (i_L^*) in a digital form.

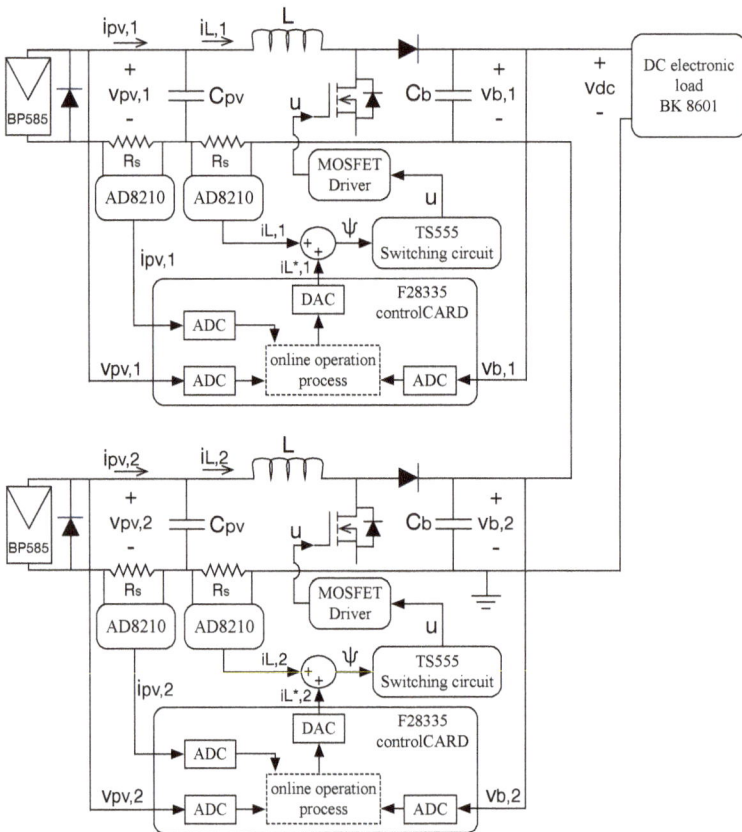

Figure 23. Circuital scheme of the experimental prototype.

The grid-connected inverter reported in Figure 2 was emulated using a BK8601 DC electronic load. Such an electronic load, configured in constant voltage mode, emulates the input voltage control imposed by a traditional grid-connected inverter. Figure 24 shows the experimental setup, which depicts the two DMPPT converters in series connection. Moreover, the figure also shows the controlCARDs, the TS555 switching circuits, and the connections to both the PV panels and electronic load. Finally, the experimental setup includes a voltage supply used to power the DSP, DAC and switching circuits.

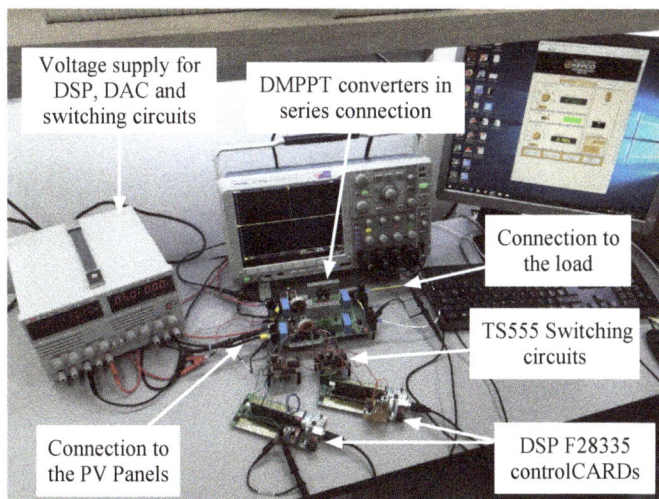

Figure 24. Experimental setup.

The electrical elements used in the platform are: 2218-H-RC inductors from from Bourns Inc. with $L = 330\ \mu H$, MKT1813622016 capacitors from Vishay BC with $C_{pv} = 22\ \mu F$ and $C_b = 44\ \mu F$, IRF540N MOSFETs from International Rectifier and MOSFET drivers A3120 from Vishay Semiconductors. The shunt-resistors used to measure the currents were WSL12065L000FEA18 from Vishay Dale with $R_s = 5\ m\Omega$. Finally, the SMC parameters were the same ones adopted for the simulations. However, the MPPT parameters were changed to $\Delta v_{mppt} = 1\ V$ and $T_a = 1\ s$ due to dynamic limitations of the BK8601 DC electronic load.

Figure 25 reports the experimental measurements of the prototype. The experiment starts with both BP585 PV panels under uniform conditions, which makes both DMPPT-U operate at the same MPP voltage and power. Therefore, the output voltages of both series-connected DMPPT converters are equal to 40 V, which is under the overvoltage limit $V_{max} = 50\ V$. Such conditions force the proposed SMC to operate in MPPT mode, which is evident from the three-point behavior described by both PV voltage profiles $v_{pv,1}$ and $v_{pv,2}$. This is also confirmed by signal Pr, which is equal to 0 at the start of the experiment.

To emulate a mismatched condition, the first PV panel is partially shaded using an obstacle as it is shown at the top of Figure 25. Therefore, from 4.8 s the first PV panel produces less power than the second PV panel, which forces the output voltage of the second DMPPT converter to grow. Subsequently, the SMC of the second DMPPT-U enters in Protection mode to prevent an overvoltage condition, i.e., $Pr = 1$, while the SMC of the first DMPPT-U keeps working in MPPT mode. The experiments confirm the correct protection of the second DMPPT converter provided by the proposed SMC.

The obstacle is removed at 14.2 s, which imposes uniform conditions again. Therefore, the SMC of the second DMPPT-U tracks the MPP voltage of the second PV panel by returning to MPPT mode.

In conclusion, the experiment reports a correct operation of the proposed SMC, in both Protection and MPPT modes, under the series-connection.

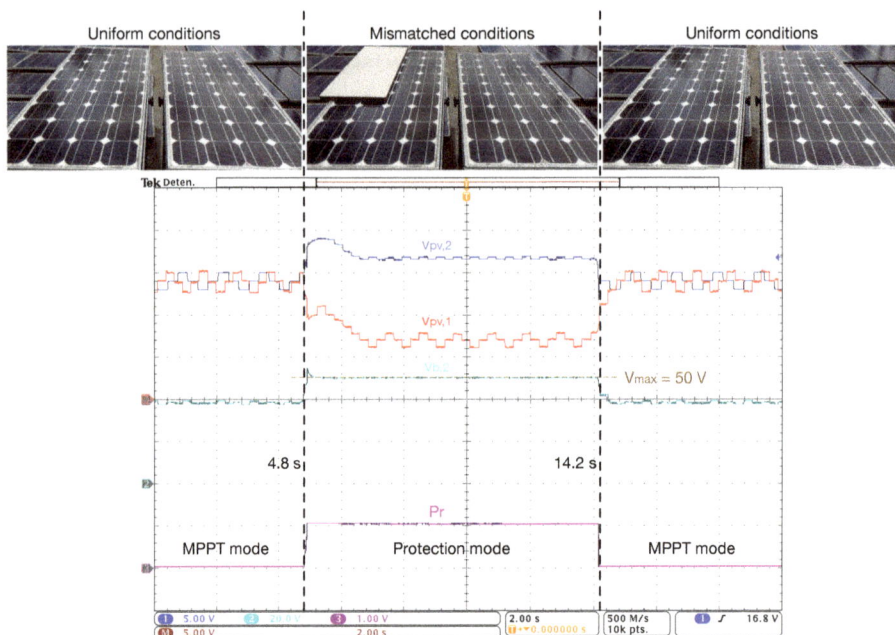

Figure 25. Experimental measurements.

9. Conclusions

A control strategy based on sliding-mode theory, for DMPPT-Us in double-stage DMPPT architectures, has been presented. The proposed controller is able to perform the MPPT on each PV panel when $v_b < V_{max}$ (MPPT mode), and to avoid the DMPPT-Us overvoltage under mismatching conditions (Protection mode). The SMC has a single sliding surface able to regulate v_{pv} and v_b in MPPT and Protection modes, respectively, including i_L into the switching function to provide a soft transition between the two operation modes. Moreover, a detailed design procedure for the SMC parameters and hardware implementation have been provided.

Simulations demonstrate the stability of the DMPPT-Us operating in both MPPT and Protection modes, and also during the transitions between both modes. Moreover, the dynamic performance reported by the simulations fulfills the design restrictions in terms of maximum setting time and overshoot. Furthermore, an experimental platform was developed to show a practical implementation of this new solution. The experimental measurements put into evidence the correct behavior of the practical SMC under real operation conditions.

The proposed control strategy ensures the stability and the dynamic performance of the DMPPT-Us in the entire operation range without a centralized controller or a communication link. Moreover, the control strategy can be implemented using low-cost hardware, which is an important characteristic for commercial DMPPT architectures. This solution can be further improved by implementing observes for both the PV and inductor currents, which will reduce the number of current sensors. Such an approach will reduce both the implementation costs and conduction losses, and it is currently under investigation.

Author Contributions: C.A.R.-P. conceived and developed the theory of the proposed solution; D.G. conceived and performed the experiments; J.D.B.-R. analyzed the data; C.A.R.-P., J.D.B.-R. and D.G. wrote the paper.

Funding: This work was supported by the Universidad Nacional de Colombia, the Instituto Tecnológico Metropolitano, the Universidad Industrial de Santander and Colciencias (Fondo nacional de financiamiento para ciencia, la tecnología y la innovación Francisco José de Caldas) under the projects UNAL-ITM-39823/P17211, "Estrategia de transformación del sector energético Colombiano en el horizonte de 2030—Energetica 2030"—"Generación distribuida de energía eléctrica en Colombia a partir de energía solar y eólica" (Code: 58838, Hermes: 38945) and the doctoral scholarship 2012-567 from Colciencias.

Conflicts of Interest: The authors declare no conflict of interest.

Appendix A. Step Response of the Second Order System with a Zero

This appendix reports the time-domain expressions for the step response and performance criteria for a canonical second-order system with a zero:

$$v(t) = \mathcal{L}^{-1} \left\{ \frac{a \cdot s + b}{(s + p) \cdot (s + q)} \cdot \frac{1}{s} \right\} \tag{A1}$$

Appendix A.1. Overdamped System (p and q Are Real and Different)

Time response expression:

$$v(t) = \frac{b}{p\,q} + \frac{e^{-p\,t}\,(b - a\,p)}{p\,(p - q)} - \frac{e^{-q\,t}\,(b - a\,q)}{q\,(p - q)} \tag{A2}$$

Derivative of the time response expression:

$$\frac{dv(t)}{dt} = \frac{e^{-q\,t}\,(b - a\,q)}{p - q} - \frac{e^{-p\,t}\,(b - a\,p)}{p - q} \tag{A3}$$

Time value at which the *MO* occurs $\left(\frac{dv(t)}{dt} = 0 \right)$:

$$t_{MO} = \frac{\ln\left(\frac{b - a\,p}{b - a\,q} \right)}{p - q} \tag{A4}$$

Appendix A.2. Critically Damped System (p = q Are Real)

Time response expression:

$$v(t) = \frac{b}{p^2} - \frac{b\,e^{-p\,t}}{p^2} - \frac{t\,e^{-p\,t}\,(b - a\,p)}{p} \tag{A5}$$

Derivative of the time response expression:

$$\frac{dv(t)}{dt} = \frac{b\,e^{-p\,t}}{p} + t\,e^{-p\,t}\,(b - a\,p) - \frac{e^{-p\,t}\,(b - a\,p)}{p} \tag{A6}$$

Time value at which the *MO* occurs $\left(\frac{dv(t)}{dt} = 0 \right)$:

$$t_{MO} = -\frac{a}{b - a\,p} \tag{A7}$$

Appendix A.3. Under Damped System (p and q Are Complex)

Time response expression:

$$v(t) = \frac{b}{pq} - \frac{be^{-\frac{(p+q)t}{2}}\left(\cosh\left(t\sqrt{\frac{(p+q)^2}{4} - pq}\right) - \chi\right)}{pq} \tag{A8}$$

where

$$\chi = \frac{\sinh\left(t\sqrt{\frac{(p+q)^2}{4} - pq}\right)\left(\frac{(p+q)}{2} + \frac{apq - b(p+q)}{b}\right)}{\sqrt{\frac{(p+q)^2}{4} - pq}}$$

Derivative of the time response expression:

$$\frac{dv(t)}{dt} = \frac{b(p+q)e^{-\frac{(p+q)t}{2}}\left(\cosh\left(\chi t\right) - \frac{\gamma\sinh(\chi t)}{\chi}\right)}{2pq} - \frac{be^{-\frac{(p+q)t}{2}}\left(\chi\sinh\left(\chi t\right) - \gamma\sinh\left(\chi t\right)\right)}{pq} \tag{A9}$$

where

$$\chi = \sqrt{\frac{(p+q)^2}{4} - pq}$$

$$\gamma = \frac{(p+q)}{2} + \frac{apq - b(p+q)}{b}$$

Time value at which the *MO* occurs $\left(\frac{dv(t)}{dt} = 0\right)$:

$$t_{MO} = \frac{\ln\left(\mathrm{abs}\left(\chi\right)\right)}{\sqrt{\frac{(p+q)^2}{4} - pq}}$$

where

$$\chi = \frac{2\sqrt{pqa^2 - (p+q)ab + b^2}}{2b - a(p+q) + 2a\sqrt{\frac{(p+q)^2}{4} - pq}}$$

References

1. IEA. *2018 Snapshot of Global Photovoltaic Markets*; Technical Report; International Energy Agency: St. Ursen, Switzerland, 2018.
2. Sharkh, S.M.; Abusara, M.A.; Orfanoudakis, G.I.; Hussain, B. *Power Electronic Converters for Microgrids*; John Wiley & Sons, Singapore Pte. Ltd.: Singapore, 2014. [CrossRef]
3. Mehrasa, M.; Adabi, M.E.; Pouresmaeil, E.; Adabi, J. Passivity-Based Control Technique for Integration of DG Resources into the Power Grid. *Int. J. Electr. Power Energy Syst.* **2014**, *58*, 281–290. [CrossRef]
4. Mehrasa, M.; Ebrahim Adabi, M.; Pouresmaeil, E.; Adabi, J.; Jørgensen, B.N. Direct Lyapunov Control (DLC) Technique for Distributed Generation (DG) Technology. *Electr. Eng.* **2014**, *96*, 309–321. [CrossRef]
5. Bastidas-Rodríguez, J.D.; Ramos-Paja, C. Types of Inverters and Topologies for Microgrid Applications Tipos de Inversores y Topologías Para Aplicaciones de Microrredes. *UIS Ingenierías* **2017**, *16*, 8.

6. Mehrasa, M.; Godina, R.; Pouresmaeil, E.; Vechiu, I.; Rodriguez, R.L.; Catalao, J.P.S. Synchronous Active Proportional Resonant-Based Control Technique for High Penetration of Distributed Generation Units into Power Grids. In Proceedings of the PES Innovative Smart Grid Technologies Conference Europe (ISGT-Europe), Torino, Italy, 26–29 September 2017; pp. 1–6. [CrossRef]

7. Mehrasa, M.; Rezanejhad, M.; Pouresmaeil, E.; Catalao, J.P.S.; Zabihi, S. Analysis and Control of Single-Phase Converters for Integration of Small-Scaled Renewable Energy Sources into the Power Grid. In Proceedings of the 7th Power Electronics and Drive Systems Technologies Conference (PEDSTC), Tehranm, Iran, 16–18 February 2016; pp. 384–389.

8. Petrone, G.; Ramos-Paja, C.A.; Spagnuolo, G. *Photovoltaic Sources Modeling*; John Wiley & Sons, Ltd: Chichester, UK, 2017. [CrossRef]

9. Serna-Garcés, S.; Bastidas-Rodríguez, J.; Ramos-Paja, C. Reconfiguration of Urban Photovoltaic Arrays Using Commercial Devices. *Energies* **2016**, *9*, 2, doi:10.3390/en9010002. [CrossRef]

10. Pendem, S.R.; Mikkili, S. Modelling and Performance Assessment of PV Array Topologies under Partial Shading Conditions to Mitigate the Mismatching Power Losses. *Sol. Energy* **2018**, *160*, 303–321. [CrossRef]

11. Bastidas-Rodriguez, J.D.; Franco, E.; Petrone, G.; Ramos-Paja, C.A.; Spagnuolo, G. Maximum power point tracking architectures for photovoltaic systems in mismatching conditions: A review. *IET Power Electron.* **2014**, *7*, 1396–1413. [CrossRef]

12. Belhachat, F.; Larbes, C. A Review of Global Maximum Power Point Tracking Techniques of Photovoltaic System under Partial Shading Conditions. *Renew. Sustain. Energy Rev.* **2018**, *92*, 513–553. [CrossRef]

13. Das, S.K.; Verma, D.; Nema, S.; Nema, R. Shading Mitigation Techniques: State-of-the-Art in Photovoltaic Applications. *Renew. Sustain. Energy Rev.* **2017**, *78*, 369–390. [CrossRef]

14. Kasper, M.; Bortis, D.; Kolar, J.W. Classification and comparative evaluation of PV panel-integrated DC-DC converter concepts. *IEEE Trans. Power Electron.* **2014**, *29*, 2511–2526. [CrossRef]

15. Khan, O.; Xiao, W. Review and Qualitative Analysis of Submodule-Level Distributed Power Electronic Solutions in PV Power Systems. *Renew. Sustain. Energy Rev.* **2017**, *76*, 516–528. [CrossRef]

16. Femia, N.; Lisi, G.; Petrone, G.; Spagnuolo, G.; Vitelli, M. Distributed Maximum Power Point Tracking of Photovoltaic Arrays: Novel Approach and System Analysis. *IEEE Trans. Ind. Electron.* **2008**, *55*, 2610–2621. [CrossRef]

17. Femia, N.; Giovanni, P.; Giovanni, S.; Massimo, V. Distributed Maximum Power Point Tracking of Photovoltaic Arrays. In *Power Electronics and Control Techniques for Maximum Energy Harvesting in Photovoltaic Systems*; Industrial Electronics; CRC Press: Boca Raton, FL, USA, 2012; pp. 139–249. [CrossRef]

18. Huusari, J.; Suntio, T. Interfacing Constraints of Distributed Maximum Power Point Tracking Converters in Photovoltaic Applications. In Proceedings of the 15th International Power Electronics and Motion Control Conference (EPE/PEMC), Novi Sad, Serbia, 4–6 September 2012. [CrossRef]

19. Wang, F.; Zhu, T.; Zhuo, F.; Yang, Y. Analysis and Comparison of FPP and DPP Structure Based DMPPT PV System. In Proceedings of the 8th International Power Electronics and Motion Control Conference (IPEMC-ECCE Asia), Hefei, China, 22–26 July 2016; pp. 207–211. [CrossRef]

20. Wang, F.; Lee, F.C.; Yue, X.; Zhuo, F. Quantified Evaluation and Criteria Analysis for DMPPT PV System. In Proceedings of the 17th European Conference on Power Electronics and Applications (EPE'15 ECCE-Europe), Geneva, Switzerland, 8–10 September 2015; pp. 1–6. [CrossRef]

21. Lyden, S.; Haque, M. Maximum Power Point Tracking Techniques for Photovoltaic Systems: A Comprehensive Review and Comparative Analysis. *Renew. Sustain. Energy Rev.* **2015**, *52*, 1504–1518. [CrossRef]

22. Mohapatra, A.; Nayak, B.; Das, P.; Mohanty, K.B. A Review on MPPT Techniques of PV System under Partial Shading Condition. *Renew. Sustain. Energy Rev.* **2017**, *80*, 854–867. [CrossRef]

23. Bratcu, A.I.; Munteanu, I.; Bacha, S.; Picault, D.; Raison, B. Power Optimization Strategy for Cascaded DC-DC Converter Architectures of Photovoltaic Modules. In Proceedings of the 2009 IEEE International Conference on Industrial Technology, Gippsland, VIC, Australia, 10–13 February 2009; pp. 1–8. [CrossRef]

24. Bratcu, A.I.; Munteanu, I.; Bacha, S.; Picault, D.; Raison, B. Cascaded DCDC converter photovoltaic systems: Power optimization issues. *IEEE Trans. Ind. Electron.* **2011**, *58*, 403–411. [CrossRef]

25. Renaudineau, H.; Donatantonio, F.; Fontchastagner, J.; Petrone, G.; Spagnuolo, G.; Martin, J.P.; Pierfederici, S. A PSO-based global MPPT technique for distributed PV power generation. *IEEE Trans. Ind. Electron.* **2015**, *62*, 1047–1058. [CrossRef]

26. Balato, M.; Vitelli, M. A new control strategy for the optimization of Distributed MPPT in PV applications. *Int. J. Electr. Power Energy Syst.* **2014**, *62*, 763–773. [CrossRef]

27. Balato, M.; Vitelli, M. Optimization of distributed maximum power point tracking PV applications: The scan of the Power vs. Voltage input characteristic of the inverter. *Int. J. Electr. Power Energy Syst.* **2014**, *60*, 334–346. [CrossRef]

28. Sitbon, M.; Leppäaho, J.; Suntio, T.; Kuperman, A. Dynamics of Photovoltaic-Generator-Interfacing Voltage-Controlled Buck Power Stage. *IEEE J. Photovolt.* **2015**, *5*, 633–640. [CrossRef]

29. Ramos-paja, C.A.; Vitelli, M. Distributed Maximum Power Point Tracking with Overvoltage Protection for Pv Systems Seguimiento Distribuido Del Punto De Maxima Potencia Con Proteccion De Sobrevoltaje. *Dyna* **2013**, *80*, 141–150.

30. Ramos-Paja, C.A.; Saavedra-Montes, A.J. Overvoltage protection for distributed maximum power point tracking converters in series connection. In *Applied Computer Sciences in Engineering, Proceedings of the Third Workshop on Engineering Applications (WEA 2016), Bogotá, Colombia, 21–23 September 2016*; Springer International Publishing: Berlin, Germany, 2016; pp. 308–319.

31. Bastidas, J.D.; Franco, E.; Petrone, G.; Ramos-Paja, C.A.; Spagnuolo, G. A model of photovoltaic fields in mismatching conditions featuring calculation speed. *Electr. Power Syst. Res.* **2013**, *96*, 81–90. [CrossRef]

32. Accarino, J.; Petrone, G.; Ramos-Paja, C.A.; Spagnuolo, G. Symbolic algebra for the calculation of the series and parallel resistances in PV module model. In Proceedings of the 2013 International Conference on Clean Electrical Power (ICCEP), Alghero, Italy, 11–13 June 2013; pp. 62–66. [CrossRef]

33. Balato, M.; Vitelli, M.; Femia, N.; Petrone, G.; Spagnuolo, G. Factors limiting the efficiency of DMPPT in PV applications. In Proceedings of the 3rd International Conference on Clean Electrical Power: Renewable Energy Resources Impact (ICCEP), Ischia, Italy, 14–16 June 2011; pp. 604–608. [CrossRef]

34. Mastromauro, R.A.; Liserre, M.; Dell'Aquila, A. Control issues in single-stage photovoltaic systems: MPPT, current and voltage control. *IEEE Trans. Ind. Inf.* **2012**, *8*, 241–254. [CrossRef]

35. Sira-Ramirez, H. Sliding Motions in Bilinear Switched Networks. *IEEE Trans. Circuits Syst.* **1987**, *34*, 919–933. [CrossRef]

36. Tan, S.C.; Lai, Y.; Tse, C.K. General Design Issues of Sliding-Mode Controllers in DC-DC Converters. *IEEE Trans. Ind. Electron.* **2008**, *55*, 1160–1174. [CrossRef]

37. Chatrenour, N.; Razmi, H.; Doagou-Mojarrad, H. Improved double integral sliding mode MPPT controller based parameter estimation for a stand-alone photovoltaic system. *Energy Convers. Manag.* **2017**, *139*, 97–109. [CrossRef]

38. Montoya, D.G.; Ramos-Paja, C.A.; Giral, R. Improved Design of Sliding-Mode Controllers Based on the Requirements of MPPT Techniques. *IEEE Trans. Power Electron.* **2016**, *31*, 235–247. [CrossRef]

39. Nicola, F.; Giovanni, P.; Giovanni, S.; Massimo, V. Maximum Power Point Tracking. In *Power Electronics and Control Techniques for Maximum Energy Harvesting in Photovoltaic Systems*; Industrial Electronics; CRC Press: Boca Raton, FL, USA, 2012; pp. 35–87. [CrossRef]

40. Erickson, R.W.; Maksimovic, D. *Fundamentals of Power Electronics*, 2nd ed.; Kluwer Academic Publishers: New York, NY, USA, 2001.

41. Petrone, G.; Ramos-Paja, C.A.; Spagnuolo, G.; Vitelli, M. Granular control of photovoltaic arrays by means of a multi-output Maximum Power Point Tracking algorithm. *Prog. Photovolt. Res. Appl.* **2012**, *21*, doi:10.1002/pip.2179. [CrossRef]

42. Golnaraghi, F.; Kuo, B.C. *Automatic Control Systems*, 9th ed.; John Wiley & Sons, Inc.: Hoboken, NJ, USA, 2010.

43. Ramos-Paja, C.A.; González, D.; Saavedra-Montes, A.J. Accurate calculation of settling time in second order systems: A photovoltaic application. *Revista Facultad de Ingenieria* **2013**, 104–117.

44. Press, W.H.; Teukolsky, S.A.; Vetterling, W.T.; Flannery, B.P. *Numerical Recipes in C: The Art of Scientific Computing*, 2nd ed.; Cambridge University Press: Cambridge, UK, 1988. [CrossRef]

45. Matworks. Rise Time, Settling Time, and Other Step-Response Characteristics—MATLAB Stepinfo. Available online: http://www.mathworks.com/help/control/ref/stepinfo.html (accessed on 24 August 2018).

46. Hammersley, J.M.; Handscomb, D.C.; Weiss, G. *Monte Carlo Methods*, 2nd ed.; Wiley-VCH: Hoboken, NJ, USA, 2008.

47. Kapat, S.; Krein, P.T. Formulation of PID control for DC-DC converters based on capacitor current: A geometric context. *IEEE Trans. Power Electron.* **2012**, *27*, 1424–1432. [CrossRef]

48. Serna-Garcés, S.I.; Gonzalez-Montoya, D.; Ramos-Paja, C.A. Sliding-mode control of a charger/discharger DC/DC converter for DC-bus regulation in renewable power systems. *Energies* **2016**, *9*, 245, doi:10.3390/en9040245. [CrossRef]

49. Montoya, D.G.; Paja, C.A.R.; Giral, R. Maximum power point tracking of photovoltaic systems based on the sliding mode control of the module admittance. *Electr. Power Syst. Res.* **2016**, *136*, 125–134. [CrossRef]

energies

MDPI

Article

A Comprehensive Strategy for Accurate Reactive Power Distribution, Stability Improvement, and Harmonic Suppression of Multi-Inverter-Based Micro-Grid

Henan Dong [1,2], Shun Yuan [1,3], Zijiao Han [4,*], Zhiyuan Cai [1], Guangdong Jia [1] and Yangyang Ge [2]

[1] Institute of Electrical Engineering, Shenyang University of Technology, Shenyang 110023, China; an1an2_nan@163.com (H.D.); yuanshun@serc.gov.cn (S.Y.); mashcaizy@hotmail.com (Z.C.); guangdongjia0302@163.com (G.J.)

[2] Liaoning Electric Power Company Electric Power Research Institute, State Grid Corporation of China, Shenyang 110006, China; 15942302722@163.com

[3] National Energy Administration, Beijing 100085, China

[4] Liaoning Electric Power Company, State Grid Corporation of China, Shenyang 110004, China

* Correspondence: thuwhatever@163.com; Tel.: +86-139-4047-6382

Received: 10 February 2018; Accepted: 6 March 2018; Published: 26 March 2018

Abstract: Among the issues of accurate power distribution, stability improvement, and harmonic suppression in micro-grid, each has been well studied as an individual, and most of the strategies about these issues aim at one inverter-based micro-grid, hence there is a need to establish a model to achieve these functions as a whole, aiming at a multi-inverter-based micro-grid. This paper proposes a comprehensive strategy which achieves this goal successfully; since the output voltage and frequency of micro-grid all consist of fundamental and harmonic components, the strategy contains two parts accordingly. On one hand, a fundamental control strategy is proposed upon the conventional droop control. The virtual impedance is introduced to solve the problem of accurate allocation of reactive power between inverters. Meanwhile, a secondary power balance controller is added to improve the stability of voltage and frequency while considering the aggravating problem of stability because of introducing virtual impedance. On the other hand, the fractional frequency harmonic control strategy is proposed. It can solve the influence of nonlinear loads, micro-grid inverters, and the distribution network on output voltage of inverters, which is focused on eliminating specific harmonics caused by the nonlinear loads, micro-grid converters, and the distribution network so that the power quality of micro-grid can be improved effectively. Finally, small signal analysis is used to analyze the stability of the multi-converter parallel system after introducing the whole control strategy. The simulation results show that the strategy proposed in this paper has a great performance on distributing reactive power, regulating and stabilizing output voltage of inverters and frequency, eliminating harmonic components, and improving the power quality of multi-inverter-based micro-grid.

Keywords: micro-grid; droop control; virtual impedance; harmonic suppression; power quality

1. Introduction

In recent years, distributed generations, e.g., wind and solar power, have been developing rapidly. Comparing with traditional power generation forms, distributed generations are environment-friendly technologies, and they often form micro-grids via inverters, which is an important complementary of bulk power network. However, since there are so many distributed generations in the micro-grid system, power electronic inverters are also widely used. In addition, various kinds of nonlinear loads such as electric vehicles are increasingly integrated into the micro-grid, thus a power quality problem inside micro-grid occurs and increases problems of heating, incremental losses, voltage and current distortion, which threaten our daily life. It is not only the micro grid itself that can be broken down by such problems, but the voltage and frequency of the distributed power system can also be influenced through the point of common coupling (PCC) [1–3]. Meanwhile, power management strategies play an increasingly important role in power quality regulation for micro-grids [4–6].

As distributed generations are connected to the micro-grid via inverter, hence the control strategy of inverters influence system stability and power quality. Among all the inverter control strategies, droop control is considered as the best strategy at present, which can distribute the output power of inverters properly under the island mode of the micro-grid, even if there is no common communication line among distributed generations (DGs); meanwhile, the voltage and frequency can be controlled within related national standards by this strategy [7–9]. However, there are some shortcomings in the traditional droop control strategy. Firstly, system reactive power cannot be distributed accurately while the equivalent impedance of micro sources is different [10]. Secondly, system voltage and frequency will not maintain their stability under abrupt load variation [11,12]. In addition, the magnitude of harmonic power varies with the amount of non-linear loads integrated into the micro-grid. The existence of harmonic power will have an effect on system devices, including transformers, capacitors, and electric rotating machines. It is also shown that harmonic power will influence the voltage amplitude and waveform of PCC [13], which may cause resonance and eventually lead to the loss of system stability.

At present, there are less researches on multi-inverter-based micro-grid power quality. Traditional methods for limiting waveforms distortion are to make direct compensation by installing passive or centralized active power filters, but the high cost of these kinds of devices should not be ignored, and these methods can only improve harmonic components, while, as to fundamental component power quality and accurate reactive power distribution, they neither make tense research nor give an integral solution [14–17]. Reference [18] proposed a compensation method based on a unified power quality conditioner (UPQC) to improve the power quality index of PCC, but though this device is widely used in large capacity and high voltage power grids, it is difficult to generalize the use of this device in low voltage distributed grids and micro-grids. Many experts considered designing an inverter control strategy to improve accurate reactive power distribution and govern harmonic components [19–28]. Reference [19] proposed a method to govern harmonic components in the micro-grid that combines the technology of active filter and inverter controller, so the utilization of inverter is obviously improved and the cost of active filter is effectively decreased; however, the design of that controller is too complex to popularize. The general method to distribute reactive power accurately is to add virtual impedances into inverters [20–22]. By measuring the output voltage and current of inverters and adjusting their output impedances, i.e., introducing virtual impedances to the multi-inverter system, the reactive power of system can be distributed accurately, although this method needs to know the parameters of lines in advance and is short of consideration about load variation [23]. Reference [24] concluded that harmonic droop control can be used for distributing harmonic power among inverters and decreasing voltage waveform distortion of PCC, whereas the calculation of non-linear loads is too complex and this method does not make an obvious function on distributing active power. Reference [25] proposed a control strategy to suppress harmonic and negative sequence current in island mode. The strategy mentioned is composed of two controllers: one is a multi-proportional resonance controller which is used to regulate load voltage, and the other is a harmonic impedance controller which is used to distribute harmonic current among micro sources. These two controllers are so intricate that they may lead to high-order feedforward transfer function. In references [26–28], experts considered reactive power distribution, voltage, and frequency stability under load abrupt variation respectively, but the influence from non-linear loads was not taken into integral consideration and given focus in the research.

Above all, this paper proposes a comprehensive strategy for accurate power distribution, stability improvement and harmonic suppression of a multi-inverter-based micro-grid, which aims at further improving the power quality of the micro-grid. Based on previous research, it mainly introduces the improved fundamental control strategy and the fractional frequency harmonic control strategy into a multi-inverter-based micro-grid, which are two components of the comprehensive strategy. The key point is to add the fractional frequency harmonic control strategy to improve the power quality of the multi-inverter-based micro-grid under the premise of ensuring the stability of the improved fundamental control strategy. In addition, the stability of this micro-grid control system is analyzed by the small signal analysis method. Following this, a micro-gird system simulation model is built in MATLAB to verify the effectiveness of the proposed strategy.

The main contributions of this work are described as follows: Section 2 proposes the comprehensive strategy for accurate power distribution, stability improvement, and harmonic suppression of the multi-inverter-based micro-grid. Section 3 analyzes the stability of the multi-converter parallel system after introducing the comprehensive control strategy. Simulation and experimental results to prove the effectiveness of the strategy are demonstrated in Section 4. Section 5 concludes the paper.

2. Comprehensive Strategy for Accurate Reactive Power Distribution, Stability Improvement, and Harmonic Suppression of a Multi-Inverter-Based Micro-Grid

It is difficult to accurately distribute reactive power and effectively improve the stability of voltage and frequency under abrupt load variation depending on conventional droop controllers, let alone suppress harmonic components in a micro-grid which are caused by many reasons. Therefore, a comprehensive strategy for accurately distributing reactive power, improving stability, and suppressing harmonics of a multi-inverter-based micro-grid is proposed in this paper. Upon the conventional droop control, an adaptive virtual impedance control loop is introduced to achieve the accurate distribution of reactive power of inverters in fundamental frequency. Considering this process may add the problem of voltage stability, so a secondary power balance controller is added to improve the stability of voltage and frequency, and the fundamental problems are settled completely so far. Next, the control strategy this paper proposed is further refined by introducing a fractional frequency harmonic suppression strategy, which can solve the harmonic problem perfectly, therefore, the power quality of the micro-grid is improved eventually.

The main circuit is mainly composed of the following parts: inverter, LC filter, line impedance, linear load, nonlinear load, and so on. The inverter is controlled by a control strategy to adjust the output voltage and frequency. The comprehensive control strategy is composed of the improved fundamental control strategy and fractional frequency harmonic control strategy.

The improved fundamental control strategy is proposed upon the conventional droop control, and the problem of accurate reactive power distribution is solved by introducing virtual impedance to inverters. The secondary power balance controller is added to improve the stability of voltage and frequency. The reference value of the frequency is 50 Hz and the reference value of the output voltage is 311 V in secondary power balance controller. The voltage value and frequency of the PCC follow a given value through the closed loop control strategy.

The fractional frequency harmonic control strategy is proposed to solve the influence of nonlinear loads, micro-grid inverters, and the distribution network on output voltage of inverters, which is focused on eliminating specific harmonics caused by the nonlinear loads, micro-grid converters, and the distribution network, so the power quality of micro-grid can be improved effectively. The 5th harmonic suppression is taken as an example. Firstly, the 5th harmonic voltage and 5th harmonic current of the PCC are detected, and the harmonic power is calculated. The reference value of the output voltage of the inverter and frequency are generated according to the droop curve of the harmonic. The closed loop control strategy is adopted in the reference value and the detection value of PCC, so that the output value of the harmonic is followed by a given value, which is 0, and the 5th harmonic is eliminated. The way to suppress other harmonics is the same as the 5th harmonic. The main control block diagram is shown in Figure 1.

In Figure 1, L_f is the filter inductance; R_f is the filter resistance; C_f is the filter capacitance; R_{line} is the line resistance; L_{line} is the line inductance; R_{ref} is the reference resistance; X_{ref} is the reference reactance; R_i is the calculated value of line resistance, X_i is the calculated value of line inductance; P_{line} is the line active power; Q_{line} is the line reactive power; P_{i_LPF} is the active power that passes through the low pass filter; Q_{i_LPF} is the reactive power that passes through the low pass filter; E^* is an reference voltage; f^* is an reference frequency; E_{nh} is the nth harmonic voltage; P_{nh} is the active power of the nth harmonic; and Q_{nh} is the reactive power of the nth harmonic.

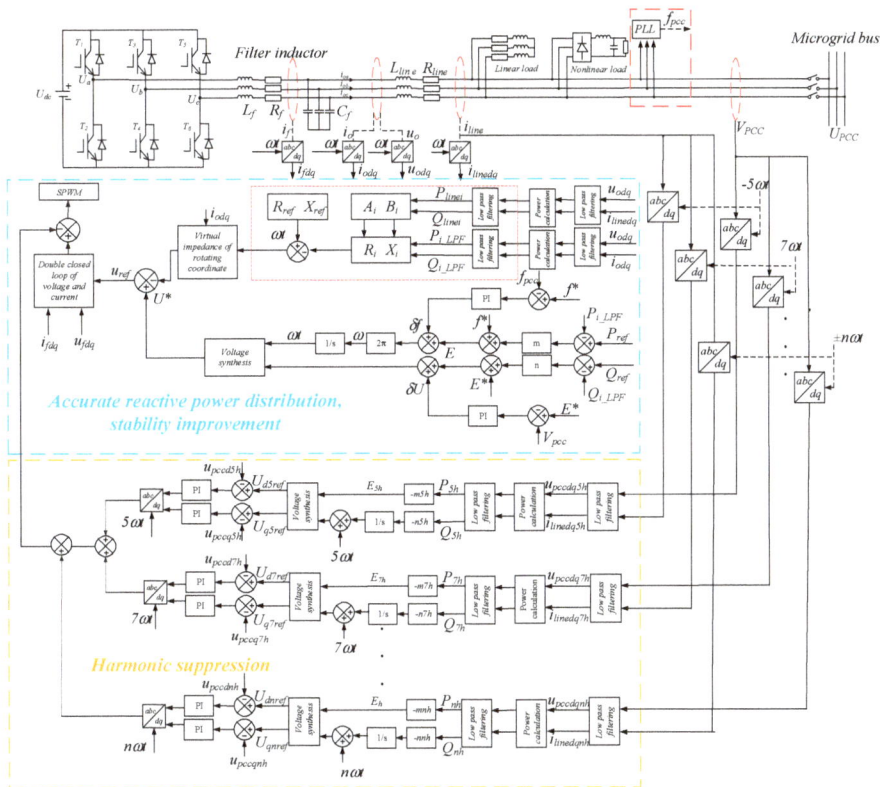

Figure 1. Block diagram of comprehensive control strategy.

2.1. Combination of Adaptive Virtual Impedance Drooping Control and Secondary Power Balance Control Strategy

In order to achieve an accurate distribution of reactive power of inverters, an adaptive virtual impedance control loop is introduced upon conventional droop control. The control block diagram is shown in Figure 2.

The output voltage function of the inverter with virtual impedance is:

$$U_0(s) = G(s)U_{ref}(s) - [G(s)Z_v(s) + Z_{eq}(s)]I_0(s). \tag{1}$$

The output power values of the inverter can be calculated by

$$P_i = \frac{U_i}{R_i^2 + X_i^2}[R_i(U_i - Ecos\delta_i) + X_iEsin\delta_i] \tag{2}$$

$$Q_i = \frac{U_i}{R_i^2 + X_i^2}[-R_i E sin\delta_i + X_i(U_i - E cos\delta_i)] \tag{3}$$

where: R_i and X_i are the power equivalent output resistance and reactance, respectively, which is defined by inverter output power and the power injected to the PCC. Z_v is virtual impedance.

Based on the above circuit, the transmission line parameters and loads fluctuation will affect the value of power equivalent impedance, then influence the reactive power distribution. The power equivalent impedance can be calculated through line power.

$$U_i(U_i - U cos\delta_i) = P_{linei} R_{linei} + Q_{linei} X_{linei} = A_i \tag{4}$$

$$U_i U sin\delta_i = P_{linei} X_{linei} - Q_{linei} R_{linei} = B_i \tag{5}$$

and the inverter equivalent output impedance can be obtained by

$$R_i = \frac{P_i A_i + Q_i B_i}{P_i^2 + Q_i^2} \tag{6}$$

$$X_i = \frac{P_i B_i - Q_i A_i}{P_i^2 + Q_i^2} \tag{7}$$

then the virtual impedance of each micro source is calculated by

$$R_{vi} = R_{ref} - R_i \tag{8}$$

$$X_{vi} = X_{ref} - X_i. \tag{9}$$

The adaptive virtual impedance control strategy can accurately distribute reactive power, meeting the requirement of the power decoupling and stability margin. The meaning of adaptive is that the virtual impedance varies along with the power equivalent output impedance when the loads change, and it can also be changed if the micro-grid central controller changes reference impedance.

The traditional droop control can realize automatic regulation of P/f and Q/V, the essence of which, however, is a kind of deviating regulation. Considering the addition of the virtual impedance will increase the degree of deviation, which is harmful to the voltage stability. Therefore, a secondary power balance control strategy is added to stabilize the output voltage of converters, as well as improve the stability of frequency. The compensation frequency and voltage of this strategy can be derived by Equations (10) and (11):

$$\delta f = k_{p\omega}(f^* - f) + k_{i\omega}\int (f^* - f)dt \tag{10}$$

$$\delta U = k_{pE}(U^* - U) + k_{iE}\int (U^* - U)dt \tag{11}$$

where: f^* and f are rated frequency and operational frequency of the microgrid, respectively; U^* and U are rated voltage and operational voltage, respectively. The frequency compensation signal δf and voltage compensation signal δU obtained from the secondary power balance controller are sent to each micro source, then the droop curves of micro sources will turn to be appropriate translations, then the voltage and frequency can be stabilized by adding this strategy, which is similar to the secondary frequency modulation of thermal power unit. Until now, not only can the reactive power be distributed accurately, but also the stability of voltage and frequency can be improved effectively, hence the combination of adaptive virtual impedance drooping control and secondary power balance control strategy in fundamental frequency is achieved completely.

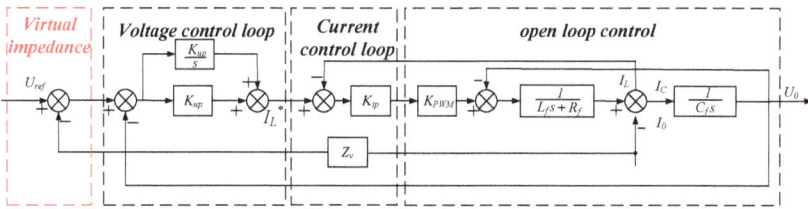

Figure 2. Inverter control block diagram with virtual impedance.

2.2. Fractional Frequency Harmonic Drooping Control Strategy

According to the circuit superposition theorem, a linear circuit with different frequencies can be analyzed separately at each frequency. Reference [29] confirmed that any harmonic, e.g., the hth harmonic can be extracted for separate analysis and control when the whole system enters steady state, so the harmonic droop control strategy is proposed to eliminate the 5th, 7th, 11th, and 13th harmonics generated by nonlinear loads. In order to reduce the harmonic voltage of the system, the harmonic voltage compensation value is calculated by the harmonic droop control strategy. The system equivalent schematic is shown in Figure 3.

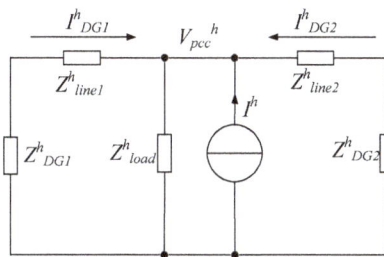

Figure 3. System equivalent circuit containing different frequencies.

In order to make the harmonic voltage of the inverter reach or close to zero, the V_{oh} should be zero, and the load part of the harmonic equivalent circuit can be equivalent to a current source [30], then the output voltage of inverter can be obtained:

$$\overline{V}_o = E\angle\delta - Z_oI\angle\theta = E\cos\delta - Z_oI\cos\theta + j(E\sin\delta - Z_oI\sin\theta). \tag{12}$$

The phase angle difference δ is the phase angle difference between the voltage source and the current source. When it is very small, Equations (13) and (14) can be obtained:

$$P \approx EI - Z_oI^2\cos\theta \tag{13}$$

$$Q \approx EI\delta - Z_oI^2\sin\theta. \tag{14}$$

From Equations (13) and (14), it can be found that whether the impedance of the line is inductive, resistive, or capacitive, the correlation between P and E, and Q and δ, can be concluded. Thus the hth harmonic droop controller can be shown as:

$$E_h = E^* - n_h P_h \tag{15}$$

$$\omega_h = \omega^* - m_h Q_h \tag{16}$$

where: P_h and Q_h are the calculated values of active power and reactive power under the hth harmonic frequency; n_h and m_h are the corresponding hth harmonic droop coefficients; E_h is the RMS of the hth harmonic voltage; and ω_h is the hth harmonic voltage angle frequency. The amplitude and angular frequency of specific harmonic voltage can be eliminated by this control strategy, and the reference value of harmonic voltage on the dq axis can be obtained through voltage synthesis and coordinate transformation. The modulated wave is used to turn on and turn off the inverter switch tube so as to produce the appropriate output voltage, so the PCC harmonic voltage can be suppressed effectively. The structure diagram of the harmonic drooping control is shown in Figure 4.

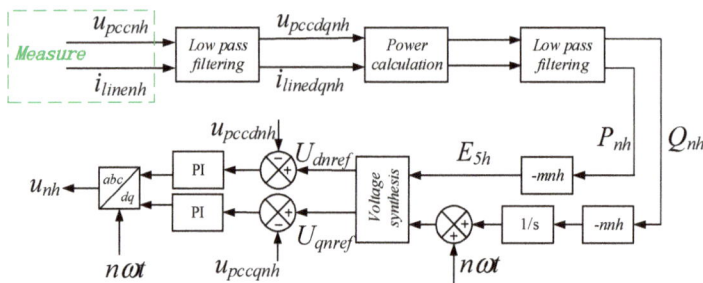

Figure 4. Harmonic drooping control structure.

3. Stability Analysis of the Multi-Converter Parallel System after Introducing the Comprehensive Control Strategy

In this section, a small signal analysis method is used to verify the effectiveness of the control strategy. The small signal analysis model of the single inverter is shown as follows (17):

$$
\begin{aligned}
[\Delta \dot{x}_{invi}] &= A_{invi}[\Delta x_{invi}] + B_{invi}[\Delta v_{bDQi}] + B_{iwcom}[\Delta \omega_{com}] \\
\begin{bmatrix} \Delta \omega_i \\ \Delta i_{odqi} \end{bmatrix} &= \begin{bmatrix} C_{INV\omega i} \\ C_{INVci} \end{bmatrix}[\Delta x_{invi}]
\end{aligned}
\tag{17}
$$

where:

$$
A_{invi} = \begin{bmatrix}
A_{pi} & 0 & 0 & 0 & B_{pi} \\
0 & A_{vi} & 0 & 0 & B_{vi} \\
B_{u1i}D_{vi}C_{pui} & B_{u1i}C_{vi} & 0 & 0 & B_{u2i} \\
B_{c1i}D_{u1i}D_{vi}C_{pui} & B_{c1i}D_{u1i}C_{vi} & B_{c1i}C_{ui} & 0 & B_{c1i}D_{u2i}+B_{c2i} \\
B_{LC1i}D_{C1i}D_{u1i}D_{vi}C_{pui}+B_{LC3i}T_u+B_{LC3i}C_{pwi} & B_{LC1i}D_{c1i}D_{u1i}C_{ui} & B_{LC1i}D_{c1i}D_{u1i}C_{ui} & B_{LC1i}+C_{ci} & A_{LCi}+B_{LC1i}D_{c1i}D_{u2i}
\end{bmatrix}
$$

$$
B_{invi} = \begin{bmatrix} 0 & 0 & 0 & B_{LC3i}T_s^{-1} \end{bmatrix}^T_{15 \times 2}
\tag{18}
$$

$$
B_{iwcom} = \begin{bmatrix} B_{pwcom} & 0 & 0 & 0 \end{bmatrix}^T_{15 \times 1}
$$

$$
C_{inv\omega i} = \begin{cases} \begin{bmatrix} C_{pw} & 0 & 0 & 0 \end{bmatrix}_{1 \times 15} & i = 1 \\ \begin{bmatrix} 0 & 0 & 0 & 0 \end{bmatrix}_{1 \times 15} & i \neq 1 \end{cases}
$$

$$
C_{invci} = \begin{bmatrix} T_i & 0 & 0 & T_s \end{bmatrix}
$$

A single inverter model is augmented in $i = 2$. The first inverter is used as the angle reference, so that the small signal model of two inverters can be obtained:

$$
\begin{aligned}
[\Delta \dot{x}_{inv}] &= A_{inv}[\Delta x_{inv}] + B_{inv}[\Delta v_{bdq}] \\
[\Delta i_{odq}] &= C_{invci}[\Delta x_{inv}]
\end{aligned}
\tag{19}
$$

where:

$$A_{inv} = \begin{bmatrix} A_{inv1} + B_{1wcom}C_{invw1} & 0 \\ 0 & A_{inv2} + B_{2wcom}C_{invw1} \end{bmatrix}$$

$$B_{inv} = \begin{bmatrix} B_{inv1} \\ B_{inv2} \end{bmatrix}; \quad C_{invc} = \begin{bmatrix} C_{invc1} & 0 \\ 0 & C_{invc2} \end{bmatrix}; \quad C_{invw} = \begin{bmatrix} C_{invw1} & C_{invw2} \end{bmatrix} \tag{20}$$

where: Δx_{inv} is the state variable of the inverter; A_{inv} is the state matrix of the inverter; Δv_{bdq} is the outlet voltage of the inverter; $\Delta \omega_{com}$ is a common angular frequency; B_{inv} and B_{wcom} are the input matrix of the inverter; $\Delta \omega$ is the angle frequency of the inverter; Δi_{odq} is the output current of the inverter; and C_{invw} and C_{invc} are the output matrices of the inverter.

The load is equivalent to the form of RL series for the micro-grid model, and its equivalent model is:

$$\left[\Delta \dot{i}_{loaddq} \right] = A_{load} \left[\Delta i_{loaddq} \right] + B_{1load} \left[\Delta u_{bdq} \right] + B_{2load} \Delta \omega \tag{21}$$

in which:

$$A_{load} = \begin{bmatrix} -\frac{R_{load}}{L_{load}} & \omega_0 \\ -\omega_0 & -\frac{R_{load}}{L_{load}} \end{bmatrix}$$

$$B_{1load} = \begin{bmatrix} \frac{1}{L_{load}} & 0 \\ 0 & \frac{1}{L_{load}} \end{bmatrix}; \quad B_{2load} = \begin{bmatrix} I_{loadq} \\ I_{loadd} \end{bmatrix} \tag{22}$$

where: R_{load} is load equivalent resistance, L_{load} is load equivalent reactance, and I_{loaddq} is load current.

In order to express the load-containing microgrid system in the form of state matrix, the virtual parameter r_n is introduced to replace the input of the original state space expression.

$$\begin{cases} u_{bd} = r_n(i_{od} - i_{loadd}) \\ u_{bq} = r_n(i_{oq} - i_{loadq}) \end{cases} \tag{23}$$

Therefore, the small signal model of the fundamental wave of the inverter model is as follows:

$$\Delta \dot{x}_{mg} = A_{mg} \Delta x_{mg} \tag{24}$$

in which:

$$A_{mg} = \begin{bmatrix} A_{inv} + B_{inv}R_nC_{invc} & -B_{inv}R_n \\ B_{1load}R_nC_{invc} + B_{2load}C_{invw} & A_{load} - B_{1load}R_n \end{bmatrix} \tag{25}$$

$$\Delta x_{mg} = \begin{bmatrix} \Delta \delta_1 & \Delta P_1 & \Delta Q_1 & \Delta X_{1dq} & \Delta \phi_{1dq} & \Delta \gamma_{1dq} & \Delta i_{l1dq} & \Delta v_{o1dq} & \Delta i_{o1dq} & \Delta \delta_2 \\ \Delta P_2 & \Delta Q_2 & \Delta X_{2dq} & \Delta \phi_{2dq} & \Delta \gamma_{2dq} & \Delta i_{l2dq} & \Delta v_{o2dq} & \Delta i_{o2dq} & \Delta i_{loaddq} \end{bmatrix} \tag{26}$$

where: Δx_{mg} is the state variable of the inverter, and ΔA_{mg} is the state matrix of the inverter.

The block diagram of the harmonic droop control is shown in Figure 5.

The harmonic droop control structure contains the harmonic power calculation module, the low pass filter module, and the harmonic droop module, which is similar to the fundamental droop control structure. The small signal model of harmonic power droop controller can be written as the form of state space function, and the expression of state space expression are:

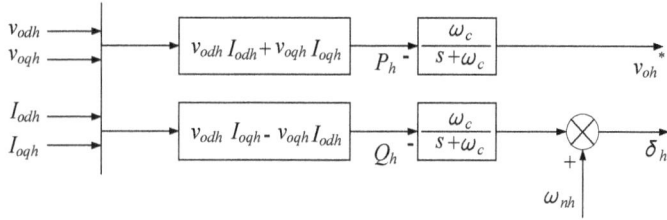

Figure 5. Harmonic droop control block diagram.

$$
\begin{bmatrix} \dot{\Delta\delta_h} \\ \Delta P_h \\ \Delta Q_h \end{bmatrix} = A_{mgh} \begin{bmatrix} \Delta\delta_h \\ \Delta P_h \\ \Delta Q_h \end{bmatrix} + B_{mgh} \begin{bmatrix} \Delta i_{ldqh} \\ \Delta v_{odqh} \\ \Delta i_{odqh} \end{bmatrix} + B_{mgwh}\Delta\omega_{comh} \tag{27}
$$

$$
\begin{bmatrix} \Delta\omega_h \\ \Delta v^*_{odqh} \end{bmatrix} = \begin{bmatrix} C_{mgwh} \\ C_{mgvh} \end{bmatrix} \begin{bmatrix} \Delta\delta_h \\ \Delta P_h \\ \Delta Q_h \end{bmatrix} \tag{28}
$$

in which:

$$
A_{mgh} = \begin{bmatrix} 0 & 0 & -m_h \\ 0 & -\omega_c & 0 \\ 0 & 0 & -\omega_c \end{bmatrix} \quad B_{mgwh} = \begin{bmatrix} -1 \\ 0 \\ 0 \end{bmatrix}
$$

$$
B_{mgh} = \begin{bmatrix} 0 & 0 & 0 & 0 & 0 & 0 \\ 0 & 0 & \omega_c I_{odh} & \omega_c I_{oqh} & \omega_c V_{odh} & \omega_c V_{oqh} \\ 0 & 0 & \omega_c I_{oqh} & -\omega_c I_{odh} & -\omega_c V_{oqh} & \omega_c V_{odh} \end{bmatrix} \tag{29}
$$

$$
C_{mgw} = \begin{bmatrix} 0 & 0 & -m_h \end{bmatrix} \quad C_{mgvh} = \begin{bmatrix} 0 & -n_h & 0 \\ 0 & 0 & 0 \end{bmatrix}
$$

where v_{odh} and v_{oqh} represent the output harmonic voltages of inverter under dq coordinate system; i_{odh} and i_{oqh} represent the output harmonic currents of inverter under dq coordinate system; P_h is the hth harmonic active power; Q_h is the hth harmonic reactive power; ω_c represents the cut-off frequency of the low pass filter; m_h and n_h represent the reactive and active power drooping coefficients; ω_h is the hth harmonic angular frequency; ω_{comh} is the common angular frequency under the hth harmonic coordinate system; and δ_h is the hth harmonic angle.

The linearization of harmonic drooping voltage loop is the same as fundamental linearization.

The state space model of the harmonic droop control strategy in the multi-inverter-based micro-grid is:

$$
\dot{\Delta x_{mgh}} = A_{mgh}\Delta x_{mgh} \tag{30}
$$

$$
\Delta x_{mgh} = \begin{bmatrix} \Delta\delta_h & \Delta P_h & \Delta Q_h & \Delta\phi_{dq} & \Delta i_{ldqh} & \Delta v_{odqh} & \Delta i_{odqh} & \Delta i_{loaddqh} \end{bmatrix}. \tag{31}
$$

The state space model of the comprehensive control strategy in the multi-inverter-based micro-grid is:

$$
\dot{\Delta x_{mga}} = A_{mga}\Delta x_{mga} \tag{32}
$$

$$
\dot{\Delta x_{mga}} = \begin{bmatrix} \dot{\Delta x_{mg}} \\ \Delta x_{mgh} \end{bmatrix} \quad \Delta x_{mga} = \begin{bmatrix} \Delta x_{mg} \\ \Delta x_{mgh} \end{bmatrix} \quad A_{mga} = \begin{bmatrix} A_{mg} & 0 \\ 0 & A_{mgh} \end{bmatrix} \tag{33}
$$

where: x_{mg} is the state variable under fundamental frequency; i_{loaddq} is a load fundamental current; A_{mg} is a fundamental state matrix; x_{mgh} is the state variable under the hth harmonic frequency; $i_{loaddqh}$ is the hth harmonic load current; A_{mgh} is the hth harmonic state matrix; x_{mga} is the comprehensive state variable under all frequencies; and A_{mga} is the comprehensive state matrix.

The stability simulation of the control strategy is carried out, in which the line resistance is 0.0428 Ω and the line inductance is 2.8 mH. The active droop coefficient $m_p = 1 \times 10^{-5}$ and the reactive droop coefficient $n_q = 3 \times 10^{-5}$. In the improved fundamental control strategy, the filter inductance is 0.6 mH, the filter capacitor is 1.5 mF, P = 1, I = 2 in PI parameters of voltage, P = 5 in PI parameters of current. In the fractional frequency harmonic control strategy, the active droop coefficient is −0.01361, the reactive power droop coefficient is −02609, KI = 10, PI = 50 in the voltage closed loop.

The stability analysis of the proposed comprehensive strategy is shown in Figures 6–9.

From Figure 6a, when the active power drooping coefficient m_p varies from 0 to 1, the overall trend of the system root locus moves towards the right side with the increase of m_p, and reaches the right half plane eventually, and the system gradually loses its stability. Namely, the system becomes unstable if the m_p is too large. From Figure 6b, when the reactive power drooping coefficient n_q varies from 0 to 1, the overall trend of the system root locus moves towards the right side with the increase of n_q, and reaches the right half plane eventually, and the system gradually loses its stability. Namely, the system also becomes unstable if n_q is too large.

From Figure 7a, when m_{ph} varies from 0 to 1, the overall trend of the system root locus moves towards the right side with the increase of m_{ph}, and the system stability is weakened gradually. From Figure 7b, when n_{qh} varies from 0 to 1, the overall trend of the system root locus moves towards the right side with the increase of n_{qh}, and reaches the right half plane eventually, and the system stability is also weakened gradually.

Figure 8 shows the root locus of the proposed multi-inverter-based micro-grid system with comprehensive control strategy, which proves that all eigenvalues are located in the left half plane, meaning that the system controlled by the comprehensive strategy is stable.

In Figure 9, the root locus of the proposed comprehensive control strategy is represented by the red "•", and the root locus of the conventional droop control strategy is represented by the blue "×". The number of poles in the low frequency area of the proposed comprehensive control strategy is 13, and the conventional droop control strategy is 6 under the same parameters. The overall trend of the characteristic root of the proposed comprehensive control strategy is on the left side of the conventional droop control strategy, and the addition of three pairs of conjugate poles can enhance the rapidity of the system. The characteristic root of the proposed comprehensive control strategy and the conventional droop control strategy are all negative real roots, so the system is stable.

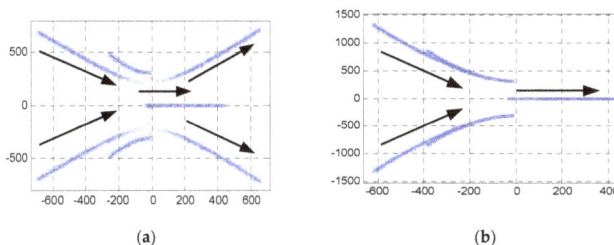

(a)   (b)

Figure 6. (**a**) The system root locus varies along with active power drooping coefficient m_p variation under fundamental frequency; (**b**) The system root locus varies along with reactive power drooping coefficient n_q variation under fundamental frequency.

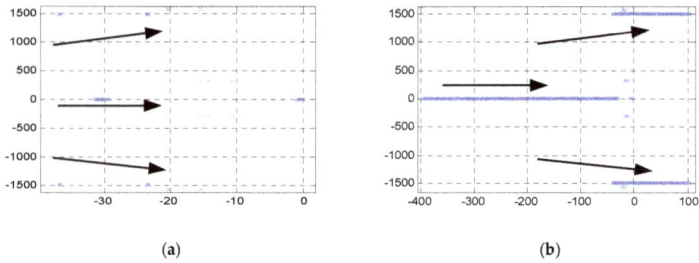

(a)   (b)

Figure 7. (a) The system root locus varies along with active power drooping coefficient m_{ph} variation under the *h*th harmonic frequency; (b) The system root locus varies along with reactive power drooping coefficient n_{qh} variation under the *h*th harmonic frequency.

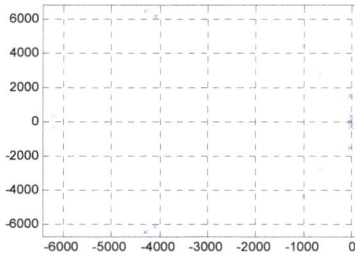

Figure 8. The root locus of the proposed multi-inverter-based micro-grid system.

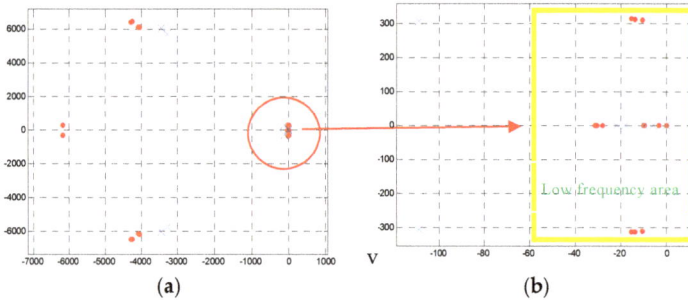

(a)   (b)

Figure 9. (a) The characteristic root contrast diagram of the proposed comprehensive control strategy and the conventional droop control strategy; (b) Part of the A graph is magnified.

4. Simulation Results

Simulations are built in MATLAB2014B, and the 5th, 7th, 11th, and 13th harmonics components are injected in the model. The simulation model is shown in Figure 10. The inverter output phase voltage amplitude is 311 V, and the frequency is 50 Hz. During simulations, the load fluctuates abruptly at 3 s, whose active power changes from 20 kW to 40 kW, and reactive power 0 kW to 20 kW. In order to verify the effectiveness of the proposed control strategy, the simulation analysis is carried out according to the following steps:

Step 1: the conventional droop control strategy of the micro-grid is simulated. The load is changed with different line impedances while the capacities of the inverters are the same, then the simulation results can be used as a reference for the following simulation.

Step 2: the comprehensive control strategy proposed in this paper is simulated and analyzed. The load is changed with different line impedances while the capacities of the inverters are the same.

Step 3: the proposed comprehensive control strategy is also proved with different inverters capacities, the load is changed in next simulations with different line impedances, while the capacity proportion of two inverters are 2:1.

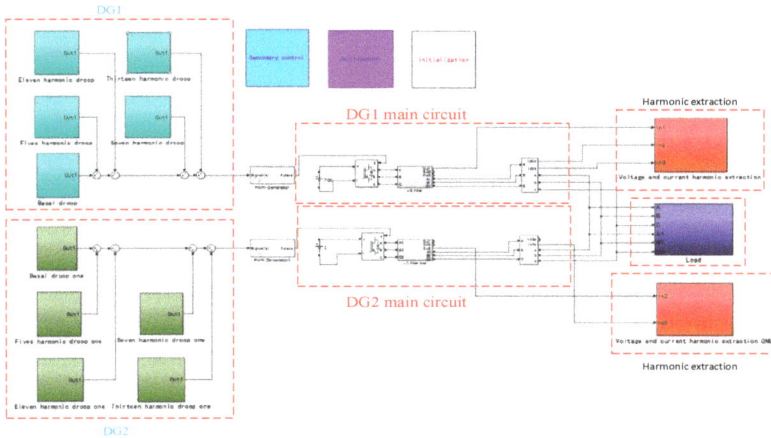

Figure 10. The simulation model in MATLAB2014B.

The simulation parameters used in this article are shown in Table 1.

Table 1. Information of simulation parameters.

Name	Parameter	Value	Parameter	Value	Parameter	Value
Main circuit	L_f	0.6×10^{-3} H	R_f	1×10^{-2} Ω	U_{ab}	700 V
	L_{line}	2.8×10^{-3} H	R_{line}	4.28×10^{-2} Ω		
Virtual impedance	R_{ref1}	2.14×10^{-3} Ω	R_{ref}	2.14×10^{-3} Ω	X_{ref1}	1.284×10^{-1}
	X_{ref2}	1.284×10^{-1}				
Fundamental wave droop control strategy	f^*	50 Hz	U^*	311 V	$m1{:}m2$	1:1
	$m1$	1×10^{-5}	$m2$	1×10^{-5}	$n1{:}n2$	1:1
	$n1$	3×10^{-5}	$n2$	3×10^{-5}	k_{up}	10
	k_{ui}	100	k_{ip}	5		
Harmonic wave droop control strategy	f^*	0 Hz	U^*	0 V	$m1{:}m2$	1:1
	$m1$	-1.361×10^{-2}	$m2$	-1.361×10^{-2}	$n1{:}n2$	1:1
	$n1$	2.609×10^{-2}	$n2$	-2.609×10^{-2}	k_{up}	10
	k_{ui}	50				

The results of the Step 1 simulations are shown in Figure 11.

From Figure 11a,b, the active powers of two inverters can be accurately distributed while the reactive powers cannot. The difference of reactive power between DG1 and DG2 is 7000 Var. In Figure 11c,d, the voltage and frequency fluctuate when the load changes. The values of voltage and frequency cannot be stabilized at 311 V and 50 Hz. The phase voltage reduction is 215 V, and the frequency reduction is 49.82 Hz. Meanwhile, the voltage waveform distortion in the 5th, 7th, 11th, and 13th harmonics are obvious. The 5th harmonic is 1.8%, 7th harmonic is 1.7%, 11th harmonic is 1.2%, and 13th harmonic is 0.7%.

The results of the Step 2 simulations are shown in Figure 12.

With the comprehensive strategy, the improvement of reactive power distribution is obvious in Figure 12b. In Figure 12c,d, the voltage and frequency can be stabilized at 311 V and 50 Hz while the load changes. The waveform has been significantly improved in Figure 12e, and the 5th, 7th, 11th, and 13th harmonic components are suppressed in Figure 12f. The 5th harmonic is 0.02%, 7th harmonic is 0.02%, 11th harmonic is 0.01%, and 13th harmonic is 0.02%.

The results of the Step 3 simulations are shown in Figure 13.

The experimental results of simulation 3 are similar to those of simulation 2. The 5th, 7th, 11th, and 13th harmonics are almost completely eliminated by simulation 2 and simulation 3. The effect is shown in Table 2. There is no doubt that the proposed strategy is also effective for the system with different micro sources capacity proportions.

According to these three simulations, it can be seen that the comprehensive strategy this paper proposed can guarantee the accurate distribution of active power and reactive power between micro sources. The voltage and frequency can be stabilized at 311 V and 50 Hz while the load changes. Meanwhile, the 5th, 7th, 11th, and 13th harmonics components can be effectively suppressed. The effects of these three simulations are shown in Table 3.

Table 2. The 5th, 7th, 11th, and 13th harmonic variation test data.

Simulation Conditions	5th	7th	11th	13th	THD
Simulation 1	1.74%	1.61%	1.23%	0.74%	2.84%
Simulation 2	0.08%	0.13%	0.09%	0.02%	0.24%
Simulation 3	0.36%	0.24%	0.23%	0.02%	0.44%

Table 3. The contrast effects of different control strategy simulations.

Simulation Conditions	Simulation 1	Simulation 2	Simulation 3
Active power accurate distribution	No	Yes	Yes
Reactive power accurate distribution	No	Yes	Yes
Voltage and frequency stability of PCC while load changes	No	Yes	Yes
The 5th, 7th, 11th, and 13th harmonic suppression	No	Yes	Yes

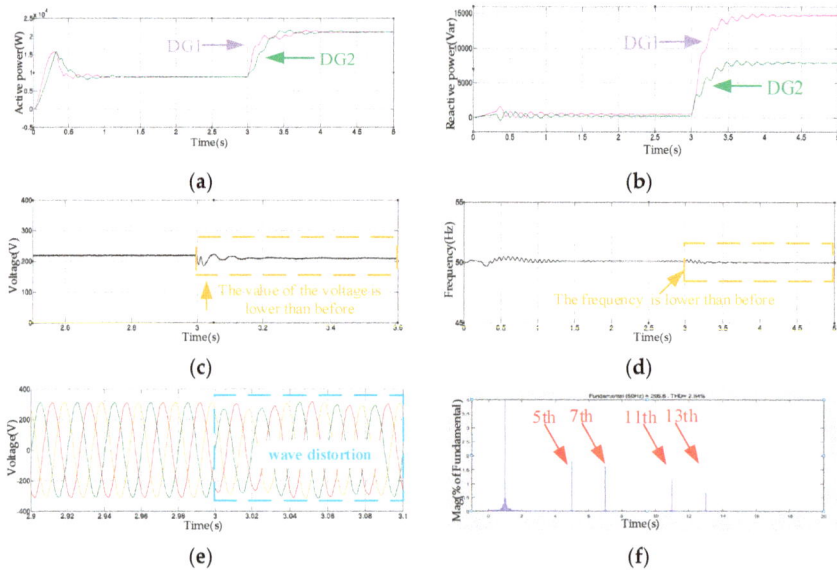

Figure 11. (**a**) Active power of DG1 and DG2 Inverters; (**b**) Reactive power of DG1 and DG2 Inverters; (**c**) The voltage value of PCC; (**d**) The frequency value of PCC; (**e**) Voltage waveform of PCC; (**f**) Voltage FFT analysis of PCC.

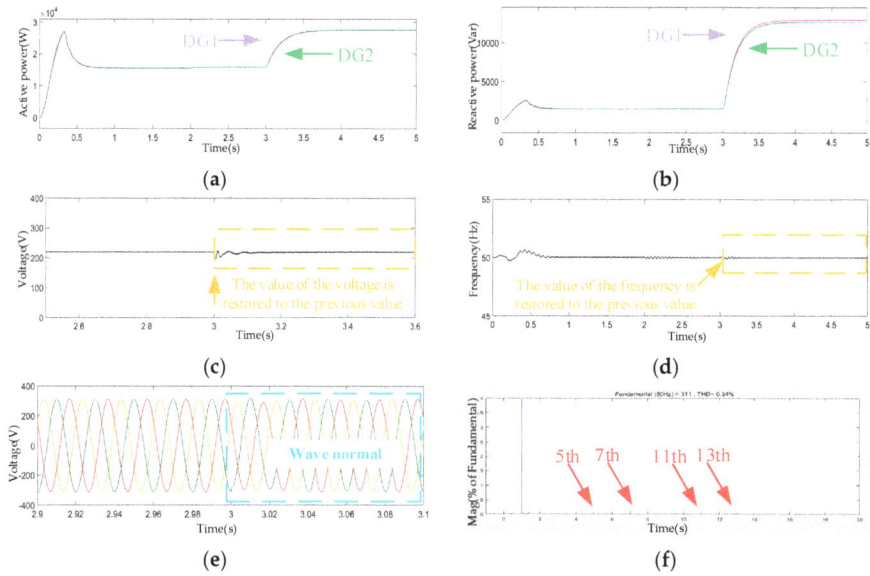

Figure 12. (**a**) Active power of DG1 and DG2 Inverters; (**b**) Reactive power of DG1 and DG2 Inverters; (**c**) The voltage value of PCC; (**d**) The frequency value of PCC; (**e**) Voltage waveform of PCC; (**f**) Voltage FFT analysis of PCC.

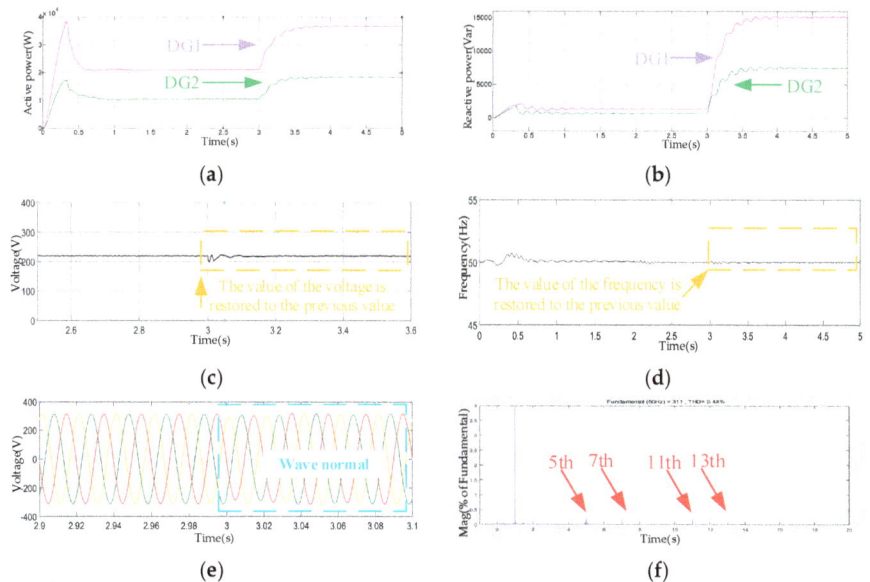

Figure 13. (**a**) Active power of DG1 and DG2 Inverters; (**b**) Reactive power of DG1 and DG2 Inverters; (**c**) The voltage value of PCC; (**d**) The frequency value of PCC; (**e**) Voltage waveform of PCC; (**f**) Voltage FFT analysis of PCC.

5. Conclusions

This paper proposes a comprehensive strategy for accurate reactive power distribution, stability improvement, and harmonic suppression of a multi-inverter-based micro-grid. With the combination of virtual impedance droop control and secondary power balance control, the active and reactive can be distributed accurately, meanwhile the stability of voltage and frequency are obviously improved. Next, a harmonic suppression control strategy is introduced to suppress harmonic components in the micro-grid. Furthermore, small signal analysis is used to analyze the stability of the proposed multi-converter parallel system after introducing the comprehensive theory. The results of MATLAB simulations certify that proposed the strategy has great performance on accurate reactive power distribution, stability improvement, and harmonic suppression.

More and more experts and scholars pay attention to the problems related to the power quality of the micro-grid. We will research on more projects related to the power quality of the micro-grid in the future, such as the power quality problems of the micro-grid with mixed loads and the transient problems of power quality in different operation modes.

Acknowledgments: This work was supported in part by the National Science and technology support project (2015BAA01B2) of China and Liaoning Electric Power Co., Ltd. science and technology project (2017YF-31) of State Grid.

Author Contributions: The paper was a collaborative effort among the authors. Henan Dong carried out relevant theoretical research, performed the simulation, analyzed the data, and wrote the paper. Shun Yuan and Zhiyuan Cai provided critical comments. Zijiao Han and Guangdong Jia performed the experiments. Yangyang Ge recorded the data.

Conflicts of Interest: The authors declare no conflict of interest.

References

1. Zhong, Q.C.; Hornik, T. *Control of Power Inverters in Renewable Energy and Smart Grid Integration*, 1st ed.; China Machine Press: Beijing, China, 2016; pp. 57–69, ISBN 978-7-11-154010-6.
2. Li, Y.; Feng, B.; Li, G.; Qi, J.; Zhao, D. Optimal distributed generation planning in active distribution networks considering integration of energy storage. *Appl. Energy* **2018**, *210*, 1073–1081. [CrossRef]
3. Zhao, B. *Key Technology and Application of Microgrid Optimal Configuration*, 1st ed.; Science Press: Beijing, China, 2015; pp. 154–196, ISBN 978-7-03-044745-6.
4. Hosseinzadeh, M.; Salmasi, F.R. Power management of an isolated hybrid AC/DC micro-grid with fuzzy control of battery banks. *IET Renew. Power Gener.* **2015**, *9*, 484–493. [CrossRef]
5. Hosseinzadeh, M.; Salmasi, F.R. Robust optimal power management system for a hybrid AC/DC micro-grid. *IEEE Trans. Sustain. Energy* **2015**, *6*, 675–687. [CrossRef]
6. Baghaee, H.R.; Mirsalim, M.; Gharehpetian, G.B.; Talebi, H.A. A decentralized power management and sliding mode control strategy for hybrid AC/DC microgrids including renewable energy resources. *IEEE Trans. Ind. Inf.* **2017**, *PP*. [CrossRef]
7. Li, D.; Zhao, B.; Wu, Z.; Zhang, L. An improved droop control strategy for low-voltage microgrids based on distributed secondary power optimization control. *Energies* **2017**, *10*, 1347. [CrossRef]
8. Zheng, Z.; Shao, W.H.; Zhao, W.F.; Ran, L.; Yang, H. Coordination control of multiple multi-functional grid-tied inverters to share power quality issues for grid-connected micro-grid. *Proc. CSEE* **2015**, *35*, 4947–4955.
9. Li, Y.; Li, Y.; Li, G.; Zhao, D.; Chen, C. Two-stage multi-objective OPF for AC/DC grids with VSC-HVDC: Incorporating decisions analysis into optimization process. *Energy* **2018**, *147*, 286–296. [CrossRef]
10. Hassan, M.; Ahmed, S.; Martin, J.P. Harmonic power sharing with voltage distortion compensation of droop controlled islanded microgrids. *IEEE Trans. Smart Grid* **2017**, *26*, 1949–1963.
11. Li, Y.W.; Kao, C.N. An accurate power control strategy for power-electronics-interfaced distributed generation units operating in a low-voltage multi-bus microgrid. *IEEE Trans. Power Electron.* **2009**, *24*, 2977–2988.
12. Lazzarin, T.B.; Bauer, G.A.T.; Barbi, I. A Control Strategy by Instantaneous Average Values for Parallel Operation of Single Phase Voltage Source Inverters Based in the Inductor Current Feedback. In Proceedings of the Energy Conversion Congress and Exposition, San Jose, CA, USA, 20–24 September 2009; pp. 495–502.

13. Jin, P.; Li, Y.; Li, G.; Chen, Z.; Zhai, X. Optimized hierarchical power oscillations control for distributed generation under unbalanced conditions. *Appl. Energy* **2017**, *194*, 343–352. [CrossRef]

14. Routimo, M.; Salo, M.; Tuusa, H. Comparison of voltage-source and current-source shunt active power filters. *IEEE Trans. Power Electron.* **2007**, *22*, 636–643. [CrossRef]

15. Grino, R.; Cardoner, R.; Costa-Castello, R.; Fossas, E. Digital repetitive control of a three-phase four-wire shunt active filter. *IEEE Trans. Power Electron.* **2007**, *54*, 1495–1503. [CrossRef]

16. Garcia-Cerrada, A.; Pinzon-Ardila, O.; Feliu-Batlle, V. Application of a repetitive controller for a three-phase active power filter. *IEEE Trans. Power Electron.* **2007**, *22*, 237–246. [CrossRef]

17. Costa-Castello, R.; Grino, R.; Parpal, C.R.; Fossas, E. High performance control of a single-phase shunt active filter. *IEEE Trans. Control Syst. Technol.* **2009**, *17*, 1318–1329. [CrossRef]

18. Ghosh, A.; Ledwich, G. A unified power quality conditioner (UPQC) for simultaneous voltage and current compensation. *Electron. Power Syst. Res.* **2001**, *59*, 55–63. [CrossRef]

19. Strzelecki, R.; Benysek, G. Power quality conditioners with minimum number of current sensors requirement. *Prz. Elektrotech.* **2008**, *11*, 295–298.

20. Sreekumar, P.; Khadkikar, V. A New Virtual Harmonic Impedance Scheme for Harmonic Power Sharing in an Islanded Microgrid. *IEEE Trans. Power Deliv.* **2016**, *31*, 936–945. [CrossRef]

21. Savaghebi, M.; Vasquez, J.C.; Jalilian, A.; Guerrero, J.M.; Lee, T.-L. Selective Harmonic Virtual Impedance for Voltage Source Inverters with LCL Filter in Microgrids. In Proceedings of the 2012 IEEE Energy Conversion Congress and Exposition (ECCE), Raleigh, NC, USA, 15–20 September 2012; pp. 1960–1965.

22. He, J.; Li, Y.W.; Guerrero, J.M.; Blaabjerg, F.; Vasquez, J.C. An islanding Microgrid power sharing approach using enhanced virtual impedance control scheme. *IEEE Trans. Power Electron.* **2013**, *28*, 5272–5282. [CrossRef]

23. Hamzeh, M.; Karimi, H.; Mokhtari, H. Harmonic and negative-sequence current control in an islanded multi-bus MV microgrid. *IEEE Trans. Smart Grid* **2014**, *5*, 167–176. [CrossRef]

24. Zhong, Q.C. Harmonic droop controller to reduce the voltage harmonics of inverters. *IEEE Trans. Ind. Electron.* **2013**, *60*, 936–945. [CrossRef]

25. Ananda, S.A.; Gu, J.C.; Yang, M.T.; Wang, J.M.; Chen, J.D.; Chang, Y.R.; Lee, Y.D.; Chan, C.M.; Hsu, C.H. Multi-Agent system fault protection with topology identification in microgrids. *Energies* **2016**, *10*, 28. [CrossRef]

26. Ren, B.Y.; Zhao, X.R.; Sun, X.D. Improved Droop Control Based Three-Phase Combined Inverters for Unbalanced Load. *Power Syst. Technol.* **2016**, *40*, 1163–1168.

27. Ziadi, Z.; Yona, A.; Senjyu, T. Optimal Scheduling of Voltage Control Resources in Unbalanced Three-Phase Distribution Systems. In Proceedings of the IEEE International Conference on Power and Energy, Kota Kinabalu, Malaysia, 2–5 December 2012; pp. 227–232.

28. Xiao, Z.; Li, T.; Huang, M.; Shi, J.; Yang, J.; Yu, J.; Wu, W. Hierarchical MAS based control strategy for microgrids. *Energies* **2010**, *3*, 8–9. [CrossRef]

29. Tu, C.M.; Yang, Y.; Xiao, F. The output side power quality control strategy for microgrid main inverter under nonlinear load. *Trans. China Electrotech. Soc.* **2017**, *32*, 53–62.

30. Guerrero, J.; Matas, J.; de Vicuna, L.G. Decentralized control for parallel operation of distributed generation inverters using resistive output impedance. *IEEE Trans. Ind. Electron.* **2007**, *54*, 994–1004. [CrossRef]

![energies logo] *energies*

MDPI

Article

Development of a Data-Driven Predictive Model of Supply Air Temperature in an Air-Handling Unit for Conserving Energy

Goopyo Hong [1] and Byungseon Sean Kim [2],*

[1] SH Urban Research Center, Seoul Housing & Communities Corporation, 621, Gaepo-ro, Gangnam-gu, Seoul 06336, Korea; goopyoh@gmail.com
[2] Department of Architectural Engineering, Yonsei University, 50 Yonsei Street, Seodaemun-gu, Seoul 03722, Korea
* Correspondence: sean@yonsei.ac.kr; Tel.: +82-2-2123-2791

Received: 29 December 2017; Accepted: 29 January 2018; Published: 9 February 2018

Abstract: The purpose of this study was to develop a data-driven predictive model that can predict the supply air temperature (SAT) in an air-handling unit (AHU) by using a neural network. A case study was selected, and AHU operational data from December 2015 to November 2016 was collected. A data-driven predictive model was generated through an evolving process that consisted of an initial model, an optimal model, and an adaptive model. In order to develop the optimal model, input variables, the number of neurons and hidden layers, and the period of the training data set were considered. Since AHU data changes over time, an adaptive model, which has the ability to actively cope with constantly changing data, was developed. This adaptive model determined the model with the lowest mean square error (MSE) of the 91 models, which had two hidden layers and sets up a 12-hour test set at every prediction. The adaptive model used recently collected data as training data and utilized the sliding window technique rather than the accumulative data method. Furthermore, additional testing was performed to validate the adaptive model using AHU data from another building. The final adaptive model predicts SAT to a root mean square error (RMSE) of less than 0.6 °C.

Keywords: data-driven; prediction; neural network; air-handling unit (AHU); supply air temperature

---

1. Introduction

The prediction of energy use and indoor environmental quality, such as temperature and relative humidity in buildings, is important for achieving energy conservation and is also employed in heating, ventilating, and air-conditioning (HVAC) applications [1].

The energy consumption and indoor air temperature of buildings can be estimated using one of two models: (1) physical or white-box models; and (2) empirical or black-box models. Both models use weather parameters and actuators' manipulated variables as inputs, with the predicted energy or temperature as their output [2,3]. The heating and cooling demand can be estimated by means of an energy simulation program, which can calculate the system capacity and zone temperature; this is what is referred to as physical modelling or white-box modelling. It is necessary to be aware of a building's physical properties in order to create accurate models of the type mentioned above [4]. Physical models can be used to provide forecasts of indoor climate variables before an actual building is constructed. However, black-box modeling is a data mining technique used to extract information from models. Black-box models, such as neural networks, need experimental data, which can be obtained after the actual building is constructed and measurements are made available. Some combination models which use energy simulation and genetic algorithms can also be utilized to select optimal design parameters and conserve energy [5].

Several articles have been published on the incorporation of black-box artificial neural networks to predict room temperature and relative humidity. This data can then be used to design and operate HVAC control systems [6–8]. Mustafaraj et al. used autoregressive linear and nonlinear neural network models to predict the room temperature and relative humidity of an open office. External and internal climate data from over three months were used to validate the models, and results showed that both models provided reasonable predictions; however, the nonlinear neural network model was superior in predicting the two variables. These predictions can be used in control strategies to save energy where air-conditioning systems operate [4]. Indoor temperature data, predicted using neural networks by the application of the Levenberg-Marquardt algorithm, was used for the control of an air-conditioned system [2]. Additionally, Kim et al. were able to make quite accurate predictions for each zone's temperature using accumulated building operational data [9]. Zhao et al. reviewed many of the techniques used to predict building energy consumption. Of the many techniques used, the ones relevant to this study are: the engineering method, which uses computer simulations; neural networks, which use artificial intelligence concepts; Support Vector Machine (SVM), which produces relatively accurate results despite small quantities of training data; and grey models, which are used when there is incomplete or uncertain data [1].

Neto et al. compared building energy consumption predicted by using an artificial neural network (ANN) and a physical model for a simulation program. Their results are suitable for energy consumption prediction with error ranges for ANNs and simulation at 10% and 13%, respectively [10]. Kreider et al. concluded that, using a neural network, operation data alone can estimate electrical demand for assessing HVAC systems [11]. Miller et al. used ANNs to predict the optimum start time for a heating plant during a setback and compared the results of ANN prediction with the conventional recursive least-squares method [12,13]. Nakahar used three different load prediction models (Kalman filter, group method of data handling (GMDH), and neural network) to estimate load for the following day and hours, to install optimal thermal storage [14]. Using general regression neural networks, Ben-Nakhi successfully predicted the time of the end of a thermostat setback to restore the designated temperature inside a building, in time for the start of business hours [6]. According to Yang et al., when the ANN prediction model is used to research optimal start and stop operation methods for cooling systems, energy consumption can be successfully reduced by 3% and 18%, respectively. Based on past building operational data, learning concepts have been introduced with optimal start and stop points discovered through analysis [15]. Nabil developed self-tuning HVAC component models based on ANNs and validated against data collected from an existing HVAC system. Errors of these models are within 2–8% in terms of coefficient variance (CV). It has been shown that the optimization process can provide total energy savings of 11% [16]. Kuisak et al. developed a data-driven approach for a daily steam load model. They used a neural network and researched 10 different data-mining algorithms to develop a predictive model [17]. Predictive neural network models have also been developed for a chiller and ice thermal storage tank of a central plant HVAC system [18]. The machine learning algorithm of an ANN is often adapted to complicated and nonlinear systems such as HVAC systems [19]. The popularity of ANNs for various applications related to online control of HVAC systems and energy management in buildings, has been increasing in the past few years [2].

The purpose of the present study was to develop a model to predict supply air temperature (SAT) of an air-handling unit (AHU) by applying AHU-historical data to a neural network, one of the black box approaches.

This current study details the process of finding various methods for improving prediction performances. Therefore, the model developed in this study was considered an adaptive model using recent hourly-operational data. When the SAT of an AHU can be estimated, it is possible to calculate the accurate amount of energy either to be eliminated or to be supplied into a building. In addition, this model can be used to calculate the heating and cooling coil load. Results of this study can be applied to control strategies and operational management of AHUs in order to reduce energy consumption.

2. Development of Models

2.1. Neural Network

The present study used a typical neural network as shown in Figure 1. Since an ANN has the ability to learn and analyze mapping relations, including nonlinear ones, its application to resolve various difficult problems has been increasing rapidly [14].

There are three layers of neurons: an input, a hidden, and an output layer. In Figure 1, neurons are placed in multiple layers. The first layer (input layer) receives inputs from the outside. The third layer (output layer) supplies the result assessed by the network and organizes the responses obtained [10]. One or more layers, called hidden layers are positioned between the first layer and the third layer. The ANN has the ability to produce output which goes through a neuron's network function. It can also match the produced output value to target value by modifying weights of interconnections. An ANN involves interconnected neurons. Each neuron or node is an independent computational unit. The current study used resilient backpropagation with weight backtracking for supervised learning in feedforward ANNs. Training a neural network is a process of setting the most suitable weights on the input of each unit. Backpropagation is the most used method for calculating error gradient for a feedforward network [20]. Connection weights and bias values initially selected as random numbers can be fixed as a result of the training process.

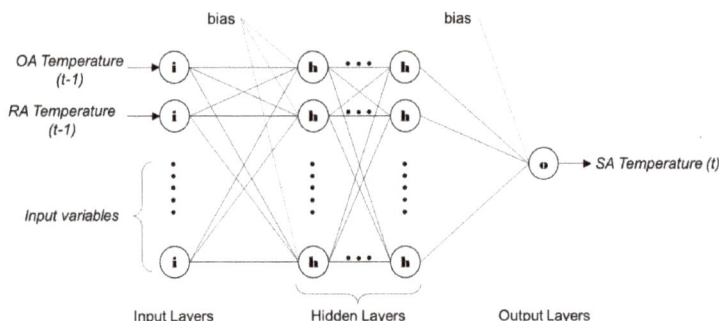

Figure 1. Typical feed-forward network (OA: Outdoor air, RA: Return air, and SA: Supply air).

2.2. Data

The case study is a fifteen-story hospital with six basement floors. The building is about 8600 m^2 and approximately 80 m high. Figure 2 shows the typical organization of AHUs in patient rooms. An AHU consists of a chilled water cooling coil, steam heating coil, two fans, filters, dampers, and sensors. Hospital AHUs require more components than AHUs serving other types of buildings [21]. Table 1 shows the specification of the sensors for temperature and relative humidity. By using sensors mounted onto the AHU and ducts, the variables listed below were monitored (as shown in Figure 3) and the data was recorded by an interval period specified by the user.

The variables that were monitored include:

- Temperature (indoor air, outdoor air, supply air, return air, mixed air)
- Relative humidity (indoor air, outdoor air, supply air, return air)
- Air flow rate (supply air, return air)
- Pressure difference
- Coil valve opening ratio (heating, cooling)
- Indoor carbon dioxide

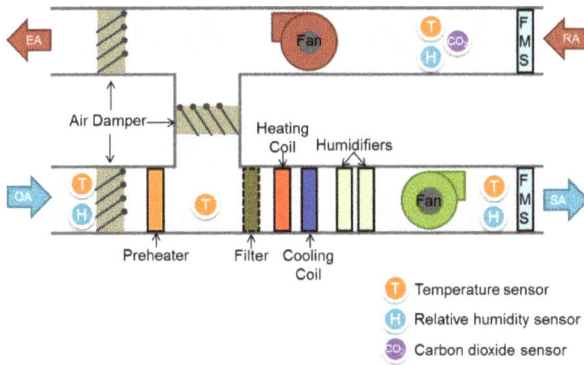

Figure 2. Schematic of case study air-handling unit (AHU) (EA: Exhaust air).

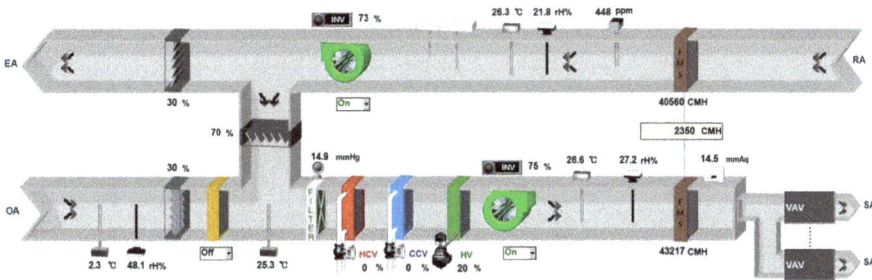

Figure 3. Monitoring of the case study AHU.

Table 1. Specification of Temperature and Humidity Sensors in the AHU.

Feature	Temperature Sensor	Humidity Sensor
Model	HTE200B12E1	HRH200A02
Range	−40 °C–70 °C	0–100% RH
Accuracy	±0.2 °C at 25 °C	±2% RH

The data collection period was one year (from December 2015 to November 2016), with a measurement interval of an hour. In order to develop a suitable predictive model, it is important to select appropriate input data. A variety of input variables such as outdoor air temperature (OAT), mixed air temperature (MAT), return air temperature (RAT), and others which were monitored in the AHU, were used to develop a predictive model for SAT output. The data set was normalized by scaling data between 0 and 1.

Input variables, output variables, statistical indices, and data sets used in the existing literature were scrutinized and are summarized in Table 2. As shown in Table 2, data collected from measurements or from simulation programs were used to develop the predictive model. Mostly, input variables were outdoor air, supply air, indoor temperature, and humidity, all of which are measurable. Using this data, energy demand, room temperature, and relative humidity can be derived as outputs. This current study examined various methods to generate proper input variables.

Table 2. Input and output variables from the literature survey.

Authors	Input Variables	Output	Statistical Indices	Data Set
Jin Yang et al. [22]	Outdoor dry-bulb Temperature Outdoor wet-bulb Temperature Temperature of water leaving the chiller	Chiller electric demand	CV, RMSE	Synthetic data (DOE 2.1E software)
Mustafaraj et al. [4]	Outside temperature Outside relative humidity Supply air temperature Supply air relative humidity Supply air flow rate Chilled water temperature Hot water temperature Room Carbon dioxide concentration	Room temperature Room relative humidity	MSE, MAE, G (goodness of fit) r^2	Measurement data (summer season)
Ruano et al. [2]	Outside solar radiation Outside air temperature Outside relative humidity	Room temperature	RMSE	Measurement data
Yang et al. [9]	Room air temperature Variation rate of room air temperature Outdoor air temperature Variation rate outdoor air temperature	The decent time of room air temperature	r^2	Simulation data
Krider et al. [18]	Outdoor Temperature Evaporator Inlet temperature Evaporator Exit temperature Ice Valve	Chiller thermal load & power usage	RMSE, CV(RMSE)	By operating a full-scale HVAC laboratory
Jeannette et al. [23]	Hot water supply temperature Boiler outlet temperature Boiler stage controller output Three-way valve controller output	Hot water supply temperature, Boiler outlet temperature	Coefficient of variation (CV)	Measurement data

As displayed in Table 2, statistical indicators were used to compare predicted values and actual values. The following statistical indicators were used: CV (coefficient of variation), mean square error (MSE), mean absolute error (MAE), root mean square error (RMSE), and coefficient of determination (r^2). The present study used MSE, RMSE, and CVRMSE in order to assess and verify the results of the predictive model.

MSE means the standard error value and uses the measure of fitness for predictive values. The neural network algorithm seeks to minimize this MSE [6].

$$\text{MSE} = \frac{1}{n} \sum_{i=1}^{n} \left(Y_{pred,i} - Y_{data,i} \right)^2 \tag{1}$$

where, Y_{pred} is the predicted value, Y_{data} is the actual value, n is the number of data.

The prediction accuracy was also measured by the root mean square error (RMSE). RMSE is a frequently used measure of the differences between a model's predicted values and the actual values. It follows Equation (2):

$$\text{RMSE} = \sqrt{\text{MSE}} \tag{2}$$

The CV (Coefficient of Variation) of the RMSE is a non-dimensional measure calculated by dividing the RMSE by the mean value of the actual temperature. Using CVRMSE makes it easier to determine the error range. The closer its value gets to 0%, the higher the accuracy.

The R program was used to predict the SAT of AHUs in this current study. R is a statistical tool that has a programming language built in. It can be used for a wide range of statistical analyses [22–24]. R offers many linear and nonlinear statistical models, classical statistical tests, time series analysis, classification, clustering, and graphical techniques. Therefore, it is very extensible. In addition, R is an analytical tool that supplies statistics and visualizations for language and the development environment. It can also provide statistical techniques, modeling, new data mining approaches, simulations, and numerical analysis methods [25].

2.3. Initial Model

The process of finding an optimal predictive model is shown in Figure 4. The figure on the left shows the initial model. To predict the SAT of the AHU, operation data such as OAT, SAT, RAT, and MAT were selected as input variables for the initial model.

Data from the monitored AHU systems were collected only up to the present time. To predict the SAT for the future 24 h, input variables for the future 24 h are needed. However, there was no future input data. Therefore, time-lagged 24 h SAT (lag 24 SAT) should be considered as output in order to estimate SAT for the future 24 h. The lag 24 SAT was included as output and trained itself to estimate. Therefore, 24 h data collected from existing SAT data was used as lag 24 SAT data. The lag 24 SAT were collections of previous 24 h data.

Predicting results should be SAT(t + 1), SAT(t + 2), ... , SAT(t + 24). However, at the present time (t), known data is input variables such as (t), (t − 1), (t − 2), ... , (t − n). By adding lag 24 SAT as output variables for 24 h, it is possible to predict SAT. In other words, input variables are: OAT (t = training period), SAT (t = training period), RAT (t = training period), and MAT (t = training period). All input values up to the present time (t) exist except for the lag 24 SAT for the recent 24 h. This makes the prediction possible. The current study used time-lagged output variables to develop a step-ahead prediction model [4,11,22].

Initial models consist of one hidden layer with nine neurons, where the number of neurons was $2n$ + 1 (n = the number of input variables). The number of neurons was inferred from the survey of exiting literature [26–28]. To validate the initial model, data from December 2015 to January 2016 was used as training data.

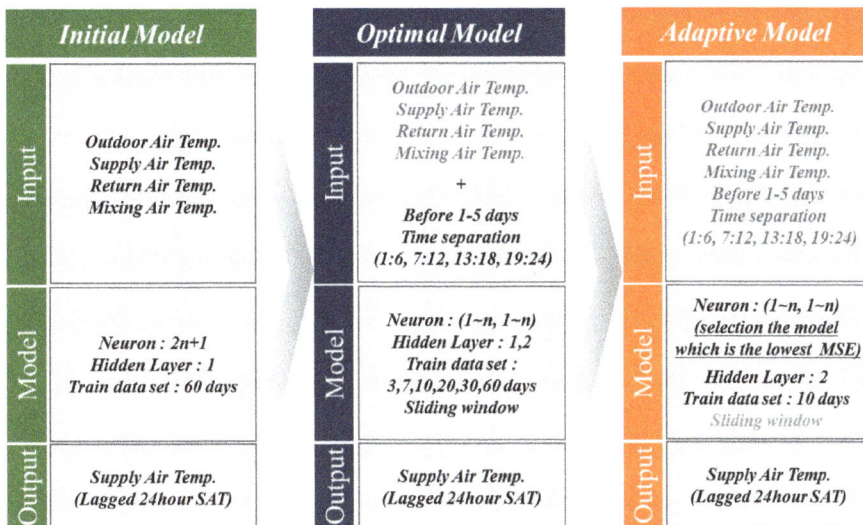

Figure 4. Evolving model process and artificial neural network (ANN) architecture.

2.4. Optimal Model

The model in the middle of Figure 4 is the optimal model. The diagram explains the ANN architecture. Different configurations of the input variables, number of hidden layers, number of neurons, and training periods were tested in order to derive the optimal model.

Since the initial model's results were not perfect, additional input variables were considered. In order to take SAT changes occurring in previous days into consideration, a model with five more input variables was created. The five extra variables were collections of previous SAT from day 1 to day 5. Moreover, four additional input variables (dawn, morning, afternoon, and evening) were included. These four variables were used to divide 24 h. These variables were referred to the time series analysis method. The performance of the prediction was analyzed by adding the input variables mentioned above.

The optimal model positioned a different number of hidden layers and neurons in order to check the accuracy of the prediction. The number of hidden layers was one or two. The number of neurons was increased from one to the number of input variables. Training times were: day 3, day 7, day 10, day 20, day 30, and day 60 because the amount of training data could affect the accuracy of the model [29].

An ANN can periodically be retrained using an augmented data set filled with new measurements. This method is also referred to as accumulative training. It can help an ANN to recognize daily and seasonal trends in predicted values. However, one drawback is that a large amount of data continuously accumulating, may become too difficult to control. Larger amounts of data mean longer training times for the ANN. Data volume can be set so that the oldest data is removed as new measurements are added. This can be achieved using a graph and periodically sliding a time window across a time series of measurements when choosing training data. In comparison with accumulative training, the sliding window technique can provide better results for real measurements [22]. The current study used the sliding window training technique.

The present study identified the optimal model by allocating the same number of neurons to the number of hidden layers. In this case, input variables were 13 and hidden layers were 2. Therefore, neuron values (number of neurons in the first hidden layer, number of neurons in the second hidden layer) were (1,1), (2,1), (2,2), (3,1), ... , (13,13), resulting in a total of 91 models.

2.5. Adaptive Model

Basic ANN architecture was created through the process of finding the optimal model. More efficient models such as an adaptive model had to be considered. Configuration of the adaptive model is shown on the right of Figure 4.

In order to find the adaptive model, a test data set was set up to examine the accuracy of the prediction. However, when actually conducting predictions, there were no test data set. It was unclear how well the prediction matched the actual value. Learning period data changed continuously because hourly data was stored. The optimal model as described in Section 2.4 might lower the prediction accuracy. To overcome this issue, data from the previous 12 h of predicted values for 24 h was used as test data and models with various number of neurons were tested again to achieve better prediction performance. A total of 91 models with neuron numbers of (1,1), (2,1), (2,2), (3,1), ... , and (13,13) were analyzed to find the most appropriate model. Outputs of these models were compared with the 12-hour test data and the model with the lowest MSE value was selected as an adaptive model.

As training data changed, an adaptive model was developed to improve the prediction performance by selecting the best model among 91 models using various neuron numbers.

For validation of the final adaptive model, AHU data from another building was used. The data measuring period was from 6 March 2017 to 23 April 2017 for each hour.

3. Results

3.1. Initial Model

Results of the comparison between predicted temperature and actual temperature derived from the initial model are shown in Figure 5. The predicted temperature followed a pattern similar to the actual temperature, with an MSE of 1.54, RMSE of 1.2 °C, and CVRMSE of 5%. Although the accuracy error was large, the initial model showed great potential in predicting SAT.

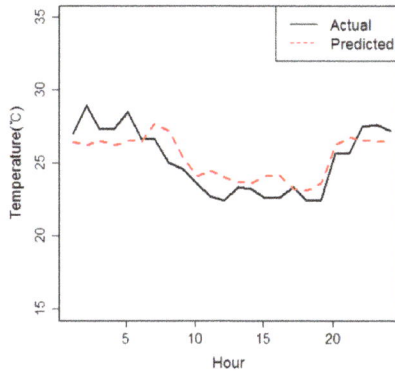

Figure 5. Comparison of actual temperature and predicted temperature using the initial model (24 February 2016).

3.2. Optimal Model

3.2.1. Kinds of Input Variables

The *basic model* had two hidden layers with the same input variables as the initial model. In order to develop an optimal model, the prediction performance of models with added input variables was evaluated. The *before model* included five added variables—from before 1 day to before 5 days. The other model was called the *hour model*. It had four added input variables derived by dividing a 24 h period into dawn, morning, afternoon, and night.

The composition and results of each model are summarized in Table 3. Since the basic model had four inputs, according to the number of neurons with two hidden layers at (1,1), (2,1), (2,2), (3,1), ... , (4,4), there were a total of 10 models. Among these 10 models, the basic model with neuron number (4,3) showed an MSE of 1.02, RMSE of 1.01 °C, and CVRMSE of 4.0%. The hour model with 13 input variables and two hidden layers had an MSE of 0.61, RMSE of 0.78 °C, and CVRMSE of 3.1%. These results indicate that the predictive performance is better when more input variables are involved in the model.

Results of comparison between actual temperature and predicted temperature among the basic model, before model, and hour model are shown in Figure 6. These are images that were expressed in program R. The predicted temperature showed a distribution pattern similar to the actual temperature. The difference between predicted temperature and the actual temperature ranged from 0.8 °C to 1.0 °C.

Table 3. Models by kinds of input variables and performance results.

Parameter	Basic Model	Before Model	Hour Model
Number of inputs	4	9	13
Hidden layer/Neurons	2/(1,1), (2,1), (2,2), (3,1), ... , (4,4)	2/(1,1), (2,1), (2,2), (3,1), ... , (9,9)	2/(1,1), (2,1), (2,2), (3,1), ... , (13,13)

Table 3. *Cont.*

Parameter	Basic Model	Before Model	Hour Model
MSE	1.02	0.87	0.61
RMSE	1.01 °C	0.93 °C	0.78 °C
CVRMSE	4.0%	3.7%	3.1%

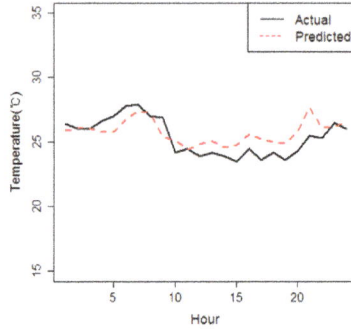

(**a**) *basic model*

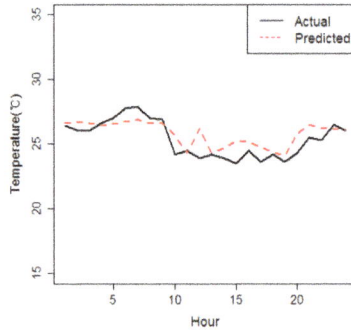

(**b**) *before model*

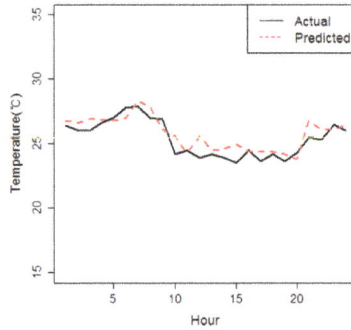

(**c**) *hour model*

Figure 6. Comparison of actual and predicted temperature by kinds of input variables (20 February 2016).

3.2.2. Number of Hidden Layers

As described in Section 3.2.1, the number of input variables was determined to be 13. Since the hour model had two hidden layers, 13 neurons were considered in one hidden layer. Results of the prediction performance of 13 models are summarized in Table 4. RMSE values ranged from 1.0 °C to 2.1 °C. CVRMSE values ranged from 3.9% to 8.3%. The model with nine neurons showed an MSE of 1.0, RMSE of 1.0 °C, and CVRMSE of 3.9%. It had the highest prediction performance among these 13 single hidden layer models.

Table 4. Model performance according to the number of neurons with one hidden layer.

Neuron	MSE	RMSE	CVRMSE
1	1.0	1.0	4.0%
2	1.9	1.4	5.5%
3	1.4	1.2	4.7%
4	1.3	1.2	4.5%
5	1.8	1.3	5.2%
6	1.3	1.1	4.4%
7	4.4	2.1	8.3%
8	1.5	1.2	4.8%
9	1.0	1.0	3.9%
10	1.1	1.0	4.1%
11	3.4	1.8	7.3%
12	2.2	1.5	5.8%
13	1.5	1.2	4.8%

When there were two hidden layers, the number of neurons was 13 for each hidden layer. A total of 91 models were obtained. RMSE values ranged from 0.8 °C to 22.7 °C. The model with the number of neurons of (5,2) showed an MSE of 0.6, RMSE of 0.8 °C, and CVRMSE of 3.1%. Therefore, the model with two hidden layers has a better prediction performance than the model with one hidden layer.

3.2.3. Period of Training Data Set

Based on the results shown above, the number of hidden layers and neurons of the optimal model were determined to be 2 and (5,2), respectively. Results for the predictive performance of models with various time periods of data training are shown in Table 5. The optimal duration was 10 days. It had an MSE of 0.59, RMSE of 0.77 °C, and CVRMSE of 3.0%. On the other hand, the MSE showed an increase when the learning period was much longer (20 days and 30 days). The learning period of 60 days also had a low MSE value at 0.77. However, data for the learning period of 60 days took much longer to obtain results. The longer the time period used for training data, the longer the processing time to obtain result. In the present study, the data training period was set to 10 days for the optimal model.

Table 5. Prediction performance according to time period used for data training.

Indicators	3 Days	7 Days	10 Days	20 Days	30 Days	60 Days
MSE	0.95	1.03	0.59	1.16	1.12	0.77
RMSE (°C)	0.97	1.01	0.77	1.08	1.06	0.88
CVRMSE	3.8%	4.0%	3.0%	4.2%	4.2%	3.5%

Graphs of actual temperature and predicted temperature by each time period of data training are shown in Figure 7.

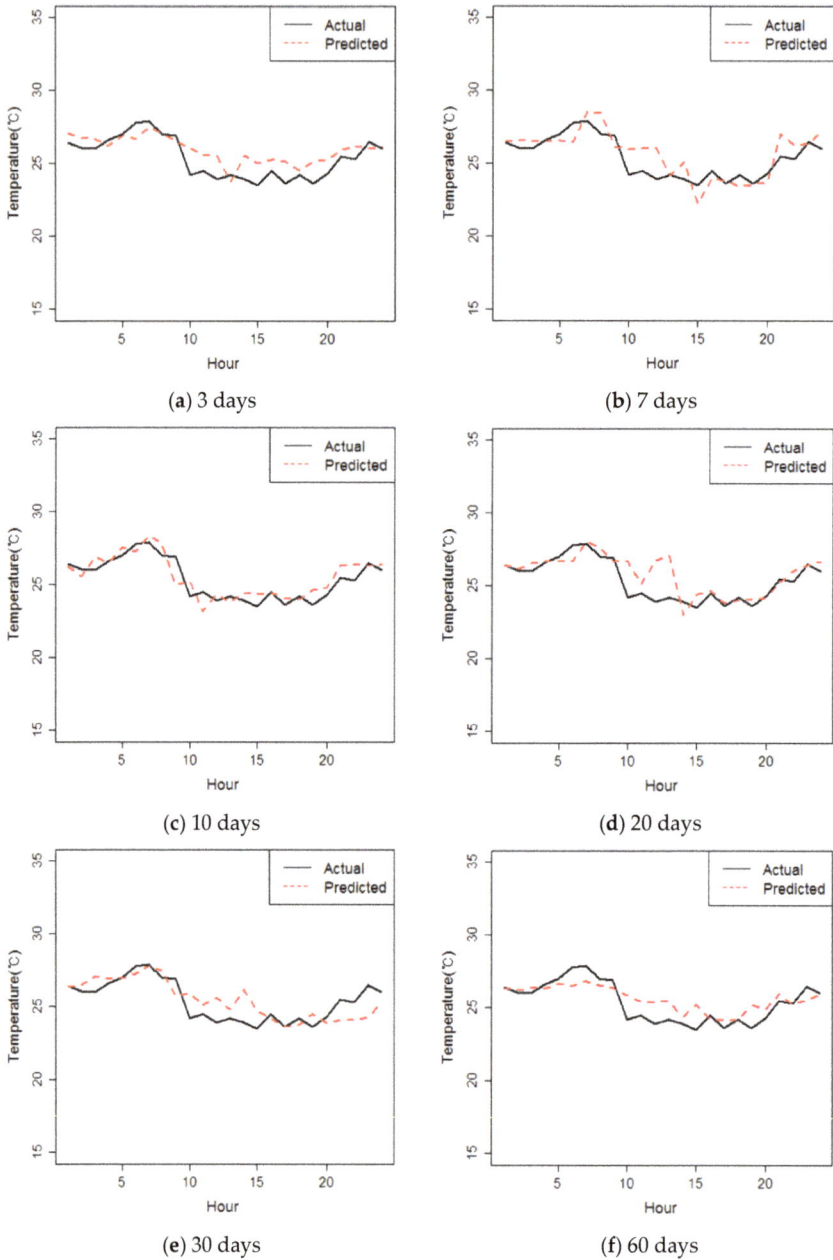

(**a**) 3 days

(**b**) 7 days

(**c**) 10 days

(**d**) 20 days

(**e**) 30 days

(**f**) 60 days

Figure 7. Comparison of actual temperature and predicted temperature by time period of data training (20 February 2016).

3.3. Fixed Optimal Model

The optimal model consisted of 13 input variables, neuron numbers of (5,2), two hidden layers, and a learning period of 10 days. It had an MSE of 0.59, RMSE of 0.77 °C, and CVRMSE of 3.0%.

Results from the comparison between actual temperature and predicted temperature using the optimal model are shown in Figure 8. The structure of the model expressed in the R program is shown in Figure 9. The weight of each connection in the optimal model is also shown in Figure 9. Black lines represent the connection between each layer and weights of each connection. Blue lines represent the bias term added in each step.

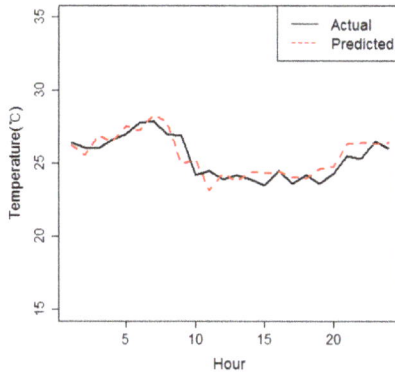

Figure 8. Comparison of actual and predicted temperatures using the optimal model (20 February 2016).

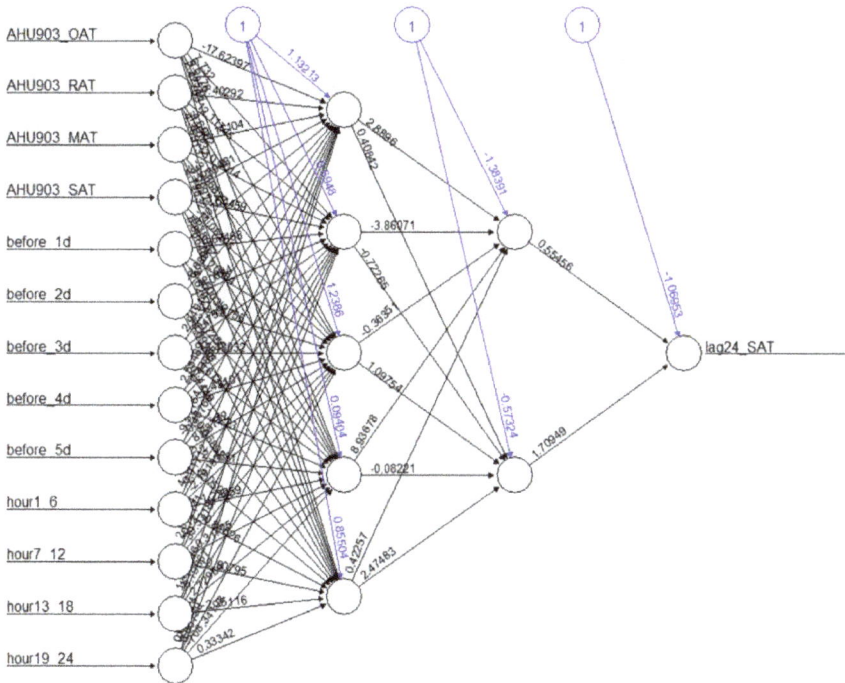

Figure 9. Structure of optimal model using R.

3.4. Final Adaptive Model

Results of the comparison between actual temperature and predicted temperature using the developed adaptive model are shown in Figure 10. The predicted temperature was learned at different time periods by using the adaptive model. Model (a) had (13,9) neurons. Training data used for model (a) was obtained from April 1 to April 10. It had an MSE of 0.66, RMSE of 0.81 °C, and CVRMSE of 3.3%. Model (b) had (5,5) neurons. Its training data was obtained from July 17 to July 26. It had an MSE of 0.52, RMSE of 0.72 °C, and CVRMSE of 3.5%.

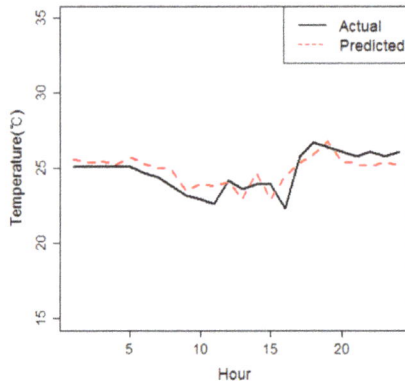

(**a**) Training data: 1–10 April (11 April 2016)

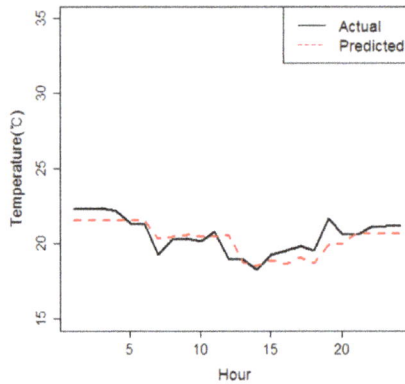

(**b**) Training data: 17–26 July (27 July 2016)

Figure 10. Comparison between actual and predicted temperature using developed adaptive model.

It could be seen that the model changed instantaneously depending on the training data. Therefore, it was an adaptive model through learning and seeking a model with the lowest MSE.

3.5. Validation of the Adaptive Model

By using another building's AHU data, the predictive performance of the adaptive model was verified. The actual temperature and predicted temperature are shown in Figure 11. In Figure 11, (a) predicts the SAT on 24 March. The difference between the predicted value and the actual value is MSE = 0.26, RMSE = 0.51 °C, and CV = 2.2%; (b,c) with a RMSE of 0.63°C and 0.55 °C, respectively, also show a better prediction performance than the original case study. Predicted temperature distribution is similar to actual temperature.

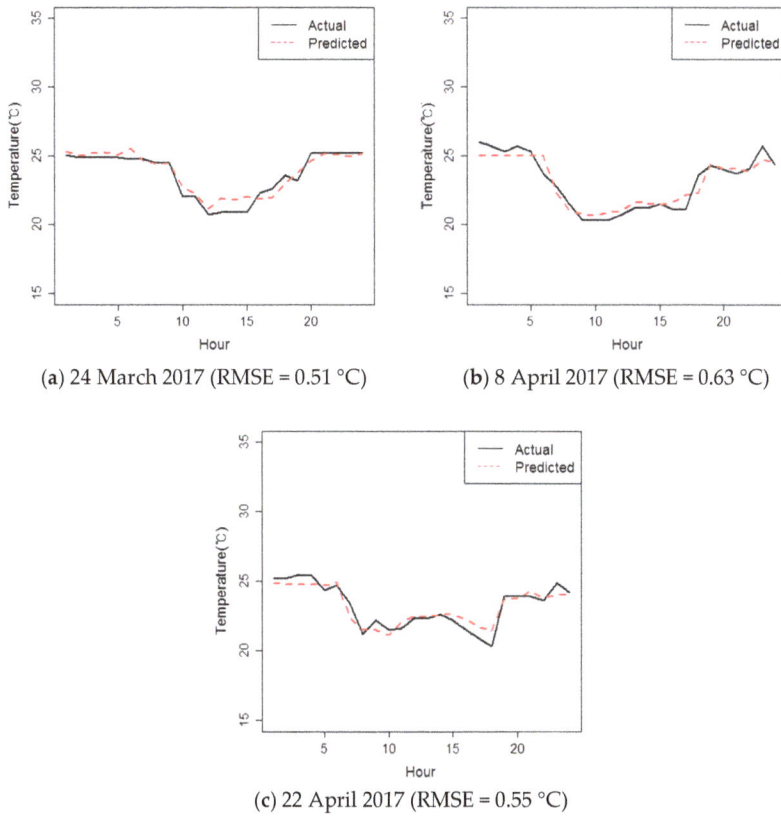

(a) 24 March 2017 (RMSE = 0.51 °C)   (b) 8 April 2017 (RMSE = 0.63 °C)

(c) 22 April 2017 (RMSE = 0.55 °C)

Figure 11. Comparison between actual and predicted temperature for validation.

4. Discussion and Conclusions

This study predicted the SAT of an AHU using an ANN. In addition, a model with good predictive performance was developed using an initial model, an optimal model, and an adaptive model.

The potential of SAT prediction was found through the initial model. Parameters and number of input variables are important in the process of finding an optimal model. The hour model with various input variables showed a 25% improvement in prediction performance than the basic model which used temperature measured within the AHU.

A total of 91 models with two hidden layers were tested. The prediction performance was improved 20% (RMSE-based) when the number of hidden layers was two instead of one.

Two hidden layers and (5,2) neurons were selected for the optimal model. It is difficult to determine the number of hidden layers and the number of neurons [9]. The number of neurons should be made to correspond to the number of input and output variables and should also follow some simple rules [30]. Unfortunately, few studies have provided guidelines for selecting the best layer or neuron numbers. Therefore, factors such as the number of hidden layers and neurons should be determined based on the characteristics of the application and data [31].

Based on the results of performance using various periods for data learning, 10 days was determined to be the best for the optimal model developed in the present study. When the learning period was increased to 20 or 30 days, the MSE value increased. Therefore, it is important to review the length of the data learning period according to data characteristics. When the learning period

Energies **2018**, *11*, 407

was increased to 60 days, the prediction performance was better than that with learning period of 20 or 30 days. However, such a long learning period was ultimately inefficient because its execution took a much longer time [32]. Even though a learning period of 3 days also showed good results, such a short time period might not completely capture the trends for the predicted values [22].

The optimal model developed in this current study is a model that can use recently collected data. It was developed by using a data training approach with a sliding window method rather than using an accumulative data method. Applying the sliding window technique makes it possible to maintain a training data set of a relatively small and constant size and retrain it quickly. However, annual and seasonal changes may not be accurately reflected in the prediction results.

Although the optimal model was developed, learning data changes over time. A fixed optimal model dependent on changing training data does not show uniformly good prediction performance results.

In conclusion, an adaptive model was developed by selecting a model with the lowest MSE. A total of 91 models were evaluated after setting up a 12-h test set at every prediction. The adaptive model can learn from training data that changes in real time. It seeks the model that has the best prediction performance. The prediction performance of the adaptive model is similar to that of the optimal model. However, it has the advantage of being able to actively cope with changing training data.

Acknowledgments: This research was supported by Basic Science Research Program through the National Research Foundation of Korea (NRF) funded by the Ministry of Education (NRF-2015R1D1A1A01057928).

Author Contributions: Goopyo Hong initiated the research idea and wrote the manuscript. Byungseon Sean Kim supervised the study and provided advice on the data analysis.

Conflicts of Interest: The authors declare no conflict of interest.

References

1. Zhao, H.-X.; Magoulès, F. A review on the prediction of building energy consumption. *Renew. Sustain. Energy Rev.* **2012**, *16*, 3586–3592. [CrossRef]
2. Ruano, A.E.; Crispim, E.M.; Conceição, E.Z.E.; Lúcio, M.M.J.R. Prediction of building's temperature using neural networks models. *Energy Build.* **2006**, *38*, 682–694. [CrossRef]
3. ASHRAE. *ASHRAE Handbook Fundamental—Chapter 19 Energy Estimating and Modeling Method*; ASHRAE: Atlanta, CA, USA, 2013.
4. Mustafaraj, G.; Lowry, G.; Chen, J. Prediction of room temperature and relative humidity by autoregressive linear and nonlinear neural network models for an open office. *Energy Build.* **2011**, *43*, 1452–1460. [CrossRef]
5. Ferdyn-Grygierek, J.; Grygierek, K. Multi-Variable Optimization of Building Thermal Design Using Genetic Algorithms. *Energies* **2017**, *10*, 1570. [CrossRef]
6. Abdullatif, E.; Ben-Nakhi, M.A.M. Energy conservation in buildings through efficient AC control using neural networks. *Appl. Energy* **2002**, *73*, 5–23.
7. Kang, I.-S.; Lee, H.-E.; Park, J.-C.; Moon, J.-W. Development of an Artificial Neural Network Model for a Predictive Control of Cooling Systems. *KIEAE J.* **2017**, *17*, 69–76. [CrossRef]
8. Baik, Y.K.; Moon, J.W. Development of Artificial Neural Network Model for Predicting the Optimal Setback Application of the Heating Systmes. *KIEAE J.* **2016**, *16*, 89–94. [CrossRef]
9. Yang, I.-H.; Kim, K.-W. Prediction of the time of room air temperature descending for heating systems in buildings. *Build. Environ.* **2004**, *39*, 19–29. [CrossRef]
10. Neto, A.H.; Fiorelli, F.A.S. Comparison between detailed model simulation and artificial neural network for forecasting building energy consumption. *Energy Build.* **2008**, *40*, 2169–2176. [CrossRef]
11. Kreider, J.F.; Blanc, S.L.; Kammerud, R.C.; Curtiss, P.S. Operational data as the basis for neural network prediction of hourly electrical demand. *ASHRAE Trans.* **1997**, *103*, 926.
12. Underwood, C.P. *HVAC Control Systems Modelling Analysis and Design*; Taylor and Francis: London, UK; New York, NY, USA, 1999.
13. Miller, R.; Seem, J. Comparison of artificial neural networks with traditional methods of predicting return time from night or weekend setback. *ASHRAE Trans.* **1991**, *97 Pt 1*, 500–508.

14. Nakahara, N.; Zheng, M.; Pan, S.; Nishitani, Y. Load Prediction for Optimal Thermal Storage—Comparison of Three Kinds of Model Application. In Proceedings of the IBPSA Building Simulation Conference, Kyoto, Japan, 13–15 September 1999; pp. 519–526.

15. Park, D.H.; N, H.M.; Chung, H.G.; Yang, I.H. *Analysis of Energy Saving Effect of Optimal Start Stop with ANN on Heating and Cooling System*; Korean Journal of Air-Conditioning and Refrigerating Engineering; SAREK: Seoul, Korea, 2014.

16. Nabil Nassif, P.E. Modeling and optimization of HVAC systems using artificial intelligence approaches. *ASHRAE Trans.* **2012**, *118*, 133.

17. Kusiak, A.; Li, M.; Zhang, Z. A data-driven approach for steam load prediction in buildings. *Appl. Energy* **2010**, *87*, 925–933. [CrossRef]

18. Massie, D.D.; Curtiss, P.S.; Kreider, J.F.; Dodier, R. Predicting Central Plant HVAC Equipment Performance Using Neural Networks Laboratory System Test Results. *ASHRAE Trans.* **1998**, *104*, 221.

19. Yao, Y.; Yu, Y. *Modeling and Control in Air Conditioning Systems*; Springer: Berlin/Heidelberg, Germany, 2016.

20. Tso, G.K.F.; Yau, K.K.W. Predicting electricity energy consumption: A comparison of regression analysis, decision tree and neural networks. *Energy* **2007**, *32*, 1761–1768. [CrossRef]

21. ASHRAE. *HVAC Design Manual for Hospitals and Clinics*, 2nd ed.; ASHRAE: Atlanta, CA, USA, 2013.

22. Yang, J.; Rivard, H.; Zmeureanu, R. On-line building energy prediction using adaptive artificial neural networks. *Energy Build.* **2005**, *37*, 1250–1259. [CrossRef]

23. Jeannette, E.; Curtiss, P.S.; Assawamartbunlue, K.; Kreider, J.F. Experimental results of a predictive neural network HVAC controller. *ASHRAE Trans.* **1998**, *104*, 192.

24. The R Project for Statistical Computing. Available online: https://www.r-project.org/ (accessed on 6 February 2017).

25. Kim, S.H.; S, H.S.; Son, S.H. A Study on Large-Scale Traffic Information Modeling using R. *J. KIISE* **2015**, *41*, 151–157.

26. Hecht-Nielsen, R. Theory of the Back propagation Neural Network. In *International Joint Conference on Neural Networks*; IEEE: Washington, DC, USA, 1989.

27. Yang, I.-H.; Yeo, M.-S.; Kim, K.-W. Application of artificial neural network to predict the optimal start time for heating system in building. *Energy Convers. Manag.* **2003**, *44*, 2791–2809. [CrossRef]

28. Argiriou, A.A.; Bellas-Velidis, I.; Kummert, M.; André, P. A neural network controller for hydronic heating systems of solar buildings. *Neural Netw.* **2004**, *17*, 427–440. [CrossRef] [PubMed]

29. Wang, S.; Jin, X. Model-based optimal control of VAV air-conditioning system using genetic algorithm. *Build. Environ.* **2000**, *35*, 471–487. [CrossRef]

30. Barga, R.; Fontama, V.; Tok, W.H.; Cabrera-Cordon, L. *Predictive Analytics with Microsoft Azure Machine Learning*; Apress: Berkley, CA, USA, 2016.

31. Zurada, J.M. *Introduction to Artificial Neural Systems*; West Publishing Company: West St. Paul, MN, USA, 1992; ISBN 10: 0314933913.

32. Anstett, M.; Kreider, J. Application of artificial neural networks to commercial energy use prediction. *ASHRAE Trans.* **1993**, *99*, 505–517.

energies

MDPI

Article

Nonlinear Modeling and Inferential Multi-Model Predictive Control of a Pulverizing System in a Coal-Fired Power Plant Based on Moving Horizon Estimation

Xiufan Liang, Yiguo Li *, Xiao Wu and Jiong Shen

Key Laboratory of Energy Thermal Conversion and Control of Ministry of Education, Southeast University, Nanjing 210096, China; lxf@seu.edu.com (X.L.); wux@seu.edu.cn (X.W.); shenj@seu.edu.cn (J.S.)
* Correspondence: lyg@seu.edu.cn; Tel.: +86-139-1397-0596

Received: 28 January 2018; Accepted: 5 March 2018; Published: 7 March 2018

Abstract: Fuel preparation is the control bottleneck in coal-fired power plants due to the unmeasurable nature or inaccurate measurement of key controlled variables. This paper proposes an inferential multi-model predictive control scheme based on moving horizon estimation for the fuel preparation system in coal-fired power plants, i.e., the pulverizing system, aimed at improving control precision of key operating variables that are unmeasurable or inaccurately measured, and improving system tracking performance across a wide operating range. We develop a first principle model of the pulverizing system considering the nonlinear dynamics of primary air, and then employ the genetic algorithm to identify the unknown model parameters. The outputs of the identified first principle model agree well with measured data from a real pulverizing system. Thereafter we derive a moving horizon estimation approach to estimate the desired, but unmeasurable or inaccurately measured, controlled variables. Estimation constraints are explicitly considered to reduce the influence of measurement uncertainty. Finally, nonlinearity of the pulverizing system is analyzed and a multi-model inferential predictive controller is developed using the extended input-output state space model to achieve offset-free performance. Simulation results show that the proposed soft sensor can provide improved estimates than conventional extended Kalman filter, and the proposed inferential control scheme can significantly improve performance of the pulverizing system.

Keywords: pulverizing system; soft sensor; inferential control; moving horizon estimation; multi-model predictive control

1. Introduction

The pulverizing system is one of the most important auxiliary parts in coal-fired power plants, and has two main functions: to grind crushed coal lumps of several cm in diameter to very fine powder (~50–100 μm in diameter), and sending the pulverized coal into the furnace and provide oxygen for its combustion [1]. The operation performance of the pulverizing system can strongly affect the fuel combustion in the furnace, and thus improving its control performance is of great significance to achieve flexible power plant operation. There are three fundamental control requirements in the pulverizing system.

(1) Pulverized coal flow into the furnace should rapidly track the power plant fuel demand, allowing power generation to be adjusted in a timely way, as required by power grids.
(2) The air to coal ratio (the ratio of primary air mass flow to raw coal mass flow) should be kept close to optimal to maintain coal combustion efficiency and reduce generation of nitrogen oxide pollutants [2].

(3) The coal mill outlet temperature must be controlled within the safe operation region to avoid wet coal conditions and coal firing [3].

However, most power plants are unable to measure pulverized coal flow into the furnace in real-time, which can significantly reduce control precision of the pulverizing system and power plant load [4]. Although hardware sensors, such as digital holography techniques [5], provide some options to solve this problem, they requires very high equipment investment, retrofitting, and maintenance costs, which make their widespread use difficult. The raw coal feed rate and primary air mass flow are also only measured approximately due to measurement technology limitations [6], which introduces many disturbances to the control system and causes fluctuation of the controlled variables.

A practical way to control unmeasurable or inaccurately measured process variables is to apply inferential control schemes, where the desired controlled variables are first estimated by a soft sensor, and subsequently employed as the feedback signal for the controller [7]. Modeling the pulverizing system provides a theoretical foundation to develop soft sensors. Agrawal et al. developed a unified thermal-mechanical model of the pulverizing system that divided coal mill internal regions into four zones and coal particles into ten size groups to consider the fineness of the pulverized coal flow into the furnace [4]; however the model is quite complex and unsuitable for control system design. Niemczyk et al. constructed and validated a dynamic pulverizing system model for different coal mill types under various operating conditions [8], and discussed the influence of classifier speed on pulverized coal flow into the furnace; however many plant details are required in their model, such as the roller breakage rate and flow parameters of the pulverized coal flow. Jin et al. established the dynamic relation between coal mill differential pressure and pulverized coal stored in the mill [9], and Zeng et al. modeled moisture content in pulverized coal by energy balance [10]; however their models ignored the nonlinear dynamics of primary air. Wei et al. developed a multi-segment model that considers coal mill dynamics from startup to shutdown separately [11], however they did not consider the moisture content and grindability of raw coal. In summary, current pulverizing system models are either too simple, too complex, or require many internal plant details, which limit their application for designing soft sensors or control systems. Most previous research has focused on simulation or fault detection of the grinding process, with primary air system dynamics simplified to linear steady-state. In practice, the primary air system is controlled via two air baffles, which have typically nonlinear dynamics. Ignoring these effects will significantly reduce accuracy when the model is used for control system or soft sensor design.

The pulverizing system is a nonlinear multi-variable system with large process inertia and measurement uncertainty, which is difficult for conventional proportional-integral-derivative (PID) controllers to control. Hence, various advanced control techniques have been proposed to improve operational performance. Lu et al. designed a fuzzy PID controller to control outlet temperature [12], however the fuzzy PID cannot handle the large process inertia well. Internal stability and tracking performance of the pulverizing system can be guaranteed with a Lyapunov function, and Fei et al. developed a robust fuzzy tracking control method [13]; however their control scheme cannot achieve the decoupling control of pulverizing system. Cortinovis et al. designed a nonlinear model predictive controller (NMPC) based on a nonlinear pulverizing system model, and updated the model parameters online with an extended Kalman Filter [14]. Although the simulation results show their control strategy is effective, NMPC is generally unable to be solved in real-time. Gao et al. designed a multi-model predictive controller for different operating points, explicitly considering the moisture content of raw coal [6], and developed an optimization control scheme for pulverized coal flow into the furnace; however they did not consider the inaccurate measurements of the key controlled variables. Zeng et al. proposed an economic control method to improve coal combustion efficiency by controlling the moisture content in the pulverized coal to an optimized set point [10]; however they did not discuss the control of pulverized coal flow into the furnace.

Although control problems associated with nonlinearities, coupling effects, and large process inertia have been widely studied for the pulverization system in previous research, few have focused on development of a soft sensor to address issues caused by the fact key controlled variables, i.e., pulverized coal flow into the furnace and primary air mass flowrate, are either unmeasurable or inaccurately measured. The most direct method to estimate desired controlled variables is to solve the model differential equations given measured inputs [4,6,10]. However, this can produce unreliable results. As discussed earlier, raw coal feed rate and primary air mass flow are only measured approximately, and using them directly to estimate pulverized coal flow into the furnace will lead to large errors. Other process measurements, such as mill electric current and outlet temperature, which could reflect the operating status of the grinding process, have not been considered for estimating pulverized coal flow into the furnace.

Considering these issues, this study develops an inferential multi-model predictive control scheme for pulverizing systems. A first principle model of the pulverizing system considering primary air nonlinear dynamics was developed, with model complexity and accuracy balanced by combining physical and empirical relationships. Based on the established model, a soft sensor was derived to estimate desired controlled variables using a moving horizon estimation (MHE) approach, where estimation constraints were explicitly considered to reduce the influence of measurement uncertainty. Finally, the pulverizing system nonlinearity was analyzed, and an inferential multi-model predictive controller designed using the extended input-output state space model to achieve offset-free performance.

The current study has two major contributions:

(1) A first principle model of the pulverizing system was developed that explicitly considered the nonlinear dynamics of primary air, which is suitable for designing a system controller and soft sensor.
(2) An inferential multi-model predictive control scheme was established based on MHE that provided improved pulverizing system control precision and tracking performance.

The main content of this paper is organized as follows: Section 2 presents the first principle model of the pulverizing system. Section 3 derives the soft sensor using MHE, and Section 4 discusses the formulation of the inferential multi-model predictive controller. Section 5 presents simulation results, including accuracy validation of the soft sensor and performance validation of the proposed inferential control scheme. Finally, Section 6 concludes the paper.

2. Dynamic Model of the Pulverizing System

2.1. Pulverizing System Description

Figure 1 shows a typical pulverizing system consisting of coal mill and the primary air systems. In the coal mill, raw coal enters the grinding region from the coal chute and is crushed. Primary air then enters the coal mill through the air ring, drying the pulverized coal and transporting it to the coarse classifier in the upper grinding zone for separation. Suitably pulverized coal is transported by the primary air to the furnace for combustion, whereas unsuitable coal falls back into the coal chute for grinding. The air pre-heater is deployed at the rear of the flue gas tunnel of the boiler, and can heat cold air to ~220 °C. Primary air is generated by mixing cold and hot air, controlled by two air baffles. The primary air fan maintains constant pressure at the entrance of the air baffles. Since the pressure has very fast dynamics, and generally can be well controlled by the primary air fan, the primary air fan has little influence on the pulverizing system operation.

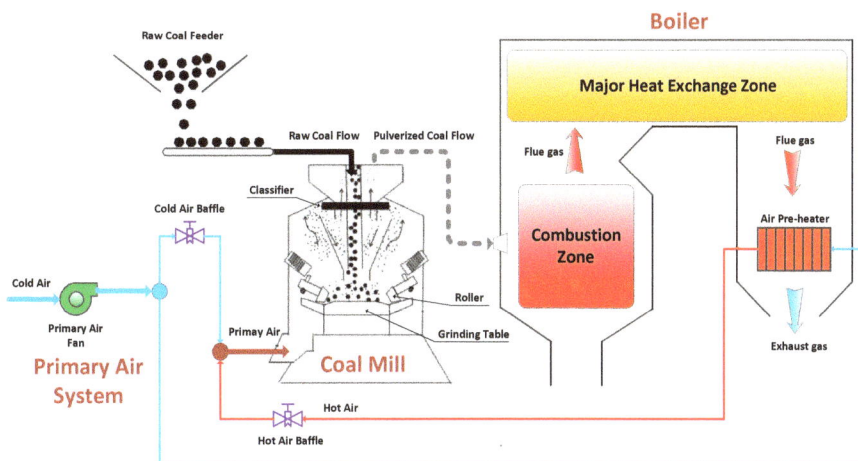

Figure 1. Simplified diagram of a typical pulverizing system.

2.2. First Principle Model of the Pulverizing System

The pulverizing system parameter model was established with the following assumptions:

(1) Raw coal grinding and pulverized coal delivery are separate processes;
(2) Pulverized coal fineness is neglected, and the coal is categorized into raw and pulverized coal only;
(3) The classifier operates at its designed rotating speed;
(4) Primary air is regarded as an ideal gas.

The pulverizing system has 24 unknown parameters to be identified: 8 in the primary air system (S_i, $i = 1, 2, \ldots, 8$), and 16 in the coal mill system (K_i, $i = 1, 2, \ldots, 16$).

The dynamics of the primary air system can be described as:

$$q_{air,cold} = S_1 (\mu_{cold})^{S_2}, \tag{1}$$

$$q_{air,hot} = S_3 (\mu_{hot})^{S_4}, \tag{2}$$

$$S_5 \frac{dt_{air}}{dt} = \frac{q_{air,cold}}{q_{air,hot} + q_{air,cold}} \cdot t_{cold} + \frac{q_{air,hot}}{q_{air,hot} + q_{air,cold}} \cdot t_{hot} - S_6 t_{air}, \tag{3}$$

and:

$$S_7 \frac{dq_{air}}{dt} = q_{air,cold} + q_{air,hot} - S_8 q_{air}, \tag{4}$$

where t_{air} is the primary air temperature, q_{air} is the primary air mass flow, μ_{cold} is the cold air baffle opening, μ_{hot} is the hot air baffle opening, t_{cold} is the cold air temperature, and t_{hot} is the hot air temperature.

Remark 1. *The air baffle has similar characteristics to valves [15]. Figure 2 shows typical valve inherent flow characteristics, and all of the curves can be well approximated by power functions with different exponents. Thus, we used (1) and (2) to identify air baffle flow characteristics.*

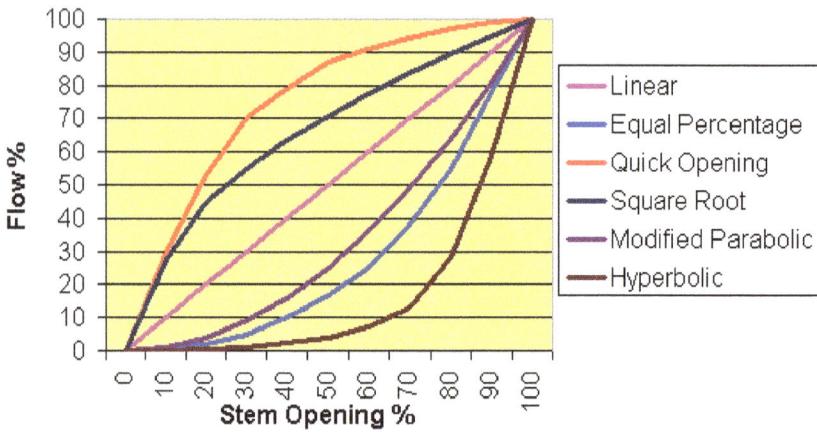

Figure 2. Typical inherent valve characteristics [16].

The mass balance of raw and pulverized coal is:

$$\frac{dm_{raw}}{dt} = q_{raw} - K_1 \cdot m_{raw} \tag{5}$$

and:

$$\frac{dm_{pul}}{dt} = K_1 \cdot m_{raw} - q_{pul}, \tag{6}$$

where q_{raw} is the mass of raw coal provided by the coal feeder per unit time, q_{pul} is mass flowrate of pulverized coal into the furnace, m_{raw} is the mass of raw coal stored in the mill, and m_{pul} is the pulverized coal stored in the mill.

The primary air blows part of the pulverized coal to the furnace, which is proportional to the differential pressure of primary air (Δp_{air}) and pulverized coal stored in the mill [6]:

$$q_{pul} \propto \Delta p_{air} \cdot m_{pul}, \tag{7}$$

and from Bernoulli's equation:

$$\Delta p_{air} = \lambda \cdot \rho_{air} \cdot \frac{v_{air}^2}{2g}, \tag{8}$$

where ρ_{air} and v_{air} are the primary air density and flow speed, and λ is the flow resistance. Since the primary air is assumed to be an ideal gas, ρ_{air} is proportional to the air temperature air, hence:

$$q_{pul} = K_2(273.15 + t_{air}) \cdot q_{air}^2 m_{pul}. \tag{9}$$

Using conservation laws, the total energy balance in the mill is:

$$\Delta E_{mill} = Q_{in} - Q_{out}, \tag{10}$$

$$Q_{in} = Q_{air,in} + Q_{coal,in} + Q_I, \tag{11}$$

and:

$$Q_{out} = Q_{air,out} + Q_{coal,out} + Q_{vapor} + Q_{loss}, \tag{12}$$

where ΔE_{mill} is the increment of inner energy; $Q_{air,in}$ and $Q_{coal,in}$ are the energy brought to the mill by the primary air and raw coal, respectively; Q_I is the heat generated by the mill electric current;

$Q_{air,out}$ and $Q_{coal,out}$ are the energy removed by primary air and pulverized coal flow to the furnace, respectively; Q_{vapor} is the heat loss from evaporation; and Q_{loss} is the heat loss to the environment. Various terms in (10), (11) and (12) can be expressed as follows:

$$\Delta E_{mill} = C_{mill}\frac{d(M_{mill}t_m)}{dt} = C_{mill}t_m\frac{dM_{mill}}{dt} + C_{mill}M_{mill}\frac{dt_m}{dt}$$
$$= K_3 t_m\left(\frac{dm_{raw}}{dt} + \frac{dm_{pul}}{dt}\right) + K_3(K_4 + m_{raw} + m_{pul})\frac{dt_m}{dt} \tag{13}$$

$$Q_{air,in} = C_a q_{air} t_{air}, \tag{14}$$

$$Q_{coal,in} = K_5 q_{raw} t_{envi}, \tag{15}$$

$$Q_I = K_6 I, \tag{16}$$

$$Q_{air,out} = C_a q_{air} t_m, \tag{17}$$

$$Q_{coal,out} = K_7 q_{pul} t_m, \tag{18}$$

$$Q_{vapor} = K_8 q_{water}, \tag{19}$$

and:

$$Q_{loss} = K_9(t_m - t_{envi}), \tag{20}$$

where t_{envi} is the environment temperature, I is the coal mill electric current, t_m is the outlet temperature, q_{water} is the mass flow rate of evaporated water, C_a is the heat capacity of air, C_{mill} is the heat capacity of the pulverizing system, and M_{mill} is the total mass of pulverizing system.

Evaporation mainly occurs inside the coal mill, hence moisture evaporation speed depends on the raw and pulverized coal stored in the coal mill, and is also exponentially related to the air mass flow [10]. Thus:

$$q_{water} = \theta(m_{raw} + m_{pul})t_m(1 - \exp(-\frac{q_{air}}{K_{10}})), \tag{21}$$

where θ is the moisture content in raw coal.

Mill differential pressure, Δp_{mill}, depends on the amount of pulverized coal carried by the primary air and flow resistance, which is assumed to be linearly related to the raw coal stored in the mill [4]. Thus:

$$\Delta p_{mill} = (K_{11} + K_{12}m_c)q_{air}^2 + K_{13}q_{pul}. \tag{22}$$

The pulverizing system electric current is determined by the raw and pulverized coal stored in the mill, and the no-load current, K_{16}:

$$I = K_{14} \cdot \eta \cdot m_{raw} + K_{15}m_{pul} + K_{16}, \tag{23}$$

where η is the grindability of raw coal.

Thus, the model has six measurable inputs, q_{raw}, μ_{cold}, μ_{hot}, t_{envi}, t_{cold}, t_{hot}; two unmeasurable inputs, θ, η; five model states, t_{air}, q_{air}, m_{raw}, m_{pul}, t_m; and five measurable outputs, t_{air}, q_{air}, I, Δp_{mill}, t_m. The desired controlled variables are q_{pul}, q_{air}, and t_m and the manipulated variables are q_{raw}, μ_{cold}, and μ_{hot}, i.e., a three input, three output control system.

2.3. Parameter Identification

The data set to identify the unknown parameters was collected from a historical database of a 660 MW power plant in China. The output prediction error was employed to evaluate the model accuracy:

$$E(\{S_i\}_{i=1,2,\dots,8}, \{K_j\}_{j=1,2,\dots,17}) = \sum_{t=1}^{N}\left\{ \begin{array}{l} w_1\|\frac{\hat{q}_{air}(t)-q_{air}(t)}{q_{air}(t)}\| + w_2\|\frac{\hat{t}_{air}(t)-t_{air}(t)}{t_{air}(t)}\| + w_3\|\frac{\hat{I}(t)-I(t)}{I(t)}\| \\ +w_4\|\frac{\Delta\hat{p}_{mill}(t)-\Delta p_{mill}(t)}{\Delta p_{mill}(t)}\| + w_5\|\frac{\hat{t}_m(t)-t_m(t)}{t_m(t)}\| \end{array}\right\}, \tag{24}$$

where N is number of data points; $q_{air}(i)$, $t_{air}(i)$, $I(i)$, $\Delta p_{mill}(i)$, and $t_m(i)$ are the ith measured outputs; $\hat{q}_{air}(i)$, $\hat{t}_{air}(i)$, $\hat{I}(i)$, $\Delta \hat{p}_{mill}(i)$, and $\hat{t}_m(i)$ are the ith model outputs; and w_1, w_2, w_3, w_4 and w_5 are the output weights. A genetic algorithm (GA) was used to minimize (24) and obtain optimal unknown parameters. Compared with more recently developed particle swarm optimization (PSO), GA has a better chance of finding a more qualified solution, since the mutation operation can make the population cluster around several "good" solutions instead of one "good" solution [17]. Moreover it has been demonstrated that GA is robust in the parameter identification problem and can achieve good results [18,19]. GA processes are well explained elsewhere [20], and we present the identification process of GA in Figure 3. Tables 1 and 2 show the GA tuning parameters and final optimal parameters, respectively. The tuning parameters are set based on the simulation parameters proposed in [21].

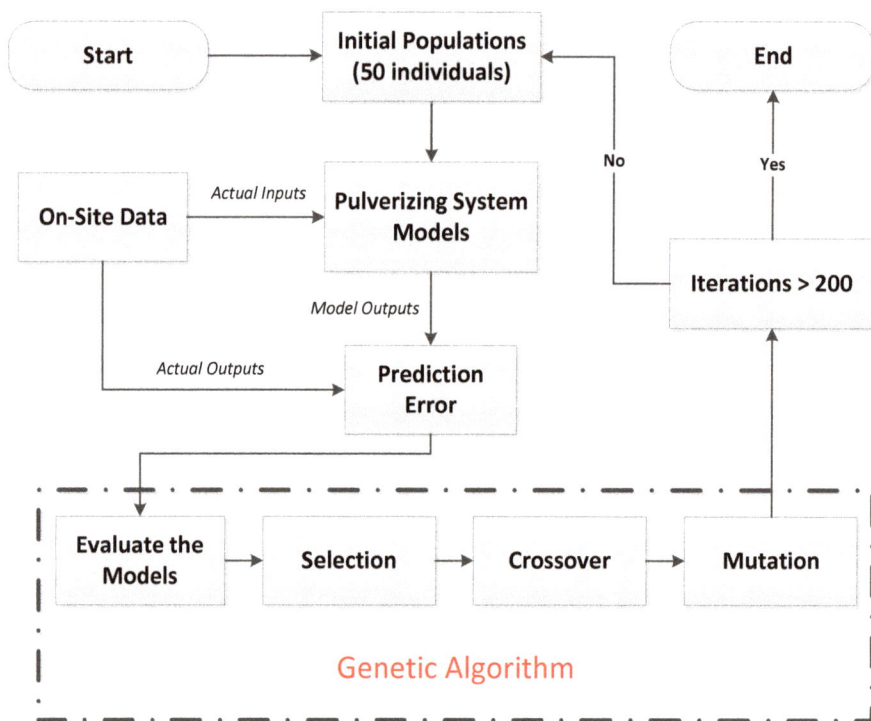

Figure 3. Identification process of GA.

Table 1. Genetic algorithm tuning parameters.

Population Size	Probability of Mutation	Probability of Crossover	Termination Iterations	Generation Gap	w_1	w_2	w_3	w_4	w_5
50	0.3	0.9	200	0.8	2	1	1	1.5	1

Table 2. Final optimal model parameters.

Parameter	S_1	S_2	S_3	S_4	S_5	S_6	S_7	S_8
Value	0.70	0.66	0.75	0.77	150.1	1.06	22.5	1.08
Parameter	K_1	K_2	K_3	K_4	K_5	K_6	K_7	K_8
Value	0.053	1.47×10^{-6}	1423	10,530	1309	2398	1306	5893
Parameter	K_9	K_{10}	K_{11}	K_{12}	K_{13}	K_{14}	K_{15}	K_{16}
Value	7037	95	5.29	0.0095	10.26	0.114	0.0292	19.11

2.4. Model Validation

The proposed model derived in Section 2.3 was validated using a different historical data set where the pulverizing system had a wide operating range (47.03–90.97% load rate), as shown in Figure 4. The real process trends and time constant were well captured by the proposed model. Thus, the model can be employed as the simulation platform for design of the soft sensor and control system. Table 3 shows the cumulative relative fitting error for the five outputs, defined as:

$$\sum_{i=1}^{N} \left| \frac{y^i_{model} - y^i_{real}}{y^i_{real}} \right|, \tag{25}$$

where N is the number of data samples, and y^i_{model} and y^i_{real} are the model output and process measurement, respectively. The primary air temperature is accurately predicted, whereas the primary air mass flowrate has significantly higher fitting error than other outputs due to the primary air mass flowrate being inaccurately measured in the real plant, as discussed above, and we cannot improve this prediction accuracy by adjusting the model parameters. However, the primary air temperature is accurately measured and the model shows high prediction accuracy.

Table 3. Cumulative relative fitting error.

Primary Air Temperature	Primary Air Mass Flowrate	Electric Current	Outlet Temperature	Differential Pressure
664	1806	852	799	995

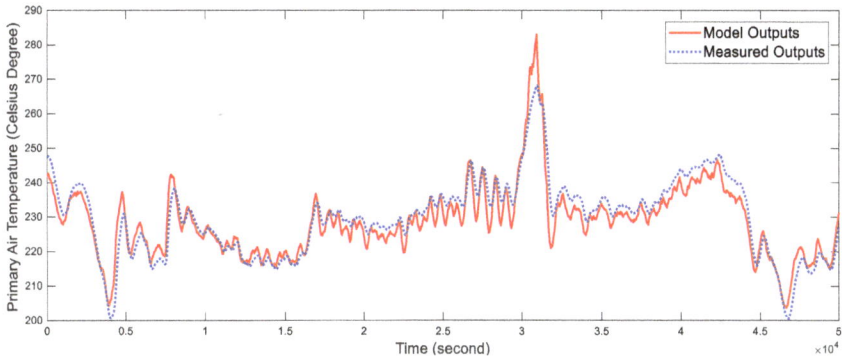

(a)

Figure 4. *Cont.*

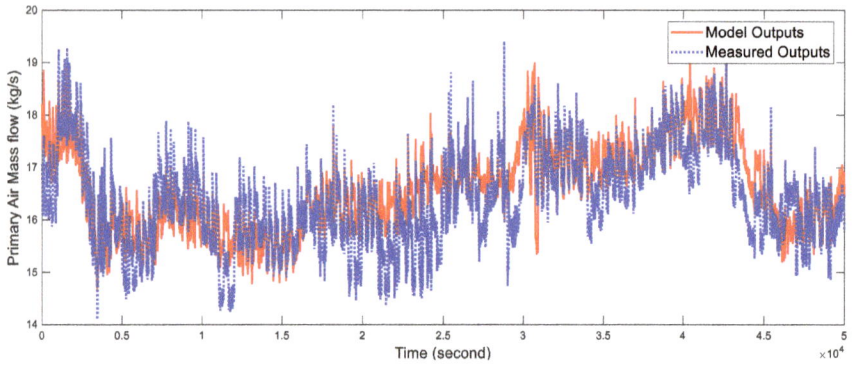

(**b**)

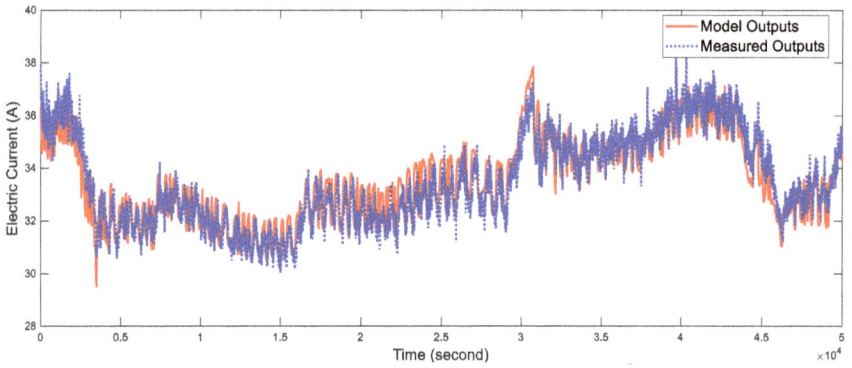

(**c**)

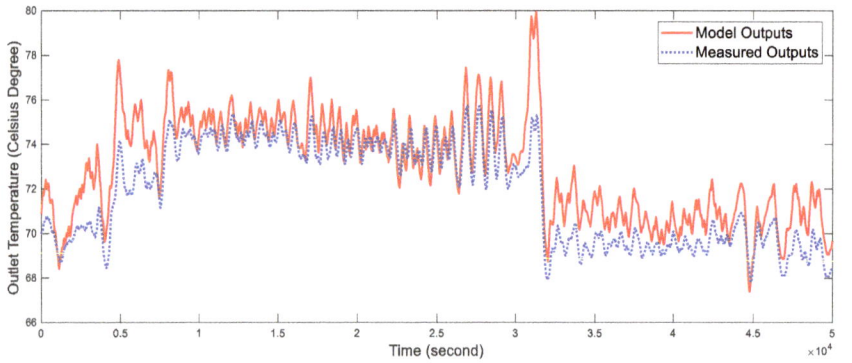

(**d**)

Figure 4. *Cont.*

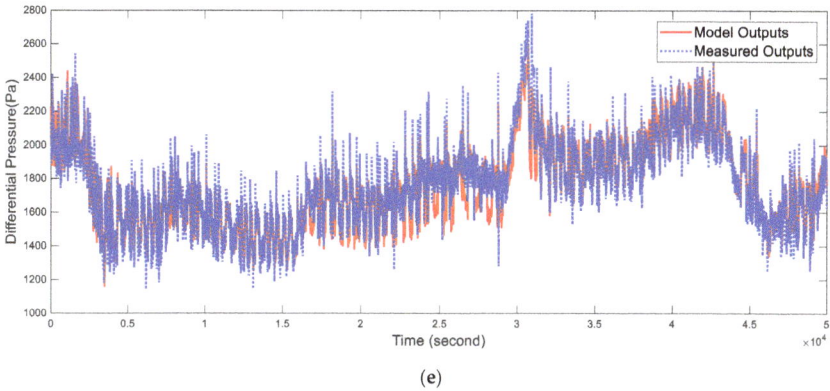

(e)

Figure 4. Validation of the proposed pulverizing system model. (**a**) Primary air temperature; (**b**) Primary air mass flow; (**c**) Electric current; (**d**) Outlet temperature; (**e**) Differential pressure.

3. Formulation of the Soft Sensor

The soft sensor to estimating the desired controlled variables was developed using MHE. We first derived the general MHE problem for the pulverizing system, and then discuss updating the arrival cost.

Artificial neural networks have been employed to develop soft sensors for many industrial processes to control unmeasurable variables [22–26]. Although such soft sensors can exhibit high fitting precision on the test data sets, they cannot explain process mechanisms, and hence can lack of robustness in the presence of process uncertainty.

Therefore, we developed the soft sensor using MHE. Moving horizon estimation is a model based optimization method to estimate the states and unknown parameters online and was originally derived as an approximation for the full-information maximum likelihood estimator (FIE) to avoid issues with FIE dimensionality [27]. Similar to model predictive control, MHE solves a finite horizon optimization problem dynamically at each sample time. Hence, the latest measurements available are employed to calculate current estimates. An important advantage of MHE over other soft sensor types is that the estimate constraints can be explicitly considered. Therefore, we can set the operating variable constraints based on prior knowledge of the pulverizing system to improve estimation accuracy.

The pulverizing system model can be expressed as:

$$\begin{cases} \frac{dx}{dt} = f(x, u, p) \\ y = h(x, u, p) \end{cases},$$ (26)

where:

$$x = \begin{bmatrix} t_{air} & q_{air} & m_{raw} & m_{pul} & t_m \end{bmatrix}^T,$$ (27)

$$p = \begin{bmatrix} \theta & \eta \end{bmatrix}^T,$$ (28)

$$u = \begin{bmatrix} \mu_{hot} & \mu_{cold} & q_{raw} & t_{envi} & t_{cold} & t_{hot} \end{bmatrix}^T,$$ (29)

$$y = \begin{bmatrix} t_{air} & q_{air} & I & \Delta p_{mill} & t_m \end{bmatrix}^T,$$ (30)

$$f(x, u, p) = \begin{bmatrix} (\dfrac{S_1(\mu_{cold})^{S_2}}{S_3(\mu_{hot})^{S_4}+S_1(\mu_{cold})^{S_2}} \cdot t_{cold} + \dfrac{S_3(\mu_{hot})^{S_4}}{S_3(\mu_{hot})^{S_4}+S_1(\mu_{cold})^{S_2}} \cdot t_{hot} - S_6 t_{air})/S_5 \\ (S_1(\mu_{cold})^{S_2} + S_3(\mu_{hot})^{S_4} - S_8 q_{air})/S_7 \\ q_{raw} - K_1 \cdot m_{raw} \\ K_1 \cdot m_{raw} - K_2(273.15 + t_{air}) \cdot q_{air}^2 m_{pul} \\ \dfrac{(Q_{in}-Q_{out})-K_3 t_m(q_{raw}-K_2(273.15+t_{air})\cdot q_{air}^2 m_{pul})}{K_3(K_4+m_{raw}+m_{pul})} \end{bmatrix}, \quad (31)$$

$$Q_{in} = C_a q_{air} t_{air} + K_5 q_{raw} t_{envi} + K_6 I, \quad (32)$$

$$Q_{out} = C_a q_{air} t_m + K_7 K_2(273.15 + t_{air}) \cdot q_{air}^2 m_{pul} t_m + K_8 \theta(m_{raw} + m_{pul}) t_m (1 - e^{-\frac{q_{air}}{K_{10}}}) + K_9(t_m - t_{envi}), \quad (33)$$

and:

$$h(x, u, p) = \begin{bmatrix} t_{air} \\ q_{air} \\ K_{14} \cdot \eta \cdot m_{raw} + K_{15} m_{pul} + K_{16} \\ (K_{11} + K_{12} m_c) q_{air}^2 + K_{13} q_{pul} \\ t_m \end{bmatrix}. \quad (34)$$

Then the MHE soft sensor is formulated as a nonlinear least squares optimization problem:

$$\min_{\substack{\hat{x}_{k-N+1},\ldots,\hat{x}_k \\ \hat{p}_{k-N+1},\ldots,\hat{p}_k}} (\| \begin{matrix} \hat{x}_{k-N+1} - \overline{x}_L \\ \hat{p}_{k-N+1} - \overline{p}_L \end{matrix} \|_{P_L}^2 + \sum_{i=k-N+1}^{k-1} \| \begin{matrix} \hat{x}_{i+1} - \phi(\hat{x}_i, \hat{p}_i, u_i) \\ \hat{p}_{i+1} - \hat{p}_i \end{matrix} \|_W^2 + \sum_{i=k-N+1}^{k} \|y_i - h(\hat{x}_i, \hat{p}_i, u_i)\|_V^2), \quad (35)$$

where:

$$\phi(\hat{x}_i, \hat{p}_i, u_i) = \int_0^T f(\hat{x}_i, \hat{p}_i, u_i) dt; \quad (36)$$

k represents the present time instance; N is the estimation horizon; T is the sampling time; $\hat{x}_{k-N+1}, \ldots, \hat{x}_k$ are the state estimates from time $k - N+1$ to k; $\hat{p}_{k-N+1}, \ldots, \hat{p}_k$ are the parameter estimates from time $k - N+1$ to k; y_i is the measured outputs at time i; P_L, V, and W are constant positive definite weighting matrixes; and $\overline{x}_L$ and $\overline{p}_L$ are constant scalars representing the influence from past measurements. The first term in the cost function (35) is typically called the arrival cost, and is important for MHE stability [28]. $\overline{x}_L$, $\overline{p}_L$, and P_L are updated when the MHE calculates a new estimate.

The analytical solution for $\phi(\hat{x}_i, \hat{p}_i, u_i)$ is difficult to find, and we approximate it using forward difference:

$$\phi(\hat{x}_i, \hat{p}_i, u_i) \approx \hat{x}_i + T \cdot f(\hat{x}_i, \hat{p}_i, u_i), \quad (37)$$

where T should be as small as possible to avoid large approximation error, or it may reduce estimation precision and possibly make the soft sensor unstable. However, since the pulverizing system has large inertia, $f(\hat{x}_i, \hat{p}_i, u_i)$ cannot change sharply during the sampling interval, hence (37) will not cause significant approximation error.

Conventionally, $\overline{x}_L$, $\overline{p}_L$ and P_L are updated using the Kalman filter. However, this introduces large errors for nonlinear systems in the approximation of the full information estimator, which necessitates a large estimation horizon, and increases the online computational burden [29]. Considering this problem, we propose an efficient arrival cost update, based on Kuhl et al. [30]. Arrival cost updating was derived for the discretized pulverizing system model as follows.

The ideal arrival cost can be expressed as:

$$C(x_L, p_L) = \min_{x_{L-1}, p_{L-1}} (\| \begin{matrix} x_{L-1} - \overline{x}_{L-1} \\ p_{L-1} - \overline{p}_{L-1} \end{matrix} \|_{P_{L-1}}^2 + \|y - h(x_{L-1}, p_{L-1})\|_V^2 + \| \begin{matrix} x_L - \phi(x_{L-1}, p_{L-1}) \\ p_L - p_{L-1} \end{matrix} \|_W^2), \quad (38)$$

where $\overline{x}_{L-1}$ and $\overline{p}_{L-1}$ are the states and parameters in the arrival cost term at the previous sampling time. To approximate $C(x_L, p_L)$ using a linear quadratic expression, nonlinear mappings $f(x_{L-1}, p_{L-1})$ and $h(x_{L-1}, p_{L-1})$ are approximated using Taylor expansion:

$$f(x_{L-1}, p_{L-1}) \approx f(x^*, p^*) + f_x \cdot (x_{L-1} - x^*) + f_p \cdot (p_{L-1} - p^*), \tag{39}$$

and:

$$h(x_{L-1}, p_{L-1}) \approx h(x^*, p^*) + h_x \cdot (x_{L-1} - x^*) + h_p \cdot (p_{L-1} - p^*), \tag{40}$$

where:

$$f_x = \left. \frac{\partial f(x, p)}{\partial x} \right|_{x_{L-1}=x^*, p_{L-1}=p^*}, \tag{41}$$

$$f_p = \left. \frac{\partial f(x, p)}{\partial p} \right|_{x_{L-1}=x^*, p_{L-1}=p^*}, \tag{42}$$

$$h_x = \left. \frac{\partial h(x, p)}{\partial x} \right|_{x_{L-1}=x^*, p_{L-1}=p^*}, \tag{43}$$

$$h_p = \left. \frac{\partial f(x, p)}{\partial p} \right|_{x_{L-1}=x^*, p_{L-1}=p^*}, \tag{44}$$

and x^* and p^* are the best available estimate at time $k - N$. Then:

$$\phi(x_{L-1}, p_{L-1}) \approx x_{L-1} + T \cdot f(x_{L-1}, p_{L-1}). \tag{45}$$

Substituting (39), (40), and (45) into (38):

$$C(x_L, p_L) \approx \min_{X_{L-1}} \left\| A \begin{bmatrix} X_{L-1} \\ X_L \end{bmatrix} - b \right\|_2^2, \tag{46}$$

where:

$$A = \begin{bmatrix} -V h_x \middle| -V h_p & O \\ -W \begin{bmatrix} I + T \cdot f_x & T \cdot f_p \\ O & I \end{bmatrix} & W \\ P_{L-1} & O \end{bmatrix}, \tag{47}$$

$$b = \begin{bmatrix} V(f(x^*, p^*) - f_x \cdot x^* - f_p \cdot p^* - y) \\ W \begin{bmatrix} h(x^*, p^*) - h_x \cdot x^* - h_p \cdot p^* \\ O \end{bmatrix} \\ P_{L-1} \begin{bmatrix} \overline{x}_{L-1} \\ \overline{p}_{L-1} \end{bmatrix} \end{bmatrix}, \tag{48}$$

$$X_L = \begin{bmatrix} x_L \\ p_L \end{bmatrix}, \tag{49}$$

$$X_{L-1} = \begin{bmatrix} x_{L-1} \\ p_{L-1} \end{bmatrix}, \tag{50}$$

and O and I are zero and unit matrices, respectively, with appropriate dimensions.
Equation (46) can be transformed using QR factorization of A to:

$$C(x_L, p_L) \approx \min_{X_{L-1}} \left\| \begin{bmatrix} Q_1 & Q_2 & Q_3 \end{bmatrix} \begin{bmatrix} R_1 & R_{12} \\ O & R_2 \\ O & O \end{bmatrix} \begin{bmatrix} X_L \\ X_{L-1} \end{bmatrix} - b \right\|_2^2, \tag{51}$$

which has the analytic solution:

$$C(x_L, p_L) \approx \|Q_3 \cdot b\|_2^2 + \left\|Q_2 \cdot b + R_2 \begin{bmatrix} x_L \\ p_L \end{bmatrix}\right\|_2^2, \tag{52}$$

where:

$$A = \begin{bmatrix} Q_1 & Q_2 & Q_3 \end{bmatrix} \begin{bmatrix} R_1 & R_{12} \\ O & R_2 \\ O & O \end{bmatrix}, \tag{53}$$

$$P_L = R_2, \tag{54}$$

$$\begin{bmatrix} \bar{x}_L \\ \bar{p}_L \end{bmatrix} = R_2^{-1} \cdot Q_2 \cdot b, \tag{55}$$

and $\bar{x}_L, \bar{p}_L$ and P_L are employed to update the MHE arrival cost.

4. Inferential Multi-Model Predictive Controller Design

Nonlinearity of the pulverizing system was analyzed to select proper local models for the multi-model controller, then the predictive controller was designed based on an extended input-output state space model to achieve offset-free performance in the presence of modeling error and unknown disturbances. Figure 5 shows an overall view of the inferential control system.

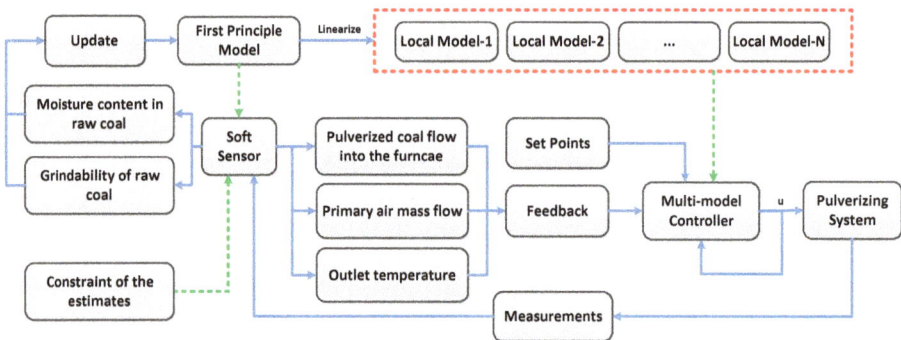

Figure 5. Inferential control system structure.

The soft sensor can not only estimate desired controlled variables but can also detect a change of raw coal. Since different raw coal types have different grindability and moisture content, pulverizing system outputs can change significant when the power plant uses a new raw coal type. Therefore, the soft sensor can be used to update model parameters online.

4.1. Nonlinearity Analysis

The basic control task for the pulverizing system is to track power plant coal demand. Hence raw coal feed rate was selected as the scheduling variable to analyze process nonlinearity. In practice, the setpoint of primary air mass flow is set according to the desired air to coal ratio, and is proportional to the raw coal feed rate. There is also a lower limit on primary air mass flow, to avoid coal jamming, and in this case the lower limit = 10 kg/s. Table 4 shows the selected operating points.

Table 4. Selected operating points.

Raw Coal Feed Rate (kg/s)	Primary Air Mass Flow (kg/s)	Primary Air Temperature (K)	Cold Air Baffle Position (%)	Hot Air Baffle Position (%)	Outlet Temperature (°C)	Mill Electric Current (A)	Mill Differential Pressure (kPa)
3	10.0	195.6	10.7	16.5	70	26.77	0.5995
5	12.5	218.0	9.3	26.2	70	31.10	0.9646
7	17.5	224.2	14.3	46.9	70	35.00	1.9944
9	22.5	235.1	16.6	74.1	70	39.06	3.5074

Local linear models at typical operating points can be obtained by linearizing the first principle model of the pulverizing system. Then the gap metric was employed to quantitatively measure nonlinearity between local models. The gap metric between two local linear systems P_1 and P_2 is defined as [31]:

$$\delta(P_1, P_2) = \max \left\{ \inf_{Q \in H_\infty} \left\| \begin{bmatrix} M_1 \\ N_1 \end{bmatrix} - \begin{bmatrix} M_2 \\ N_2 \end{bmatrix} Q \right\|_\infty , \inf_{Q \in H_\infty} \left\| \begin{bmatrix} M_2 \\ N_2 \end{bmatrix} - \begin{bmatrix} M_1 \\ N_1 \end{bmatrix} Q \right\|_\infty \right\}, \tag{56}$$

where $P_1 = N_1 M_1^{-1}$ and $P_2 = N_2 M_2^{-1}$ are the normalized right coprime factorization on P_1 and P_2, respectively.

If the $\delta(P_1, P_2) \approx 1$, dynamic behavior between the local linear models is significantly different and process nonlinearity is strong between the two operating points. In contrast, if the $\delta(P_1, P_2) \approx 0$, dynamic behavior between the two local models is similar, and process nonlinearity is weak. Figure 6 shows the gap metric between all the local models.

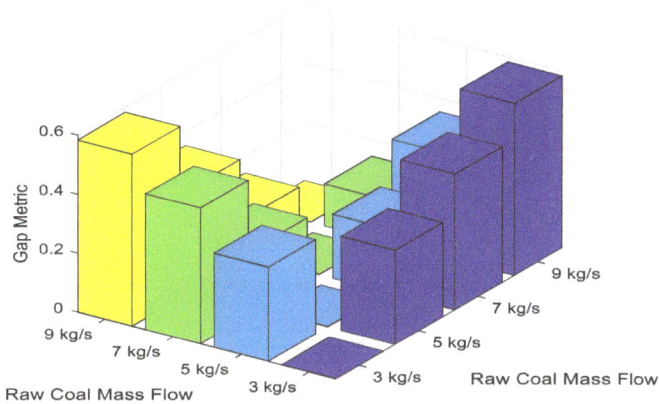

Figure 6. Gap metric between local linear models.

The gap metric is approximately linear with local linear model distance, i.e., the difference of the raw coal feed rate. Therefore, we divided the operating range uniformly by selecting local models with 5 and 9 kg/s raw coal feed rate and employed the selected local models for controller design. When $\delta(P_1, P_2) < 0.3$ between any operating point and one of the selected operating points, nonlinearity within the local controller working range is not strong. The proposed division of the operating range can satisfy this condition. Although we can select all four models to set up the multi-model controller, this will lead to heavy online computation overhead, for insignificant improvement in control performance.

4.2. Multi-Model Predictive Controller Based on Extended Input-Output State Space Model

Modeling error and unknown disturbances always exist in practice. Therefore, integration must be included in the control algorithm. To achieve this, we can transform the original local linear models into the equivalent extended input-output state space for offset-free tracking performance [32,33]. In this control scheme, past values of the manipulated and controlled variables together with the tracking error form the new state variables. Therefore the method is free from the difficulties of observer based control techniques, such as convergence rate and observer robustness [32]. When the pulverizing system operates over a wide range, a single linear model for the MPC design will cause model discrepancies due to nonlinearities, with consequential control performance degradation. Therefore, two local MPC controllers were assigned with different operating ranges according the nonlinearity analysis. The proposed controller algorithm for the pulverizing system is as follows:

The selected local linear models can be described using the input-output linear difference model:

$$
\begin{aligned}
&y(k+1) + F_1 y(k) + F_2 y(k-1) + \ldots + F_n y(k-n+1) \\
&= H_1 u(k) + H_2 u(k-1) + \ldots + H_n u(k-n+1)
\end{aligned}
\tag{57}
$$

where $F_i \in R^{3 \times 3}$, $H_i \in R^{3 \times 6}$ $(i = 1, 2, \ldots, n)$, $y = \begin{bmatrix} q_{pul} & q_{air} & t_m \end{bmatrix}^T$ is the controlled variables, $u = \begin{bmatrix} u_{mpc}^T & u_d^T \end{bmatrix}^T$ is the input variables, $u_{mpc} = \begin{bmatrix} q_{raw} & \mu_{cold} & \mu_{hot} \end{bmatrix}^T$ is the manipulated variables, and $u_d = \begin{bmatrix} t_{envi} & t_{cold} & t_{hot} \end{bmatrix}^T$ is the feed forward signal of measured disturbances. The local linear models are continuous and can be obtained by linearizing the model differential equations using first-order Taylor expansion.

Equation (57) can be transformed into the differenced form using the backshift operator, Δ:

$$
\begin{aligned}
&\Delta y(k+1) + F_1 \Delta y(k) + F_2 \Delta y(k-1) + \ldots + F_n \Delta y(k-n+1) \\
&= H_1 \Delta u(k) + H_2 \Delta u(k-1) + \ldots + H_n \Delta u(k-n+1)
\end{aligned}
\tag{58}
$$

where $\Delta y(i) = y(i) - y(i-1)$, $\Delta u(i) = u(i) - u(i-1)$.

We define the input–output states as:

$$
\Delta x_m = \begin{bmatrix} \Delta y(k)^T & \Delta y(k-1)^T & \cdots & \Delta y(k-n+1)^T & \Delta u(k-1)^T & \Delta u(k-2)^T & \cdots & \Delta u(k-n+1)^T \end{bmatrix}^T
\tag{59}
$$

Thus, the corresponding state space model can be expressed as:

$$
\begin{cases}
\Delta x_m(k+1) = A_m \Delta x_m(k) + B_m \Delta u(k) \\
\Delta y(k) = C_m \Delta x_m(k)
\end{cases}
\tag{60}
$$

where:

$$
A_m = \begin{bmatrix}
-F_1 & -F_2 & \cdots & -F_{n-1} & -F_n & H_2 & \cdots & H_{n-1} & H_n \\
I & O & \cdots & O & O & O & \cdots & O & O \\
O & I & \cdots & O & O & O & \cdots & O & O \\
\vdots & \vdots & \cdots & \vdots & \vdots & \vdots & \cdots & \vdots & \vdots \\
O & O & \cdots & I & O & O & \cdots & O & O \\
O & O & \cdots & O & O & O & \cdots & O & O \\
O & O & \cdots & O & O & I & \cdots & O & O \\
\vdots & \vdots & \cdots & \vdots & \vdots & \cdots & \vdots & \vdots \\
O & O & \cdots & O & O & O & \cdots & I & O
\end{bmatrix},
\tag{61}
$$

$$
B_m = \begin{bmatrix} H_1^T & O & O & \cdots & O & I & O & O \end{bmatrix}
\tag{62}
$$

and:

$$C_m = \begin{bmatrix} I & O & O & \cdots & O & O & O & O \end{bmatrix} \tag{63}$$

Since the states are formed using input and output variables, the MPC controller does not require the design of state observers.

The output tracking error is defined as:

$$e(k) = y(k) - r(k), \tag{64}$$

where $r(k)$ is the reference signal. Combining (60) and (64):

$$e(k+1) = e(k) + C_m A_m \Delta x_m(k) + C_m B_m \Delta u(k) - \Delta r(k+1), \tag{65}$$

by augmenting $e(k)$ into the state variables and:

$$z(k) = \begin{bmatrix} \Delta x_m(k) \\ e(k) \end{bmatrix}. \tag{66}$$

The extended input–output state space model can be expressed as:

$$z(k+1) = Az(k) + B\Delta u(k) + C\Delta r(k+1), \tag{67}$$

where:

$$A = \begin{bmatrix} A_m & 0 \\ C_m A_m & I \end{bmatrix}, \tag{68}$$

$$B = \begin{bmatrix} B_m \\ C_m B_m \end{bmatrix} \tag{69}$$

and:

$$C = \begin{bmatrix} 0 \\ -I \end{bmatrix}. \tag{70}$$

Note that when the system is in steady-state, the elements in $z(k)$ must be zero and hence can guarantee $y(k) = r(k)$, which indicates, using the extended input–output state space model as the prediction model in MPC, the desired controlled variables can track the reference signal with no offset.

The optimal control moves can be calculated by minimizing the objective function:

$$\underset{\{\Delta u_{mpc}(k+i)i=1,2...,M\}}{\text{argmin}} \quad J = \sum_{j=1}^{P} z^T(k+j)Q_j z(k+j) + \sum_{j=1}^{M} \Delta u^T(k+j)R_j\Delta u(k+j)$$

$$\text{s.t.} \begin{cases} \Delta u_{mpc}^{max} \leq \Delta u_{mpc}(k+i) \leq \Delta u_{mpc}^{min} 0 \leq j < M \\ u_{mpc}^{max} \leq u_{mpc}(k+i) \leq u_{mpc}^{min} 0 \leq j < M \\ \Delta u_{mpc}(k+i) = 0 j \geq M \end{cases}, \tag{71}$$

where:

$$Q_j = diag\{q_{j,y1}, q_{j,y2}, q_{j,y3}, q_{j,u1}, q_{j,u2}, \ldots, q_{j,u6}, q_{j,e1}, q_{j,e2}, q_{j,e3}\}; \tag{72}$$

$$R_j = diag\{ r_{j,u1} \quad r_{j,u2} \quad r_{j,u3} \}; \tag{73}$$

P and M are the prediction and control horizons, respectively; and Q_j and R_j are the weighting matrices. Generally, $q_{j,ei}$ ($i = 1, 2, 3$) and $r_{j,ui}$ ($i = 1, 2, 3$) cannot be set to zero, because the tracking error and control effort must be considered in the cost function.

Tuning of the MPC parameters is actually a compound problem owing to the lack of agreement on what satisfactory controller performance is [34]. Generally the weighting matrixes should be tuned based on practical needs. In the pulverizing system, since safe operation is the primary concern, the controller cannot take aggressive moves and hence $r_{j,ui}$ should be large enough to avoid overshot or oscillation of the controlled variables. To achieve this, we first fix $q_{j,ei}$ and then gradually increase $r_{j,ui}$ until overshot or oscillation disappears. In practice, the error weights $q_{j,ei}$ can be tuned empirically: if one or more process variables are more important than others, larger weights should be set on them to ensure the tracking performance [35]. We put more weights on the tracking error of primary air mass flow to maintain the economic air to coal ratio. The prediction and control horizons can be determined using empirical formulas proposed in [35].

Solving the optimization problem (71) for the two local controllers provides their control inputs, U_1 and U_2. Then the control move of the multi-model predictive controller can be expressed as:

$$u = \varphi_1 U_1 + \varphi_2 U_2, \tag{74}$$

where φ_1 and φ_2 are the weighting functions, and Figure 7 shows their relationship with the scheduling variable (raw coal mass flow). Trapezoidal relationship is employed owing to its simplicity in design. The switching points are placed at the 1/4 points on the line segment between the adjacent selected operating points, i.e., the 6 kg/s and 8 kg/s raw coal mass flowrate, so that the local controllers can switch smoothly.

The design procedures of the proposed MMPC are summarized in Figure 8.

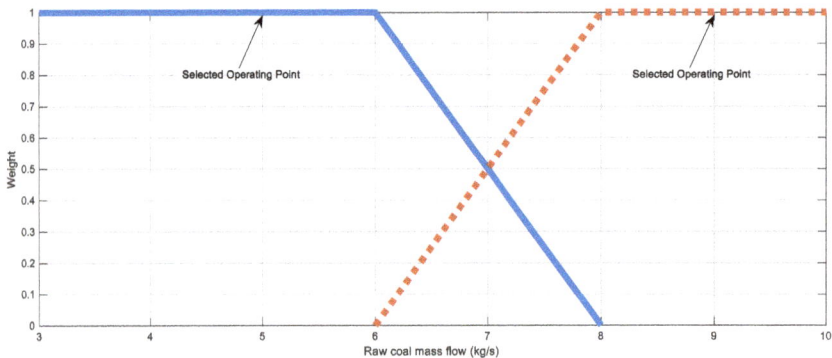

Figure 7. Weighting functions and scheduling variable: φ_1 = solid line, φ_2 = dotted line.

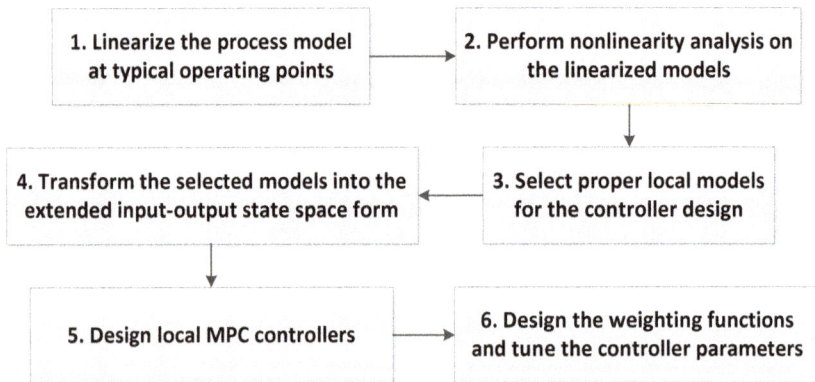

1. Linearize the process model at typical operating points	2. Perform nonlinearity analysis on the linearized models
4. Transform the selected models into the extended input-output state space form	3. Select proper local models for the controller design
5. Design local MPC controllers	6. Design the weighting functions and tune the controller parameters

Figure 8. Design procedures of proposed MMPC.

5. Simulation Results

We tested the proposed inferential multi-model predictive control performance. The soft sensor effectiveness is presented first, and then the inferential multi-model control system is compared with proportional-integral (PI) control strategy for a real power plant.

5.1. Soft Sensor Test

The proposed soft sensor was compared with a conventional extended Kalman filter (EKF), with the same weighting matrixes for states and outputs both cases. Sampling time for the soft sensor = 1 s, the same as the power plant DCS sampling time. Weighting matrixes in (27) were W = diag(0.5, 0.3, 0.1, 0.5, 0.5, 1), V = diag(1, 1, 10, 0.1, 5), which was a trade-off between model prediction and measurement data. Estimation horizon N = 10. Table 5 shows the input and state estimate constraints, where k represents the present sample time, i represents the ith estimate in (27) ($i = k - N+1, \ldots , k$), and Δ means the difference between estimates at time k and $k - 1$. State constraints can be determined from the input constraints by simulating the first principle model.

Table 5. State estimate constraints.

State Constraints	$|\Delta t_{air}(k,i)|$	$|\Delta q_{air}(k,i)|$	$|\Delta m_{raw}(k,i)|$	$|\Delta m_{pul}(k,i)|$	$|\Delta t_m(k,i)|$	$|\Delta\theta(k,i)|$	$|\Delta\eta(k,i)|$
Value	0.9 K/s	0.3 kg/s	0.5 kg/s	0.3 kg/s	0.2 K/s	0.01	0.01
Input Constraints	$|\Delta q_{raw}|$	$|\Delta\mu_{cold}|$	$|\Delta\mu_{hot}|$	$|\Delta t_{envi}|$	$|\Delta t_{cold}|$	$|\Delta t_{hot}|$	
Value	0.05 kg/s	2%/s	2%/s	0.1 °C/s	0.1 °C/s	1.5 °C/s	

As discussed earlier, raw coal and primary air mass flow cannot be accurately measured. Therefore, we set ±5% measurement uncertainty in the simulation, and ±1% measurement uncertainty for other input and output signals. Additionally, at 200 s we increased the raw coal moisture content and grindability to simulate the power plant changing raw coal type. Figure 9 shows unmeasurable states and parameters estimates, and Figure 10 shows controlled variables estimates. Since pulverized coal flow into the furnace is unmeasurable, the measured raw coal feed rate was also regarded as the pulverized coal flow for the simulation. Table 6 shows the root-mean-square (RMS) errors and 3-sigma error bounds of the estimates for MHE and EKF.

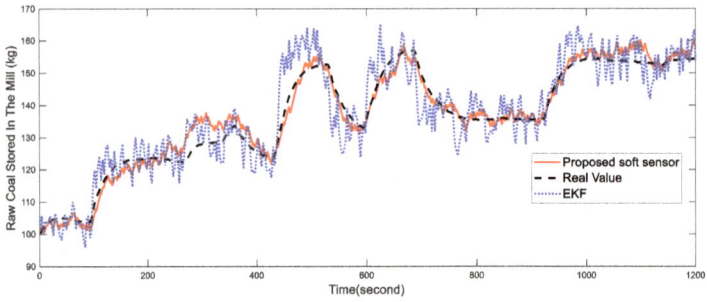

(a)

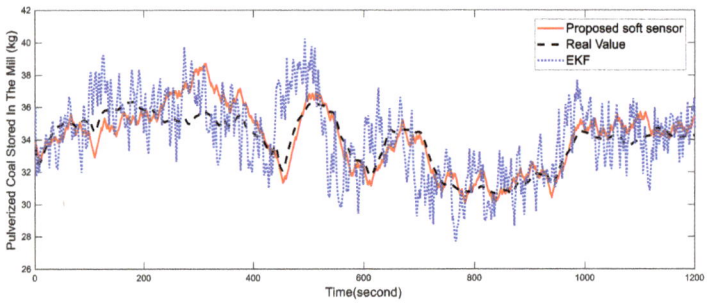

(b)

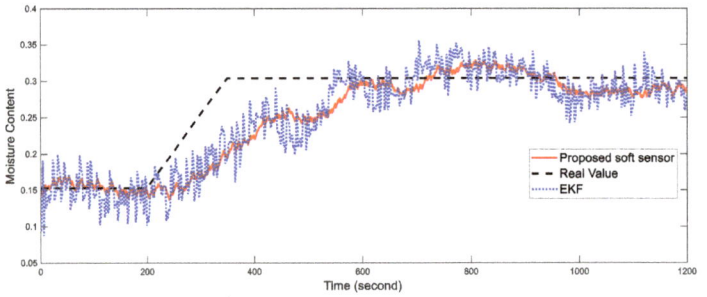

(c)

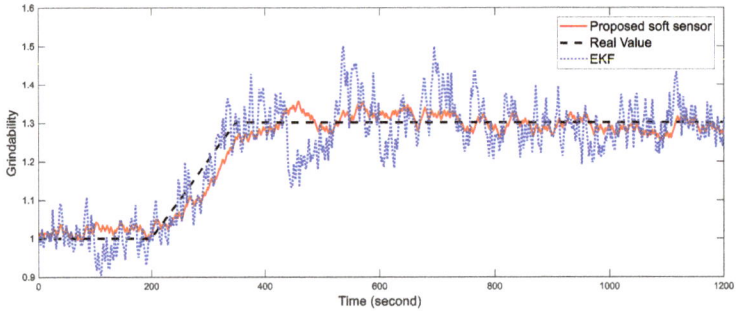

(d)

Figure 9. Estimates of (**a**,**b**) unmeasurable states and (**c**,**d**) raw coal parameters. (**a**) Raw coal stored in the mill; (**b**) Pulverized coal stored in the mill; (**c**) Moisture content; (**d**) Grindability.

(a)

(b)

(c)

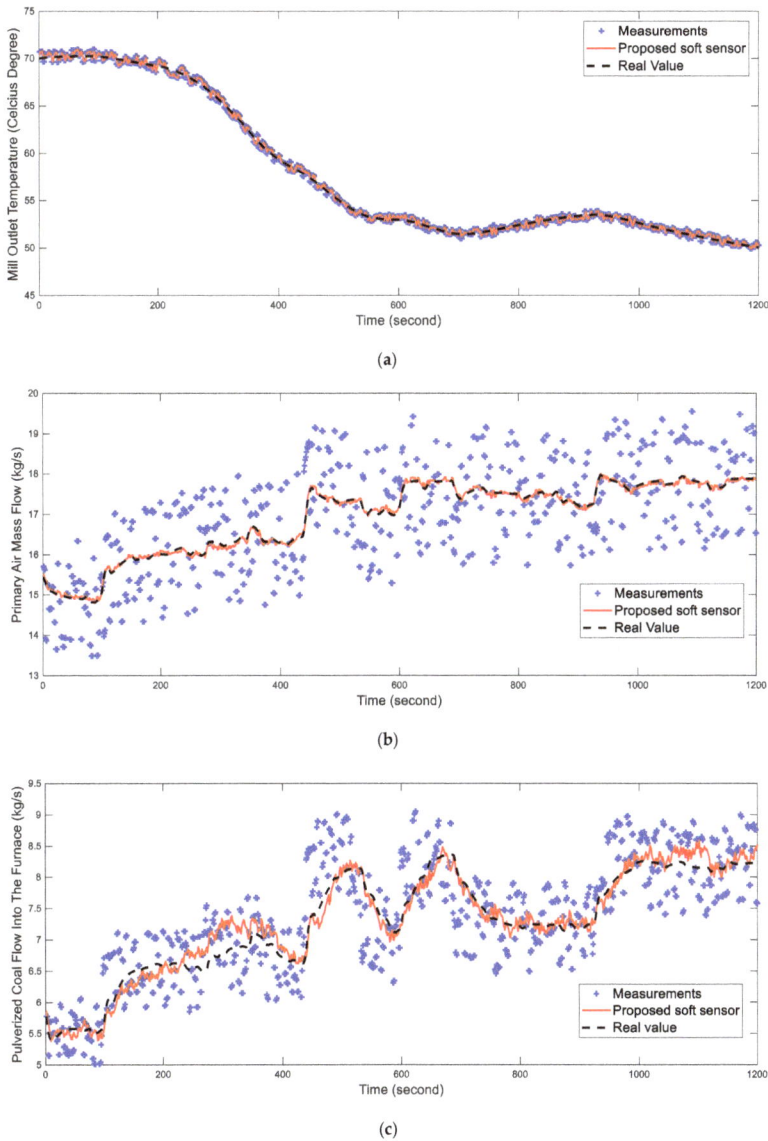

Figure 10. Controlled variable estimates. (**a**) Mill outlet temperature; (**b**) Primary air mass flow; (**c**) Pulverized coal flow into the furnace.

Table 6. RMS errors and 3-sigma error bounds of the estimates.

	Raw Coal Stored in the Mill	Pulverized Coal Stored in the Mill	Moisture Content	Grindability
RMS of MHE	2.8493	0.9084	0.0391	0.028
RMS of EKF	6.0084	1.8175	0.0417	0.0585
Error bound of EKF	±18.0326	±5.4549	±0.1253	±0.1757
Error bound of MHE	±8.5516	±2.7263	±0.1172	±0.0839

Figure 9 and Table 6 show that the proposed soft sensor provides satisfactory unmeasurable states and parameter estimates in the presence of measurement uncertainty. Since the state constraint is considered, which represents prior knowledge of the process, the proposed soft sensor is less affected by measurement uncertainty than EKF. Previous studies have shown that, given the same tuning parameters, MHE can provide improved estimates and greater robustness than EKF [36], which is verified by the current simulation.

Raw coal property changes were successfully detected by the soft sensor. Therefore, when the power plant changes raw coal type, we can slowly update the model parameters online rather then re-identifying the model parameters. There was a large delay between real and estimated moisture content, since the changed moisture content only influences outlet temperature slowly due to the large energy balance inertia, hence the true value cannot be immediately estimated.

Figure 10 shows that pulverized coal flow into the furnace and primary air flow estimates are significantly closer to the real values than were the measurements, and outlet temperature estimates had similar precision to the measurements. Since the outlet temperature is already measured accurately, the soft sensor cannot significantly improve its measurement accuracy. However, the other two controlled variables are only approximately measured, and the soft sensor can significantly improve their measurement quality because it employs accurately measured signals to reconstruct measurement signals based on the first principle model. Therefore, using estimates rather than measurements as the control system feedback signal can significantly enhance control precision of the desired controlled variables.

5.2. Inferential Control Strategy Test

We tested tracking performance of the proposed inferential multi-model predictive controller. Measurement uncertainty was set the same as the previous simulation, and sample time for the controller = 5 s due to the large process inertia. We set $q_{j,yi} = 0$ ($i = 1, 2, 3$) and $q_{j,ui} = 0$ ($i = 1, 2, \ldots, 8$) to simplify (71), which also means that only tracking error and control effort were considered. The weights for tracking error and control effort were $q_{j,y1} = 4$, $q_{j,y2} = 8$, $q_{j,y3} = 1$, $q_{j,u1} = 60$, $q_{j,u2} = 10$, and $q_{j,u3} = 10$. Prediction horizon = 100, long enough to cover key pulverizing system dynamics. Tuning the control horizon was a trade-off between computation cost and control performance [37], and was set = 5.

In real power plants, the pulverizing system is controlled via three independent single PI control loops, which are tuned conservatively to ensure safe and reliable operation [6]. Hence the PI controllers were employed to compare with proposed control system. Figure 11 shows the PI control structure used for comparison, and Figure 12 shows the simulation results. Note that, in the PI control scheme, the pulverized coal flow into the furnace is estimated by solving the model differential equations given the input signals.

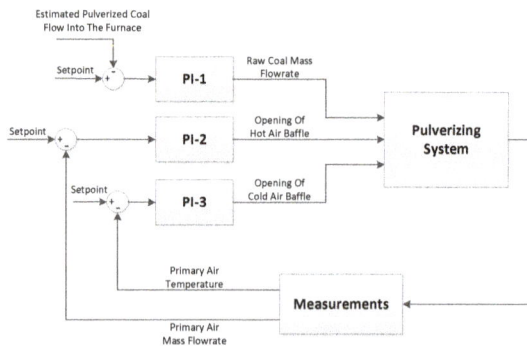

Figure 11. Conventional PI control structure.

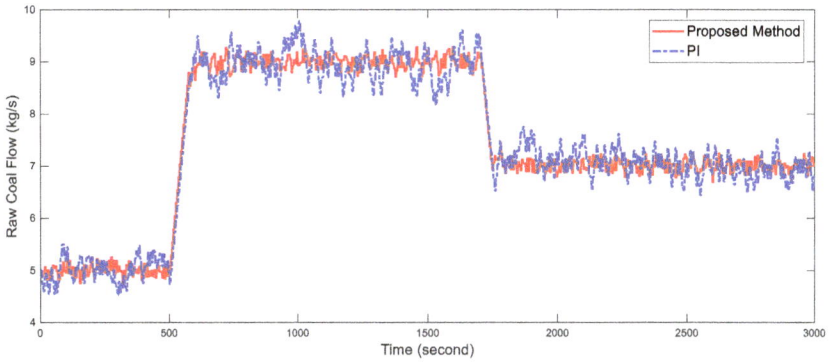

(**a**)

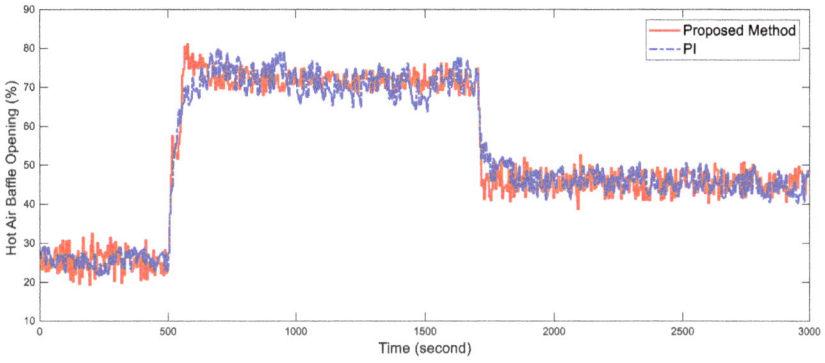

(**b**)

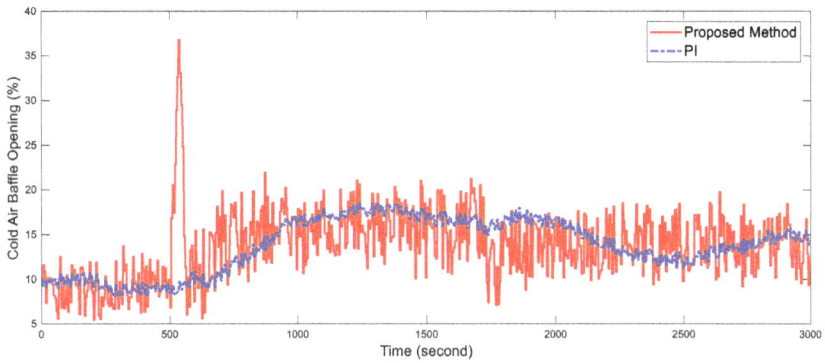

(**c**)

Figure 12. *Cont.*

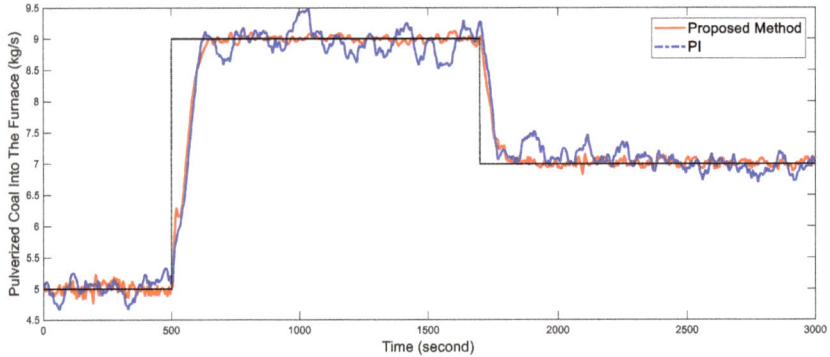

(d)

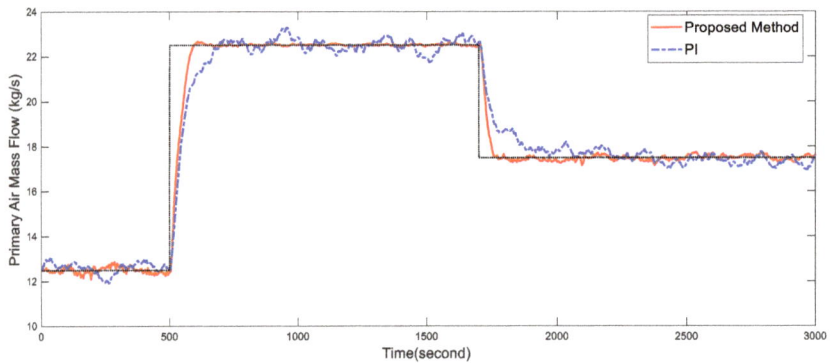

(e)

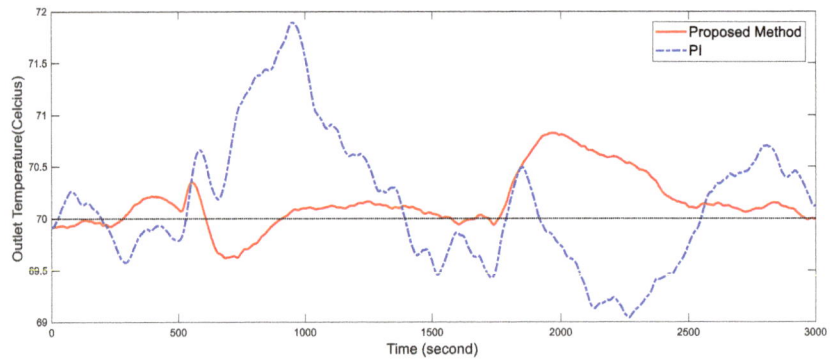

(f)

Figure 12. Proposed and PID control method performances for (**a–c**) control inputs, and (**d–f**) controlled outputs. (**a**) Raw coal flow; (**b**) Hot air baffle opening; (**c**) Cold air baffle opening; (**d**) Pulverized coal into the furnace; (**e**) Primary air mass flow; (**f**) Outlet temperature.

To investigate control performance quantitatively, we introduce the cumulative tracking error:

$$\sum_{i=1}^{T} \left| \frac{y_{real}^i - y_{ref}^i}{y_{ref}^i} \right|, \tag{75}$$

where T is the total simulation time, and subscripts *ref* denotes the reference signal and *real* denotes the real controlled variable value. Figure 12 shows the cumulative tracking error for the proposed and PI controllers.

Figures 12 and 13 show that the proposed multi-model inferential controller can significantly improve pulverizing system control precision and tracking performance over a wide operating range. The reasons for this good performance are summarized as follows.

(1) The desired controlled variables are more accurately "measured" by the soft sensor, hence their control precision is significantly improved. The proposed control scheme produces fewer fluctuations around its set point for mass flowrate of primary air and pulverized coal into the furnace, which indicates that the inferential controller is less sensitive to measurement uncertainty.

(2) The multi-model MPC controller can automatically handle nonlinearity, large inertia, and coupling effects of the pulverizing system. At 500 s, the power plant coal demand increased to 9 kg/s. Since the predictive controller can foresee the future outlet temperature increment, it opens the cold air baffle in advance to compensate for the excess energy input by the hot air. Hence temperature is successfully maintained around 70 °C. A similar result is observed at 1700 s, where coal demand falls to 7 kg/s. The PI controller cannot predict the influence from other control loops and handle it timely, resulting in poorly controlled outlet temperature. The PI controller can also easily result in oscillatory performance, due to the large energy balance inertia.

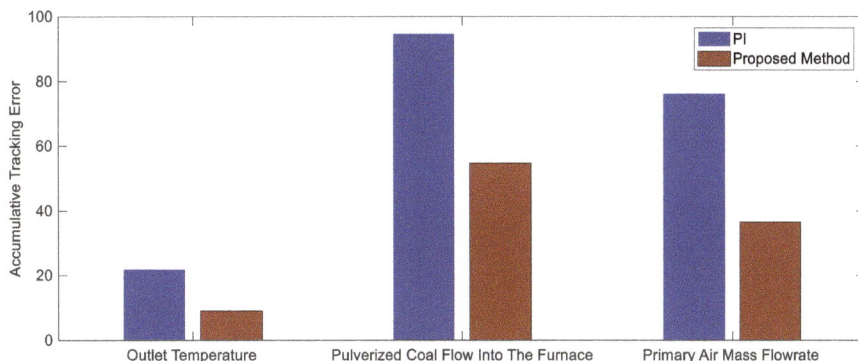

Figure 13. Cumulative tracking error.

Since the pulverized coal flow into the furnace is more accurately controlled within the proposed control scheme, the power plant load will have fewer fluctuations caused by measurement uncertainty. Primary air also tracks the set point faster than the PI controller, which indicates that the air to coal ratio is better controlled. The outlet temperature exhibits almost no oscillations, showing that safe operation of the system has been improved.

6. Conclusions

This paper proposed an inferential multi-model predictive control method to improve pulverizing system control precision and tracking performance. A first principle model of the pulverizing system was developed considering primary air nonlinear dynamics. The proposed model also considered the grindability and moisture content of raw coal to adapt to the change of raw coal type. The unknown parameters in the pulverizing system model were identified using a genetic algorithm. Model validation showed that the proposed model agreed well with measurement data from a real plant, and hence it was employed as the simulation platform for the design of soft sensor and inferential controller.

A soft sensor was developed based on the established model using an MHE approach to estimate desired controlled variables that are unmeasurable or inaccurately measured. The proposed soft sensor can reconstruct signals of the desired controlled variables from more accurately measured variables and thus can improve their "measurement" accuracy. Moreover constraints in the estimates were explicitly considered in the MHE, such that the influence of measurement uncertainty can be significantly reduced. To improve accuracy and computation speed of the MHE, we derived an efficient arrival cost update based on the pulverizing system model. Simulation results showed that the proposed soft sensor can give improved estimates compared with conventional EKF.

Estimated outputs of the soft sensor were employed as feedback signals for an inferential multi-model predictive controller, because, as shown in simulation results, the estimates were much closer to the real value than measurements. We analyzed nonlinearity of the pulverizing system using gap metric and then selected two linear models to construct the local MPC controller based on the analysis. To achieve offset free performance in the presence of unknown disturbances and modeling error, the local linear models were transformed into the extended input-output state space model for controller design. The proposed controller was compared with conventional PI controllers applied in real power plants. Simulation results showed that the proposed inferential method could significantly improve control precision and tracking performance of pulverized coal flow into the furnace, primary air mass flow and outlet temperature.

Acknowledgments: This study was supported by the National Natural Science Foundation of China (NSFC) (grants 51476027, 51576041, and 51506029); the Natural Science Foundation of Jiangsu Province, China (grant BK20150631); and China Postdoctoral Science Foundation.

Author Contributions: Each author has contributed to the present paper. Yiguo Li conceived the idea and directed the simulations. Xiufan Liang wrote the paper and performed the simulations. Jiong Shen directed the simulations. Xiao Wu analyzed the data.

Conflicts of Interest: The authors declare no conflict of interest.

References

1. Flynn, D. *Thermal Power Plant Simulation and Control*; IET: London, UK, 2003.
2. Bhatt, M.S. Effect of air ingress on the energy performance of coal fired thermal power plants. *Energy Convers. Manag.* **2007**, *48*, 2150–2160. [CrossRef]
3. Agrawal, V.; Panigrahi, B.K.; Subbarao, P.M.V. Review of control and fault diagnosis methods applied to coal mills. *J. Process Control* **2015**, *32*, 138–153. [CrossRef]
4. Agrawal, V.; Panigrahi, B.K.; Subbarao, P.M.V. A unified thermo-mechanical model for coal mill operation. *Control Eng. Pract.* **2015**, *44*, 157–171. [CrossRef]
5. Wu, Y.; Wu, X.; Wang, Z.; Chen, L.; Cen, K. Coal powder measurement by digital holography with expanded measurement area. *Appl. Opt.* **2011**, *50*, 22–29. [CrossRef] [PubMed]
6. Gao, Y.; Zeng, D.; Liu, J.; Jian, Y. Optimization control of a pulverizing system on the basis of the estimation of the outlet coal powder flow of a coal mill. *Control Eng. Pract.* **2017**, *63*, 69–80. [CrossRef]
7. Iii, F.J.D. Nonlinear inferential control for process applications. *J. Process Control* **1998**, *8*, 339–353.
8. Niemczyk, P.; Bendtsen, J.D.; Ravn, A.P.; Andersen, P.; Pedersen, T.S. Derivation and validation of a coal mill model for control. *Control Eng. Pract.* **2012**, *20*, 519–530. [CrossRef]
9. Jin, A.; Hitotumatu, S.; Sato, I. Modeling and Parameter Identification of Coal Mill. *J. Power Electron.* **2009**, *9*, 700–707.

10. Zeng, D.L.; Hu, Y.; Gao, S.; Liu, J.Z. Modelling and control of pulverizing system considering coal moisture. *Energy* **2015**, *80*, 55–63. [CrossRef]

11. Wei, J.L.; Wang, J.; Wu, Q.H. Development of a Multisegment Coal Mill Model Using an Evolutionary Computation Technique. *IEEE Trans. Energy Convers.* **2007**, *22*, 718–727. [CrossRef]

12. Lu, J.; Chen, L.; Shen, J.; Wu, Y.; Lu, F. A study of control strategy for the bin system with tube mill in the coal fired power station. *ISA Trans.* **2002**, *41*, 215–224. [CrossRef]

13. Fei, M.; Zhang, J. Robust Fuzzy Tracking Control Simulation of Medium-speed Pulverizer. In *Systems Modeling and Simulation*; Springer: Heidelberg, Germany, 2007.

14. Cortinovis, A.; Mercangöz, M.; Mathur, T.; Poland, J.; Blaumann, M. Nonlinear coal mill modeling and its application to model predictive control. *Control Eng. Pract.* **2013**, *21*, 308–320. [CrossRef]

15. Song, G.L.; Zhou, J.H.; Weng, W.G. Experimental Research on Cold Aerodynamic Field of 75t/h Boiler with Tangential Firing. *Power Syst. Eng.* **2005**, *21*, 14–16.

16. The Engineering ToolBox. Available online: https://www.engineeringtoolbox.com/control-valves-flow-characteristics-d_485.html (accessed on 23 January 2018).

17. Kachitvichyanukul, V. Comparison of three evolutionary algorithms: GA, PSO, and DE. *Ind. Eng. Manag. Syst.* **2012**, *11*, 215–223. [CrossRef]

18. Rashtchi, V.; Rahimpour, E.; Rezapour, E.M. Using a genetic algorithm for parameter identification of transformer RLCM model. *Electr. Eng.* **2006**, *88*, 417–422. [CrossRef]

19. Wang, J.; Wang, J.; Daw, N.; Wu, Q. Identification of pneumatic cylinder friction parameters using genetic algorithms. *IEEE/ASME Trans. Mechatron.* **2004**, *9*, 100–107. [CrossRef]

20. Zhang, Y.G.; Wu, Q.H.; Wang, J.; Oluwande, G. Coal Mill Modeling by Machine Learning Based on on-Site Measurements. *IEEE Trans. Energy Convers.* **2002**, *17*, 549–555. [CrossRef]

21. Ghosh, A. *Evolutionary Computation in Data Mining*; Springer: Heidelburg, Germany, 2004.

22. Pachauri, N.; Singh, V.; Rani, A. Two degree of freedom PID based inferential control of continuous bioreactor for ethanol production. *ISA Trans.* **2017**, *68*, 235–250. [CrossRef] [PubMed]

23. Darko, S.I.; Nikola, J.; Nikola, P.; Velimir, O. Soft sensor for real-time cement fineness estimation. *ISA Trans.* **2015**, *55*, 250–259.

24. Rani, A.; Singh, V.; Gupta, J.R. Development of soft sensor for neural network based control of distillation column. *ISA Trans.* **2013**, *52*, 438–449. [CrossRef] [PubMed]

25. Zhai, Y.J.; Yu, D.L.; Qian, K.J.; Lee, S.; Theera-Umpon, N. A Soft Sensor-Based Fault-Tolerant Control on the Air Fuel Ratio of Spark-Ignition Engines. *Energies* **2017**, *10*, 131. [CrossRef]

26. Zhang, D.; Liu, G.; Zhao, W.; Miao, P.; Jiang, Y.; Zhou, H. A Neural Network Combined Inverse Controller for a Two-Rear-Wheel Independently Driven Electric Vehicle. *Energies* **2014**, *7*, 4614–4628. [CrossRef]

27. Rao, C.V.; Rawlings, J.B.; Lee, J.H. Constrained linear state estimation—A moving horizon approach. *Automatica* **2001**, *37*, 1619–1628. [CrossRef]

28. Rao, C.V.; Rawlings, J.B.; Mayne, D.Q. Constrained state estimation for nonlinear discrete-time systems: Stability and moving horizon approximations. *IEEE Trans. Autom. Control* **2003**, *48*, 246–258. [CrossRef]

29. Lopez-Negrete, R.; Patwardhan, S.C.; Biegler, L.T. Constrained particle filter approach to approximate the arrival cost in moving horizon estimation. *J. Process Control* **2011**, *21*, 909–919. [CrossRef]

30. Kühl, P.; Diehl, M.; Kraus, T.; Schlöder, J.P.; Bock, H.G. A real-time algorithm for moving horizon state and parameter estimation. *Comput. Chem. Eng.* **2011**, *35*, 71–83. [CrossRef]

31. Du, J.; Song, C.; Yao, Y.; Li, P. Multilinear model decomposition of MIMO nonlinear systems and its implication for multilinear model-based control. *J. Process Control* **2013**, *23*, 271–281. [CrossRef]

32. Zhang, R.; Xue, A.; Wang, S.; Ren, Z. An improved model predictive control approach based on extended non-minimal state space formulation. *J. Process Control* **2011**, *21*, 1183–1192. [CrossRef]

33. Wang, L.; Young, P.C. An improved structure for model predictive control using non-minimal state space realisation. *J. Process Control* **2006**, *16*, 355–371. [CrossRef]

34. Gous, G.Z.; De Vaal, P.L. Using MV overshoot as a tuning metric in choosing DMC move suppression values. *ISA Trans.* **2012**, *51*, 657–664. [CrossRef] [PubMed]

35. Garriga, J.L.; Soroush, M. Model predictive control tuning methods: A review. *Ind. Eng. Chem. Res.* **2010**, *49*, 3505–3515. [CrossRef]

36. And, E.L.H.; Rawlings, J.B. Critical Evaluation of Extended Kalman Filtering and Moving-Horizon Estimation. *Ind. Eng. Chem. Res.* **2005**, *44*, 2451–2460.

37. Wu, X.; Shen, J.; Li, Y.; Lee, K.Y. Fuzzy modeling and stable model predictive tracking control of large-scale power plants. *J. Process Control* **2014**, *24*, 1609–1626. [CrossRef]

![energies logo] *energies*

MDPI

Article

Modeling and Control of a Combined Heat and Power Unit with Two-Stage Bypass

Yaokui Gao [1], Yong Hu [1,*], Deliang Zeng [1], Jizhen Liu [1] and Feng Chen [2]

[1] School of Control and Computer Engineering, North China Electric Power University, Beijing 102206, China; gaoyaokui05@126.com (Y.G.); zdl@ncepu.edu.cn (D.Z.); ljz@ncepu.edu.cn (J.L.)

[2] Beijing Guodian Zhishen Control Technology CO., Ltd., Beijing 102200, China; chenfeng@kh.cgdc.com.cn

*   Correspondence: ncepu_hu@yahoo.com; Tel.: +86-10-6177-2840

Received: 10 May 2018; Accepted: 25 May 2018; Published: 29 May 2018

Abstract: This paper presents a non-linear dynamic model of a combined heat and power (CHP) unit with two-stage bypass for the first time. This model is derived through an analysis of the material and energy balance of the CHP unit. The static parameters are determined via the design data of the CHP unit, and the dynamic parameters refer to model parameters of same type of units in other references. Based on the model, an optimized control scheme for the coordination system of the unit is proposed. This scheme introduces a stair-like feedforward-feedback predictive control algorithm to solve the control problem of large delays in boiler combustion, and integrates decoupling control to reduce the effect of external disturbance on the main steam pressure. Simulation results indicate that the model effectively reflects the dynamics of the CHP unit and can be used for designing and verifying its coordinated control system; the control scheme can achieve decoupling control of the CHP unit; the fluctuation of main steam pressure is considerably reduced; and the adjustment of coal feed flow is stable. In this case, the proposed scheme can guarantee the safe, stable and flexible operation of the unit and lay the foundation for decoupling the heat load-based constraint of CHP units, thereby expanding the access space of wind power in northern China.

Keywords: combined heat and power unit; two-stage bypass; dynamic model; coordinated control system; predictive control; decoupling control

1. Introduction

In recent years, the installed capacity of wind turbines in China has increased at an alarming rate [1,2]; however, the phenomenon whereby wind power is abandoned is very serious, especially in northern China. This condition is attributed to the particularly scarce peak-load regulation power (hydropower and condensing unit) in these areas compared with the numerous combined heat and power (CHP) units. The CHP units have considerable environmental and financial benefits when compared to conventional energy generation [3–7]. However, a CHP unit is subject to a heat load-based constraint, which causes its power output to be high in the heating season and limits the access space of wind power in the grid [8]. On the basis of this issue, the North-East Energy Regulatory Bureau promulgated the "Special Reform Program for the North-East Electric Power Auxiliary Service Market [9]" and the "North-East Electric Power Auxiliary Service Market Operation Rules (Trial) [10]" in November 2016. The policy aims to give full play to the economic leverage, optimize peaking resources through marketization, and allow operators to maximize their effectiveness. The main idea of this policy is to compensate thermal power units with high peaking rates. The compensation funds are shared equally by wind power, nuclear power, and thermal power units with low peak peaking rates. In order to significantly reduce the phenomenon of the abandonment of wind power and increase its access space in the grid, the National Energy Administration officially launched pilot projects to improve the flexibility of thermal power units in 2016 [11]. After comprehensive comparison and selection, 22 thermal power plants in areas with prominent problems of renewable energy consumption were selected as pilot projects. It can be seen that the flexible operation of thermal power units meets the major needs of the national energy industry and is supported by the policies of the National Energy Administration. For the special situation in northern China, wind power would have significant access space to the power grid if the heat load-based constraints are decoupled during the heating season. In this case, decoupling the heat load-based constraints of CHP units is an important means to solve the problem of wind power consumption in northern China.

Currently, the main methods for decoupling the heat load-based constraint of CHP units are bypass heating, electric heating, and heat storage methods [12]. In bypass heating, part of the main steam is cooled and decompressed directly to heat circulating water in the heat supply network. This method does not meet the designed operating conditions of CHP units. However, all related equipment has a certain degree of anti-fatigue capacity at a design time, and a small deviation from design conditions has a slight effect on equipment wear and service life. In electric heating [13], part of the electricity produced by a CHP unit is directly used to heat the circulating water in the heat supply network. This method is equivalent to using excess wind power rather than CHP units for heating from the point of view of power grid. Thus, significant coal-saving benefits are gained. However, the renovation costs of electric heating equipment (electric boiler) are extremely high. In heat storage heating [14–17], the storage tank begins to store heat when the heating capacity of the CHP unit is sufficient and then releases heat when the heating capacity of the CHP unit is insufficient. The coal-saving benefits are obvious, considering that no conversion of high-quality energy to low-quality energy occurs. The work presented in this paper focuses on the first aforementioned method, bypass heating, and mainly focuses on its effect on the energy balance of a CHP unit. Such a study provides a solid foundation for the safe operation of the CHP unit with two-stage bypass.

Recently, research of the bypass system has mainly focused on the process of start-up, shut-down, and rapid load changes. In order to study the dynamics of bypass temperature, Zhou, Y et al. established a dynamic model for a high-pressure bypass system, verified by fast cut back (FCB) field test data. The results show that the model has high degree of accuracy. Moreover, an improved control technology is proposed to solve the bypass over-temperature problem during FCB. Simulation results show that the improved control technology is better than the traditional controller. However, this model is only a partial model of the bypass system and cannot demonstrate the effect of the bypass on the energy balance of the unit. Therefore, this model cannot be used to design and verify the coordinated control system (CCS) of the unit [18]. Considering that people are increasingly interested in the optimization of bypass controllers and actuators, Pugi et al. presented a model for real-time simulation of a steam plant, and on this basis, they developed a modular Simulink library of components such as heaters, turbines, and valves. This model has been used for closed-loop testing of hardware such as bypass controllers and valve positioners [19]. Considering the dynamics of the bypass system, the feedwater system and the feedwater heater during FCB, Wang et al. established a dynamic model for a coal-fired unit. The effectiveness of the model was verified by FCB field tests [20]. Although there are many studies on the bypass system, the research content only focuses on special working conditions of the unit and has not considered bypass heating. In order to fundamentally analyze the effect of the bypass heating on the energy balance of CHP units, it is necessary to further study the dynamics of CHP units. In recent years, researchers have conducted extensive studies on the drum-boiler model of CHP units. On the basis of the dynamic model of a drum-boiler condensing unit [21–23], Liu et al. established a three input, three output, and nonlinear dynamic model for a drum-boiler CHP unit. The inputs of the model are coal feed flow, valve position of turbine, and valve position for heating. The outputs of the model are the main steam pressure, electrical load, and heating steam flow. The simulation results indicate that the control methods of CHP and condensing units are basically the same and there is a more flexible way to improve the load ramp rate of CHP units (valve throttling for heating). However, heating steam pressure is generally used as the controlled variable for heat load rather than heating steam flow [24]. Considering that the problem exists in [24], Liu et al.

presented a mathematical model for a drum-boiler CHP unit, a model that differs from that described in [24] in which heating steam pressure is used as the controlled variable for heat load. The simulation results indicate that the model effectively reflects the dynamics of a CHP unit. However, the effect of bypass heating method on the CHP unit is not considered [25]. To deeply analyze the effect of bypass heating on the energy balance of a CHP unit and ensure the safe operation of the unit, a dynamic model of a CHP unit with two-stage bypass should be established.

On the basis of [24,25], a five input, three output, and nonlinear dynamic model of a CHP unit with two-stage bypass is proposed in the current study. The effect of bypass heating on the energy balance of the CHP unit is considered. Based on the model, an optimized control scheme for the CCS of the unit is proposed. In this scheme, a stair-like feedforward-feedback predictive control algorithm is taken as key to solving the control problem of large delays in boiler combustion, and the decoupling control is integrated into the scheme to reduce the effect of external disturbance on main steam pressure. Simulation results indicate that the model effectively reflects the dynamics of the CHP unit and can be used for designing and verifying its coordinated control system. The control scheme can achieve decoupling control of the CHP unit, the fluctuation of main steam pressure is considerably reduced, and the adjustment of coal feed flow is stable. In this case, the proposed scheme can guarantee the safe, stable, and flexible operation of the CHP unit and lay the foundation for decoupling the heat load-based constraint of CHP units, thereby expanding the access space of wind power in northern China.

This paper is organized as follows. Section 2 presents a brief introduction of the bypass heating method. Section 3 deduces and establishes a nonlinear dynamic model for a CHP unit with two-stage bypass and contains a simple verification of the model dynamics. Section 4 designs an optimized control scheme for the unit. Section 5 simulates and verifies the control scheme proposed in the former section. Section 6 presents the conclusion of this paper.

2. Working Principle of the Bypass Heating Method

Compared with a traditional CHP unit, the CHP unit with two-stage bypass is different in several aspects (Figure 1). In the latter, a high-pressure bypass is installed in front of the main steam valve, and part of the main steam is cooled, decompressed, and sent to reheat the steam pipe (in the cold section) to mix with the exhaust steam from the high-pressure cylinder (HPC), and then the mixed steam is fed to the reheater for reheating. Moreover, a low-pressure bypass is installed on the reheat steam pipe (in the hot section), and part of the reheat steam is cooled, decompressed, and sent to the heating steam pipe to mix with the extraction steam for heating, and then the mixed steam is fed to the heater in the heat supply network. Considering that the extraction steam from the intermediate pressure cylinder (IPC) is generally insufficient when the unit is involved in peak regulation, the bypass can be opened at this point to assist heating, which allows decoupling of the heat load-based constraint of the CHP unit.

In the bypass heating, since there is an extraction system in the turbine, the steam flow through the turbine decreases stepwise, while the steam flow through the bypass increases with the increase of the desuperheating water flow. Therefore, the reheat steam flow increases when the high-pressure bypass is opened. Given that the resistance of the reheater is constant, the reheat steam pressure (in the hot section) increases rapidly when the low-pressure bypass is not opened in a timely manner. In this case, the exhaust temperature of HPC increases simultaneously due to the compression effect of the steam, which will increase the thermal stress damage of the turbine. In addition, the original axial thrust of the turbine will be destroyed when the adjustment of the two-stage bypass mismatches, which will affect the safe operation of the turbine. Therefore, establishing a dynamic model for a CHP unit with two-stage bypass is of great significance.

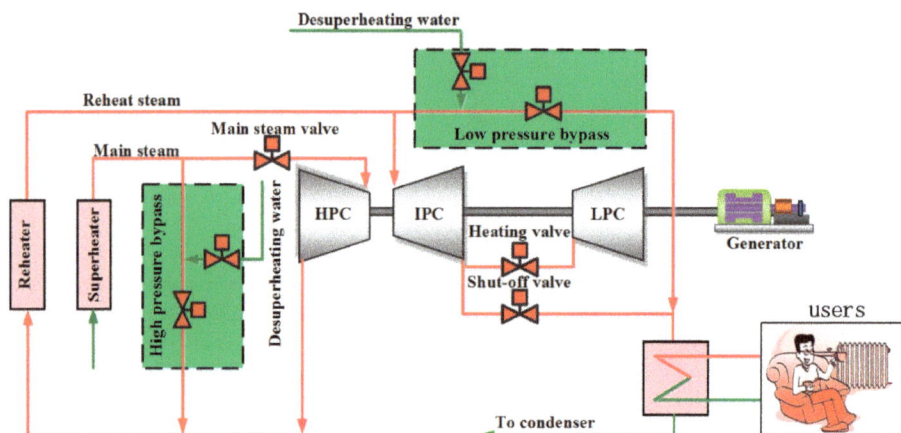

Figure 1. Schematic of a combined heat and power (CHP) unit with two-stage bypass.

3. Modeling and Verification of a Combined Heat and Power (CHP) Unit with Two-Stage Bypass

3.1. Modeling of a CHP Unit with Two-Stage Bypass

3.1.1. Modeling of the Milling Processes

A positive-pressure, direct-fired milling system is mainly composed of a coal feeder, a coal mill, a separator, a primary air tube, and a burner. The dynamics of the coal feeder can be approximated as a pure delay link [25]; the dynamics of the milling process mainly depends on the coal mill and separator, which can be approximated as an inertia link [25]. The residence time of the coal powder in the primary air tube and burner is short, and the combustion process is fast. Therefore, this part of the dynamics can be ignored.

The dynamic model of the coal feeder is established in Equation (1):

$$q_{m,m} = q_{m,b}(t - \tau),\qquad(1)$$

The dynamic model of the milling process is established in Equation (2):

$$T_f \frac{dq_{m,f}}{dt} = -q_{m,f} + q_{m,m}\qquad(2)$$

where $q_{m,b}$ is the coal feed flow, t/h; $q_{m,m}$ is the amount of coal entering the coal mill per unit time, t/h; $q_{m,f}$ is the amount of coal entering the boiler per unit time, t/h; τ is the delay time from coal feeder to coal mill, s; and T_f is the inertia time of the milling process. In the above model, τ and T_f are pending dynamic parameters.

3.1.2. Modeling of the Drum

As the dynamics of drum can be accurately reflected by drum pressure [21], the drum pressure is selected as the state variable to establish the following differential equation (Equation (3)):

$$C_b \frac{dp_b}{dt} = K_1 q_{m,f} Q_{net,ar} - K_2 \sqrt{p_b - p_t}\qquad(3)$$

where C_b is the energy storage coefficient of the drum, t/MPa; p_b is the drum pressure, MPa; $Q_{net,ar}$ is the low calorific value of coal, MJ/kg; and p_t is the main steam pressure, MPa. In the above model, K_1 and K_2 are pending static parameters, and C_b is a pending dynamic parameter.

3.1.3. Modeling of the Main Steam Pipe

As the dynamics of the main steam pipe can be accurately reflected by the main steam pressure, the main steam pressure is selected as the state variable to establish the following differential equation (Equation (4)):

$$C_t \frac{dp_t}{dt} = K_2 \sqrt{p_b - p_t} - K_3 p_t u_t - K_4 p_t u_H \tag{4}$$

where C_t is the energy storage coefficient of the main steam pipe, t/MPa, u_t is the valve position of turbine, %; and u_H is the valve position of high-pressure bypass, %. In the above model, K_3 and K_4 are pending static parameters, and C_t is a pending dynamic parameter.

3.1.4. Modeling of the Reheat Steam Pipe

As the dynamics of the reheat steam pipe can be accurately reflected by the reheat steam pressure, the reheat steam pressure is selected as the state variable to establish the following differential equation (Equation (5)):

$$C_r \frac{dp_r}{dt} = K_5 K_3 p_t u_t + K_6 K_4 p_t u_H - K_7 p_r u_L - 100 K_8 p_r \tag{5}$$

where C_r is the energy storage coefficient of reheat steam pipe, t/MPa, p_r is the reheat steam pressure, MPa; and u_L is the valve position of low-pressure bypass, %. In the above model, K_5, K_6, K_7 and K_8 are pending static parameters, and C_r is a pending dynamic parameter.

3.1.5. Modeling of Intermediate Pressure Cylinder (IPC) Extraction Steam Pipe

As the dynamics of the IPC extraction steam pipe can be accurately reflected by the extraction steam pressure, the extraction steam pressure is selected as the state variable to establish the following differential equation (Equation (6)):

$$C_{IPC} \frac{dp_{IPC}^{out}}{dt} = 100 K_9 K_8 p_r + K_{10} K_7 p_r u_L - K_{11} p_{IPC}^{out} u_{LPC}^{in} - K_{12} q_{m,x} (\theta_o - \theta_i), \tag{6}$$

where C_{IPC} is the extraction heat storage coefficient of IPC, t/MPa; p_{IPC}^{out} is the extraction pressure of IPC, MPa; $q_{m,x}$ is the circulating water flow in the heat supply network, t/h; θ_o is the supply water temperature in the heat supply network, °C; θ_i is the return water temperature in the heat supply network, °C; and u_{LPC}^{in} is the valve position for heating, %. In the above model, K_9, K_{10}, K_{11} and K_{12} are pending static parameters, and C_{IPC} is a pending dynamic parameter.

The condensate in the heater is saturated water according to the characteristics of the heater in the heat supply network. Considering that a one-to-one correspondence exists between saturation temperature and saturation pressure, the saturation pressure is equal to the extraction pressure (ignoring the extraction pressure drop). On the basis of the above assumptions, the supply water temperature should be scientifically expressed as a function of extraction pressure (ignoring the heater terminal temperature difference) [25]. The extraction parameters of the IPC in the normal operating condition are shown in the Table 1.

Table 1. Extraction parameters of intermediate pressure cylinder (IPC) in normal operating condition.

p_{IPC}^{out}/MPa	0.2	0.3	0.4	0.5	0.6
θ_o/°C	120.2	133.5	143.6	151.8	158.8

The function of extraction pressure can be obtained by fitting (Equation (7)).

$$\theta_o = 95.5 p_{IPC}^{out} + 103.3 \tag{7}$$

where the goodness of fit $R^2 = 0.9834$.

3.1.6. Modeling of the Turbine

The exhaust steam pressure of IPC is different between condensing and heating conditions. Although the inlet parameters of the turbine are the same, the power output among HPC, IPC, and low pressure cylinder (LPC) are different. Assume that the power outputs of HPC, IPC, and LPC account for 30%, 35%, and 35% of the total power. This assumption does not affect the calculation of extraction pressure and extraction temperature. The power output of the turbine can be regarded as the sum of the power outputs of HPC, IPC, and LPC and can be approximated as an inertial link (Equation (8)).

$$T_t \frac{dN_E}{dt} = -N_E + 0.3K_{13}K_3 p_t u_t + 0.35K_{14}K_8 p_r \times 100 + 0.35K_{15}K_{11}p_{IPC}^{out}u_{LPC}^{in}, \tag{8}$$

where N_E is the electric power, MW; T_t is the inertia time of turbine, s. In the above model, K_{13}, K_{14} and K_{15} are pending static parameters, and T_t is a pending dynamic parameter.

In summary, the proposed model of the CHP unit is expressed as follows:

$$
\begin{cases}
q_{m,m} = q_{m,b}(t - \tau) \\
T_f \frac{dq_{m,f}}{dt} = -q_{m,f} + q_{m,m} \\
C_b \frac{dp_b}{dt} = K_1 q_{m,f} Q_{net,ar} - K_2\sqrt{p_b - p_t} \\
C_t \frac{dp_t}{dt} = K_2\sqrt{p_b - p_t} - K_3 p_t u_t - K_4 p_t u_H \\
C_r \frac{dp_r}{dt} = K_5 K_3 p_t u_t + K_6 K_4 p_t u_H - K_7 p_r u_l - 100K_8 p_r \\
C_{IPC} \frac{dp_{IPC}^{out}}{dt} = 100K_9 K_8 p_r + K_{10}K_7 p_r u_L - K_{11}p_{IPC}^{out}u_{LPC}^{in} - K_{12}q_{m,x}(96p_{IPC}^{out} - \theta_i + 103) \\
T_t \frac{dN_E}{dt} = -N_E + 0.3K_{13}K_3 p_t u_T + 0.35K_{14}K_8 p_r \times 100 + 0.35K_{15}K_{11}p_{IPC}^{out}u_{LPC}^{in} \\
\theta_o = 95.5p_{IPC}^{out} + 103.38
\end{cases}
\tag{9}
$$

where the inputs of the model are $q_{m,b}$, u_t, u_H, u_L and u_{LPC}^{in}; the outputs are p_t, N_E and θ_o; the time-varying parameters are $Q_{net,ar}$ and θ_i; the pending static parameters are $K_1, \cdots, K_{15}$; and the pending dynamic parameters are τ, T_f, C_t, C_b, C_r and C_{IPC}.

3.2. Model Parameter Determination

The static parameters of the model are determined via the designed data of a CHP unit in China (Table 2) where the subscript RG donates the rated generation condition, and the subscript RH donates the rated heating condition. The electric power output $N_{E(RG)}$, main steam flow $D_{E(RG)}$, reheat steam flow $D_{R(RG)}$, and exhaust steam flow from the IPC $D_{IPC(RG)}^{out}$ are determined via the heat balance diagram in the RG condition; in this paper, the turbine heat acceptance (THA) condition is adopted as the RG condition (Appendix A, Figure A1); the main steam flow $D_{E(RH)}$, reheat steam flow $D_{R(RH)}$, exhaust steam flow from IPC $D_{IPC(RH)}^{out}$, extraction steam flow for heating $D_{H(RH)}$, inlet steam flow of low-pressure cylinder $D_{LPC(RH)}^{in}$, main steam pressure $p_{t(RH)}$, governing stage pressure $p_{1(RH)}$, reheat steam pressure in the cold section $p_{r(RH)}$, exhaust steam pressure from the IPC $p_{IPC(RH)}^{out}$, and inlet steam pressure of the LPC $P_{LPC(RH)}^{in}$ are determined via the heat balance diagram in the RH condition (Appendix A, Figure A1); the drum pressure $p_{b(RH)}$ is determined via the performance parameters of the boiler in Appendix A, Table A1; the steam flow via high-pressure bypass $D_{HTDPR(RH)}^{in}$, steam flow via low-pressure bypass $D_{LTDPR(RH)}^{in}$, cooling water flow via high-pressure bypass $q_{HTDPR(RH)}^{in}$, and cooling water flow via low-pressure bypass are obtained from the design data of the two-stage bypass in Appendix A, Table A2; the low calorific value of coal $Q_{net,ar}$ is determined via the data of designed coal in Appendix A, Table A3; In addition, the designed coal feed flow $q_{m,b(RG)}$ of the boiler in the THA condition is 207.74 t/h; the designed circulating water flow in heat supply network

$q_{m,x(RH)}$ is 12,000 t/h, and the return water temperature from heat supply network $\theta_{i(RH)}$ is 40 °C according to the boiler operation regulations.

Table 2. Designed data of a CHP unit in China.

Parameter	Value	Parameter	Value
Electric power output, $N_{E(RG)}$ (MW)	330	Steam flow via low-pressure bypass, $D^{in}_{LTDPR(RH)}$ (t/h)	208.50
Coal feed flow, $q_{m,b(RG)}$ (t/h)	207.74	Drum pressure, $p_{b(RH)}$ (MPa)	18.57
Low calorific value of coal, $Q_{net,ar}$ (MJ/kg)	14.522	Main steam pressure, $p_{t(RH)}$ (MPa)	16.70
Main steam flow, $D_{E(RG)}$ (t/h)	997.56	Governing stage pressure, $p_{1(RH)}$ (MPa)	13.887
Reheat steam flow, $D_{R(RG)}$ (t/h)	829.81	Reheat steam pressure in cold section, $p_{r(RH)}$ (MPa)	3.699
Exhaust steam flow from IPC, $D^{out}_{IPC(RG)}$ (t/h)	695.00	Exhaust steam pressure from IPC, $p^{out}_{IPC(RH)}$ (MPa)	0.490
Main steam flow, $D_{E(RH)}$ (t/h)	1043.26	Inlet steam pressure of LPC, $p^{in}_{LPC(RH)}$ (MPa)	0.157
Reheat steam flow, $D_{R(RH)}$ (t/h)	860.27	Cooling water flow via high-pressure bypass, $q^{in}_{HTDPR(RH)}$ (t/h)	33.5
Exhaust steam flow from IPC, $D^{out}_{IPC(RH)}$ (t/h)	726.67	Cooling water flow via low-pressure bypass, $q^{in}_{LTDPR(RH)}$ (t/h)	33.0
Extraction steam flow for heating, $D_{H(RH)}$ (t/h)	500	Circulating water flow in heat supply network, $q_{m,x(RH)}$ (t/h)	12,000
Inlet steam flow of low-pressure cylinder, $D^{in}_{LPC(RH)}$ (%)	226.67	Return water temperature from heat supply network, $\theta_{i(RH)}$ (°C)	40
Steam flow via high-pressure bypass, $D^{in}_{HTDPR(RH)}$ (t/h)	175	-	-

The calculation formulas of K_1–K_{15} are given as follows:

$$q_{m,b(RH)} = q_{m,b(RG)} \frac{D_{E(RH)}}{D_{E(RG)}} \tag{10}$$

$$K_1 = \frac{D_{E(RH)}}{q_{m,b(RH)} Q_{net,ar}}, \tag{11}$$

$$K_2 = \frac{D_{E(RH)}}{\sqrt{p_{b(RH)} - p_{t(RH)}}} \tag{12}$$

$$K_3 = \frac{D_{E(RH)}}{p_{t(RH)} u_{T(RH)}} = \frac{D_{E(RH)}}{100 p_{1(RH)}} \tag{13}$$

$$K_4 = \frac{D^{in}_{HTDPR(RH)}}{p_{t(RH)} u_{H(RH)}} \tag{14}$$

$$K_5 = \frac{D_{R(RH)}}{D_{E(RH)}} \tag{15}$$

$$K_6 = \frac{D^{in}_{HTDPR(RH)} + q^{in}_{HTDPR(RH)}}{D^{in}_{HTDPR(RH)}} \tag{16}$$

$$K_7 = \frac{D^{in}_{LTDPR(RH)}}{p_{r(RH)} u_{L(RH)}} \tag{17}$$

$$K_8 = \frac{D_{R(RH)}}{100 p_{r(RH)}} \tag{18}$$

$$K_9 = \frac{D^{out}_{IPC(RH)}}{D_{R(RH)}} = \frac{D^{in}_{LPC(RH)} + D_{H(RH)}}{D_{R(RH)}} \tag{19}$$

$$K_{10} = \frac{D^{in}_{LTDPR(RH)} + q^{in}_{LTDPR(RH)}}{D^{in}_{LTDPR(RH)}} \tag{20}$$

$$K_{11} = \frac{D^{in}_{LPC(RH)}}{u^{in}_{LPC(RH)} p^{out}_{IPC(RH)}} = \frac{D^{in}_{LPC(RH)}}{100 \frac{p^{in}_{LPC(RH)}}{p^{out}_{IPC(RH)}} p^{out}_{IPC(RH)}} = \frac{D^{in}_{LPC(RH)}}{100 p^{in}_{LPC(RH)}} \tag{21}$$

$$K_{12} = \frac{D_{H(RH)}}{q_{m,x(RH)} \left(96 p^{out}_{IPC(RH)} - \theta_{i(RH)} + 103\right)}, \tag{22}$$

$$K_{13} = \frac{N_{E(RG)}}{D_{E(RG)}} \tag{23}$$

$$K_{14} = \frac{N_{E(RG)}}{D_{R(RG)}} \tag{24}$$

$$K_{15} = \frac{N_{E(RG)}}{D^{out}_{IPC(RG)}} \tag{25}$$

The model dynamic parameters refer to model parameters of same type of units in [25]. The model parameters obtained are shown in Table 3.

Table 3. Model parameters of a CHP unit with two-stage bypass.

$K_1 = 0.3307$	$K_6 = 1.1914$	$K_{11} = 14.4375$	$\tau = 15$	$C_{IPC} = 160$
$K_2 = 800.1323$	$K_7 = 0.5637$	$K_{12} = 3.7865 \times 10^{-4}$	$T_f = 120$	$T_t = 12$
$K_3 = 0.7512$	$K_8 = 2.3257$	$K_{13} = 0.3308$	$C_b = 3300$	-
$K_4 = 0.1050$	$K_9 = 0.8447$	$K_{14} = 0.3977$	$C_t = 20$	-
$K_5 = 0.8246$	$K_{10} = 1.1785$	$K_{15} = 0.4748$	$C_r = 10$	-

Let

$x_1 = q_{m,f}; x_2 = p_b; x_3 = p_t; x_4 = p_r; x_5 = p^{out}_{IPC}; x_6 = N_E.$

$u_1 = q_{m,b}; u_2 = u_T; u_3 = u_H; u_4 = u_L; u_5 = u^{in}_{LPC}.$

$y_1 = p_t; y_2 = N_E; y_3 = \theta_o.$

$Q_{net,ar} = 14.522; \theta_i = 40.$

The nonlinear state equation of the CHP unit is obtained as follows:

$$\begin{cases} \dot{x}_1 = -0.00833x_1 + 0.00833u_1(t-15) \\ \dot{x}_2 = 0.00146x_1 - 0.24246(x_2 - x_3)^{0.5} \\ \dot{x}_3 = 40.00662(x_2 - x_3)^{0.5} - 0.03756x_3u_2 - 0.00525x_3u_3 \\ \dot{x}_4 = 0.06194x_3u_2 + 0.01251x_3u_3 - 0.05637x_4u_4 - 23.257x_4 \\ \dot{x}_5 = 1.22782x_4 + 0.00415x_4u_4 - 0.09023x_5u_5 - 2.3665e^{-6}(96x_5 + 63) \\ \dot{x}_6 = -x_6 + 0.00621x_3u_2 + 2.69772p_r + 0.19994x_5u_5 \end{cases} \tag{26}$$

The output equation of the CHP unit is expressed as follows:

$$\begin{cases} y_1 = x_3 \\ y_2 = x_6 \\ y_3 = 95.5x_5 + 103.38 \end{cases} \tag{27}$$

3.3. Verification of Model Dynamics

In order to verify the dynamics of the model, the step disturbance is applied to each input of the model, the simulation results are shown in Figure 2.

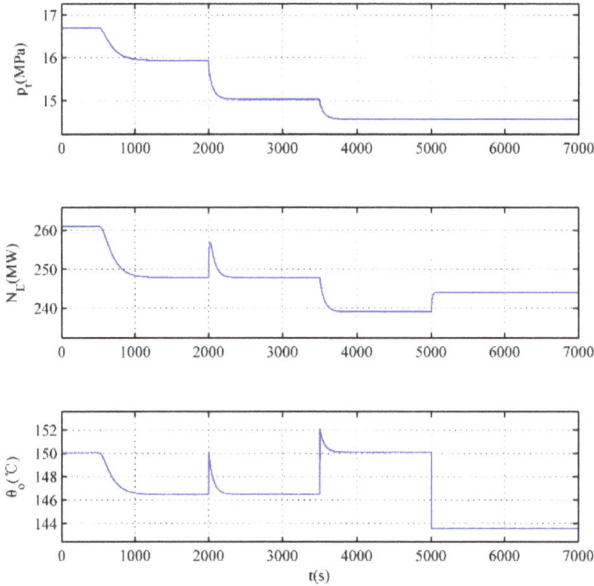

Figure 2. Curves for model outputs.

- When a step decreasing signal is applied to coal feed flow $q_{m,b}$ at 500 s, the main steam pressure p_t, power output of unit N_E, and supply water temperature θ_o are decreased simultaneously due to the energy balance between the boiler and turbine.
- When a step increasing signal is applied to main steam valve u_t at 2000 s, the main steam pressure p_t is rapidly decreased, while the power output of unit N_E and the supply water temperature θ_o are increased first and then restored; this is due to the energy storage of the unit.
- When a step increasing signal is applied to high-pressure bypass u_H and low-pressure bypass u_L at 3500 s, the main steam pressure p_t and the power output of unit N_E are rapidly decreased, while the supply water temperature is also increased rapidly, this verifies the effectiveness of the bypass heating.
- When a step increasing signal is applied to heating valve u_{LPC}^{in} at 5000 s, the power output of unit N_E is increased, while supply water temperature θ_o is decreased, This is due to the fact that more steam is sent to the low-pressure cylinder to generate electricity, instead of heating the return water in the heat supply network.

4. Optimized Control of a CHP Unit with Two-Stage Bypass

In order to solve fundamentally the control problem of large delay in boiler combustion, this paper takes a feedforward-feedback predictive control algorithm as the core. This control algorithm incorporates the concept of feedforward control to preserve the traditional control experience. Moreover, the idea of stair-like control is used to solve the optimal control law in order to avoid the problem of matrix inversion.

4.1. Stair-Like Feedforward-Feedback Generalized Predictive Control

For ease of understanding, a predictive control algorithm with only one feedforward is derived in this study, and predictive control algorithms with multiple feedforwards are similar.

4.1.1. Model Prediction

Consider the following controlled auto-regressive integrated moving average (CARIMA) model [26,27]:

$$A(q^{-1})y(t) = B(q^{-1})u(t-1) + C(q^{-1})v(t-1) + \frac{D(q^{-1})\xi(t)}{\Delta} \tag{28}$$

where $A(q^{-1})$, $B(q^{-1})$, $C(q^{-1})$ and $D(q^{-1})$ are n-, n_b-, n_c- and n_d-order polynomials of q^{-1}, respectively; $u(t)$ is the input of the system; $v(t)$ is the feedforward input of the system; $y(t)$ is the output of the system; $\xi(t)$ denotes the white noise; Δ is a difference operator, $\Delta = 1 - q^{-1}$.

Introduce the following Diophantine equation:

$$1 = E_j(q^{-1})A(q^{-1})\Delta + q^{-j}F_j(q^{-1}), \tag{29}$$

where $E_j(q^{-1}) = e_{j,0} + e_{j,1}q^{-1} + \cdots + e_{j,j-1}q^{-(j-1)}$, $F_j(q^{-1}) = f_{j,0} + f_{j,1}q^{-1} + \cdots + f_{j,n}q^{-n}$. Multiply Equation (28) by $E_j(q^{-1})\Delta q^j$, and combine it with Equation (29). Then the prediction equation can be obtained as,

$$y(t+j) = E_j(q^{-1})B(q^{-1})\Delta u(t+j-1) + E_j(q^{-1})C(q^{-1})\Delta v(t+j-1) + \\ F_j(q^{-1})y(t) + E_j(q^{-1})D(q^{-1})\xi(t+j). \tag{30}$$

Since the noise in the future is unknown, the best prediction equation is

$$\hat{y}(t+j) = E_j(q^{-1})B(q^{-1})\Delta u(t+j-1) + E_j(q^{-1})C(q^{-1})\Delta v(t+j-1) + F_j(q^{-1})y(t) \tag{31}$$

4.1.2. Rolling Optimization

Assume that the objective function is

$$J = \sum_{j=N_1}^{N_2} [w(t+j) - y(t+j)]^2 + \lambda \sum_{j=1}^{N_u} [\Delta u(t+j-1)]^2 \tag{32}$$

where w is the desired setting value; N_1 and N_2 are the initial and terminal values of the optimization horizon; N_u is the control horizon; and λ is the weight of control variable.

The desired setting value in Equation (32) is

$$w(t+j) = \alpha^j y(t) + (1 - \alpha^j)y_r, \, j = 1, 2, \cdots, \tag{33}$$

where $y(t)$ is the actual output of the system; y_r is the real setting value of the system; and α is a softening factor, $0 \le \alpha < 1$.

Let $G_j(q^{-1}) = E_j(q^{-1})B(q^{-1}) = g_{j,0} + g_{j,1}q^{-1} + \cdots + g_{j,n_b+j-1}q^{-(n_b+j-1)}$ and $H_j(q^{-1}) = E_j(q^{-1})C(q^{-1}) = h_{j,0} + h_{j,1}q^{-1} + \cdots + h_{j,n_c+j-1}q^{-(n_c+j-1)}$. Equation (32) can be simply written as,

$$\hat{y}(t+j) = G_j(q^{-1})\Delta u(t+j-1) + H_j(q^{-1})\Delta v(t+j-1) + F_j(q^{-1})y(t) \tag{34}$$

Since the feedforward increment in the future is unknown, the future output in Equation (32) can be written as

$$\hat{y}(t+N_1|t) = G_{N_1}(q^{-1})\Delta u(t+N_1-1) + H_j(q^{-1})\Delta v(t+j-1) + F_{N_1}(q^{-1})y(t)$$
$$= g_{N_1,0}\Delta u(t+N_1-1) + \cdots + g_{N_1,N_1-1}\Delta u(t) + f_{N_1}(t)$$

$$\vdots$$

$$\hat{y}(t+N_2|t) = G_{N_2}(q^{-1})\Delta u(t+N_2-1) + H_j(q^{-1})\Delta v(t+j-1) + F_{N_2}(q^{-1})y(t)$$
$$= g_{N_2,0}\Delta u(t+N_2-1) + \cdots + g_{N_2,N_2-1}\Delta u(t) + f_{N_2}(t)$$

where

$$f_{N_1}(t) = q^{N_1-1}[G_{N_1}(q^{-1}) - g_{N_1,0} - \cdots - g_{N_1,N_1-1}q^{-(N_1-1)}]\Delta u(t)$$
$$+q^{N_1-1}[H_{N_1}(q^{-1}) - h_{N_1,0} - \cdots - h_{N_1,N_1-1}q^{-(N_1-1)}]\Delta v(t) + F_{N_1}(q^{-1})y(t)$$

$$\vdots$$

$$f_{N_2}(t) = q^{N_2-1}[G_{N_2}(q^{-1}) - g_{N_2,0} - \cdots - g_{N_2,N_2-1}q^{-(N_2-1)}]\Delta u(t)$$
$$+q^{N_2-1}[H_{N_2}(q^{-1}) - h_{N_2,0} - \cdots - h_{N_2,N_2-1}q^{-(N_2-1)}]\Delta v(t) + F_{N_2}(q^{-1})y(t)$$

All values involved in $f_j(t)$ are known at time t. Given $i \geq 0$, $\Delta u(t+N_u-i) = 0$. When $N_1 < N_u$, $g_{N_1,N_1-N_u} = \cdots = g_{N_1,-1} = 0$, then the future output can be written as

$$\hat{y}(t+N_1|t) = g_{N_1,N_1-N_u}\Delta u(t+N_u-1) + \cdots + g_{N_1,N_1-1}\Delta u(t) + f_{N_1}(t)$$

$$\vdots$$

$$\hat{y}(t+N_2|t) = g_{N_2,N_2-N_u}\Delta u(t+N_u-1) + \cdots + g_{N_2,N_2-1}\Delta u(t) + f_{N_2}(t)$$

Let $\hat{y} = \begin{bmatrix} \hat{y}(t+N_1|t) \\ \vdots \\ \hat{y}(t+N_2|t) \end{bmatrix}$, $\Delta u = \begin{bmatrix} \Delta u(t) \\ \vdots \\ \Delta u(t+N_u-1) \end{bmatrix}$, and $f = \begin{bmatrix} f_{N_1}(t) \\ \vdots \\ f_{N_2}(t) \end{bmatrix}$.

Then, the following formula can be obtained:

$$\hat{y} = G\Delta u + f \tag{35}$$

where G is a $(N_2 - N_1 + 1) \times N_u$ dimensional matrix.

$$G = \begin{bmatrix} g_{N_1,N_1-1} & \cdots & g_{N_1,N_1-N_u} \\ \vdots & \ddots & \vdots \\ g_{N_2,N_2-1} & \cdots & g_{N_2,N_2-N_u} \end{bmatrix}$$

By using the idea of stair-like control [28–30], the future increment of the control variable can be explicitly planned as $\Delta u(t) = \delta$, $\Delta u(t+j) = \beta\Delta u(t+j-1) = \beta^j\delta$, $1 \leq j \leq N_u$. Then, $\Delta u(t) = (\Delta u(t)\Delta u(t+1)\cdots\Delta u(t+N_u-1))^T = (\delta\ \beta\delta\cdots\beta^{N_u-1}\delta)^T = (1\ \beta\cdots\beta^{N_u-1})^T\delta$.

$$G\Delta u = \begin{bmatrix} g_{N_1,N_1-1} & \cdots & g_{N_1,N_1-N_u} \\ \vdots & \ddots & \vdots \\ g_{N_2,N_2-1} & \cdots & g_{N_2,N_2-N_u} \end{bmatrix}\begin{bmatrix} 1 \\ \vdots \\ \beta^{N_u-1} \end{bmatrix}\delta = \begin{bmatrix} g_{N_1,N_1-1} + \cdots + \beta^{N_u-1}g_{N_1,N_1-N_u} \\ \vdots \\ g_{N_2,N_2-1} + \cdots + \beta^{N_u-1}g_{N_2,N_2-N_u} \end{bmatrix} = \tilde{G}\delta$$

Therefore, the prediction model in Equation (35) can be written as

$$\hat{y} = \tilde{G}\delta + f. \tag{36}$$

The objective function in Equation (32) can be written as

$$\min_{\delta} J = (\tilde{G}\delta + f - w)^T(\tilde{G}\delta + f - w) + \lambda(1 + \beta^2 + \cdots + \beta^{2(N_u-1)})\delta^2 \tag{37}$$

Minimize the objective function $\frac{\partial J}{\partial \delta} = 0$. Then, the optimal control law can be obtained as

$$\delta = \frac{\tilde{G}^T(w - f)}{\tilde{G}^T\tilde{G} + \lambda(1 + \beta^2 + \cdots + \beta^{2(N_u-1)})\delta^2}. \tag{38}$$

In the actual control process, only the current control law is implemented, that is, $\Delta u(t) = \Delta u(t-1) + \delta$. As shown in Equation (38), there is no matrix inversion problem in the control law; therefore, the algorithm can be directly applied to engineering.

4.1.3. Feedback Correction

In the rolling optimization of the generalized predictive control (GPC), the optimization starting point is emphasized to be consistent with the actual output of the system (Equation (34)), which can achieve the function of feedback correction and no difference adjustment of controlled variables. Therefore, feedback correction is not necessary in the case of low control accuracy requirements.

4.2. Optimized Design of the Control Scheme

An optimized control scheme for the CHP unit with two-stage bypass is designed as shown in Figure 3.

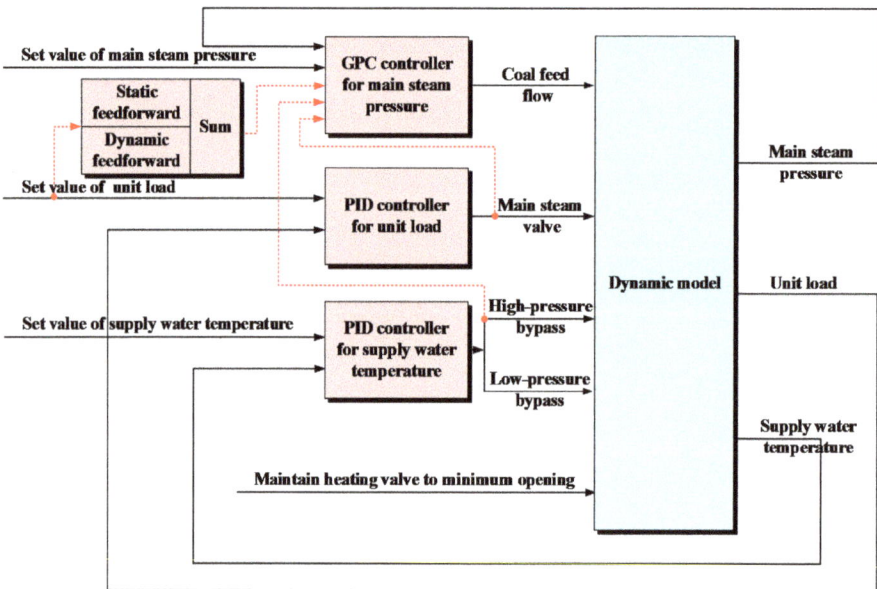

Figure 3. Optimized control scheme for the CHP unit with two-stage bypass.

As shown in Figure 3, the optimized control scheme retains the feedforward-feedback control concept, but unlike the traditional control scheme, the control loop for main steam pressure adopts the stair-like GPC algorithm. Since the response rate of the main steam valve is relatively fast and the control task for unit load is generally completed independently in the digital electro-hydraulic control system, the control loop for unit load retains the traditional proportional-integral-derivative (PID) control method. For the same reason, the control loop for supply water temperature also retains the traditional PID control method. The heating valve will be kept at the minimum opening when the heating capacity of the unit fails to meet the heat load demand.

The inputs of the GPC controller for main steam pressure include: (1) static and dynamic feedforward controls from the set value of the unit load; (2) decoupling feedforward control from the main steam valve; (3) decoupling feedforward control from the high-pressure bypass; and (4) the set and feedback values of the main steam pressure. Among them, the static feedforward control can convert linearly unit load into coal feed flow; therefore, the coal feed flow can be roughly quantify and the adjustment burden of the GPC controller can be reduced simultaneously. The dynamic feedforward control can pre-feed coal according to the deviation of the actual and target loads; in this case, the problem of slow adjustment at the initial period of variable load can be overcome. The decoupling feedforward control from the main steam valve and high-pressure bypass can reduce fluctuation of the main steam pressure caused by external disturbances. In addition, the set and feedback values of the main steam pressure are mainly used for the no-difference adjustment of the main steam pressure. The details of the control loop for the main steam pressure is as shown in Figure 4.

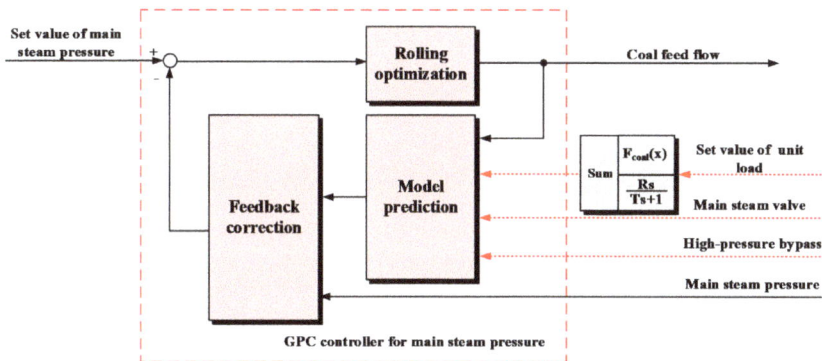

Figure 4. Control loop for main steam pressure.

5. Simulation and Validation

In order to verify the effectiveness of the control scheme, the constant pressure operation and sliding pressure operation of the CHP unit were simulated respectively for this paper, and the control effect in the decoupling and non-decoupling modes were compared (decoupling and non-decoupling of high-pressure bypass). All simulation and validation were performed in MATLAB (R2014a MathWorks, Natick, MA, USA) environment. During the simulation, the ramp rate of the unit was 6 MW/min, which accounted for 1.8% of the rated load; the sampling time $T_s = 0.1$ s, the initial value of the optimization horizon $N_1 = 300$, the terminal value of the optimization horizon $N_2 = 500$, the control horizon $N_u = 10$, the weight of control variable $\lambda = 0.001$, the softening factor $\alpha = 0.998$, the stair-like factor $\beta = 0.1$; the proportional gain and integral gain of the PID controller for unit load were 1.2 and 0.04 respectively; and the proportional gain and integral gain of the PID controller for supply water temperature were 1.2 and 0.3 respectively. The parameters for feedforward controller $R = 20$, $T = 10$. In addition, for ease of simulation, this study simplifies the dynamic feedforward $F_{coal}(x)$ into a proportional coefficient $F = 0.835$.

Figure 5 shows the curves for controlled variables. Under the decoupling control mode of high-pressure bypass, the fluctuating amplitude of controlled variables is reduced considerably, especially for that of the main steam pressure. It can be seen from the figure that although there is a certain delay in the response of the main steam pressure, it has almost no overshoot in the decoupled control mode, this is due to the compensation effect of the decoupling control of the high-pressure bypass, which allows the GPC controller to adjust the coal feed flow in advance, thereby increasing the capability of the control system to overcome external disturbances of the high-pressure bypass.

In addition, since the response time for the other two controlled variables is short, the response values of the unit load and the supply water temperature can be closely matched to their set values.

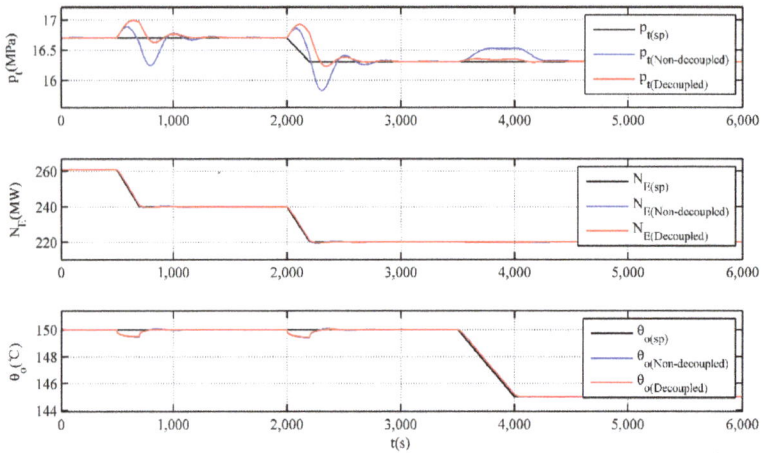

Figure 5. Curves for controlled variables.

Figure 6 shows the curves for the control variables. It can be seen from the figure that the adjustment of each control variable is stable under the decoupling control mode of high-pressure bypass. In addition, compared with the non-decoupling control mode, the decoupling control can reduce the coal feed flow rapidly at the initial period of variable load and pull back it rapidly at the terminal period of variable load. This just validates the compensation effect of feedforward control and decoupling control.

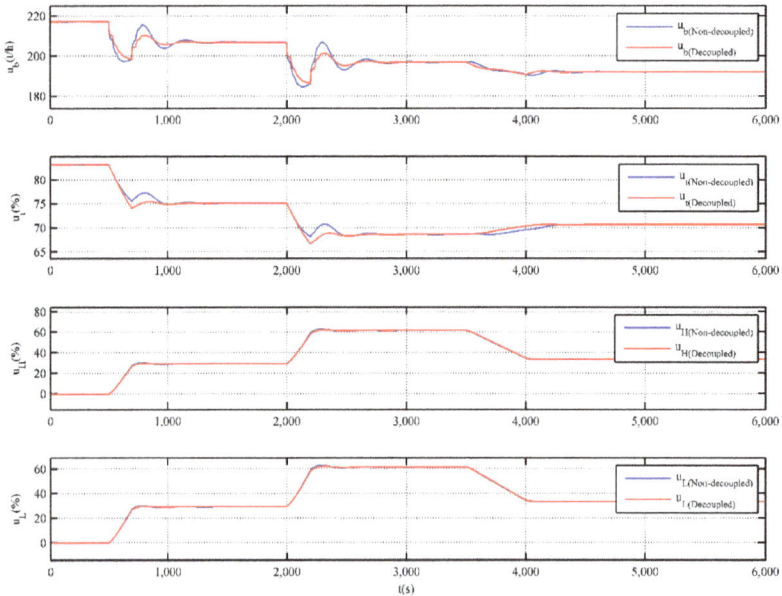

Figure 6. Curves for control variables.

In order to further illustrate the effectiveness of the decoupling control of the high-pressure bypass, the curves of the main steam pressure and coal feed flow are enlarged and plotted in the same figure when the setting value of the supply water temperature changes at 3500 s (Figure 7). Since the setting values of the unit load and main steam pressure are unchanged during this period of simulation, the simulation verifies the effectiveness of the decoupling control well. It can be seen from the figure, due to the effect of the decoupling control, the coal feed flow drops rapidly at the initial period of the dynamic process, and in this case, the overshoot of the main steam pressure is greatly reduced, which fully proves that the decoupling control of the high-pressure bypass can improve the anti-disturbance capacity of the main steam pressure.

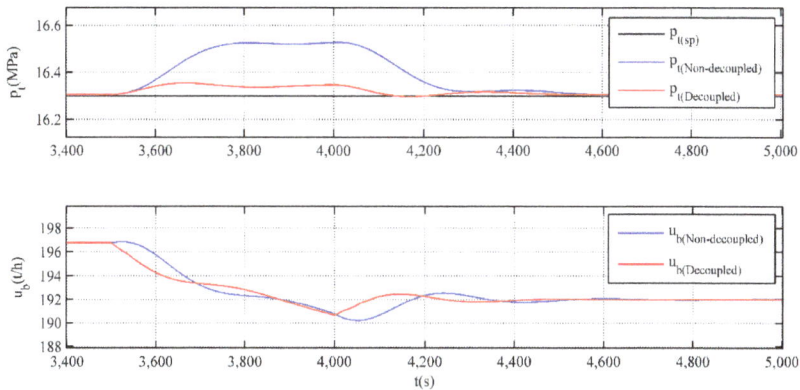

Figure 7. Amplification curves of p_t and u_b.

In addition, the integrated time and absolute error (ITAE) for each controlled variable are calculated (Table 4). It can be seen from the table that the ITAE indicators of the controlled variables are reduced to different degrees under the decoupling control mode, especially for that of main steam pressure. In this case, the proposed scheme can guarantee the safe, stable, and flexible operation of the unit.

Table 4. Integrated time and absolute error (ITAE) indicator for each controlled variable.

Indicator	Main Steam Pressure (MPa)	Unit Load (MW)	Supply Water Temperature (°C)
Non-decoupling control	3699.61	6290.31	3282.26
Decoupling control	2209.68	5707.16	3112.17

6. Conclusions

Considering the effect of bypass heating on the energy balance of a CHP unit, a nonlinear dynamic model for a CHP unit with a two-stage bypass is proposed. The static parameters are determined via the design data of a CHP unit in northern China, and the dynamic parameters refer to model parameters of the same type of units in other references. On the basis of the model, an optimized control scheme for the coordination system of the unit is proposed. In this scheme, a stair-like feedforward-feedback predictive control algorithm is adopted to solve the control problem of large delays in boiler combustion, and the decoupling control is integrated to reduce the effect of external disturbance (the main steam valve and high-pressure bypass) on the main steam pressure. Simulation results indicate that the model effectively reflects the dynamics of the CHP unit with a two-stage bypass and can be used for designing and verifying its CCS; the control scheme can achieve optimal control of the CHP unit with a two-stage bypass; the compensation effect of feedforward and decoupling control allows the GPC controller to adjust the coal feed flow in advance; and in this case the coal feed flow can be reduced rapidly at the initial period of variable load and pulled back rapidly at the terminal period of variable load. Moreover, in the decoupling control of the high-pressure bypass mode, the fluctuation of the main steam pressure is considerably reduced and the adjustment of the coal feed flow is stable, which proves that the decoupling control of the high-pressure bypass can improve the anti-disturbance capacity of the main steam pressure. In this case, the proposed scheme can guarantee the safe, stable, and flexible operation of the unit and lay the foundation for decoupling the heat load-based constraint of CHP units, thereby expanding the access space of wind power in northern China.

Author Contributions: Conceptualization, Y.G. and Y.H.; methodology, Y.G.; Software, Y.G. and D.Z.; validation, Y.H., F.C. and J.L.; formal analysis, Y.G.; investigation, Y.H.; resources, D.Z.; data curation, F.C.; writing original draft, Y.G.; writing, review and editing, Y.H.; visualization, Y.G.; supervision, Y.G.; project administration, Y.H.; funding acquisition, D.Z.

Acknowledgments: This paper is supported by National Natural Science Foundation of China (51776065) and the Fundamental Research Funds for the Central Universities (2018ZD05).

Conflicts of Interest: The authors declare no conflict of interest.

Nomenclature

C_b	energy storage coefficient of drum, t/MPa	P_{LPC}^{in}	inlet steam pressure of low-pressure cylinder, MPa
C_t	energy storage coefficient of main steam pipe, t/MPa,	$Q_{net,ar}$	low calorific value of coal, MJ/kg
C_r	energy storage coefficient of reheat steam pipe, t/MPa,	$q_{m,b}$	coal feed flow, t/h
C_z	energy storage coefficient of extraction pipe for intermediate-pressure cylinder, t/MPa,	$q_{m,m}$	amount of coal entering the coal mill per unit time, t/h
D_E	main steam flow, t/h	$q_{m,f}$	amount of coal entering the boiler per unit time, t/h
D_R	reheat steam flow, t/h	q_{HTDPR}^{in}	cooling water flow via high-pressure bypass, t/h
D_{IPC}^{out}	exhaust steam flow from intermediate-pressure cylinder, t/h	q_{LTDPR}^{in}	cooling water flow via low-pressure bypass, t/h
D_H	extraction steam flow for heating, t/h	$q_{m,x}$	circulating water flow in heat supply network, t/h
D_{LPC}^{in}	inlet steam flow of low-pressure cylinder, t/h	T_f	inertia time of the milling process, s
D_{HTDPR}^{in}	steam flow via high-pressure bypass, t/h	T_t	inertia time of turbine, s
D_{LTDPR}^{in}	steam flow via low-pressure bypass, t/h	u_H	valve position of high-pressure bypass, %
N_E	electric power, MW	u_L	valve position of low-pressure bypass, %
p_b	drum pressure, MPa	u_{LPC}^{in}	valve position for heating, %
p_t	main steam pressure, MPa	u_t	valve position of turbine, %
p_1	governing stage pressure, MPa	θ_i	return water temperature in heat supply network, °C
p_r	reheat steam pressure, MPa	θ_o	supply water temperature in heat supply network, °C
p_{HPC}^{out}	exhaust steam pressure from high-pressure cylinder, MPa	τ	inertia time from coal feeder to coal mill, s
p_{IPC}^{out}	exhaust steam pressure from intermediate-pressure cylinder, MPa	K_i	pending static parameters, $i = 1, \cdots, 15$

Appendix A

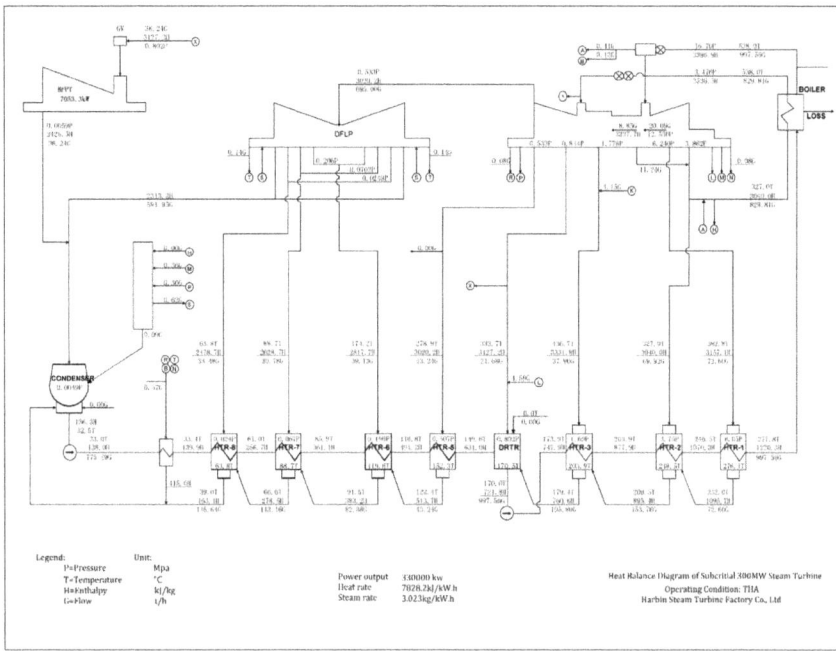

Figure A1. Heat balance diagram in turbine heat acceptance (THA) condition.

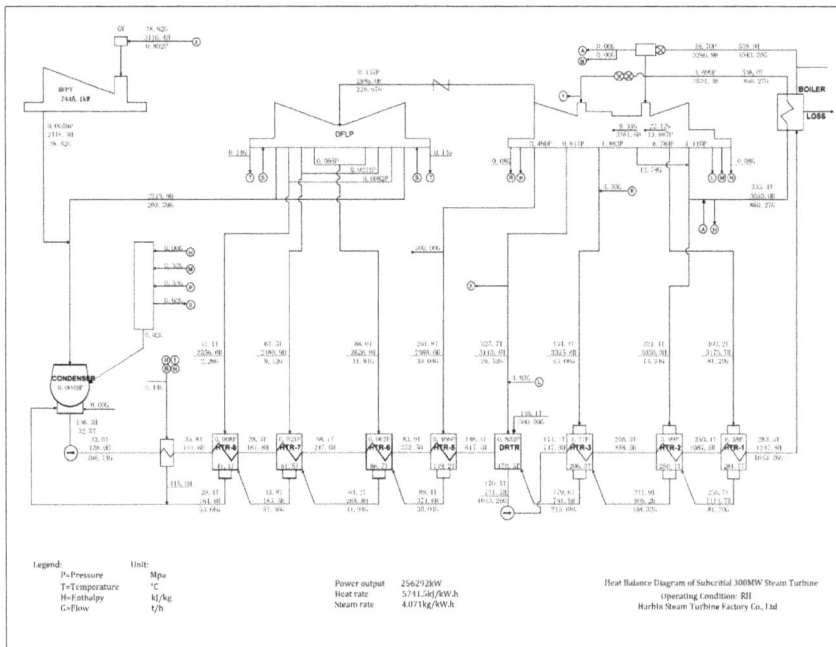

Figure A2. Heat balance diagram in rated heating (RH) condition.

Table A1. Performance parameters of boiler (boiler maximum continuous rating (BMCR); turbine rated load (TRL); turbine heat acceptance (THA)).

Items	Unit	BMCR	TRL	THA	HTO	75% THA	40% THA
Main steam flow	t/h	1100	1043.26	997.56	775.27	714.46	400.01
Outlet steam pressure of superheater	MPa	17.50	17.41	17.34	17.05	16.98	8.08
Outlet steam temperature of superheater	°C	541	541	541	541	541	530.1
Boiler pressure	MPa	19.0	18.77	18.59	17.83	17.65	8.62
Reheat steam flow	t/h	909.19	865.47	829.81	761.85	606.49	349.74
Inlet/outlet steam pressure of reheater	MPa	4.053/3.833	3.653/3.443	3.699/3.498	3.447/3.263	2.689/2.542	1.487/1.402
Inlet/outlet steam temperature of reheater	°C	336.8/541	325.5/541	327/541	328.3/541	302.1/541	316.7/503
Feed water temperature	°C	284.2	276.2	277.8	173.1	257.1	226.6

Table A2. Designed data of the two-stage bypass.

Device Name		High-Pressure Bypass	Low-Pressure Bypass
Primary steam	Designed flow t/h	175	182
	Designed pressure MPa	17.6	4.8
	Designed temperature °C	546	546
	Operating pressure MPa	7.85	1.35
	Operating temperature °C	523.8	493.5
Secondary steam	Designed flow t/h	208.5	215
	Designed pressure MPa	4.8	0.49
	Designed temperature °C	355	280
	Operating pressure MPa	1.5	0.2452
	Operating temperature °C	308.6	240
Desuperheating water;	Designed flow t/h	33.5	33
	Designed pressure MPa	24.6	4.0
	Designed temperature °C	285	140
	Operating pressure MPa	9.5	2
	Operating temperature °C	222.6	36.1

Table A3. Parameters of designed coal.

Items		Symbol	Unit	Designed Coal	Checked Coal
Elemental analysis	Carbon (received base)	Car	%	38.54	39.434
	Hydrogen (received base)	Har	%	3.25	3.53
	Oxygen (received base)	Oar	%	9.92	8.846
	Nitrogen (received base)	Nar	%	0.73	0.672
	Sulfur (received base)	St,ar	%	0.43	0.402
Industrial analysis	Ash (received base)	Aar	%	15	19.802
	Water (received base)	Mt	%	32.4	27.54
	Water (air-drying base)	Mad	%	14.20	12.142
	Water (dry ash-free base)	Vdaf	%	49.28	43.512
Low calorific value of coal (received base)		Qnet,ar	kcal/kg	3228	3472
			MJ/kg	13.50	14.522
Wearable coefficient		HGI	-	56	58.8
...		...	...	...	...

References

1. National Energy Administration. Development of Wind Power Industry in 2015. Available online: http://www.nea.gov.cn/2016-02/02/c_135066586.htm (accessed on 10 May 2018).
2. Tan, Z.F.; Ju, L.W. Review of China's Wind Power Development: History, Current Status, Trends and Policy. *J. North China Electr. Power Univ.* **2013**, *2*, 2.
3. Salem, R.; Bahadori-Jahromi, A.; Mylona, A.; Godfrey, P.; Cook, D. Comparison and Evaluation of the Potential Energy, Carbon Emissions, and Financial Impacts from the Incorporation of CHP and CCHP Systems in Existing UK Hotel Buildings. *Energies* **2018**, *11*, 1219. [CrossRef]

4. Helin, K.; Zakeri, B.; Syri, S. Is District Heating Combined Heat and Power at Risk in the Nordic Area?—An Electricity Market Perspective. *Energies* **2018**, *11*, 1256. [CrossRef]
5. Cozzolino, R. Thermodynamic Performance Assessment of a Novel Micro-CCHP System Based on a Low Temperature PEMFC Power Unit and a Half-Effect Li/Br Absorption Chiller. *Energies* **2018**, *11*, 315. [CrossRef]
6. Zhang, G.; Cao, Y.; Cao, Y.; Li, D.; Wang, L. Optimal Energy Management for Microgrids with Combined Heat and Power (CHP) Generation, Energy Storages, and Renewable Energy Sources. *Energies* **2017**, *10*, 1288. [CrossRef]
7. Rong, S.; Li, W.; Li, Z.; Sun, Y.; Zheng, T. Optimal Allocation of Thermal-Electric Decoupling Systems Based on the National Economy by an Improved Conjugate Gradient Method. *Energies* **2015**, *9*, 17. [CrossRef]
8. Lv, Q.; Chen, T.; Wang, H.; Lv, Y.; Liu, R.; Li, W.D.; Huaneng, D.P.P. Review and Perspective of Integrating Wind Power into CHP Power System for Peak Regulation. *Electr. Power* **2013**, *11*, 129–136+.
9. Northeast China Energy Regulatory Bureau of the National Energy Administration Special Reform Program for the Northeast Electric Power Auxiliary Service Market. Available online: http://dbj.nea.gov.cn/nyjg/hyjg/201611/t20161124_2580781.html (accessed on 24 May 2018).
10. Northeast China Energy Regulatory Bureau of the National Energy Administration Northeast Electric Power Auxiliary Service Market Operation Rules (Trial). Available online: http://dbj.nea.gov.cn/nyjg/hyjg/201708/t20170817_3015219.html (accessed on 24 May 2018).
11. National Energy Administration. The Pilot Project for Improving the Flexibility of Thermal Power Units Was Launched by the National Energy Administration. Available online: http://www.nea.gov.cn/2016-06/22/c_135456540.htm (accessed on 24 May 2018).
12. Chen, X.; Kang, C.; O'Malley, M.; Xia, Q.; Bai, J.; Liu, C.; Sun, R.; Wang, W.; Li, H. Increasing the Flexibility of Combined Heat and Power for Wind Power Integration in China: Modeling and Implications. *IEEE Trans. Power Syst.* **2015**, *30*, 1848–1857. [CrossRef]
13. Mathiesen, B.V.; Lund, H. Comparative Analyses of Seven Technologies to Facilitate the Integration of Fluctuating Renewable Energy Sources. *IET Renew. Power Gener.* **2009**, *3*, 190–204. [CrossRef]
14. Streckienė, G.; Martinaitis, V.; Andersen, A.N.; Katz, J. Feasibility of CHP-plants with Thermal Stores in the German Spot Market. *Appl. Energy* **2009**, *86*, 2308–2316. [CrossRef]
15. Lund, H.; Andersen, A.N. Optimal Designs of Small CHP Plants in a Market with Fluctuating Electricity Prices. *Energy Convers. Manag.* **2005**, *46*, 893–904. [CrossRef]
16. Christidis, A.; Koch, C.; Pottel, L.; Tsatsaronis, G. The Contribution of Heat Storage to the Profitable Operation of Combined Heat and Power Plants in Liberalized Electricity Markets. *Energy* **2012**, *41*, 75–82. [CrossRef]
17. Li, P.; Wang, H.; Lv, Q.; Li, W. Combined Heat and Power Dispatch Considering Heat Storage of Both Buildings and Pipelines in District Heating System for Wind Power Integration. *Energies* **2017**, *10*, 893. [CrossRef]
18. Zhou, Y.; Wang, D.; Qi, T. Modelling, Validation and Control of Steam Turbine Bypass System of Thermal Power Plant. *J. Control Eng. Appl. Inform.* **2017**, *19*, 41–48.
19. Pugi, L.; Galardi, E.; Lucchesi, N.; Carcasci, C. Hardware-in-the-loop Testing of Bypass Valve Actuation System: Design and Validation of a Simplified Real Time Model. *Proc. Inst. Mech. Eng. Part E J. Process Mech. Eng.* **2017**, *231*, 212–235. [CrossRef]
20. Wang, D.; Zhou, Y.; Li, X. A Dynamic Model Used for Controller Design for Fast Cut Back of Coal-fired Boiler-turbine Plant. *Energy* **2018**, *144*, 526–534. [CrossRef]
21. Zeng, D.L.; Zhao, Z.; Chen, Y.Q.; Liu, J.Z. A Practical 500MW Boiler Dynamic Model Analysis. *Proc. CSEE* **2003**, *5*, 33. [CrossRef]
22. Åström, K.J.; Bell, R.D. Drum-boiler dynamics. *Automatica* **2000**, *36*, 363–378. [CrossRef]
23. Tian, L.; Zeng, D.L.; Liu, J.Z.; Zhao, Z. A Simplified Nonlinear Dynamic Model of 330MW Unit. *Proc. CSEE* **2004**, *24*, 180–184. [CrossRef]
24. Liu, J.Z.; Wang, Q.; Tian, L.; Liu, X.P. Simplified Model and Characteristic Analysis of Load-Pressure Object in Heat Supply Units. *J. Chin. Soc. Power Eng.* **2012**, *3*, 3.
25. Liu, X.; Tian, L.; Wang, Q.; Liu, J. Simplified Nonlinear Dynamic Model of Generating Load-Throttle Pressure-Extraction Pressure for Heating Units. *J. Chin. Soc. Power Eng.* **2014**, *34*, 115–121.
26. Clarke, D.W.; Mohtadi, C.; Tuffs, P.S. Generalized Predictive Control—Part I. The Basic Algorithm. *Automatica* **1987**, *23*, 137–148. [CrossRef]

27. Clarke, D.W.; Mohtadi, C.; Tuffs, P.S. Generalized Predictive Control—Part II Extensions and interpretations. *Automatica* **1987**, *23*, 149–160. [CrossRef]

28. Wu, G.; Peng, L.X.; Sun, D.M. Application of Stair-like Generalized Predictive Control to Industrial Boiler. In Proceedings of the IEEE International Symposium on Industrial Electronics, Xi'an, China, 25–29 May 1992; pp. 218–221.

29. Xuefeng, Q.; Meisheng, X.; Demin, S.; Gang, W. The Stair-like Generalized Predictive Control for Main-steam Pressure of Boiler in Steam-power Plant. In Proceedings of the 3rd World Congress on Intelligent Control and Automation, Hefei, China, 26 June–2 July 2000; pp. 3165–3167.

30. Li, X.; Wang, X.; Wang, Z.; Qian, F. A stair-like generalized predictive control algorithm based on multiple models switching. *CIESC J.* **2012**, *1*, 30. [CrossRef]

![energies logo] *energies*

MDPI

Review

Review of Electromagnetic Vibration in Electrical Machines

Xueping Xu, Qinkai Han and Fulei Chu *

Department of Mechanical Engineering, Tsinghua University, Beijing 100084, China;
xuxueping@mail.tsinghua.edu.cn (X.X.); hanqinkai@hotmail.com (Q.H.)
* Correspondence: chufl@mail.tsinghua.edu.cn

Received: 5 June 2018; Accepted: 27 June 2018; Published: 6 July 2018

Abstract: Electrical machines are important devices that convert electric energy into mechanical work and are widely used in industry and people's life. Undesired vibrations are harmful to their safe operation. Reviews from the viewpoint of fault diagnosis have been conducted, while summaries from the perspective of dynamics is rare. This review provides systematic research outlines of this field, which can help a majority of scholars grasp the ongoing progress and conduct further investigations. This review mainly generalizes publications in the past decades about the dynamics and vibration of electrical machines. First the sources of electromagnetic vibration in electrical machines are presented, which include mechanical and electromagnetic factors. Different types of air gap eccentricity are introduced and modeled. The analytical methods and numerical methods for calculating the electromagnetic force are summarized and explained in detail. The exact subdomain analysis, magnetic equivalent circuit, Maxwell stress tensor, winding function approach, conformal mapping method, virtual work principle and finite element analysis are presented. The effects of magnetic saturation, slot and pole combination and load are discussed. Then typical characteristics of electromagnetic vibration are illustrated. Finally, the experimental studies are summarized and the authors give their thoughts about the research trends.

Keywords: electrical machine; electromagnetic vibration; multiphysics; rotor dynamics; air gap eccentricity; calculation method; magnetic saturation

---

1. Introduction

With the continuous development of the economy, electrical machines have been widely used in industries and people's lives. Ever higher performance requirements are being put forward for electrical machines. The magnetic field can interact with mechanical structures, which will produce unbalanced magnetic forces and excite harmful vibrations. These forces may have significant effects on the dynamic behavior and noise of rotors. The vibration and noise range of electrical machines is one of the important indicators in the manufacture. National standards also clearly prescribe the vibration and noise limits of rotating electrical machines. Vibration and noise will affect people's daily life and severe vibration will cause significant economic losses. Furthermore, the vibrations of electrical machines may decrease the efficiency of the energy conversion because the vibration and possible related temperature rise are unwanted energy losses for the system.

The vibrations of electrical machines can be divided into three categories: mechanical vibrations, electromagnetic vibrations and aerodynamic vibrations. Benefitting from the continuous improvement of design and manufacturing level, the performance of electrical machines has been greatly improved and their volume has become very small. For the widely used small and medium-sized electrical machines, electromagnetic vibrations are the main type. With the growth of living standards, people will pay more and more attention to the vibration of electrical machines. Investigating the vibration mechanism can be helpful for the design of electrical machines. Therefore, the study of the electromagnetic vibration of electrical machines has practical significance.

Electromagnetic vibrations are usually generated by the distorted air-gap field of an eccentric rotor in electrical machines. The uneven air-gap is directly related to the eccentricity, which is common in rotating electrical machines. Eccentricity can be caused by several reasons, such as relative misalignment of the rotor and stator in the fixing stage, misalignment of the load axis and rotor shaft, elliptical stator inner cross section, wrong placement or rubbing of ball bearings, mechanical resonance, and unbalanced loads [1,2]. Eccentricities can be further subdivided into two categories: circumferential unequal air gaps and axial unequal air gaps. The former can be grouped into static eccentricity and dynamic eccentricity. In the case of static eccentricity, the rotor rotates around its own geometric axis, which is not the geometric axis of the stator. In the case of dynamic eccentricity, the rotor is not concentric and rotates around the geometric axis of the stator. In reality, both static eccentricity and dynamic eccentricity tend to coexist. An inherent static eccentricity exists, even in newly manufactured machines, due to the build-up of tolerances during the manufacturing and assembly procedure, as has been reported in [3]. Unequal air gaps cause unbalanced magnetic forces (UMFs) [4] on the rotor, which lead to mechanical stress on some part of the shaft and bearing. After prolonged operation, these factors cause broken mechanical parts or even the stator to rub the rotor, causing major breakdowns of the machines [5].

The calculation of UMF is essential for the analysis of vibrations and the optimal design of electrical rotating machinery. Two common approaches are the analytical method and the finite element method (FEM). Although the FEM has been widely applied to study the UMF [6,7], the analytical method still receives much attention because insights into the origins and pivotal factors in the production of UMF is provided by this method. Earlier publications focused primarily on the theoretical formulation of UMF and linear equations were mainly adopted. Werner [8] established a dynamic model for an induction motor with eccentric excitation by taking radial electromagnetic stiffness into account. The linear expressions are convincing only for cases where the eccentricity is small enough. Therefore linear approaches are far from industrial applications. The nonlinear relationship between the UMF and eccentricity was pointed out in [9]. After that, many researchers have introduced nonlinear approaches to determine UMF in the last two decades. For instance, winding function analysis [10], conformal mapping method [11], energy conservation law [12], magnetic equivalent circuit method [13] and exact subdomain model [14] were all applied to investigate the magnetic field distribution and UMF for electrical machines with non-uniform air-gaps. The most commonly adopted analytical method is the air-gap permeance approach [15,16]. An analytical expression of UMF for different pole-pairs was obtained by expressing the air-gap permeance as a Fourier series in [15]. A calculation model for UMF was presented in [16] based on the actual position of the rotor inside the stator.

It should be noted that the design and modeling of electrical machine systems are a multidisciplinary problem because the electromagnetics, structural mechanics and heat transfer are involved, the design optimization process becomes more and more complex [17]. Therefore it is significant to pursue optimal system performance rather than optimal components such as motors or controllers, because assembling individually optimized components into a system cannot ensure an optimal performance for the whole system [18,19]. The problem is really a challenge for both the research and industrial communities since it includes not only theoretical multidisciplinary design and analysis (such as electromagnetic, thermal, mechanical analysis and power electronics) but also practical engineering manufacturing of the system. Lei et al. [20] developed a robust approach for the system-level design optimization of the electrical machine system. Khan et al. [21] presented a multilevel design optimization framework to improve the efficiency of the proposed method by combining it with several techniques, such as design of experiments and approximate models.

The electromagnetic vibration of electrical machines has always been a hot topic in the mechanical discipline and electrical discipline fields, and there exist rich research results. The existing reviews of electromagnetic vibration are basically summarized from the view of fault diagnosis [22–27], and summaries from the perspective of dynamics are rather insufficient. With the continuous progress of research, some new technologies and methods are emerging. It is necessary to generalize the latest research progress of electromagnetic vibration from the perspective of dynamics and vibration to avoid repetitive work. In addition, a review which provides systematic research outlines and references can be beneficial for the majority of scholars in this field to promote the ongoing progress and development of the investigations.

Different from the condition monitoring standpoint, this review is mainly concerned with the dynamic issues of electromagnetic vibrations in electrical machines. First, the vibration sources, which include the mechanical and electromagnetic aspects, are summarized in Section 2. Then Section 3 presents in detail different analytical and numerical calculation methods for the electromagnetic force. After that, the electromagnetic vibration characteristics and experimental investigations are demonstrated in Section 4. Finally Section 5 summarizes the authors' thoughts about the trends and prospects of this research.

2. Sources of Electromagnetic Vibration

Under ideal conditions, the air gap between the stator and rotor is uniform and the magnetic circuit is symmetrical. The rotor rotates in the uniform magnetic field and the total force of the radial electromagnetic force is zero. If mechanical or electromagnetic factors make the radial force around the rotor circumference uneven, it will produce an electromagnetic force which is also known as the unbalanced magnetic force (UMF). UMF will cause undesired electromagnetic vibration and noise, exacerbate the bearing wear, influence the stability of the rotor system and even produce rubbing between the rotor and stator. The potential hazards are prominent. Therefore, the investigation of this coupling interaction is necessary and important.

In essence, the main source of electromagnetic vibration is the asymmetry of the magnetic circuit in the electrical machine. The misalignment between the stator and rotor is the most common cause of UMF. Furthermore, the uneven magnetization of the material and the improper winding can also generate UMF. Generally speaking, the electromagnetic sources can be divided into mechanical factors and electromagnetic factors.

2.1. Mechanical Sources

The mechanical causes of the electromagnetic force are mainly the air gap eccentricity between the stator and the rotor. As Figure 1 indicates, the sources of air gap eccentricity can be divided into four categories: shape deviation, parallel eccentricity, inclined eccentricity and curved eccentricity. The surface corrugations of the outer rotor circle and inner stator circle will affect the uniformity of the air gap length. In addition, when the stator and rotor are not regular cylinders, their shape deviation can produce air gap eccentricity. Lundström et al. [28,29] studied the air gap eccentricity and electromagnetic force caused by the deviation of generator shape in detail. The characteristics of dynamic responses including the whirling frequency and amplitude were investigated. With the progress of mechanical manufacturing technology, the probability of shape deviation in the rotor and stator is gradually declining. Another very important and widely investigated air gap eccentricity type is that the rotor shaft does not coincide with the stator axis. It is assumed that the stator and rotor are ideal cylinders and have a smooth surface. Moreover the stator axis is straight. The air gap changes caused by assembly error and bearing wear etc. can be regarded as different air gap eccentricities. When the rotor shaft is straight, eccentricity can be further divided into parallel eccentricity and inclined eccentricity. When the rotor shaft is bending, the eccentricity type is curved eccentricity.

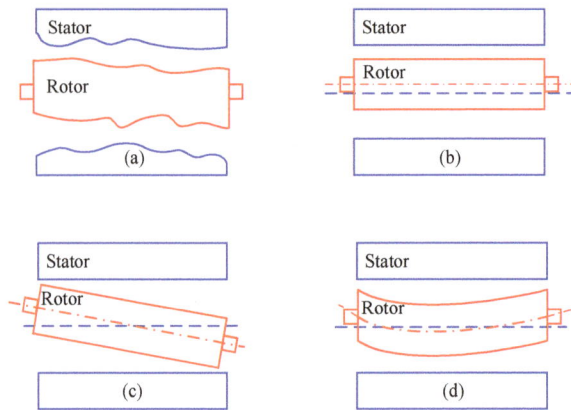

Figure 1. The main mechanical sources of air-gap eccentricity: (**a**) shape deviation, (**b**) parallel eccentricity, (**c**) inclined eccentricity and (**d**) curved eccentricity.

2.1.1. Parallel Eccentricity

As shown in Figure 2, when the rotor shaft and the stator axis are parallel, the air gap eccentricity can be divided into three categories: one is the static eccentricity which refers to that the air gap eccentricity already exists before operation and the rotor rotates with its own geometric center axis [30,31]. The second category is the dynamic eccentricity which occurs when the stator and rotor are concentric at first and the eccentricity occurs during the operation. The rotor rotates with the geometric axis of the stator [32,33]. Static eccentricity and dynamic eccentricity are the most basic eccentricity types. The third category is the static and dynamic mixed eccentricity, that is, static eccentricity and dynamic eccentricity coexist [33–35]. The dynamic eccentricity is mainly caused by the mass unbalance of the rotor. The radial centrifugal force is generated during the rotation of the electrical machine, which results in the uneven air gap between the stator and the rotor. The static eccentricity is easily caused by the installation parallel deviation and bearing wear. Static and dynamic mixed eccentricity can be regarded as the static eccentricity plus the dynamic eccentricity.

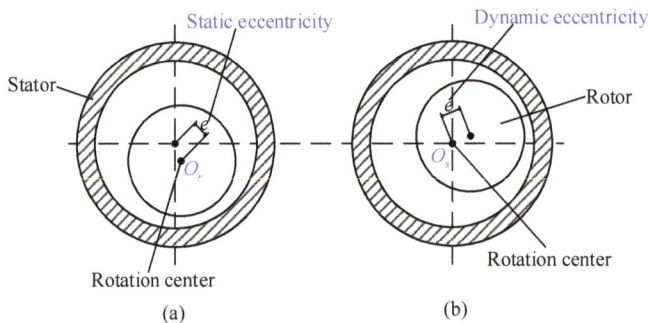

Figure 2. Two common cases of parallel eccentricity: (**a**) static eccentricity and (**b**) dynamic eccentricity.

The air gap length for parallel eccentricity is of great importance to the calculation of the UMF and the vibration analysis of the rotor system. Taking the static and dynamic mixed eccentricity (as Figure 3 indicates) as an example, the air gap length formula is derived as follows [36]:

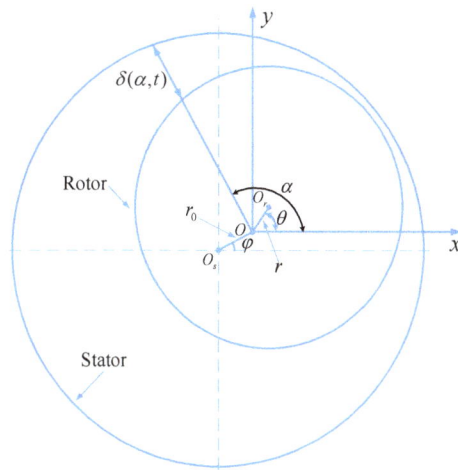

Figure 3. Schematic diagram of static and dynamic mixed eccentricity.

A two-dimensional Cartesian orthogonal coordinate system is established. O_s is the geometric center of the stator. O is initial geometric center of the rotor and O_r is geometric center of the rotor during operation. By geometric relationship derivation, the air-gap length for static and dynamic mixed eccentricity can be approximately expressed as:

$$\delta(\alpha, t) \approx \delta_0 - r_0 \cos(\alpha - \varphi) - r \cos(\alpha - \theta) \tag{1}$$

where δ_0 is the average air-gap length when the rotor is centered. r_0 is the static eccentricity and r is the dynamic eccentricity. α is the air-gap angle with respect to x-axis. φ and θ are the angles of static eccentricity and dynamic eccentricity with reference to x-axis, respectively.

If just the static eccentricity or dynamic eccentricity exists, Equation (1) can be respectively simplified as:

$$\delta(\alpha, t) \approx \delta_0 - r_0 \cos(\alpha - \varphi) \tag{2}$$

$$\delta(\alpha, t) \approx \delta_0 - r \cos(\alpha - \theta) \tag{3}$$

2.1.2. Inclined Eccentricity

In engineering practice, the height difference between the bearings on both sides of the rotor shaft and the inclination of the shaft etc. will cause air gap differences in the rotor at different positions along the axial direction. This eccentricity is named inclined eccentricity. In 1992 Akiyama [37] proposed inclined eccentricity based on the actual engineering needs. The inclined eccentricity can be further subdivided into symmetrical inclined eccentricity and mixed inclined eccentricity. Symmetric inclined eccentricity means that there is only angular deviation between the stator axis and the rotor shaft. For the mixed inclined eccentricity, there exist angular deviation and radial displacement between the stator axis and the rotor shaft. The mixed inclined eccentricity can be considered as a combination of symmetric inclined eccentricity and parallel eccentricity. Figure 4 shows the schematic diagram of symmetrical inclined eccentricity. Due to the fact that the air gap length of inclined eccentricity is related to the inclined angle and axial position, it is necessary to investigate this problem in the three-dimensional coordinate system.

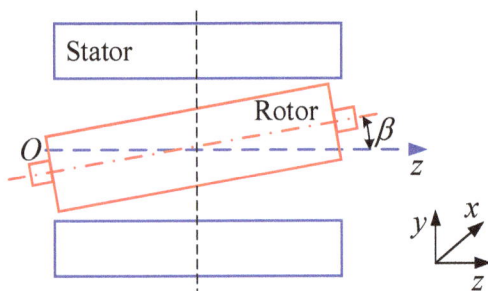

Figure 4. Schematic diagram of symmetrical inclined eccentricity.

The air gap length of mixed inclined eccentricity is further investigated in the following paragraphs. The general case of air gap changes along the axial direction can be considered as mixed inclined eccentricity [38,39]. As Figure 5 demonstrates, it is assumed that the stator is stationary and the midpoint of rotor in the axial direction is selected to be the origin of the coordinates. There exist two orthogonal coordinate systems in the three-dimensional space and they are parallel. One is the stationary coordinate system *O-XYZ* for the stator. The other is the unfixed coordinate system *o-xyz* for the rotor [40].

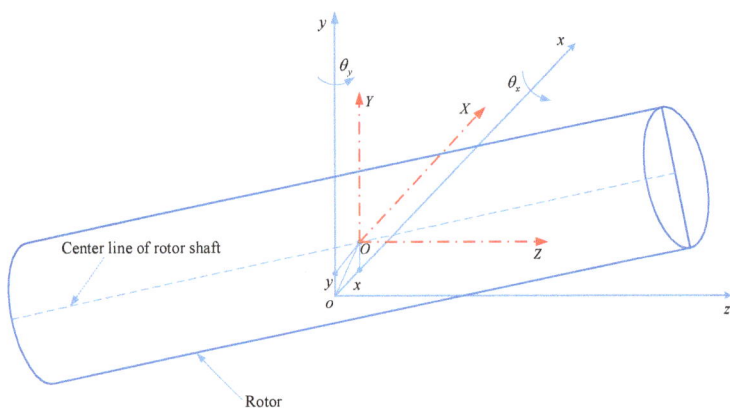

Figure 5. Schematic diagram of the mixed inclined eccentricity.

The parameters that describe the rotor state can be divided into two categories: one is the dynamic parameters (x, y and θ_x, θ_y) describing the dynamic displacement and angle responses respectively, and the other is static parameters (x_0, y_0 and θ_{x0}, θ_{y0}) representing the static displacement and angle eccentricities respectively. The dynamic parameters are determined by the dynamic responses of the rotor system. However, the static parameters need to be provided in the beginning.

The complex three-dimensional problem can be decomposed to be a two-dimensional eccentricity by cutting the axis into numerous cross sections along the axial direction. The comprehensive eccentricity of the investigated cross section is:

$$r_0 = \sqrt{[x_0 + Z\cos(\theta_{x0} + \theta_x)\sin(\theta_{y0} + \theta_y)]^2 + [y_0 - Z\sin(\theta_{x0} + \theta_x)]^2} \tag{4}$$

where the value of the coordinate Z is the intersection point between the investigated cross section and *OZ* axis.

The common range of φ is $[0, 2\pi]$, while the scope of inverse trigonometric functions is $[0, \pi]$. Hence, the following extension is conducted:

$$\varphi = \begin{cases} \arccos \frac{x_0 + Z\cos(\theta_{x0}+\theta_x)\sin(\theta_{y0}+\theta_y)}{r_0} & y_0 - Z\sin(\theta_{x0}+\theta_x) \geq 0 \\ 2\pi - \arccos \frac{x_0 + Z\cos(\theta_{x0}+\theta_x)\sin(\theta_{y0}+\theta_y)}{r_0} & y_0 - Z\sin(\theta_{x0}+\theta_x) < 0 \end{cases} \tag{5}$$

The air-gap length is a function of the air-gap angle, time and axial position [41–43]. The unified air-gap length in an arbitrary cross-section can be approximately expressed by the equation as follows:

$$\delta(\alpha, t, z) = \delta_0 - r_0 \cos(\alpha - \varphi) - r\cos(\alpha - \theta) \tag{6}$$

The effective air-gap length along the axial direction becomes short because the rotor is inclined with respect to the stator. The actual air-gap interaction length between the rotor and the stator is determined by the static angle eccentricities and the dynamic angle responses:

$$z = Z\cos(\theta_{x0}+\theta_x)\cos(\theta_{y0}+\theta_y) \quad Z \in [-L/2, L/2] \tag{7}$$

where L is the axial length of the air-gap.

2.1.3. Curved Eccentricity

The curved eccentricity always occurs to some extent in most large motors where the axis is bending. For example, a three-phase diving induction motor holds dynamic arc eccentricity. Due to the effects of load or the insufficient shaft stiffness, the shaft will also bend and form curved air gap eccentricity [44–46]. The usual way to analyze this complex situation is to treat the electrical rotor as a number of small slices. In each slice, the air gap length can be analyzed according to the pattern of the basic eccentricities (parallel eccentricity or inclined eccentricity).

As shown in Figure 6, in order to obtain the air gap length at different positions, a multi-layer model is designed. The rotor is divided into many layers along the axial direction. Each layer is small enough so that the parallel eccentricity or inclined eccentricity can be applied [47].

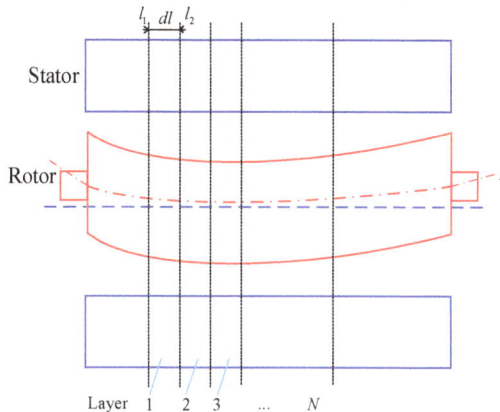

Figure 6. Schematic diagram of the curved eccentricity.

To sum up, the common types of air gap eccentricity and logical relationship between them can be summarized in Figure 7:

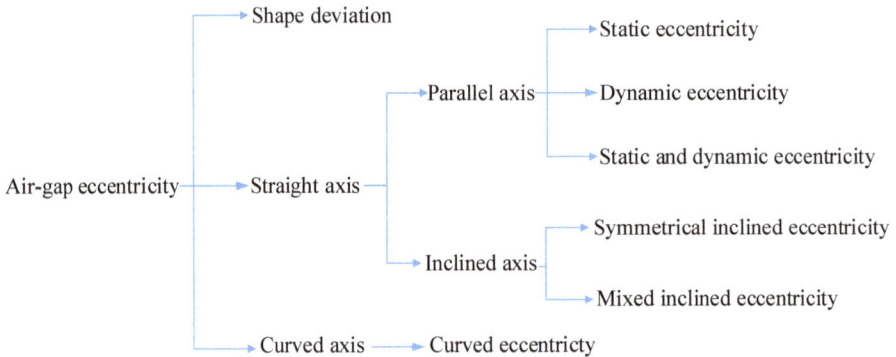

Figure 7. Summary of different air-gap eccentricity types.

2.2. Electromagnetic Sources

The sources in electromagnetic aspects can be summarized in four categories: short circuit, open circuit, magnetization unevenness and winding topology asymmetry. Under normal circumstances, the air gap flux and the electromagnetic force distribution are even and symmetrical. When a short circuit occurs in the rotor or stator slot, the magnetic flux of air gap changes. The UMF results in radial vibration [48,49]. The reasons for short circuits include the unfixed excitation winding end, winding deformation, winding manufacturing defects and foreign matter intake, etc. The current flowing through the short-circuited coil is zero, which causes a decrease in the magnetic potential of the corresponding magnetic pole and an asymmetry of the magnetic field. Thereby the UMF is generated. The common types of short circuit are turn-to-turn, coil-to coil, phase-to-phase and phase-to ground short-circuit [23]. Wan et al. [50] studied the influence of the short circuit on the force acting on the rotor and pointed out that the inter-turn short circuit in generator rotor caused thermal imbalance and magnetic imbalance. Wallin et al. [51] studied the UMF and flux density distribution resulting from winding inter turn-short circuits. Through experimental and numerical simulations, it was found that an additional unbalanced electromagnetism was generated.

The open circuit can also make the magnetic circuit unequal [52,53]. Broken rotor bars and broken end rings are the most common open circuit types in a squirrel-cage rotor [54,55]. If the broken bars are distributed over the poles, the current of the broken bars flows to the adjacent bars. This leads to unbalanced magnetic flux. If the broken bars are adjacent to each other, the current of each broken bar may not flow to its adjacent bar, therefore a more uneven magnetic flux distribution occurs. Therefore both the adjacent broken bars and distributed broken bars can result in UMF [56]. Based on the electromagnetic theory, Jung et al. [57] derived a corrosion model for a rotor-bar-fault induction motors. In addition, Baccarini et al. [58] proposed an analytical model for induction machines considered the broken rotor bars and other factors.

Moreover, when the magnetization of the motor material is not uniform, the electromagnetic force per unit area at the rotor surface is different. Therefore the electromagnetic force is not zero [59]. The factors that cause uneven magnetization can be divided into non-uniform magnetization of permanent magnet materials and magnetization of soft magnetic materials. The uniformity of the magnetization is affected by many factors, such as the aging of the magnet, the mutual repulsion of the magnetic field, the asymmetric magnetization during manufacture and the magnetic edge effect of the magnetic ring etc. The magnetization inhomogeneity will make the magnetic circuit asymmetric and produce the electromagnetic force. In addition, the asymmetry winding of the electrical machine will also cause electromagnetic force [60–63]. Zhu et al. [64] analyzed the electromagnetic force characteristics for permanent magnet motor in the case of completely asymmetrical winding.

3. Calculation Method of the Electromagnetic Force

The calculation of UMF is an important part of any electromagnetic vibration analysis. For the sake of dynamic modeling, the UMF is usually decomposed into a radial force directed to the shortest air gap and a tangential force directed perpendicular to the radial one. Based on the assumption that the magnitude of the electromagnetic force is proportional to eccentricities, Tenhunen et al. [65–67] studied the radial and tangential electromagnetic forces of the eccentric rotor at different rotational speeds. Frosini et al. [68] investigated the effects of eccentricity on the radial and tangential electromagnetic forces at open circuit and load. The authors established an analytical function of the electromagnetic force with respect to the known parameters. Wu et al. [61] studied the radial and tangential electromagnetic forces of surface-mounted permanent magnet motors under load and revealed the mechanism of increasing and decreasing the radial and tangential forces. Since the radial component of the air gap flux is much larger than the tangential component, the tangential component is generally neglected in the calculation [69–74].

As Figure 8 displays, the calculation method of UMF has undergone a complex development process. The early research mainly focused on the theoretical analysis of electromagnetic force and the linear expression was adopted [75–77]. Behrend [75] obtained a linear formula of the electromagnetic force based on the hypothesis that the UMF is proportional to the air gap eccentricity. Calleecharan et al. [76] simplified the electromagnetic force as a linear spring which holds a negative stiffness coefficient when studying industrial hydro-generators. Although linear expressions are simple and convenient, the results are only reliable when the air gap eccentricity is small enough. In 1965, Funke et al. [11] suggested that there was a nonlinear function between the electromagnetic force and the air gap eccentricity. The nonlinear calculation methods of the UMF were widely studied since then.

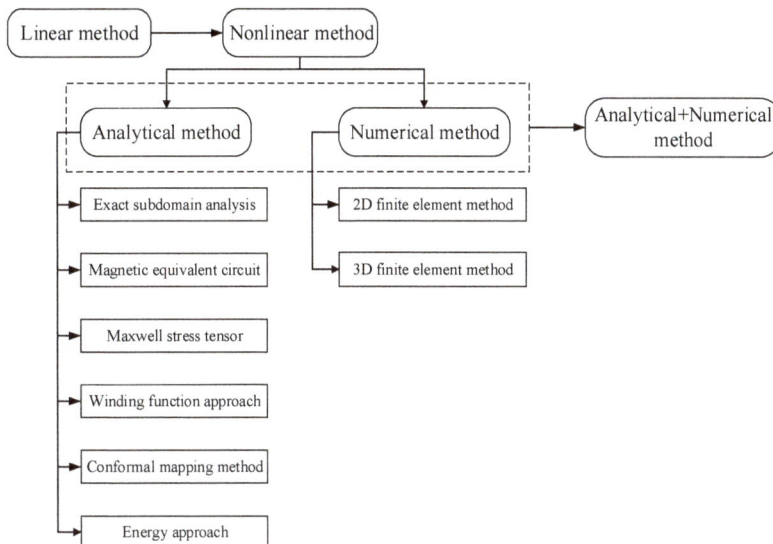

Figure 8. Calculation method for the unbalanced magnetic force.

In recent years, with the increasing reliability requirements for motor product, the nonlinear calculation of electromagnetic force has attracted extensive attention. Many calculation methods are emerging and generally there are three major categories: analytic methods, numerical methods and combinations of analytical and numerical methods.

3.1. Analytical Method

3.1.1. Exact Subdomain Analysis

Exact subdomain analysis (ESA) is a method that divides the whole domain of the electrical machine into several subdomains and each subdomain is solved precisely. As Figure 9 displays, the solution domain in the ESA method is divided into five parts: air gap domain, stator core domain, rotor core domain, outer boundary domain and permanent magnet domain. There are boundary conditions between the regions and the radial and tangential components of the electromagnetic force can be considered. To obtain the solution of the air-gap field distribution, some basic assumptions for ESA are needed. For instance, the permeability of stator/rotor iron is infinite and the saturation as well as end effect are neglected.

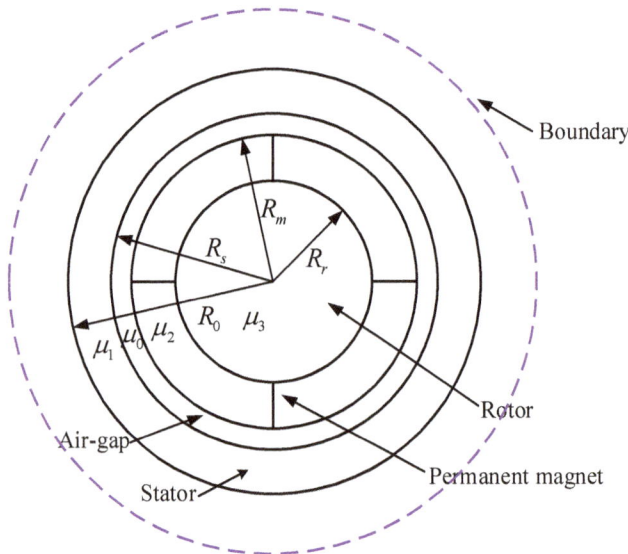

Figure 9. Schematic diagram for the exact subdomain analysis.

The scalar magnetic potential distribution in the air gap, stator iron, rotor iron and the exterior region is governed by the Laplace equation. While the scalar magnetic potential distribution in the magnet domain is governed by the quasi-Poissonian equation. These equations are expressed as follows:

$$-\frac{\partial^2 \varphi_A}{\partial r^2} + \frac{1}{r}\frac{\partial \varphi_A}{\partial r} + \frac{1}{r^2}\frac{\partial^2 \varphi_A}{\partial \theta^2} = 0 \tag{8}$$

$$-\frac{\partial^2 \varphi_M}{\partial r^2} + \frac{1}{r}\frac{\partial \varphi_M}{\partial r} + \frac{1}{r^2}\frac{\partial^2 \varphi_M}{\partial \theta^2} = \frac{1}{\mu_2}\nabla \cdot \mathbf{M} \tag{9}$$

$$-\frac{\partial^2 \varphi_S}{\partial r^2} + \frac{1}{r}\frac{\partial \varphi_S}{\partial r} + \frac{1}{r^2}\frac{\partial^2 \varphi_S}{\partial \theta^2} = 0 \tag{10}$$

$$-\frac{\partial^2 \varphi_R}{\partial r^2} + \frac{1}{r}\frac{\partial \varphi_R}{\partial r} + \frac{1}{r^2}\frac{\partial^2 \varphi_R}{\partial \theta^2} = 0 \tag{11}$$

$$-\frac{\partial^2 \varphi_O}{\partial r^2} + \frac{1}{r}\frac{\partial \varphi_O}{\partial r} + \frac{1}{r^2}\frac{\partial^2 \varphi_O}{\partial \theta^2} = 0 \qquad (12)$$

where r and θ are the radius and angle of the investigated subdomain in the polar coordinates, respectively. φ_A, φ_M, φ_S, φ_R and φ_O represent the magnetic scalar potential in the air gap, magnet, stator, rotor and the exterior region, respectively.

Equations (8)–(12) are usually solved by variable separation technique and many scholars have applied the SEA method to investigate the magnetic field distribution in the electrical machine. The ESA method works for radial or parallel magnetized magnets as well as for the overlapping or nonoverlapping stator windings [78]. In addition, the slotting effects [79,80] and different slot and pole combinations [81] can be considered in the SEA model. In a specific research, the five subdomains are often simplified. For instance, two domains including the air gap domain and permanent magnet domain were investigated [82]. Three domains consist of magnet, air gap and slots were developed [83,84]. Moreover, the ESA method is applicable to many types of electrical machines such as brushless permanent magnet machines and surface mounted permanent-magnet machines etc. Rahideh et al. [14] established a polar coordinate system to analyze the air gap/winding field and permanent magnet field distribution of slotless brushless permanent magnet motor with open circuit. Kumar et al. [85] proposed an improved analytical model by developing the instantaneous air-gap field distribution for a permanent magnet brushless DC motor.

3.1.2. Magnetic Equivalent Circuit

As Figure 10 illustrates, in the magnetic equivalent circuit (MEC) method, the rotor yoke, rotor teeth, rotor leakage, air gap, stator leakage, stator teeth and stator yoke are equivalent to a voltage loop and are superimposed along the magnetic circuit. The nodes of these voltage loop represent scalar magnetic potentials at different positions in the electrical machines. The current in the voltage loop passes through each node and it means magnetic flux goes through the magnetic unit. Figure 10 is an illustration of a permanent-magnet synchronous machine. The physical connection between the left and right ends of the magnetic circuit is modeled through common variables. A linear leakage permeance exists between every two stator teeth due to stator slot opening. The magnetic equivalent circuit method is mainly based on Kirchhoff's law and Gauss's law. The equations are solved by the Gaussian elimination method. The MEC can be regarded as a compromise between the electrical lumped-parameter models and finite element analysis. This approach shows the advantage of a close relationship with the physical field distributions in electrical machines. Moderate complexity and reasonable accuracy are reflected in the calculation. When compared with the finite element method, the disadvantages may be that the eddy current and skin effect cannot be handled perfectly.

The MEC method was formally introduced to electromechanical systems by Laithwaite [86] and Carpenter [87], respectively. In the late 1980s, Ostovic developed a series of publications about MEC modeling of induction and permanent-magnet synchronous machines [88–91]. Since then, the application of MEC method was extensively investigated. The publications are mainly about electrical machines and selected examples are as follow: Sudhoff et al. [92] used the MEC method for induction motor modeling and proposed a methodology of constructing a state-variable model. Serri et al. [93] applied the MEC method to analyze the torque and radial force of a multi-phase bearingless permanent magnet motor, which simplified the design process of the motor. Xie et al. [94] studied the air-gap flux density of a dual-rotor permanent magnet motor by the MEC method and discussed the effects of winding current harmonics, winding types and radial and parallel magnetization types. Based on Kirchhoff's second law of magnetic circuit, Hanic et al. [95] proposed an analysis approach for saturated surface-mounted permanent magnet motor with no-load by using the conformal mapping and MEC method. Fleming et al. [96] conducted real-time simulation of switched reluctance motor by MEC.

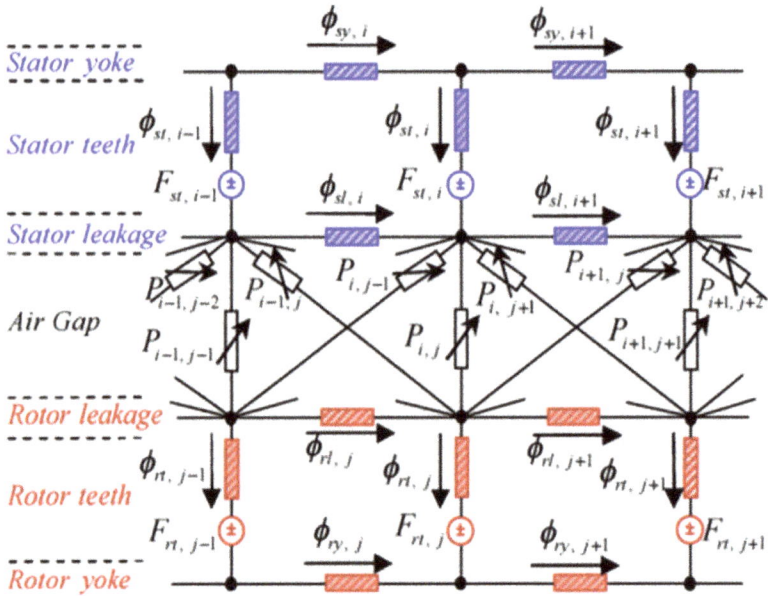

Figure 10. A portion of magnetic equivalent circuit network in electrical machines. Adapted from [97].

3.1.3. Maxwell Stress Tensor

The Maxwell stress tensor (MST) method describes the interaction between current and magnetic field as well as the distribution of magnetic flux density over the contour of the body. The basic laws of macroscopic electromagnetic phenomena can be expressed by the Maxwell equations. The total force on an object can be obtained by integrating the Maxwell stress tensor over a closed surface enclosing the object [98]. Therefore, when the air gap in electrical machine is considered, the integration surface can be the circular plane between the two bodies and has a cylindrical shape with the normal vector pointing outwards. The Maxwell stress tensor $\mathbb{K}$ is utilized to calculate electromagnetic force $\mathbf{F}$ on a moving body:

$$\mathbf{F} = \int_V \nabla \cdot \mathbb{K} dv \tag{13}$$

where volume V contains the investigated object. The Maxwell stress tensor is independent of coordinate system and defined as:

$$\mathbb{K}_{ij} = \frac{B_i B_j}{\mu_0} - \delta_{ij} \frac{|\mathbf{B}|^2}{2\mu_0} \tag{14}$$

where $\mathbf{B}$ is the magnetic flux density. i, j represent the components in the specific coordinate system and δ_{ij} is the Kronecker delta function. By considering the Gauss's theorem, the Maxwell stress tensor force can be rewritten in a more convenient form:

$$F = \oint_S \mathbb{K} \cdot \mathbf{n} ds \tag{15}$$

where S is a surface enclosing the investigated body and $\mathbf{n}$ is a unit vector normal to the surface. For the specific application of MST method in electrical machines, the basic idea is to obtain the Maxwell stress on the rotor surface and the detailed process is presented in the following.

The air-gap permeance can be calculated as:

$$\Lambda(\alpha,t) = \frac{\mu_0}{\delta(\alpha,t)} \tag{16}$$

where μ_0 is the vacuum permeability

The magnetic flux density distribution of the air-gap is [99,100]:

$$B(\alpha,t) = \Lambda(\alpha,t)F(\alpha,t) \tag{17}$$

where $F(\alpha,t)$ is the synthesis fundamental magnetomotive force (MMF).

The normal component and tangential component of Maxwell stress are:

$$\sigma_n(\alpha,t) = \frac{1}{2\mu_0}\left(B_n(\alpha,t)^2 - B_\tau(\alpha,t)^2\right) \tag{18}$$

$$\sigma_\tau(\alpha,t) = \frac{1}{\mu_0}B_n(\alpha,t)B_\tau(\alpha,t) \tag{19}$$

In general, the tangential component of the flux density is much smaller than the normal component and can be considered negligible. Then the Maxwell stress perpendicular to the core/air boundary is given by:

$$B_n(\alpha,t) = B(\alpha,t), \quad B_\tau(\alpha,t) = 0 \tag{20}$$

$$\sigma(\alpha,t) = \frac{B(\alpha,t)^2}{2\mu_0} \tag{21}$$

The two common approaches calculating the Maxwell stress are the Fourier series method [101] and direct integral method [36]. In the Fourier series method, the air-gap permeance can be expanded as a Fourier series:

$$\Lambda_n = \begin{cases} \frac{\mu_0}{\delta_0}\frac{1}{\sqrt{1-\varepsilon^2}} & (n=0) \\ \frac{2\mu_0}{\delta_0}\frac{1}{\sqrt{1-\varepsilon^2}}\left(\frac{1-\sqrt{1-\varepsilon^2}}{\varepsilon}\right)^n & (n>0) \end{cases} \tag{22}$$

where ε is the relative eccentricity.

It can be figured out that the first three harmonic components are dominant for ordinary eccentricities. Therefore, by ignoring the higher permeance harmonics and making some simplifications, the resulting electromagnetic forces in the horizontal and vertical direction are obtained as follows:

$$F_x = \begin{cases} f_1\cos\theta + f_2\cos(2\omega t - \theta) + f_3\cos(2\omega t - 3\theta) & p=1 \\ f_1\cos\theta + f_3\cos(2\omega t - 3\theta) + f_4\cos(2\omega t - 5\theta) & p=2 \\ f_1\cos\theta + f_4\cos(2\omega t - 5\theta) & p=3 \\ f_1\cos\theta & p>3 \end{cases} \tag{23}$$

$$F_y = \begin{cases} f_1\sin\theta + f_2\sin(2\omega t - \theta) - f_3\sin(2\omega t - 3\theta) & p=1 \\ f_1\sin\theta + f_3\sin(2\omega t - 3\theta) - f_4\sin(2\omega t - 5\theta) & p=2 \\ f_1\sin\theta + f_4\sin(2\omega t - 5\theta) & p=3 \\ f_1\sin\theta & p>3 \end{cases} \tag{24}$$

where p is the number of pole-pair and ω is the supply frequency of the electrical machine. It can be observed that the results change with polyphase excitations:

$$f_1 = \frac{RL\pi}{4\mu_0}F_c^2(2\Lambda_0\Lambda_1 + \Lambda_1\Lambda_2 + \Lambda_2\Lambda_3) \tag{25}$$

$$f_2 = \frac{RL\pi}{4\mu_0}F_c^2\left(\Lambda_0\Lambda_1 + \frac{1}{2}\Lambda_1\Lambda_2 + \frac{1}{2}\Lambda_2\Lambda_3\right) \tag{26}$$

$$f_3 = \frac{RL\pi}{4\mu_0} F_c^2 (2\Lambda_0\Lambda_1 + \Lambda_1\Lambda_2 + \Lambda_2\Lambda_3) \tag{27}$$

$$f_4 = \frac{RL\pi}{8\mu_0} F_c^2 \Lambda_2\Lambda_3 \tag{28}$$

R and L are the radius and length of the rotor, respectively.

For the direct integration method, the UMF can be obtained by integration of the Maxwell stress on the rotor surface. The expressions for parallel eccentricity are:

$$F_x = RL \int_0^{2\pi} \sigma(\alpha, t) \cos \alpha d\alpha \tag{29}$$

$$F_y = RL \int_0^{2\pi} \sigma(\alpha, t) \sin \alpha d\alpha \tag{30}$$

If the inclined eccentricity is taken into consideration, not only the electromagnetic force but also the electromagnetic torque need to be calculated:

$$F_x = R \int_{-l/2}^{l/2} \int_0^{2\pi} \sigma(\alpha, t, z) \cos \alpha d\alpha dz \tag{31}$$

$$F_y = R \int_{-l/2}^{l/2} \int_0^{2\pi} \sigma(\alpha, t, z) \sin \alpha d\alpha dz \tag{32}$$

$$M_x = -R \int_{-l/2}^{l/2} z \int_0^{2\pi} \sigma(\alpha, t, z) \sin \alpha d\alpha dz \tag{33}$$

$$M_y = R \int_{-l/2}^{l/2} z \int_0^{2\pi} \sigma(\alpha, t, z) \cos \alpha d\alpha dz \tag{34}$$

where l is the axial length of the air-gap and l satisfies the following equation:

$$l = L \cos(\theta_{x0} + \theta_x) \cos(\theta_{y0} + \theta_y) \tag{35}$$

The last five years have witnessed some progress in the MST method. Meessen et al. [102] selected the MST method to calculate the magnetic force components in the cylindrical coordinate system. By inserting the analytical expressions, the method can be fast and accurate. Spargo et al. [103] developed a semi-numerical method to calculate the harmonic torque components based on the MST theory which provides a simple algebraic expression. Bermúdez et al. [104] extended the MST method to consider the nonlinear magnetic media and local force distribution. The resultant electromagnetic force was verified well.

3.1.4. Winding Function Approach

The winding function approach (WFA) takes all harmonics of the magnetomotive force into consideration, with no restrictions concerning the symmetry of stator windings and rotor bars. However, the classic WFA is not suitable for the modeling of eccentricities since it cannot consider air-gap variations, although it was initially also applicable of these cases. The mutual inductances between stator phase and rotor loops (L_{sr}) are different from those between the rotor loops and stator phase (L_{rs}), and it is difficult to find a physical meaning of this asymmetry. As a result, the modified winding function approach (MWFA) for the inductance calculation considering air-gap eccentricity was proposed [105]. In this method, the air-gap constant is replaced by an air-gap function which depends on the relative position of the rotor with respect to the stator. This method has been applied to analyze static, dynamic and mixed eccentricity in induction machines [106,107]. In addition, this modification can be further extended to consider axial skewing.

The basic idea of WFA can be explained in an induction machine. Taking an induction machine which has m stator circuits and n rotor bars as an example, the cage can be regarded as n identical and equally spaced rotor loops. Voltage equations for the induction machine can be written in vector-matrix form as follows [108,109]:

$$\mathbf{V}_s = \mathbf{R}_s \mathbf{I}_s + \frac{d\lambda_s}{dt} \tag{36}$$

$$\mathbf{V}_r = \mathbf{R}_r \mathbf{I}_r + \frac{d\lambda_r}{dt} \tag{37}$$

where:

$$\mathbf{V}_s = \begin{bmatrix} v_1^s & v_2^s & \cdots & v_m^s \end{bmatrix}^T, \quad \mathbf{V}_r = \begin{bmatrix} 0 & 0 & \cdots & 0 \end{bmatrix}^T, \quad \mathbf{I}_s = \begin{bmatrix} i_1^s & i_2^s & \cdots & i_m^s \end{bmatrix}^T,$$
$$\mathbf{I}_r = \begin{bmatrix} i_1^r & i_2^r & \cdots & i_n^r \end{bmatrix}^T,$$

The stator and rotor flux linkages are given by:

$$\lambda_s = \mathbf{L}_{ss}\mathbf{I}_s + \mathbf{L}_{sr}\mathbf{I}_r \tag{38}$$

$$\lambda_r = \mathbf{L}_{rs}\mathbf{I}_s + \mathbf{L}_{rr}\mathbf{I}_r \tag{39}$$

$\mathbf{L}_{ss}$ is an $m \times m$ matrix with the stator self and mutual inductances, $\mathbf{L}_{rr}$ is an $n \times n$ matrix with the rotor self and mutual inductances, $\mathbf{L}_{sr}$ is an $m \times n$ matrix composed by the mutual inductance between the stator phases and the rotor loops. $\mathbf{L}_{rs}$ is an $n \times m$ matrix and $\mathbf{L}_{rs} = \mathbf{L}_{sr}{}^T$.

The mechanical equations for electrical machines are obtained:

$$\frac{d\omega}{dt} = \frac{1}{J_{rl}}(T_e - T_l) \tag{40}$$

$$\frac{d\theta_r}{dt} = \omega \tag{41}$$

where θ_r is the rotor position, ω is the angular speed and J_{rl} is the rotor-load inertia. T_e is the electromagnetic torque and T_l is the load torque.

The magnetic co-energy which stores in the magnetic circuits and can be written as:

$$W_{co} = \frac{1}{2}\mathbf{I}_s^T \mathbf{L}_{ss}\mathbf{I}_s + \frac{1}{2}\mathbf{I}_s^T \mathbf{L}_{sr}\mathbf{I}_r + \frac{1}{2}\mathbf{I}_r^T \mathbf{L}_{rs}\mathbf{I}_s + \frac{1}{2}\mathbf{I}_r^T \mathbf{L}_{rr}\mathbf{I}_r \tag{42}$$

The electromagnetic torque T_e can be obtained from the magnetic co-energy:

$$T_e = \left[\frac{\partial W_{co}}{\partial \theta_r}\right]_{(I_s, I_r \text{ constant})} \tag{43}$$

According to the WFA, the general expression of the mutual inductance between any winding i and j in any electrical machine is given by:

$$L_{ij}(\theta) = \mu_0 \delta_0 l \int_0^{2\pi} N_i(\varphi, \theta) N_j(\varphi, \theta) g^{-1}(\varphi, \theta) d\varphi \tag{44}$$

where $g^{-1}(\varphi, \theta)$ is the inverse air-gap length. $N_i(\varphi, \theta)$ and $N_j(\varphi, \theta)$ are the winding functions of the windings i and j. φ is the angle along the inner surface of the stator and θ is the angular position of the rotor with respect to the stator. The winding function $N_i(\varphi, \theta)$ is a function of φ and θ for a rotating coil. While it is only a function of φ for a stationary coil. This expression is inappropriate to handle arbitrary distribution windings of synchronous machines and the MWFA should be employed.

Much work has been done on the WFA method. Faiz et al. extended the winding function theory for non-uniform air gap eccentricity in rotating electric machinery [110] and applied the MWFA to calculate the time-dependent inductances of the motor with static, dynamic and mixed eccentricities in a unified manner [111]. Ghoggal et al. proposed a MWFA by taking account the skewing rotor bars effects [112] and teeth saturation due to local air-gap flux concentration [113]. Tu et al. [114] investigated the actual winding arrangement of a synchronous machine based on the WFA and Iribarnegaray et al. [115] gave a critical review of the modified winding function theory.

3.1.5. Conformal Mapping Method

The Laplace equations of the magnetic field distribution in the electrical machine are hard to solve directly. A possible solution is to convert this problem into an orthogonal coordinate system by the conformal mapping method (CMM). The CMM can maintain the solution of Laplace's equation in both the original and transformed domain. If the field distribution (e.g., rectangle, circle) of a geometrical subdomain is known, the field distribution in more complex geometrical subdomains (e.g., slotted air gap) can be calculated by the CMM. The permeance of the electrical machine was often obtained by applying the unit magnetic potential in the CMM. The magnet is ignored and infinitely deep rectilinear stator slots are assumed. The results can be acceptable when the ratio of the air gap length to slot pitch is relatively high. Such a single slot model is appropriate for the electrical machine with a relatively small number of slots. However an electrical machine with plenty of slots are common in ordinary cases. Considering that the effect of slotting on the variation of air gap is similar in an electrical machine, the single-slot approach is extensively investigated [116].

As illustrated in Figure 11, the conformal transformation is applied to transform the geometric shape into a slotless air gap in which the field distribution can be solved. The solution is then mapped back into the complex plane where the actual slot shape exists. Four conformal transformations are required to transform the slotted air gap into a slotless air gap:

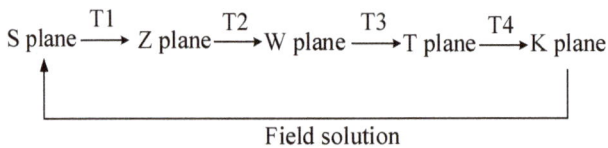

Figure 11. Basic steps for obtaining the field solution in a slotted air gap based on CMM.

The original geometry represents a single slot opening in the S plane. This geometric shape needs to be transformed into its linear model in the Z plane utilizing a logarithmic conformal transformation defined as [117,118]:

$$z = \ln(s) \tag{45}$$

where $s = m + jn = re^{j\theta}$, $z = x + jy$. The relationship between the coordinate in the S and Z planes is:

$$x = \ln(r) \quad y = \theta \tag{46}$$

The next step is to convert the geometric structure in the Z plane into the upper half of the W plane through Schwarz-Christoffel transformation. The approaches of W plane to T plane and T plane to K plane are similar to that from Z plane to W plane. The individual transformations between the planes can be written as follows:

$$\text{T1}: \frac{\partial z}{\partial s} = \frac{1}{s}$$
$$\text{T2}: \frac{\partial w}{\partial z} = -j\frac{\pi}{g'}\frac{(w-1)w}{(w-a)^{1/2}(w-b)^{1/2}}$$
$$\text{T3}: \frac{\partial t}{\partial w} = j\frac{g'}{\pi}\frac{1}{w}$$
$$\text{T4}: \frac{\partial k}{\partial t} = k$$

(47)

where $g' = \ln(R_s/R_r)$. R_s and R_r are the radius of the stator and rotor, respectively. a and b are the coefficients of w at the corner points of the slot.

Finally, the relationship between flux density in the S and K planes is given by:

$$B_s = B_k\left(\frac{\partial k}{\partial t}\frac{\partial t}{\partial w}\frac{\partial w}{\partial z}\frac{\partial z}{\partial s}\right)^*$$

(48)

$$B_s = B_k\left[\frac{k}{s}\frac{(w-1)}{(w-a)^{1/2}(w-b)^{1/2}}\right]^* = B_k\lambda^*$$

(49)

$$B_s = (B_r + jB_\theta)(\lambda_a + j\lambda_b)^*$$

(50)

where λ^* is the conjugate of the air-gap permeance with λ_a and λ_b as its real and imaginary parts. The flux density B_k with its real and imaginary parts B_r and B_θ represents the field solution in the slotless air gap.

There also exist extensive publications about CMM. Lin et al. introduced an analytical method for universal motors based on the actual air-gap field distribution from the field solutions by the CMM [119], and presented a generalized analytical field solution for spoke type permanent magnet machines [120]. Hanic et al. [95] proposed a novel method for magnetic field analysis based on the CMM and MEC. Alam et al. [121,122] took the slotting effect, winding distribution, armature reaction and saturation effect into account. They presented an improved CMM for magnetic field analysis in surface-mounted permanent magnet motors considering eccentricities.

3.1.6. Virtual Work Principle

The force in the virtual work principle (VMP) is calculated from a spatial derivative of stored energy. The VMP obtains the air gap magnetic field through the electromagnetic theory and the corresponding boundary conditions. Moreover the energy function of the air gap magnetic field is obtained. The partial differentials of energy function in horizontal direction and vertical direction are obtained respectively. By using the method of electromechanical analysis, the coupling dynamics equation is established. These strong nonlinear equations are solved by mathematical transformation.

The magnetic field energy of the air gap space is calculated first and the magnetic tube energy of air gap is:

$$d_w = \frac{R}{2}\Lambda(\alpha, t)F(\alpha, t)^2 d\alpha dz$$

(51)

The magnetic field energy of the air gap space is:

$$W = \frac{R}{2}\int_0^{2\pi}\int_0^L \Lambda(\alpha, t)F(\alpha, t)^2 d\alpha dz$$

(52)

The electromagnetic force in the x-direction and y-direction can be obtained by the derivation of energy:

$$F_x^{ump} = \frac{\partial W}{\partial x} = \frac{RL}{2}\int_0^{2\pi}\frac{\partial \Lambda}{\partial x}F(\alpha, t)^2 d\alpha$$

(53)

$$F_y^{ump} = \frac{\partial W}{\partial y} = \frac{RL}{2} \int_0^{2\pi} \frac{\partial \Lambda}{\partial y} F(\alpha, t)^2 d\alpha \tag{54}$$

The VMP has developed for a long time and the existing publications about this method mainly emerged in the late 1990s. Several authors have calculated spatial derivatives of position-dependent air gap reluctances and formulated an analytical expression for force and torque in terms of these derivatives [123,124]. Others have implemented VWP discretely by evaluating the total system energy difference and dividing through the spatial difference [125,126].

3.2. Numerical Method

3.2.1. 2D Finite Element Method

The numerical calculation of electromagnetic field is a problem about solution of partial differential equations. For many practical engineering problems, the application of finite element methods (FEM) based on variational principles is often easier than the direct solution of partial differential equations. The Maxwell stress method and virtual work method are usually used for calculating the UMF in the FEM. The FEM divides the continuous field into finite units and then expresses the solution of each unit by the interpolation function which makes it meet the boundary conditions. Then the solution of the continuous field over the entire field is obtained. Numerous scholars studied the FEM for the magnetic field distribution of electrical machines and the FEM is often applied as the contrast or verification for the analytical methods. The FEM can be generally divided into 2D FEM and 3D FEM from the perspective of investigation scope. Moreover, it can be also classified as static FEM and dynamic FEM which is also named time-stepping FEM.

The 2D FEM was adopted to investigate the electromagnetic characteristics of the switched-flux permanent magnet motor [127] and the squirrel-cage induction motor [128]. In addition, the time steeping FEM was developed to solve the transient magnetic field in induction machines [129,130]. Wang et al. [131] overcame difficulties of hundred stator slots operating under loads and proposed a method for the 2D finite element calculation of UMF in large hydro-generators. Zarko et al. [132] used the FEM to calculate the UMF of a salient-pole synchronous generator in no-load and loaded conditions. Lee et al. [133] adopted the 2D FEM to model magnetic vibration sources in two 100-kW marine fractional-slot interior permanent magnet machines with different pole and slot combinations.

3.2.2. 3D Finite Element Method

As illustrated in Figure 12, 2D FEM simplifies the electrical machine system to be a plane problem, while the 3D FEM investigates the issue in the space. Therefore the results obtaining by 3D FEM are more accurate to some extent due to the fact that less assumptions are made. Lee et al. [134] applied the 3D EFM to study the performance of traditional longitudinal flux and transverse flux permanent magnet motors. Chen et al. [135] developed a finite element solution approach for the analysis of the dynamic behavior and balancing effects of an induction motor system. Sibue et al. [136] studied the current density distribution and losses in a large air-gap transformer composed of toe cores and two windings using homogenization and 3D FEM. Ha et al. [137] investigated the coupled mechanical and magnetic forces by the FEM in the transient state solved in a step by step procedure with respect to time. Faiz et al. [129] proposed a time-stepping FEM that identifies mixed eccentricity and overcomes the difficulty of applying FEM to transient behaviors.

The current density is considered as input of 2D FEM which neglects the harmonics of the stator current. The current and rotor angular position can be appropriately considered in the 3D FEM. Generally speaking, the 3D FEM is more time consuming but accurate compared with the 2D FEM. If the size of electrical machines is large or the requirements of calculation accuracy are harsh, the 3D FEM model is needed.

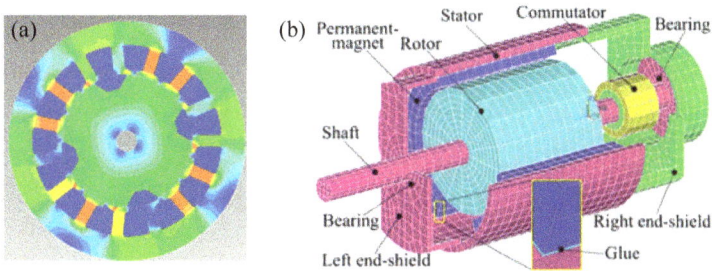

Figure 12. 2D FEM and 3D FEM of the electrical machines: (**a**) 2D FEM and (**b**) 3D FEM. Adapted from [138].

3.3. Comprehensive Method

The analytical method which can obtain the origin and frequency characteristics of the electromagnetic force is mainly used to explore the mechanism and the solution speed is fast. However the calculation is simplified based on many assumptions and the model is relatively simple. The numerical method can consider more parameters and the accuracy of calculation results is more reliable than the analytical method. However it requires massive time and the computational efficiency for large as well as complex situations is low. Based on these characteristics, some scholars develop a numerical plus analytical calculation method, which takes advantage of the merits of simple analytical method and numerical method. Numerical method is used to solve electromagnetic field distribution for obtaining accurate magnetic flux density. Analytic method is adopted to get the electromagnetic force and the computational efficiency is improved. In addition, experiments are applied to verify the calculation results and this approach is being widely promoted. He et al. [138] combined the FEM with the boundary finite element method to study the electromagnetic, mechanical and acoustic characteristics of permanent magnet DC motors. First, the MMF harmonic of the candidate winding was obtained by analytical method and then each set of harmonics was analyzed by a simplified FEM. Li et al. [139] proposed a semi-analytical method to analyze the eddy current loss of an axial flux permanent magnet motor. Tudorache et al. [140] proposed a hybrid model (numerical and analytical combination) to reduce the cogging torque of a permanent magnet synchronous motor. Compared with the FEM, the computational time was very low and the calculation accuracy remained high.

Based on the advantage combinations of the analytical and numerical methods, Chao [141] proposed a hybrid method to analyze the UMF in hard disk drive spindle motors. Hanic et al. [95] adapted the conformal mapping and magnetic equivalent circuits to calculate the back electromotive force and cogging torque of saturated surface-mounted permanent magnet machines. Sprangers et al. [142] presented a semi-analytical method based on the harmonic modeling technique and analyzed the magneto-static field distribution in the slotted structure of rotating electrical machines. Guo et al. [143] combined the analytical and FEM techniques to predict the air-gap magnetic field distribution of a permanent motor embedded salient pole.

3.4. Main Factors Considered in the Calculation

The values of UMF are influenced by many factors. The existing calculation methods are all based on some assumptions, even though the numerical method (which here refers to FEM) can take numerous situations into account. Considering that proposing an exactly perfect model is rather difficult and may be very complex, some key influential factors are needed to paid more attention. This way is efficient in improving the accuracy and will not increase the complexity much at the same time. According to the literature review, the relatively important factors are magnetic saturation, slot and pole combination and load effects.

3.4.1. Magnetic Saturation

In 1918, Rosenberg [144] discovered the effect of magnetic saturation on UMF, which attracted the attention of many researchers. With the increase of excitation current, UMF increases nonlinearly under normal circumstances, but the existence of magnetic saturation limits the infinite increase of electromagnetic force. In addition, the complex geometry structure of rotor and stator tends to cause magnetic field distortion. The magnetic saturation may also be formed in the narrow places wherein slot and pole are located. Generally speaking, the saturation phenomenon is inevitable because of two major factors. One is due to the saturation effects of magnetization characteristic in the ferromagnetic materials. The other is the existence of magnetic flux leakage caused by the distortion in the narrow air-gap space. There exists much evidence supporting the fact of magnetic saturation [145]. The magnetic saturation will have a great influence on the UMF and extensive studies have been conducted. Variable degree of saturation [146] and nonlinear magnetic saturation [147] were investigated, respectively.

Calculation methods considering the influence of magnetic saturation effect are emerging in the last decade. Covo [76] used the slope of the magnetization curve to analyze the effect of saturation on the UMF, which was verified by Tenhunen [67] experimentally. Ohishi et al. [148] made further improvements on this method by applying the magnetization curve of the ferromagnetic material and obtained the polynomial relationship between the air gap flux density and the excitation current in the non-air gap portion of the motor. A polynomial equation describing the magnetization curve of the electrical machine was obtained based on the air gap line. Perers et al. [149] studied the effect of magnetic saturation in a hydro-generator eccentric rotor on UMF, which indicated that magnetic saturation significantly influenced the UMF at high voltage and high load. Dorrell [38] proposed a flexible UMF calculation method considering the situation of axial eccentricity and magnetic saturation, and this approach can be applied in the design process effectively.

3.4.2. Slot and Pole Effects

The N-S poles are common in the electrical rotor. The stator slots with the current form winding current. Many calculation methods ignore the influence of poles and slots on the magnetic field distribution. However, the pole and slot change the density of the magnetic field lines and affect the calculation accuracy of UMF. According to the relationship between the number of rotor poles and the stator slots, the combination of poles and slots can be divided into two major categories: one is the integer combination, that is, the ratio of stator slot number to rotor pole number is a positive integer and the other is a fractional combination. More and more motors begin to adopt fractional combination and the systematic research is necessary to investigate its influence on the UMF. On the other hand, the slot skewing may also have a great influence over the harmonics of UMF [150,151].

Some studies showed that harmonics of air gap magnetic flux density generated by the slotted rotating machinery were an important part of UMF [152]. The slotting effect will cause an additional magnetic field [153]. Furthermore it was found that the vibration modes and frequency components of higher harmonics strongly depend on the combination of the pole-slot number [154,155] and shape of poles [156]. Zhu et al. [60] analyzed the influence of different pole/slot combinations on the UMF and found that UMF increases for higher fractional pole-to-slot ratio. In another article, Zhu et al. [157] further proposed the additive effects ($p = 3k + 1$, p-pole number and k-slot number) and cancelling effects ($p = 3k - 1$) of different pole/slot combinations on UMF. The slotting effects and air gap eccentricity are often investigated together in recent years. Zarko et al. [117] proposed an analytical method for magnetic field distribution in the slotted air gap of a surface-mounted permanent magnet motor. Bao et al. [158] combined the effects of eccentricity and slotting to conduct magnetic field monitoring in a submersible motor.

3.4.3. Load Effects

According to the basic electromagnetic relationship inside the electrical machine, the magnetic field generated by the stator current and the magnetic field generated by the excitation current are superposed to form the composite air-gap magnetic field under the load condition. Many studies calculated the UMF based on the open-circuit current without load [159]. This assumption is applicable and reasonable in ordinary cases because the load has little effect on the calculation of UMF. Wang et al. [2] investigated the effect of different load conditions (no load, half rated and rated load) on the UMF and obtained the conclusion that the load had little effect on the electromagnetic force. Although the load current and power factor have changed with load, the UMF has little change compared with no-load [160]. The no load cases are widely investigated and the discussions are extensive. For example, Gaussens et al. [43] proposed a general and accurate approach to determine the no-load flux of field-excited flux-switching machines. Hu et al. [161] presented an improved analytical subdomain model for predicting the no-load magnetic field and cogging force.

However, the load in electrical machines exists actually and the effects of load cannot always be ignored. When the accuracy demands for calculations are harsh and the load is high, the load factor should be taken into consideration seriously. Moreover, in the case of saturation, the existence of load may weaken the UMF. Therefore it should be investigated respectively. Perers et al. [149] pointed out that the UMF decreased with increasing load at a given terminal voltage because the inter-pole leakage flux became more pronounced with increasing load. Zhu et al. [60] established a general analytical model in a two-dimensional polar coordinate system and studied the UMF during loading. Studies showed that the additional UMF was large when the electrical load was high. Dorrell et al. [162] pointed out that there was only rotor flux component when the motor was open without load. Flux components were contained in both stator and rotor under load conditions.

4. Characteristics of Electromagnetic Vibration and Experimental Study

4.1. The Magnitude and Frequency of the Electromagnetic Force

Electromagnetic force is an electromagnetic attraction that pushes the rotor toward to the stator. As the vibration increases, the air gap length will become smaller and smaller and forms negative feedbacks which further aggravate the vibration. Therefore the UMF is always in the smallest air gap direction. The magnitude of the force is influenced by many factors such as eccentricity, current and winding structure etc. Research have shown that the magnitude of UMF has the nonlinear relationship with these factors and specific situation should be investigated respectively [163]. The UMF is rotating if the electrical machine is on operation. Otherwise the force is static.

The common source of UMF is eccentricity. In addition, the static and dynamic eccentricities tend to coexist. The equation describing the frequency components of interest is [22]:

$$f_{ecc} = \left[\left(nR\frac{1-s}{p} \pm k \right) \pm \left(n_d\frac{1-s}{p} \pm 2n_{sat} \right) \right] f_1 \qquad (55)$$

where R is the number of rotor slots, p is the number of fundamental pole pairs, f_1 is the fundamental supply frequency, s is the slip, n_d is the eccentricity order ($n_d = 0$ in case of static eccentricity and $n_d = 1$ in case of dynamic eccentricity), n is the any positive integer, n_{sat} models magnetic saturation ($n_{sat} = 0, 1, 2, \dots$) and k is the order of the stator time harmonics that are present in power supply driving the motor.

When neglecting saturation and considering only static eccentricity, the frequency components that are characteristic of a failure agree exactly with the rotor slot harmonics. Furthermore, in the case of saturation, new sidebands around dynamic-eccentricity-characteristic components will appear, according to the combinations of the feasible values of n_d and n_{sat}. Classic theory predicts that spatial harmonics of $(p \pm 1)$ pole pairs will result from pure dynamic eccentricity. The frequencies of these harmonics will be given by $(f_1 \pm f_r)$, where f_r is the rotor frequency. Nevertheless, in practical applications, it is likely that these additional fields caused by dynamic eccentricity may induce currents in stator windings because the motor is not completely electrically and magnetically symmetrical. Moreover, the low-frequency components near the fundamental given by:

$$f_l = |f \pm k f_r|, \qquad k = 1,2,3\ldots \tag{56}$$

where f_r is the rotating frequency.

The magnitude of UMF due to rotor eccentricity is relatively large because of the existence of the direct-current (DC) component in the static eccentric cases and the low fundamental frequency content in dynamic eccentric cases. The magnitude of the force decreases with the increasing of the frequency harmonic. Therefore the magnitudes of the DC component and low frequency content are large and dominate [164]. The effects of first several frequency harmonic are often investigated in detail among publications. Arkkio et al. [165] pointed out that the static eccentricity generates an additional force component varying at twice of the supply frequency. Pennacchi et al. [166] investigated the rotating component and twice of it in the UMF for a three-phase generator. Li et al. [167] found that the 3rd harmonic magnetic force plays a major role in the production of the squeaking noise in small permanent magnet DC brush motors.

4.2. Measures to Reduce Electromagnetic Vibration

Electromagnetic vibration will cause stability problems and affect the safe and stable operation of the system. Some measures are needed to reduce the vibration of the electromagnetic excitation source. Currently the equalizing winding, damping winding, parallel circuit, magnetic saturation and slot/pole combination are mainly adopted in the electromagnetic vibration suppression. The parallel circuit and damping winding methods are the effective approaches that have been extensively investigated [27].

The parallel circuit is achieved by reconnection of the stator coils groups. Magnetic field harmonics due to rotor eccentricity generate currents circulating in the parallel paths of the rotor and stator windings. These currents equalize the magnetic field distribution in the air gap and hence reduce the resultant UMF. Burakov et al. [100] compared the two approaches (stator parallel paths and rotor windings) reducing UMF in detail. The study found that parallel stator windings can reduce the UMF more effectively than damping windings. Wallin et al. [168] used parallel circuit method to reduce UMF of a synchronous motor. The research shows that the reduction of UMF strongly depends on the relative unbalanced direction of the stator current isolation line.

In the studies of damping winding, Dorrell et al. [169] found that damped windings can significantly reduce the UMF. Wallin et al. [170] studied the effect of damping winding on UMF in an eccentric salient pole synchronous motor. It was found that continuous or discontinuous damping winding produced different damping winding currents, but the influence on UMF was similar. Dorrell [70] analyzed the effect of damped windings in induction machines on reducing the UMF, and thereby reducing bearing wear on the rotor system.

In addition, many other methods are applied to reduce the UMF. For example, Nguyen et al. [171] designed the dual-stator to reduce the UMF for a wound-field flux switching machine. Bi et al. [172] revealed that lead wires can generate severe UMF in permanent magnet synchronous motor, especially at high speed and propose several ways to reduce lead wire asymmetry. Oliveira et al. [173] proposed an equipotential bonding method to change the magnetic attraction force on the circumference of a rotor, and thereby reducing the UMF.

4.3. Experimental Study

The publications on theoretical modeling and analysis about UMF are extensive, but experimental data for the measurement of UMF is sparse due to the difficulties in building suitable experimental equipment. The measurement of the air gap field is rather difficulty. Dorrell et al. [145] reviewed the experimental data available about UMF in induction motors and also put forward further data for consideration.

The experiment method of UMF can be divided into direct approach and indirect approach. Considering that the UMF in a real machine is difficult to be measured, the indirect approach is usually adopted [174]. For instance, Pennacchi et al. [16] evaluated the UMF effects indirectly by means of the vibrational behavior of the machine. Zarko et al. [132] carried out the measurement and analysis of bearing and shaft vibrations with no-load and loaded conditions. Kim et al. [175] developed an experimental device to measure the axial UMF and verify the simulated results in brushless DC motors. The experimental setup is illustrated in Figure 13 as follows:

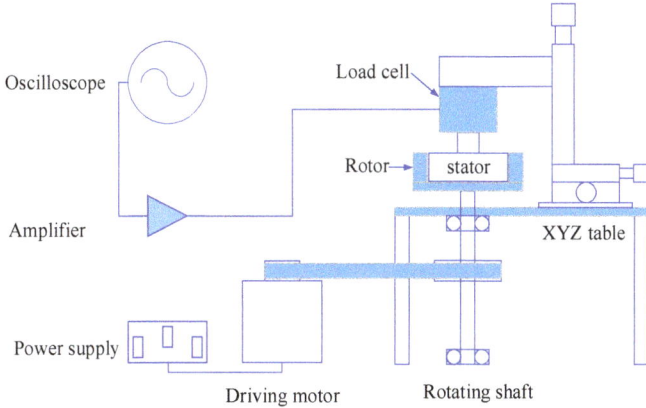

Figure 13. Experimental setup used to measure the axial UMF of brushless DC motors. Adapted from [64].

The direct experiment method was studied by Lee et al. [176]. As displayed in Figure 14, the rotor is separated from a stator and the stator eccentricity is adjustable. The disassembled rotor is clamped with a load cell and there are no bearings. This experimental device measures the variation of UMF by the load cell on the stationary rotor. The experimental setup is demonstrated as follows:

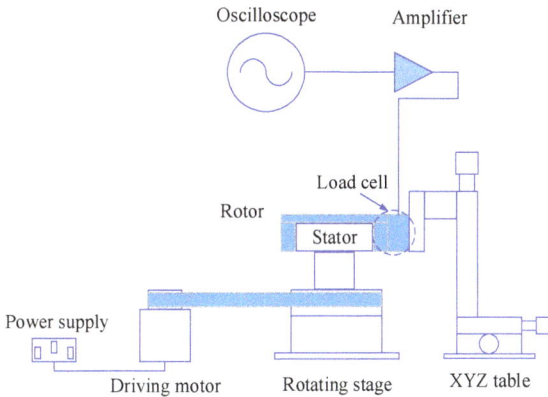

Figure 14. Experimental setup for direct measurement of UMF. Adapted from [176].

The experimental approaches are usually adopted to conduct some verifications for the analytical or numerical calculation methods. Kumar et al. [85] compared the analytical results with the experimental results based on the back electromotive force (EMF) which is obtained from the air-gap field distribution. Kim et al. [128] compared the back EMFs measured in the experiment with simulation results and verified the proposed FEM method. Wang et al. [177] conducted the experiment to validate the theoretical analyses and FE results for the surface mounted permanent-magnet machine.

5. Development Trend and Prospect

5.1. Accurate Calculation of Electromagnetic Excitation

The mechanism of UMF is complicated and further exploration is needed. Moreover the calculation of UMF is still the focus of future research. There is no universally applicable nonlinear analytical expression for computing the UMF so far. Although extensive research about the nonlinear analytical expression of UMF has been conducted, it is based on some assumptions and simplifications. The computational accuracy needs to be improved and the applicability of obtained expressions are limited. The finite element method is more accurate, but tedious steps and time-consuming highlight the limitations in practical applications. Accurate analysis and calculation of UMF under different operating conditions is an important research topic in the design. Therefore, it may be a future research direction to explore the accurate and universal nonlinear model for UMF.

5.2. Control of Electromagnetic Vibration

Electromagnetic vibration threatens the safe operation of the entire system. Measures to reduce or eliminate the UMF for the purpose of reducing vibration and noise have been studied. These measures mainly make the electromagnetic field between the stator and stator be as uniform as possible from the aspects of improving the manufacture and installation precision, carrying out multiple balancing checks and arranging the circuit structure rationally. However, as electrical machines move toward high-speed, heavy loads and subtle trends, efficiency and operability will constrain the use of these measures. If a sudden fault caused by the electromagnetic field changes affects the normal operation of the electrical machines, all possible influencing factors need to be checked. This situation cannot meet the long-time and trouble-free design requirements. The frequency components of UMF can be adopted as the basis for fault diagnosis and the fault characteristics can be extracted for targeted dynamic parameters adjustment. At the same time, the dynamic characteristics of the electrical machine can be applied as the monitoring object for fault diagnosis, and thus forming a feedback system. Real-time monitoring and adjustment of the parameters can ensure safe and stable operation of the electrical machine. Therefore, the adoption of electromagnetic vibration characteristics in the vibration control will be a worthwhile study field in the future.

5.3. Multiphysics Coupled Modeling

At present, the research of electromagnetic vibration mainly investigates the effect of UMF on solid structures, and ignores the further influence of the structural field on the electromagnetic field. The weak interaction between solid field and electromagnetic field was involved. Future studies will gradually focus on the strong interaction between solid and electromagnetic fields. The end of electrical machines will heat up during operation, which causes variations in the magnetic field. In addition, the mechanical strength and stiffness properties of the device will change. All these factors influence the vibration characteristics. Moreover the air flow of ventilation components, air flow interference during high-speed operation and the external sound field excitation may affect the dynamic characteristics of electrical machines. Coupling vibration between rotating machinery and electromagnetic field is worth exploring. Meanwhile, the rotating structure is affected by the temperature field and surrounding acoustic excitation [178]. Due to the flexibility of structure, the multiphysics interaction will produce significant effects on the dynamic properties of structures.

Therefore, the coupling of flow field, structure field, temperature field, electromagnetic field, sound fields and other physical field will be a trend in the future modeling of electrical machines. In addition, the multiobjective and multidisciplinary approach for the system-level design optimization of electrical machine systems are necessary. Moreover, material model, manufacturing condition, electromagnetic, thermal and mechanical models should be investigated for the motor-level design optimization in future work [17–20].

6. Concluding Remarks

Over the past decades, research on the dynamics and vibration of electrical machines has undergone great progress. This paper provides a comprehensive review of the electromagnetic vibration in electrical machines. The basic but important aspects such as sources, calculations, characteristics and experiments of electromagnetic vibration are summarized in detail. The mechanisms for mechanical and electromagnetic factors were revealed, especially the modeling of different air-gap. Different calculations methods which include ESA, MEC, MST, WFA, CMM, VMP, 2D FEM, 3D FEM and comprehensive approaches were demonstrated. Main factors such as magnetic saturation, slot and pole as well as load effects were involved. The magnitude, frequency and reduction measures of the electromagnetic vibration were presented. The common experiment schemes were summarized and the research prospect was provided. This paper can be benefit for the majority of scholars in this field to promote the ongoing progress and development of the investigation.

Author Contributions: Conceptualization, X.X., Q.H. and F.C.; Methodology, X.X.; Software, X.X.; Validation, X.X., Q.H. and F.C.; Formal Analysis, X.X.; Investigation, X.X., Q.H. and F.C.; Resources, F.C.; Data Curation, X.X.; Writing-Original Draft Preparation, X.X., Q.H. and F.C.; Writing-Review & Editing, X.X.; Visualization, X.X.; Supervision, F.C.; Project Administration, F.C.; Funding Acquisition, F.C.

Funding: The authors gratefully acknowledge the financial support from Natural Science Foundation of China (Grant Nos. 51335006 and 11472147).

Conflicts of Interest: The authors declare no conflicts of interest.

Abbreviations

UMF	Unbalanced magnetic force
ESA	Exact subdomain analysis
MEC	Magnetic equivalent circuit
MST	Maxwell stress tensor
MMF	Fundamental magnetomotive force
WFA	Winding Function Approach
MWFA	Modified winding function approach
CMM	Conformal mapping method
VWP	Virtual work principle
FEM	Finite element method
2D	Two degrees of freedom
3D	There degrees of freedom
DC	Direct-current
EMF	Back electromotive force

References

1. Rodriguez, P.V.J.; Belahcen, A.; Arkkio, A.; Laiho, A.; AntoninoDaviu, J. Air-gap force distribution and vibration pattern of induction motors under dynamic eccentricity. *Electr. Eng.* **2008**, *90*, 209–218. [CrossRef]
2. Xu, X.P.; Han, Q.K.; Chu, F.L. Nonlinear vibration of a rotating cantilever beam in a surrounding magnetic field. *Int. J. Non-Linear Mech.* **2017**, *95*, 59–72. [CrossRef]
3. Dorrell, D.G.; Thomson, W.T.; Roach, S. Analysis of air-gap flux, current, vibration signals as a function of the combination of static and dynamic air-gap eccentricity in 3-phase induction motors. *IEEE Trans. Ind. Appl.* **1997**, *33*, 24–34. [CrossRef]

4. Smith, A.C.; Dorrell, D.G. Calculation and measurement of unbalanced magnetic pull in cage induction motors with eccentric rotors. Part 1: Analytical model. *IEEE Proc. Electr. Power Appl.* **1996**, *143*, 202–210. [CrossRef]

5. Nandi, S.; Bharadwaj, R.M.; Toliyat, H.A. Mixed eccentricity in three phase induction machines: Analysis, simulation and experiments. In Proceedings of the 37th IAS Annual Meeting Industry Applications Conference, Pittsburgh, PA, USA, 13–18 October 2002.

6. Lundin, U.; Wolfbrandt, A. Method for modeling time-dependent nonuniform rotor/stator configurations in electrical machines. *IEEE Trans. Magn.* **2009**, *45*, 2976–2980. [CrossRef]

7. Donát, M.; Dušek, D. Eccentrically mounted rotor pack and its influence on the vibration and noise of an asynchronous generator. *J. Sound Vib.* **2015**, *344*, 503–516. [CrossRef]

8. Werner, U. Rotordynamic model for electromagnetic excitation caused by an eccentric and angular rotor core in an induction motor. *Arch. Appl. Mech.* **2013**, *83*, 1215–1238. [CrossRef]

9. Funke, H.; Maciosehek, G. Influence of unbalanced magnetic pull on the running of synchronous machine. *Electric* **1965**, *19*, 233–238.

10. Tu, X.P.; Dessaint, L.A.; Fallati, N.; Kelper, B.D. Modeling and real-time simulation of internal faults in synchronous generators with parallel-connected windings. *IEEE Trans. Ind. Electron.* **2007**, *54*, 1400–1409. [CrossRef]

11. Li, J.T.; Liu, Z.J.; Nay, L.H.A. Effect of radial magnetic forces in permanent magnet motors with rotor eccentricity. *IEEE Trans. Magn.* **2007**, *43*, 2525–2527. [CrossRef]

12. Lundström, N.L.P.; Aidanpää, J.O. Dynamic consequences of electromagnetic pull due to deviations in generator shape. *J. Sound Vib.* **2007**, *301*, 207–525. [CrossRef]

13. Kia, S.H.; Henao, H.; Capolino, G.A. Gear tooth surface damage fault detection using induction machine stator current space vector analysis. *IEEE Trans. Ind. Electron.* **2014**, *62*, 1866–1878. [CrossRef]

14. Rahideh, A.; Korakianitis, T. Analytical open-circuit magnetic field distribution of slotless brushless permanent-magnet machines with rotor eccentricity. *IEEE Trans. Magn.* **2011**, *47*, 4791–4808. [CrossRef]

15. Guo, D.; Chu, F.; Chen, D. The unbalanced magnetic pull and its effects on vibration in a three-phase generator with eccentric rotor. *J. Sound Vib.* **2002**, *254*, 297–312. [CrossRef]

16. Pennacchi, P. Computational model for calculating the dynamical behavior of generators caused by unbalanced magnetic pull and experimental validation. *J. Sound Vib.* **2008**, *312*, 332–353. [CrossRef]

17. Lei, G.; Zhu, J.G.; Guo, Y.G.; Liu, C.C.; Ma, B. A review of design optimization methods for electrical machines. *Energies* **2017**, *10*, 962. [CrossRef]

18. Lei, G.; Guo, Y.G.; Zhu, J.G.; Wang, T.S.; Chen, X.M.; Shao, K.R. System level six sigma robust optimization of a drive system with pm transverse flux machine. *IEEE Trans. Magn.* **2012**, *48*, 923–926. [CrossRef]

19. Lei, G.; Wang, T.S.; Zhu, J.G.; Guo, Y.G.; Wang, S.H. System level design optimization method for electrical drive systems-deterministic approach. *IEEE Trans. Ind. Electron.* **2014**, *61*, 6591–6602. [CrossRef]

20. Lei, G.; Wang, T.S.; Zhu, J.G.; Guo, Y.G.; Wang, S.H. System level design optimization method for electrical drive systems-robust approach. *IEEE Trans. Ind. Electron.* **2015**, *62*, 4702–4713. [CrossRef]

21. Khan, M.A.; Husain, I.; Islam, M.R.; Klass, J.T. Design of experiments to address manufacturing tolerances and process variations influencing cogging torque and back EMF in the mass production of the permanent-magnet synchronous motors. *IEEE Trans. Ind. Appl.* **2014**, *50*, 346–355. [CrossRef]

22. Nandi, S.; Toliyat, H.A.; Li, X.D. Condition monitoring and fault diagnosis of electrical motors—A review. *IEEE Trans. Energy Convers.* **2005**, *20*, 719–729. [CrossRef]

23. Faiz, J.; Ebrahimi, B.M.; Sharifian, M.B.B. Different faults and their diagnosis techniques in three-phase squirrel-cage induction motors—A review. *Electromagnetics* **2006**, *26*, 543–569. [CrossRef]

24. Bellini, A.; Filippetti, F.; Tassoni, C.; Capolino, G.A. Advances in diagnostic techniques for induction machines. *IEEE Trans. Ind. Electron.* **2008**, *55*, 4109–4126. [CrossRef]

25. Faiz, J.; Ojaghi, M. Different indexes for eccentricity faults diagnosis in three-phase squirrel-cage induction motors: A review. *Mechatronics* **2009**, *19*, 2–13. [CrossRef]

26. Singh, A.; Grant, B.; DeFour, R.; Sharma, C.; Bahadoorsingh, S. A review of induction motor fault modeling. *Electr. Power Syst. Res.* **2016**, *133*, 191–197. [CrossRef]

27. Salah, A.; Guo, Y.G.; Dorrell, D. Monitoring and damping unbalanced magnetic pull due to eccentricity fault in induction machines a review. In Proceedings of the 20th International Conference on Electrical Machines and Systems (ICEMS), Sydney, NSW, Australia, 11–14 August 2017.

28. Lundström, N.L.P.; Aidanpää, J.O. Whirling frequencies and amplitudes due to deviations of generator shape. *Int. J. Non-Linear Mech.* **2008**, *43*, 933–940. [CrossRef]

29. Lundström, N.L.P.; Grafström, A.; Aidanpää, J.O. Small shape deviations causes complex dynamics in large electric generators. *Eur. Phys. J. Appl. Phys.* **2014**, *66*, 1–15. [CrossRef]

30. Li, X.D.; Wu, Q.; Nandi, S. Performance analysis of a three phase induction machine with inclined static eccentricity. *IEEE Trans. Ind. Appl.* **2007**, *43*, 531–541. [CrossRef]

31. Li, S.L.; Li, Y.J.; Sarlioglu, B. Rotor unbalanced magnetic force in flux switching permanent magnet machines due to static and dynamic eccentricity. *Electr. Power Compon. Syst.* **2016**, *44*, 336–342. [CrossRef]

32. Babaei, M.; Faiz, J.; Ebrahimi, B.M.; Amini, S.; Nazarzadeh, J. A detailed analytical model of a salient-pole synchronous generator under dynamic eccentricity fault. *IEEE Trans. Magn.* **2011**, *47*, 764–771. [CrossRef]

33. Zhou, Y.; Bao, X.H.; Di, C.; Wang, L. Analysis of dynamic unbalanced magnetic pull in induction motor with dynamic eccentricity during starting period. *IEEE Trans. Magn.* **2016**, *52*. [CrossRef]

34. Faiz, J.; Ebrahimi, B.M. Mixed fault diagnosis in three-phase squirrel-cage induction motor using analysis of air-gap magnetic field. *Prog. Electromagn. Res. Pier* **2006**, *64*, 239–255. [CrossRef]

35. Morinigo-Sotelo, D.; Garcia-Escudero, L.A.; Duque-Perez, O.; Perez-Alonso, M. Practical aspects of mixed-eccentricity detection in PWM voltage-source-inverter-fed induction motors. *IEEE Trans. Ind. Electron.* **2010**, *57*, 252–262. [CrossRef]

36. Xu, X.P.; Han, Q.K.; Chu, F.L. Nonlinear vibration of a generator rotor with unbalanced magnetic pull considering both dynamic and static eccentricities. *Arch. Appl. Mech.* **2016**, *86*, 1521–1536. [CrossRef]

37. Akiyama, Y. Unbalanced-heating phenomenon of induction motor with eccentric rotor. In Proceedings of the Conference Record of the IEEE Industry Applications Society Annual Meeting, Houston, TX, USA, 4–9 October 1992.

38. Dorrell, D.G. Sources and characteristics of unbalanced magnetic pull in 3-phase cage induction motors with axial-varying rotor eccentricity. *IEEE Trans. Ind. Appl.* **2011**, *47*, 12–24. [CrossRef]

39. Li, Y.X.; Zhu, Z.Q. Cogging torque and unbalanced magnetic force prediction in PM machines with axial-varying eccentricity by superposition method. *IEEE Trans. Magn.* **2017**, *99*, 1–4. [CrossRef]

40. Xu, X.P.; Han, Q.K.; Chu, F.L. A four degrees-of-freedom model for a misaligned electrical rotor. *J. Sound Vib.* **2015**, *358*, 356–374. [CrossRef]

41. Nandi, S.; Ahmed, S.; Toliyat, H.A. Detection of rotor slot and other eccentricity-related harmonics in a three-phase induction motor with different rotor cages. *IEEE Trans. Energy Convers.* **2001**, *16*, 253–260. [CrossRef]

42. Yang, B.; Kim, Y.; Son, B. Instability and imbalance response of large induction motor rotor by unbalanced magnetic pull. *J. Vib. Control* **2004**, *10*, 447–460. [CrossRef]

43. Gaussens, B.; Hoang, E.; Barrière, O.; Saint-Michel, J.; Lecrivain, M.; Gabsi, M. Analytical approach for air-gap modeling of field-excited flux-switching machine: No-load operation. *IEEE Trans. Magn.* **2012**, *48*, 2505–2517. [CrossRef]

44. Tenhunen, A.; Benedetti, T.; Holopainen, T.P.; Arkkio, A. Electromagnetic forces in cage induction motors with rotor eccentricity. In Proceedings of the IEEE International Electronic Machines and Drives Conference, Madison, WI, USA, 1–4 June 2003.

45. Di Gerlando, A.; Foglia, G.M.; Perini, R. Analytical modelling of unbalanced magnetic pull in isotropic electrical machines. In Proceedings of the International Conference on Electrical Machines, Vilamoura, Portugal, 6–9 September 2008.

46. Xu, X.P.; Han, Q.K.; Chu, F.L. Electromagnetic vibration characteristics of an eccentric rotor with a static load. *J. Tsinghua Univ. Sci. Technol.* **2016**, *56*, 176–184.

47. Di, C.; Bao, X.H.; Wang, H.F.; Lv, Q.; He, Y.G. Modeling and analysis of unbalanced magnetic pull in cage induction motors with curved dynamic eccentricity. *IEEE Trans. Magn.* **2015**, *51*. [CrossRef]

48. He, Y.L.; Ke, M.Q.; Tang, G.J.; Jiang, H.C.; Yuan, X.H. Analysis and simulation on the effect of rotor interturn short circuit on magnetic flux density of turbo-generator. *J. Electr. Eng.* **2016**, *67*, 323–333. [CrossRef]

49. He, Y.L.; Wang, F.L.; Tang, G.J.; Ke, M.Q. Analysis on steady state electromagnetic characteristics and online monitoring method of stator inter turn short circuit of turbo generator. *Electr. Power Compon. Syst.* **2017**, *45*, 198–210. [CrossRef]

50. Wan, S.T.; Li, H.M.; Li, Y.G. Analysis of generator vibration characteristic on rotor winding inter-turn short circuit fault. *Proc. CSEE* **2005**, *25*, 122–126.

51. Wallin, M.; Lundin, U. Dynamic unbalanced pull from field winding turn short circuits in hydropower generators. *Electr. Power Compon. Syst.* **2013**, *41*, 1672–1685. [CrossRef]

52. Seghiour, A.; Seghier, T.; Zegnini, B. Diagnostic of the simultaneous of dynamic eccentricity and broken rotor bars using the magnetic field spectrum of the air-gap for an induction machine. In Proceedings of the 3rd International Conference on Control, Engineering & Information Technology, Tlemcen, Algeria, 25–27 May 2015.

53. Jannati, M.; Idris, N.R.N.; Salam, Z. A new method for modeling and vector control of unbalanced induction motors. In Proceedings of the IEEE Energy Conversion Congress and Exposition, Raleigh, NC, USA, 15–20 September 2012.

54. Toliyat, H.A.; Lipo, T.A. Transient analysis of cage induction machines under stator, rotor bar and end ring faults. *IEEE Trans. Energy Convers.* **1995**, *10*, 241–247. [CrossRef]

55. Milimonfared, J.; Kelk, H.M.; Nandi, S.; Minassinas, A.D.; Toliyat, H.A. A novel approach for broken-rotor-bar detection in cage induction motors. *IEEE Trans. Ind. Appl.* **1999**, *35*, 1000–1006. [CrossRef]

56. Faiz, J.; Ebrahimi, B.M.; Toliyat, H.A.; Abu-Elhaija, W.S. Mixed-fault diagnosis in induction motors considering varying load and broken bars location. *Energy Convers. Manag.* **2010**, *51*, 1432–1441. [CrossRef]

57. Jung, J.H.; Kwon, B.H. Corrosion model of a rotor-bar-under-fault progress in induction motors. *IEEE Trans. Ind. Electron.* **2006**, *53*, 1829–1841. [CrossRef]

58. Baccarini, L.M.R.; de Menezes, B.R.; Caminhas, W.M. Fault induction dynamic model, suitable for computer simulation: Simulation results and experimental validation. *Mech. Syst. Signal Proc.* **2010**, *24*, 300–311. [CrossRef]

59. Petrinic, M.; Tvoric, S.; Car, S. The effects of pole number and rotor wedge design on unbalanced magnetic pull of the synchronous generator. In Proceedings of the International Conference on Electrical Machines (ICEM), Berlin, Germany, 2–5 September 2014.

60. Zhu, Z.Q.; Ishak, D.; Howe, D.; Chen, J.T. Unbalanced magnetic forces in permanent-magnet brushless machines with diametrically asymmetric phase windings. *IEEE Trans. Ind. Appl.* **2007**, *43*, 1544–1553. [CrossRef]

61. Wu, L.J.; Zhu, Z.Q.; Chen, J.T.; Xia, Z.P. An analytical model of unbalanced magnetic force in fractional-slot surface-mounted permanent magnet machines. *IEEE Trans. Magn.* **2010**, *46*, 2686–2700. [CrossRef]

62. Wu, L.J.; Zhu, Z.Q.; Jamil, M.L.M. Unbalanced magnetic force in permanent magnet machines having asymmetric windings and static/rotating eccentricities. In Proceedings of the International Conference on Electrical Machines and Systems ICEMS, Busan, South Korea, 26–29 October 2013.

63. Krotsch, J.; Piepenbreier, B. Radial forces in external rotor permanent magnet synchronous motors with non-overlapping windings. *IEEE Trans. Ind. Electron.* **2012**, *59*, 2267–2276. [CrossRef]

64. Zhu, Z.Q.; Jamil, M.L.M.; Wu, L.J. Influence of slot and pole number combinations on unbalanced magnetic force in machines with diametrically asymmetric windings. *IEEE Trans. Ind. Appl.* **2013**, *49*, 19–30. [CrossRef]

65. Tenhunen, A.; Holopainen, T.P.; Arkkio, A. Spatial linearity of an unbalanced magnetic pull in induction motors during eccentric rotor motions. In *Compel-The International Journal for Computation and Mathematics in Electrical and Electronic Engineering*; MCB UP Ltd.: Bingley, UK, 2003.

66. Tenhunen, A.; Benedetti, T.; Holopainent, T.P.; Arkkio, A. Electromagnetic forces of the cage rotor in conical whirling motion. *IEEE Proc. Electr. Power Appl.* **2003**, *50*, 563–568. [CrossRef]

67. Tenhunen, A.; Holopainen, T.P.; Arkkio, A. Effects of saturation on the forces in induction motors with whirling cage rotor. *IEEE Trans. Magn.* **2004**, *40*, 766–769. [CrossRef]

68. Frosini, L.; Pennacchi, P. The effect of rotor eccentricity on the radial and tangential electromagnetic stresses in synchronous machines. In Proceedings of the 32nd Annual Conference on IEEE Industrial Electronics, Paris, France, 6–10 November 2006.

69. Yim, K.H.; Jang, J.W.; Jang, G.H.; Kim, M.G.; Kim, K.N. Forced vibration analysis of an IPM motor for electrical vehicles due to magnetic force. *IEEE Trans. Magn.* **2012**, *48*, 2981–2984. [CrossRef]

70. Dorrell, D.G.; Shek, J.K.; Mueller, M.A.; Hsieh, M.F. Damper windings in induction machines for reduction of unbalanced magnetic pull and bearing wear. *IEEE Trans. Ind. Appl.* **2013**, *49*, 2206–2216. [CrossRef]

71. Robinson, R.C. The calculation of unbalanced magnetic pull in synchronous and induction motors. *AIEE Trans.* **1943**, *62*, 620–624.

72. Fruchtenicht, J.; Jordan, H.; Seinsch, H.O. Running instability of cage induction-motors caused by harmonic fields due to eccentricity. 1. Electromagnetic spring constant and electromagnetic damping coefficient. *Arch. Elektrotech.* **1982**, *65*, 271–281.

73. Fruchtenicht, J.; Jordan, H.; Seinsch, H.O. Running instability of cage induction-motors caused by harmonic fields due to eccentricity. 2. Self-excited transverse vibration of the rotor. *Arch. Elektrotech.* **1982**, *65*, 283–292.

74. Belmans, R.; Geysen, W.; Jordan, H. Unbalanced magnetic pull and monopolar flux in three phase induction motors with eccentric rotors. *Int. Conf. Electr. Mach.* **1982**, *3*, 916–921.

75. Behrend, B. On the mechanical forces in dynamos caused by magnetic attraction. *Trans. Am. Inst. Electr. Eng.* **1990**, *17*, 613–633. [CrossRef]

76. Covo, A. Unbalanced magnetic pull in induction motors with eccentric rotors. *Power Appar. Syst. Part III* **1954**, *73*, 1421–1425.

77. Calleecharan, Y.; Aidanpaa, J.O. Dynamics of a hydropower generator subjected to unbalanced magnetic pull. *Proc. Inst. Mech. Eng. Part C* **2011**, *225*, 2076–2088. [CrossRef]

78. Zhu, Z.Q.; Howe, D.; Chan, C.C. Improved analytical model for predicting the magnetic field distribution in brushless permanent-magnet machines. *IEEE Trans. Magn.* **2002**, *38*, 229–238. [CrossRef]

79. Lubin, T.; Mezani, S.; Rezzoug, A. Exact analytical method for magnetic field computation in the air gap of cylindrical electrical machines considering slotting effects. *IEEE Trans. Magn.* **2010**, *46*, 1092–1099. [CrossRef]

80. Fu, J.J.; Zhu, C.S. Subdomain model for predicting magnetic field in slotted surface mounted permanent-magnet machines with rotor eccentricity. *IEEE Trans. Magn.* **2012**, *48*, 1906–1917. [CrossRef]

81. Zhu, Z.Q.; Wu, L.J.; Xia, Z.P. An accurate subdomain model for magnetic field computation in slotted surface-mounted permanent-magnet machines. *IEEE Trans. Magn.* **2010**, *46*, 1100–1115. [CrossRef]

82. Wang, X.H.; Li, Q.F.; Wang, S.H. Analytical calculation of air-gap magnetic field distribution and instantaneous characteristics of brushless DC motors. *IEEE Trans. Energy Convers.* **2003**, *18*, 424–432. [CrossRef]

83. Wu, L.J.; Zhu, Z.Q.; Staton, D.A.; Popescu, M.; Hawkins, D. Subdomain model for predicting armature reaction field of surface-mounted permanent-magnet machines accounting for tooth-tips. *IEEE Trans. Magn.* **2011**, *47*, 812–822. [CrossRef]

84. Wu, L.J.; Zhu, Z.Q.; Staton, D.A.; Popescu, M.; Hawkins, D. Comparison of analytical models of cogging torque in surface-mounted PM machines. *IEEE Trans. Ind. Electron.* **2012**, *59*, 2414–2425. [CrossRef]

85. Kumar, P.; Bauer, P. Improved analytical model of a permanent-magnet brushless DC motor. *IEEE Trans. Magn.* **2008**, *44*, 2299–2309. [CrossRef]

86. Laithwai, R. Magnetic equivalent circuits for electrical machines. *Proc Inst. Electr. Eng Lond.* **1967**, *144*, 1805–1809.

87. Carpenter, C.J. Magnetic equivalent circuits. *Proc Inst. Electr. Eng. Lond.* **1968**, *115*, 1503–1511. [CrossRef]

88. Ostovic, V. A method for evaluation of transient and steady state performance in saturated squirrel cage induction machines. *IEEE Trans. Energy Convers.* **1986**, *1*, 190–197. [CrossRef]

89. Ostoviv, V. Magnetic equivalent-circuit presentation of electric machines. *Electr. Mach. Power Syst.* **1987**, *12*, 407–432. [CrossRef]

90. Ostoviv, V. A simplified approach to magnetic equivalent-circuit modeling of induction machines. *IEEE Trans. Ind. Appl.* **1988**, *24*, 308–316. [CrossRef]

91. Ostoviv, V. A novel method for evaluation of transient states in saturated electric machines. *IEEE Trans. Ind. Appl.* **1989**, *25*, 96–100. [CrossRef]

92. Sudhoff, S.D.; Kuhn, B.T.; Corzine, K.A.; Branecky, B.T. Magnetic equivalent circuit modeling of induction motors. *IEEE Trans. Energy Convers.* **2007**, *22*, 259–270. [CrossRef]

93. Serri, S.; Tani, A.; Serra, G. A method for non-linear analysis and calculation of torque and radial forces in permanent magnet multiphase bearingless motors. In Proceedings of the International Symposium on Power Electronics Power Electronics, Electrical Drives, Automation and Motion, Sorrento, Italy, 20–22 June 2012.

94. Xie, W.; Dajaku, G.; Gerling, D. Analytical method for predicting the air-gap flux density of dual-rotor permanent-magnet (DRPM) machine. In Proceedings of the 2012 XXth International Conference on Electrical Machines, Marseille, France, 2–5 September 2012.

95. Hanic, A.; Zarko, D.; Hanic, Z. A novel method for no-load magnetic field analysis of saturated surface permanent-magnet machines using conformal mapping and magnetic equivalent circuits. *IEEE Trans. Energy Convers.* **2016**, *31*, 747–756. [CrossRef]

96. Fleming, F.E.; Edrington, C.S. Real-time emulation of switched reluctance machines via magnetic equivalent circuits. *IEEE Trans. Ind. Electron.* **2016**, *63*, 3366–3376. [CrossRef]

97. Feki, N.; Clerc, G.; Velex, P.H. Gear and motor fault modeling and detection based on motor current analysis. *Electr. Power Syst. Res.* **2013**, *95*, 28–37. [CrossRef]

98. Pennacchi, P. Nonlinear effects due to electromechanical interaction in generators with smooth poles. *Nonlinear Dyn.* **2009**, *57*, 607–622. [CrossRef]

99. Pillai, K.P.P.; Nair, A.S.; Bindu, G.R. Unbalanced magnetic pull in rain-lighting brushless alternators with static eccentricity. *IEEE Trans. Veh. Technol.* **2008**, *57*, 120–126. [CrossRef]

100. Burakov, A.; Arkkio, A. Comparison of the unbalanced magnetic pull mitigation by the parallel paths in the stator and rotor windings. *IEEE Trans. Magn.* **2007**, *43*, 4083–4088. [CrossRef]

101. Wu, B.; Sun, W.; Li, Z. Circular whirling and stability due to unbalanced magnetic pull and eccentric force. *J. Sound Vib.* **2011**, *330*, 4949–4954. [CrossRef]

102. Meessen, K.J.; Paulides, J.J.H.; Lomonova, E.A. Force calculations in 3-D cylindrical structures using fourier analysis and the maxwell stress tensor. *IEEE Trans. Magn.* **2013**, *49*, 536–545. [CrossRef]

103. Spargo, C.M.; Mecrow, B.C.; Widmer, J.D. A seminumerical finite-element postprocessing torque ripple analysis technique for synchronous electric machines utilizing the air-gap Maxwell stress tensor. *IEEE Trans. Magn.* **2014**, *50*. [CrossRef]

104. Bermudez, A.; Rodriguez, A.L.; Villar, I. Extended formulas to compute resultant and contact electromagnetic force and torque from Maxwell stress tensors. *IEEE Trans. Magn.* **2017**, *53*. [CrossRef]

105. Al-Nuaim, N.A.; Toliyat, H.A. A novel method for modeling dynamic air-gap eccentricity in synchronous machines based on modified winding function theory. *IEEE Trans. Energy Convers.* **1998**, *13*, 156–162. [CrossRef]

106. Nandi, S.; Toliyat, H.; Parlos, A. Performance analysis of a single phase induction motor under eccentric conditions. In Proceedings of the IEEE Industry Applications Conference Thirty-Second IAS Annual Meeting, New Orleans, LA, USA, 5–9 October 1997.

107. Nandi, S.; Bharadwaj, R.; Toliyat, H. Performance analysis of a three phase induction motor under mixed eccentricity condition. *IEEE Trans. Energy Convers.* **2002**, *17*, 392–399. [CrossRef]

108. Joksimovic, G.M.; Durovic, M.D.; Penman, J.; Arthur, N. Dynamic simulation of dynamic eccentricity in induction machines-winding function approach. *IEEE Trans. Energy Convers.* **2000**, *15*, 143–148. [CrossRef]

109. Bossio, G.; De Angelo, C.; Solsona, J.; García, G.; Valla, M.I. A 2-D model of the induction machine: An extension of the modified winding function approach. *IEEE Trans. Energy Convers.* **2004**, *19*, 144–150. [CrossRef]

110. Faiz, J.; Tabatabaei, I. Extension of winding function theory for nonuniform air gap in electric machinery. *IEEE Trans. Magn.* **2002**, *38*, 3654–3657. [CrossRef]

111. Faiz, J.; Ojaghi, M. Unified winding function approach for dynamic simulation of different kinds of eccentricity faults in cage induction machines. *IET Electr. Power Appl.* **2009**, *3*, 461–470. [CrossRef]

112. Ghoggal, A.; Sahraoui, M.; Aboubou, A.; Souzou, S.E.; Razik, H. An improved model of the induction machine dedicated to faults detection-extension of the modified winding function. In Proceedings of the IEEE International Conference on Industrial Technology, Hong Kong, China, 14–17 December 2005.

113. Ghoggal, A.; Zouzou, S.E.; Sahraoui, M.; Derghal, H.; Hadri-Haminda, A. A winding function-based model of air-gap eccentricity in saturated induction motors. In Proceedings of the XXTH International Conference on Electrical Machines (ICEM), Marseille, France, 2–5 September 2012.

114. Tu, X.P.; Dessaint, L.A.; El Kahel, M.; Barry, A.O. A new model of synchronous machine internal faults based on winding distribution. *IEEE Trans. Ind. Electron.* **2006**, *53*, 1818–1828. [CrossRef]

115. Serrano-Iribarnegaray, L.; Cruz-Romero, P.; Gomez-Exposito, A. Critical review of the modified winding function theory. *Prog. Electromagn. Res.* **2013**, *133*, 515–534. [CrossRef]

116. Zhu, Z.Q.; Howe, D. Instantaneous magnetic field distribution in brushless permanent magnet dc motors, Part III: Effect of stator slotting. *IEEE Trans. Magn.* **1993**, *29*, 143–151. [CrossRef]

117. Zarko, D.; Ban, D.; Lipo, T.A. Analytical calculation of magnetic field distribution in the slotted air gap of a surface permanent-magnet motor using complex relative air-gap permeance. *IEEE Trans. Magn.* **2006**, *42*, 1828–1837. [CrossRef]

118. Zarko, D.; Ban, D.; Lipo, T.A. Analytical solution for cogging torque in surface permanent-magnet motors using conformal mapping. *IEEE Trans. Magn.* **2008**, *44*, 52–65. [CrossRef]

119. Lin, D.; Zhou, P.; Stanton, S. An analytical model and parameter computation for universal motors. In Proceedings of the IEEE International Electric Machines & Drives Conference, Niagara Falls, ON, Canada, 15–18 May 2011.

120. Lin, D.; Zhou, P.; Lu, C.; Lin, S. Analytical prediction of cogging torque for spoke type permanent magnet machines. *IEEE Trans. Magn.* **2012**, *48*, 1035–1038. [CrossRef]

121. Alam, F.R.; Abbaszadeh, A. Magnetic field analysis in eccentric surface-mounted permanent-magnet motors using an improved conformal mapping method. *IEEE Trans. Energy Convers.* **2016**, *31*, 333–344. [CrossRef]

122. Rezaee-Alam, F.; Rezaeealam, B.; Faiz, J. Unbalanced magnetic force analysis in eccentric surface permanent-magnet motors using an improved conformal mapping method. *IEEE Trans. Energy Convers.* **2017**, *32*, 146–154. [CrossRef]

123. Sewell, P.; Bradley, K.J.; Clare, J.C.; Wheeler, P.W.; Ferrah, A.; Magill, R. Efficient dynamic models for induction machines. *Int. J. Numer. Model. Electron. Netw. Dev. Fields* **1999**, *12*, 449–464. [CrossRef]

124. Meshgin-Kelk, H.; Milimonfared, J.; Toliyat, H.A. A comprehensive method for the calculation of inductance coefficients of cage induction machines. *IEEE Trans. Energy Convers.* **2003**, *18*, 187–193. [CrossRef]

125. Law, J.D.; Busch, T.J.; Lipo, T.A. Magnetic circuit modelling of the field regulated reluctance machine. Part I: Model development. *IEEE Trans. Energy Convers.* **1996**, *11*, 49–55. [CrossRef]

126. Profumo, F.; Tenconi, A.; Gianolio, G. PM linear synchronous motors normal force calculation. In Proceedings of the IEEE International Electric Machines and Drives Conference (IEMDC 99), Seattle, WA, USA, 9–12 May 1999.

127. Thomas, A.S.; Zhu, Z.Q.; Wu, L.J. Novel modular-rotor switched-flux permanent magnet machines. *IEEE Trans. Ind. Appl.* **2012**, *48*, 2249–2258. [CrossRef]

128. Kim, M.J.; Kim, B.K.; Moon, J.W.; Cho, Y.H.; Hwang, D.H.; Kang, D.S. Analysis of inverter-fed squirrel-cage induction motor during eccentric rotor motion using FEM. *IEEE Trans. Magn.* **2008**, *44*, 1538–1541.

129. Faiz, J.; Ebrahimi, B.M.; Akin, B. Finite-element transient analysis of induction motors under mixed eccentricity fault. *IEEE Trans. Magn.* **2008**, *44*, 66–74. [CrossRef]

130. Vandevelde, L.; Gyselinck, J.J.C.; Melkebeek, J.A.A. Long-range magnetic force and deformation calculation using the 2D finite element method. *IEEE Trans. Magn.* **1998**, *34*, 3540–3543. [CrossRef]

131. Wang, L.; Cheung, R.W.; Ma, Z.Y. Finite-element analysis of unbalanced magnetic pull in a large hydro-generator under practice operations. *IEEE Trans. Magn.* **2008**, *44*, 1558–1561. [CrossRef]

132. Zarko, D.; Ban, D.; Vazdar, I.; Jarica, V. Calculation of unbalanced magnetic pull in a salient-pole synchronous generator using finite-element method and measured shaft orbit. *IEEE Trans. Ind. Electron.* **2012**, *59*, 2536–2549. [CrossRef]

133. Lee, S.K.; Kang, G.H.; Hur, J. Finite element computation of magnetic vibration sources in 100 KW two fractional-slot interior permanent magnet machines for ship. *IEEE Trans. Magn.* **2012**, *48*, 867–870. [CrossRef]

134. Lee, J.Y.; Hong, D.K.; Woo, B.C.; Park, D.H.; Nam, B.U. Performance comparison of longitudinal flux and transverse flux permanent magnet machines for turret applications with large diameter. *IEEE Trans. Magn.* **2012**, *48*, 915–918. [CrossRef]

135. Chen, Y.S.; Cheng, Y.D.; Liao, J.J.; Chiou, C.C. Development of a finite element solution module for the analysis of the dynamic behavior and balancing effects of an induction motor system. *Finite Elem. Anal. Des.* **2008**, *44*, 483–492. [CrossRef]

136. Sibue, J.R.; Ferrieux, J.P.; Meunier, G.; Periot, R. Modeling of losses and current density distribution in conductors of a large air-gap transformer using homogenization and 3-D FEM. *IEEE Trans. Magn.* **2012**, *48*, 763–766. [CrossRef]

137. Ha, K.H.; Hong, J.P. Dynamic rotor eccentricity analysis by coupling electromagnetic and structural time stepping FEM. *IEEE Trans. Magn.* **2001**, *37*, 3452–3455. [CrossRef]

138. He, G.; Huang, Z.; Qin, R.; Chen, D.Y. Numerical prediction of electromagnetic vibration and noise of permanent-magnet direct current commutator motors with rotor eccentricities and glue effects. *IEEE Trans. Magn.* **2012**, *48*, 1924–1931. [CrossRef]

139. Li, J.; Choi, D.W.; Son, D.H.; Cho, Y.H. Effects of MMF harmonics on rotor eddy-current losses for inner-rotor fractional slot axial flux permanent magnet synchronous machines. *IEEE Trans. Magn.* **2012**, *48*, 839–842. [CrossRef]

140. Tudorache, T.; Trifu, I. Permanent-magnet synchronous machine cogging torque reduction using a hybrid model. *IEEE Trans. Magn.* **2012**, *48*, 2627–2632. [CrossRef]

141. Chao Bi Liu, Z.J.; Low, T.S. Analysis of unbalanced magentic pull in hard disk drive spindle motors using a hybrid method. *IEEE Trans. Magn.* **1996**, *32*, 4308–4310. [CrossRef]

142. Sprangers, R.L.J.; Paulides, J.J.H.; Gysen, B.L.J.; Lomonova, E.A. Magnetic Saturation in Semi-Analytical Harmonic Modeling for Electric Machine Analysis. *IEEE Trans. Magn.* **2016**, *52*. [CrossRef]

143. Guo, Y.J.; Lin, H.Y.; Huang, Y.K.; Fang, S.H.; Yang, H.; Wang, K. Air gap magnetic field analysis of wind generator with PM embedded salient poles by analytical and finite element combination technique. *IEEE Trans. Magn.* **2014**, *50*, 777–780. [CrossRef]

144. Rosenberg, E. Magnetic pull in electric machines. *Trans. Am. Inst. Electr. Eng.* **1918**, *37*, 1425–1469. [CrossRef]

145. Dorrell, D.G. Experimental behaviour of unbalanced magnetic pull in 3-phase induction motors with eccentric rotors and the relationship with tooth saturation. *IEEE Trans. Energy Convers.* **1999**, *14*, 304–309. [CrossRef]

146. Ojaghi, M.; Faiz, J. Extension to multiple coupled circuit modeling of induction machines to include variable degrees of saturation effects. *IEEE Trans. Magn.* **2008**, *44*, 4053–4056. [CrossRef]

147. Skin, K.H.; Choi, J.Y.; Cho, H.W. Characteristic analysis of interior permanent-magnet synchronous machine with fractional-slot concentrated winding considering nonlinear magnetic saturation. *IEEE Trans. Appl. Supercond.* **2016**, *26*, 1–4. [CrossRef]

148. Ohishi, H.; Sakabe, S.; Tsumagari, K. Radial magnetic pull in salient pole machines with eccentric rotors. *IEEE Trans. Energy Convers.* **1987**, *2*, 439–443. [CrossRef]

149. Perers, R.; Lundin, U.; Leijon, M. Saturation effects on unbalanced magnetic pull in a hydroelectric generator with an eccentric rotor. *IEEE Trans. Magn.* **2007**, *43*, 3884–3890. [CrossRef]

150. Ghoggal, A.; Zouzou, S.E.; Razik, H.; Saharoui, M.; Khezzar, A. An improved model of induction motors for diagnosis purposes-Slot skewing effect and air–gap eccentricity faults. *Energy Convers. Manag.* **2009**, *50*, 1336–1347. [CrossRef]

151. Dorrell, D.G.; Knight, A.M.; Betz, R.E. Issues with the design of brushless doubly-fed reluctance machines: Unbalanced magnetic pull, skew and iron losses. In Proceedings of the IEEE International Electric Machines & Drives Conference (IEMDC), Niagara Falls, ON, Canada, 15–18 May 2011.

152. Frauman, P.; Burakov, A.; Arkkio, A. Effects of the slot harmonics on the unbalanced magnetic pull in an induction motor with an eccentric rotor. *IEEE Trans. Magn.* **2007**, *43*, 3441–3444. [CrossRef]

153. Kim, U.; Lieu, D.K. Magnetic field calculation in permanent magnet motors with rotor eccentricity with slotting effect considered. *IEEE Trans. Magn.* **1998**, *34*, 2253–2266. [CrossRef]

154. Zhu, Z.Q. Influence of slot and pole number combination on radial force and vibration modes in fractional slot PM brushless machines having single-and double-layer windings. In Proceedings of the IEEE Energy Conversion Congress and Exposition, San Jose, CA, USA, 20–24 September 2009.

155. Chen, J.T.; Zhu, Z.Q. Comparison of all- and alternate-poles-wound flux-switching PM machines having different stator and rotor pole numbers. *IEEE Trans. Ind. Appl.* **2010**, *46*, 1406–1415. [CrossRef]

156. Dajaku, G.; Gerling, D. Air-Gap flux density characteristics of salient pole synchronous permanent-magnet machines. *IEEE Trans. Magn.* **2012**, *48*, 2196–2204. [CrossRef]

157. Zhu, Z.Q.; Mohd Jamil, M.L.; Wu, L.J. Influence of slot and pole number combinations on unbalanced magnetic force in permanent magnet machines. *IEEE Energy Convers. Congr. Expo.* **2011**, *49*, 3291–3298.

158. Bao, X.H.; Wang, H.F.; Di, C.; Cheng, Z.H. Magnetic field monitoring in submersible motor under eccentricity fault considering slotting effect. *Int. J. Appl. Electromagn. Mech.* **2016**, *50*, 233–245. [CrossRef]

159. Wang, Q.W.; Chen, L.; Chai, F.; Gan, L. No-load magnetic field distribution in axial flux permanent magnet machine with static eccentricity. In Proceedings of the 20th International Conference on Electrical Machines and Systems (ICEMS), Sydney, NSW, Australia, 11–14 August 2017.

160. An, X.; Zhou, J.; Xiang, X. Dynamic response of a rub-impact rotor system under axial thrust. *Arch. Appl. Mech.* **2009**, *79*, 1009–1018. [CrossRef]

161. Hu, H.Z.; Zhao, J.; Liu, X.D.; Guo, Y.G.; Zhu, J.G. No-load magnetic field and cogging force calculation in linear permanent-magnet synchronous machines with semi closed slots. *IEEE Trans. Ind. Electron.* **2017**, *64*, 5564–5575. [CrossRef]

162. Dorrell, D.G.; Popescu, M.; Ionel, D.M. Unbalanced magnetic pull due to asymmetry and low-level static rotor eccentricity in fractional-slot brushless permanent-magnet motors with surface magnet and consequent-pole rotors. *IEEE Trans. Magn.* **2010**, *46*, 2675–2685. [CrossRef]

163. Gustavsson, R.K.; Aidanpaa, J.O. The influence of nonlinear magnetic pull on hydropower generator rotors. *J. Sound Vib.* **2006**, *297*, 551–562. [CrossRef]

164. Kim, U.; Lieu, D.K. Effects of magnetically induced vibration force in brushless permanent-magnet motors. *IEEE Trans. Magn.* **2005**, *41*, 2164–2172. [CrossRef]

165. Arkkio, A.; Antila, M.; Pokki, K.; Simon, A.; Lantto, E. Electromagnetic force on a whirling cage rotor. *IEEE Proc. Electr. Power Appl.* **2000**, *147*, 353–360. [CrossRef]

166. Pennacchi, P.; Frosini, L. Dynamical behaviour of a three-phase generator due to unbalanced magnetic pull. *IEEE Proc. Electr. Power Appl.* **2005**, *152*, 1389–1400. [CrossRef]

167. Li, Y.B.; Ho, S.L.; Fu, W.N.; Xue, B.F. Analysis and solution on squeak noise of small permanent-magnet dc brush motors in variable speed applications. *IEEE Trans. Magn.* **2009**, *45*, 4752–4755. [CrossRef]

168. Wallin, M.; Ranlof, M.; Lundin, U. Reduction of unbalanced magnetic pull in synchronous machines due to parallel circuits. *IEEE Trans. Magn.* **2011**, *47*, 4827–4833. [CrossRef]

169. Dorrell, D.G. Unbalanced magnetic pull in cage induction machines for fixed-speed renewable energy generators. *IEEE Trans. Magn.* **2011**, *47*, 4096–4099. [CrossRef]

170. Wallin, M.; Bladh, J.; Lundin, U. Damper winding influence on unbalanced magnetic pull in salient pole generators with rotor eccentricity. *IEEE Trans. Magn.* **2013**, *49*, 5158–5165. [CrossRef]

171. Nguyen, H.Q.; Jiang, J.Y.; Yang, S.M. Design of a wound-field flux switching machine with dual-stator to reduce unbalanced shaft magnetic force. *J. Chin. Inst. Eng.* **2017**, *40*, 441–448. [CrossRef]

172. Bi, C.; Phyu, H.N.; Jiang, Q. Unbalanced magnetic pull induced by leading wires of permanent magnet synchronous motor. In Proceedings of the 12th International Conference on Electrical Machines and Systems, Tokyo, Japan, 15–18 November 2009.

173. Oliveira, W. Reduction of unbalanced magnetic pull (UMP) due to equipotential connections among parallel circuits of the stator winding. In Proceedings of the IEEE International Electric Machines and Drives Conference, Miami, FL, USA, 3–6 May 2009.

174. Lee, C.I.; Jang, G.H. Experimental measurement and simulated verification of the unbalanced magnetic force in brushless DC motors. *IEEE Trans. Magn.* **2008**, *44*, 4377–4380. [CrossRef]

175. Kim, J.Y.; Sung, S.J.; Jang, G.H. Characterization and experimental verification of the axial unbalanced magnetic force in brushless DC motors. *IEEE Trans. Magn.* **2012**, *48*, 3001–3004. [CrossRef]

176. Lee, C.I.; Jang, G.H. Experimental Measurement and Simulated Verification of the Unbalanced Magnetic Force in Brushless DC Motors. *IEEE Trans. Magn.* **2008**, *44*, 4377–4380. [CrossRef]

177. Wang, K.; Zhu, Z.Q.; Ombach, G.; Koch, M.; Zhang, S.; Xu, J. Design and experimental verification of an 18-slot10-pole fractional-slot surface mounted permanent-magnet machine. In Proceedings of the IEEE International Electric Machines and Drives Conference (IEMDC), Chicago, IL, USA, 12–15 May 2013.

178. Lv, Q.; Bao, X.H.; He, Y.G. Influence of thermal expansion on eccentricity and critical speed in dry submersible induction motors. *J. Electr. Eng. Technol.* **2014**, *9*, 106–113. [CrossRef]

energies

MDPI

Article

Experimental Study of Flow-Induced Whistling in Pipe Systems Including a Corrugated Section

Hee-Chang LIM * and Faran RAZI

School of Mechanical Engineering, Pusan National University, Busandaehak-ro 63beon-gil, Geumjeong-gu, Busan 46241, Korea; ralph366@nate.com
* Correspondence: hclim@pusan.ac.kr; Tel.: +82-515-102-302; Fax: +82-515-125-236

Received: 6 June 2018; Accepted: 26 July 2018; Published: 27 July 2018

Abstract: When air flows through pipe systems that include a corrugated segment, a whistling tone is generated and increases in intensity with increasing flow velocity. This whistling sound is related to the particular geometry of corrugated pipes, which is in the form of alternating cavities. This whistling is an environmental noise problem as well as a possible structural danger because of the resulting induced vibration. This paper studies the whistling behavior of various pipe systems with a combination of smooth and corrugated pipes through a series of experiments. The considered pipe systems consist of two smooth pipes attached at the upstream and downstream ends of a corrugated segment. Experiments with smooth and corrugated pipes, which had inner diameters of 15.25 and 16.5 mm, respectively, and various lengths, were performed for flow velocities of up to approximately 30 m/s. The minimum and maximum Strouhal numbers (St) obtained during our experiments were 0.25 and 0.38, respectively. For all pipe configurations investigated in this study, the lowest Mach number at which whistling was observed was 0.017, and the maximum was 0.093. The lowest frequency at which whistling was detected in our experiments was 650 Hz, and the highest was 3080 Hz. The results presented in the form of different variables and dimensionless parameters, including the frequency, Mach number, Strouhal number, and Helmholtz number. The average mode gap and number of excited acoustic modes were also taken into account for all considered configurations. The pipe systems with longer corrugated segments had broader whistling ranges than did configurations with shorter segments, indicating that the number of cavities inside the corrugated pipe has a direct effect on whistling. Increasing the smooth pipe length (either upstream or downstream) resulted in a decrease in the average mode gap between successive modes. The number of excited acoustic modes was primarily related to the corrugated segment length, but the smooth pipe length also had a pronounced effect on the excited modes for a constant corrugation length. The highest number of excited modes (13) was seen in the case of corrugated length 450 mm and smooth pipe length (either upstream or downstream) 400 mm while the lowest number of excited modes (1) was observed for corrugated length 250 mm and smooth pipe length (downstream) 300 mm and 400 mm.

Keywords: corrugated pipe; whistling noise; Helmholtz number; excited modes

---

1. Introduction

A pipe with a periodic variation in its diameter is called a corrugated pipe. Due to the particular geometry of corrugated pipes, they possess the unique characteristic of being locally rigid while at the same time globally flexible [1]. The corrugations are basically alternating cavities and flat regions that are symmetric about the axis of the pipe [2]. Corrugated pipes have the tendency to generate strong whistling noise as a result of fluid or gas flow through the pipe. The acoustic field produced, in addition to noise pollution, can pose a threat to the structural stability in the systems they exist, as a result of strong vibrations. Corrugated pipes are extensively used industrially in heat exchangers, offshore natural gas production systems, and vacuum cleaners [3–5].

Sound generation in corrugated pipes is a result of an oscillation produced through a flow-acoustic interaction [6,7]. The generation of a shear layer is the result of flow separation taking place at the upstream edge of each cavity. This shear layer serves as a source of unsteadiness in the flow. Due to this unsteadiness, an unsteady force is being exerted on the walls. In response to this hydrodynamic force, the walls of the pipe exert a force that is basically considered as a source of sound, as shown in Figure 1 [8]. The flexibility of these pipes is not a prerequisite condition for generation of sound in them. However, the unsteady forces on the walls of the pipe thereby induce some mechanical vibrations that may have a significant effect [9]. The coupling of the shear layers with longitudinal acoustic waves in the pipe is the most commonly observed phenomenon [9–12]. The vortex shedding is controlled by the resulting high-amplitude oscillations [13,14]. These types of flow pulsations are referred to as self-sustained oscillations and result in high-amplitude sound generation, which is also called whistling. This whistling can consist of hydrodynamic and acoustic subsystems [9,15]. The instability of the shear layer, which is acting as an amplifier, is basically the hydrodynamic subsystem, and provides the system with acoustic energy. Acting as the acoustic subsystem are the longitudinal standing waves that act as a band-pass filter, which, in turn, is responsible for maintaining synchronization in this feedback mechanism. Because of this band-pass filter, there is a stepwise increment in the whistling frequency corresponding to certain flow velocities, as observed during various experimental studies [3,4,11,12,16–18].

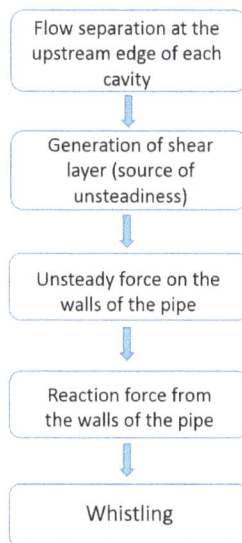

Figure 1. Whistling mechanism in corrugated pipes [8].

In comparatively shorter whistling pipes displaying very strong acoustical reflections at their ends, the vortex shedding taking place at the upstream edge of the cavities is activated as a result of the oscillating velocity, which is associated with resonant longitudinal acoustical standing waves inside the pipe. The values of the acoustic passive resonance frequency of the pipe and the oscillating frequency will often be in close proximity to each other resulting in an acoustic pipe mode [19]. The combination of the vortex shedding occurring locally at the cavities and the longitudinal acoustic waves that travel along the pipe results in whistling inside corrugated pipes [2,15,17,20–27]. When the acoustic oscillations and the source of sound are synchronized with each other, this synchronization can be described in terms of a ratio of a convection time due to the vorticity perturbations across the cavity to the oscillation time period of the acoustic field. This ratio is most commonly regarded as the Strouhal number [19] and is given as

$$St = \frac{fW}{U_{corr}} \tag{1}$$

where f is the oscillation frequency, W is the cavity width, and U_{corr} is the steady flow velocity in the corrugated pipe.

Tonon et al. [15] presented an experimental study comparing the whistling behavior of a pipe system with multiple side branches and a system consisted of corrugated pipes. They suggested that the multiple side branch system is an acceptable model for corrugated pipes. A captivating aspect of their study was that the system was found to whistle at the second hydrodynamic mode of the cavities rather than at the first. They proposed a prediction model for the whistling behavior that consisted of an energy balance, formulated on the basis of vortex sound theory.

Nakiboglu et al. [25] performed an investigative work regarding the whistling behaviors of two geometrically periodic systems, i.e., corrugated pipes and a multiple side branch system. In both systems, they observed a non-monotonic behavior in the whistling amplitude as a function of flow velocity, with local maxima corresponding to lock-in frequencies. In their effort to quantify the Strouhal number, they considered a variety of characteristic lengths. In their study, the shape of the upstream edge of the cavity also exhibited a significant effect on pressure fluctuation amplitudes for both corrugated pipes and the multiple side branch system [25]. They reported that the round upstream edges of the cavities increased the amplitude of the pressure fluctuation by up to five times in comparison to the sharp edges. Moreover, they found that the radius of the downstream edge did not have any considerable effect on the sound production. Using the same experimental setup, Nakiboglu et al. [26] studied the effects of the variation of different parameters on the whistling of corrugated pipes. They performed this study on corrugated pipe segments with different lengths and cavity geometries, and demonstrated that the peak-whistling Strouhal number, which is based on a characteristic length of the sum of the cavity width, and the upstream edge radius, was independent of the pipe length. They also indicated that the peak-whistling Strouhal number decreased with increasing confinement ratio, which was defined as the ratio of the pipe diameter to the sum of the cavity width and the upstream edge radius $D_p/(W + r_{up})$.

Nakiboglu et al. [27] conducted a study related to the aeroacoustics of a swinging corrugated tube. The main idea behind the work was that, when a short corrugated pipe segment is swung around one's head, it tends to produce a musically intriguing whistling sound; this system was named the "Hummer". Their experiments indicated that the Hummer could remain silent even when there was turbulence in the flow. Thus, they concluded that the absence of whistling was not in relation to a lack of turbulence. They anticipated that the reason for the absence of the fundamental mode in short corrugated pipes was the inability of the acoustic sources at the inlet and the outlet of the pipe to cooperate with each other, as a result of difference in their mean velocity profile.

Rudenko et al. [19] proposed a linear model for plane-wave propagation along a corrugated pipe. They considered an experiment in which a corrugated segment was placed between two smooth pipe segments, creating a system called a composite pipe. Their experiments assessed a quasi-steady model for convective acoustic losses which was dependent upon the pipe inlet geometry. Their model predicted some whistling modes that were not observed in the experiments. They reported that in various cases, they encountered a large overlap between the whistling ranges of successive modes, implying the domination of one mode and the suppression of the neighboring ones.

This paper discusses the results obtained through an experimental study performed using a combination of smooth and corrugated pipes. The pipe system under consideration consists of two smooth pipe segments attached to either end of a corrugated pipe segment. We performed experiments on different combinations of such pipe systems using various smooth and corrugated pipe lengths. The objective of this study is to analyze the effect of the variation in the lengths of corrugated and smooth pipe segments, while maintaining the same geometric specifications. For each pipe configuration, there exists a critical Mach number, M_{cr}, at which the system starts whistling. In our experiments, the Reynolds number, Re, was defined within the range of $6500 \leq \text{Re} \leq 32,000$, while the inner diameter of the smooth pipe (15.25 mm) was considered to be the characteristic length, and the kinematic viscosity of the working fluid (air) was $\nu = 1.5 \times 10^{-5} \text{ m}^2/\text{s}$.

In Section 2, we briefly describe vortex sound theory and explain the dimensionless parameters used. In Section 3, we describe the experimental set-up and procedure. We define the geometric specifications of the smooth and corrugated pipes used in our study and the considered pipe configurations. In Sections 4–6, we discuss the experimental results, and Section 7 provides the conclusion of this work.

2. Theoretical Background

2.1. Vortex Sound Theory

Vorticity as a source of sound can be demonstrated by considering the analogy provided by Howe [28–30], according to which an acoustical flow is basically the unsteady irrotational component of the total flow. Howe [29] used the decomposition of Helmholtz to divide the flow into rotational and irrotational parts for a given flow field $\mathbf{u}$:

$$\mathbf{u} = \nabla \phi + \nabla \times \boldsymbol{\psi} \tag{2}$$

where ϕ is a scalar potential and $\boldsymbol{\psi}$ is a vector stream function. Because the acoustical field should be a compressible and unsteady flow, the acoustical flow velocity $\mathbf{u}'$ is defined as

$$\mathbf{u}' = \nabla \phi' \tag{3}$$

where $\phi' = \phi - \phi_0$ is the deviation of ϕ from the steady component ϕ_0 of the potential. Since the flows with low Mach numbers and high Reynolds numbers are being dealt with here, heat transfer and friction can be neglected. Thus, for a homentropic flow, an explicit relation between vorticity and sound production is obtained using Crocco's formulation for the momentum equation [15]:

$$\frac{\overline{\partial \mathbf{u}}}{\partial t} + \nabla B = -\omega \times \mathbf{u} \tag{4}$$

where $B = \frac{1}{2}|\mathbf{u}^2| + \int \frac{dp}{\rho}$ is the total enthalpy and $\omega = \nabla \times \mathbf{u}$ is the vorticity. At low Mach numbers, the convective effects on sound wave propagation can be neglected, which results in the following relation [15],

$$\frac{1}{c_0^2}\frac{\partial^2 B}{\partial t^2} - \nabla^2 B = \nabla \cdot (\omega \times \mathbf{u}) \tag{5}$$

In Equation (5), the right-hand side corresponds to the assumption that the Coriolis force density $\mathbf{f}_{coriolis} = -\rho_0(\omega \times \mathbf{u})$, where ρ_0 is the fluid density, acts as the source of sound.

According to Howe [29], the time-averaged acoustic source power $< P_{source} >$ can be estimated as

$$< P_{source} > = -\rho_0 < \int_V (\omega \times \mathbf{u}) \cdot \mathbf{u}' dV >, \tag{6}$$

where V is the volume in which ω is not vanishing and the brackets $< \cdots >$ indicate time averaging.

2.2. Dimensional Parameters

The centre-line velocity U_{cl} relates the average velocity inside a smooth pipe U_{sp} by the empirical relation [31]

$$U_{sp} \simeq U_{cl}/(1 + 1.33\sqrt{F}), \tag{7}$$

where the friction factor F for a smooth pipe is given by the formula of Blasius [31]:

$$F \simeq 0.316 Re^{-0.25}. \tag{8}$$

For the Reynolds number considered in our experiments, we approximate the smooth pipe velocity as $U_{sp} \simeq U_{cl}/1.22$. Because of the slight difference between the smooth and corrugated pipe diameters, the average velocity inside the corrugated pipe is

$$U_{corr} = \frac{U_{sp}D_{sp}}{D_{corr}}, \tag{9}$$

where U_{corr} is the average velocity in corrugated segment, D_{sp} is the smooth pipe diameter, and D_{corr} is the corrugated pipe diameter. Thus,

$$U_{corr} \simeq U_{cl}/\alpha, \quad \alpha = 1.32. \tag{10}$$

Therefore, the steady cross-sectional average velocity inside the corrugated pipe is approximately 1.32 times lower than the centre-line velocity at the end of the downstream smooth pipe.

The Mach number (M) was calculated as

$$M = \frac{U_{cl}}{c_0}, \tag{11}$$

where c_0 is the speed of sound in air at room temperature, which is equal to 340 m/s. Moreover, the Helmholtz Number was calculated as

$$He = \frac{Lf}{c_0} = \frac{LSt_{cr}}{\alpha W}M, \tag{12}$$

where $L = L_{up} + L_{corr} + L_{dw}$ is the sum of the lengths of all three pipes considered in each configuration, as can be seen in Figure 2; f is the whistling frequency in Hz; St_{cr} is the critical Strouhal number at which the whistling begins; and W is the cavity width.

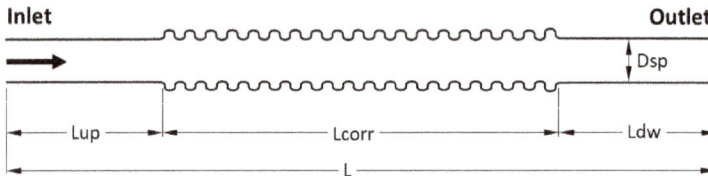

Figure 2. Complete pipe system consisting of smooth pipe segments attached to a corrugated segment (not to scale).

3. Design of Model Experiment (Set-Up and Procedure)

3.1. Test Model and Corrugated Pipe System Configurations

The pipe system consists of three pipe segments, as shown in Figure 2. The first segment is the upstream smooth pipe segment L_{up} and is followed by the corrugated tube L_{corr} and the downstream smooth segment L_{dw}. The smooth pipe segments are made of aluminium. The inner diameter of the smooth pipes is 15.25 mm, and their thickness is 4.75 mm. The corrugated pipe segment is made of plastic.

Figure 3 shows the corrugated pipe geometry (drawn not to scale). The inner diameter D_{corr} is 16.5 mm, the depth d of the cavities is estimated to be 1.3 mm, and the width W is 2.3 mm. Three different lengths of corrugated pipes, 250, 350, and 450 mm, were used. The smooth pipes used in the experiment had lengths of 100, 200, 300, and 400 mm. Table 1 shows the details of all pipe system configurations considered in this study.

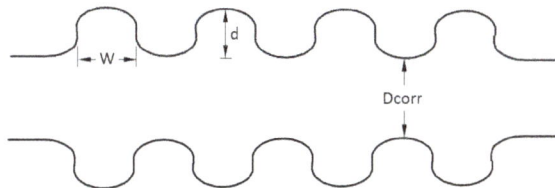

Figure 3. Corrugated pipe geometry (not to scale).

Table 1. Various combinations of smooth and corrugated pipes categorized into two cases. Case a refers to the three different L_{corr} values (250, 350, and 450 mm) for one L_{up} value (100 mm) and four L_{dw} values (100, 200, 300 and 400 mm). Case b corresponds to the same three L_{corr} values for one L_{dw} value (100 mm) and four L_{up} values (100, 200, 300 and 400 mm).

Case a	L_{up} (mm)	L_{corr} (mm)	L_{dw} (mm)	Case b	L_{up} (mm)	L_{corr} (mm)	L_{dw} (mm)
I		250		I		250	
II	100	350	100–400	II	100–400	350	100
III		450		III		450	

Experiments were performed with various combinations of smooth and corrugated pipe segments in an on-coming inlet flow, which is considered to be a uniform flow because the inlet end pipe is directly connected to the wind tunnel contraction area. Each corrugated pipe was tested with all four lengths of smooth pipes first at the downstream and then at the upstream end. The pipe configurations were divided into two cases, which are described as follows.

- Case a: $L_{dw} \geq L_{up}$. The length of the downstream smooth pipe was equal to or greater than the length of the upstream smooth pipe.
- Case b: $L_{dw} \leq L_{up}$. The length of the downstream smooth pipe was equal to or less than the length of the upstream smooth pipe.

3.2. Test Equipment and Measurement Procedure

Figure 4 shows a three-dimensional (3D) sketch of our experimental set-up. The set-up consists of a wind tunnel as the primary source of high-velocity air with a settling chamber installed. A 5-cm-thick layer of acoustic absorbing material was attached to the side walls of the settling chamber to avoid acoustic resonance in the settling chamber. The length of the wind tunnel is 270 cm, and the discharge flange has an outer cross section of 50 cm × 50 cm and an inner cross section of 40 cm × 40 cm. The air is generated by a Turbo Fan with a volumetric air flow of 250 m³/min, a total pressure of 800 mmAq, and a rated power of 7.5 kW, thus ensuring a uniform inflow condition at the upstream end of the pipe system. The air from the wind tunnel is passed through the pipe system by connecting a converging section at the end of the tunnel. The dimensions of the section are such that the upstream end had a cross section of 40 cm × 40 cm, whereas the downstream end cross section was 16 cm × 16 cm. A straight miniature wire probe (55P11) was fixed inside the pipe on the pipe axis (between the centre and the pipe wall) a few millimeters upstream from the downstream open pipe end by means of a probe support. Miniature wire probes have platinum-plated tungsten wire sensors with a diameter of 5 μm and a length of 1.25 mm. The probe body is a 1.9 mm diameter ceramic tube, equipped with gold-plated connector pins that connect to the probe supports by plug-and-socket arrangements. The output of the probe support was attached to an IFA 300 constant temperature anemometer system. It provides a frequency response of up to 300 kHz, depending on the sensor used. The output of the anemometer system was sent to a low-pass filter (dual channel programmable filter 3624, NF Corporation) for signal conditioning. In the HW measurement, the sampling rate for data acquisition was 20 kHz, while the range of low-pass filter was 10 kHz taking the Nyquist criterion into account. After being filtered, the signal was sent to the computer via a *National Instruments* data acquisition board. A pitot tube was also coaxially attached to the hotwire probe support by a clamping mechanism at the downstream pipe outlet to measure the free stream air velocity. The velocity was measured using an FC012-Micromanometer. National Instruments LabVIEW software was used for data processing and analysis. The calibrated hot wire provided a measurement of the time-dependent velocity at the axis of the pipe. Using the spectrum analysis in LabVIEW, we obtained the frequency peaks at which the whistling sound occurred.

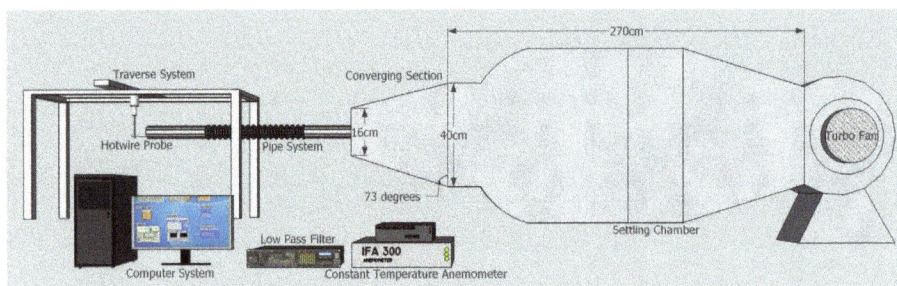

Figure 4. 3D sketch of experimental set-up and pipe system.

4. Effect of Downstream Smooth Pipe Length on the Whistling Behavior for Constant Corrugated Segment Length (Case a)

4.1. Shortest Corrugated Segment: Case a-I ($L_{corr}/D_{corr} = 15.2$)

With increasing corrugated segment length, the pipe systems for all smooth pipe lengths, both upstream and downstream, tend to whistle at higher frequencies and Mach numbers. For the corrugated pipe with a length of 250 mm (Case a-I), when the length of the downstream smooth pipe was increased from 100 to 400 mm at intervals of 100 mm, the frequency and Mach number range within which whistling occurred decreased, as shown in Figure 5 and Table 2. The maximum whistling range was observed for equal upstream and downstream pipe lengths of 100 mm. This pipe system whistled at Mach numbers ranging from $M = 0.027$ to 0.069 and with frequencies from 1080 to 2500 Hz, which respectively correspond with the minimum and maximum Mach numbers in the range. Maintaining a fixed upstream pipe length of 100 mm, when $L_{dw} = 200$ mm, a drastic decrement in the whistling range was observed. The pipe system whistled from $M = 0.028$ to 0.039, corresponding to frequencies of 880 and 1170 Hz, respectively. Pipe systems with a downstream pipe of 300 mm in length resulted in further reduction in whistling ranges; whistling occurred from $M = 0.031$ to 0.036 at a constant frequency of 1010 Hz. Finally, for the downstream smooth pipe with a length of 400 mm, the pipe system whistled very briefly at a frequency of 1100 Hz from $M = 0.033$ to 0.035.

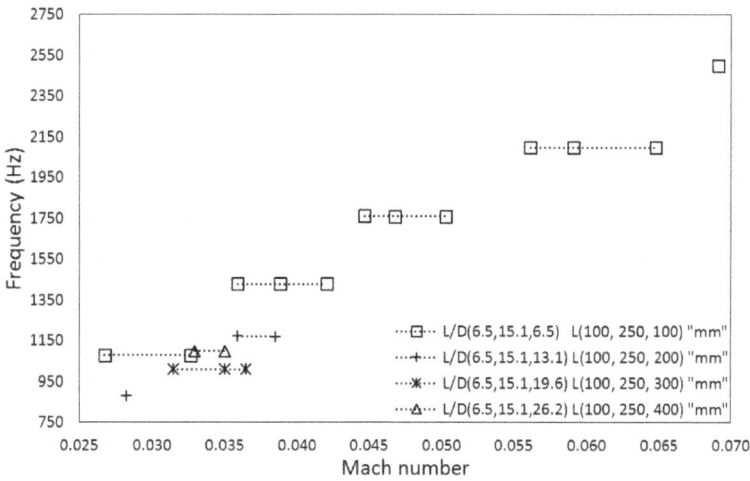

Figure 5. Frequency (Hz) plotted against Mach number M for L_{up} = 100 mm, L_{corr} = 250 mm, and L_{dw} = 100–400 mm.

Table 2. Mach number and corresponding frequency range (Hz) for pipe systems in Case a-I.

L_{up} (mm)	L_{corr} (mm)	L_{dw} (mm)	Mach Number Range	Corresponding Frequency Range (Hz)
		100	0.027–0.069	1080–2500
100	250	200	0.028–0.039	880–1170
		300	0.031–0.036	1010
		400	0.033–0.035	1100

Unlike frequency, the Strouhal number does not increase linearly with the Mach number but instead shows fluctuating behavior. For L_{dw} = 100 mm, the initial and final Strouhal numbers at M = 0.027 and 0.067 were 0.36 and 0.32, respectively. The Strouhal and Mach number ranges were the broadest of all pipe systems considered in Case a-I, as can be seen in Figure 6 and Table 3, with 0.36 as the highest Strouhal number in the Mach number range. For L_{dw} = 200 mm, the range was significantly reduced; the Strouhal numbers are estimated to be 0.28 and 0.27 at M = 0.028 and 0.039, respectively, with a maximum Strouhal number of 0.29 in this range. For the downstream smooth pipe with a length of 300 mm, the starting Strouhal number was 0.29 at M = 0.031, and the ending Strouhal number was 0.25 at M = 0.036, with a maximum Strouhal number of 0.29 in this case. For L_{dw} = 400 mm, the longest downstream smooth pipe considered in this study, the Mach numbers of M = 0.033 and 0.035 yielded Strouhal numbers of 0.30 and 0.28, respectively, with 0.30 as the highest Strouhal number in the range.

Similar to frequency, the Helmholtz number He is linearly related to the Mach number. For the smallest downstream pipe, with a length of 100 mm, the minimum He was 1.31 at M = 0.027, whereas the maximum He was 3.38 at M = 0.067, as shown in Figure 7 and Table 4. For L_{dw} = 200 mm, the minimum and maximum He were 1.33 and 1.81 at M = 0.028 and 0.039, respectively. The minimum and maximum He for L_{dw} = 300 mm were 1.75 and 2.03 at M = 0.031 and 0.033, respectively. For the longest downstream smooth pipe of 400 mm, the minimum He, which occurred at M = 0.033, was 2.28, whereas the maximum He occurred at M = 0.035 and was 2.42. This trend suggests that the range of He with respect to M continuously shrinks as the length of the downstream smooth pipe increases.

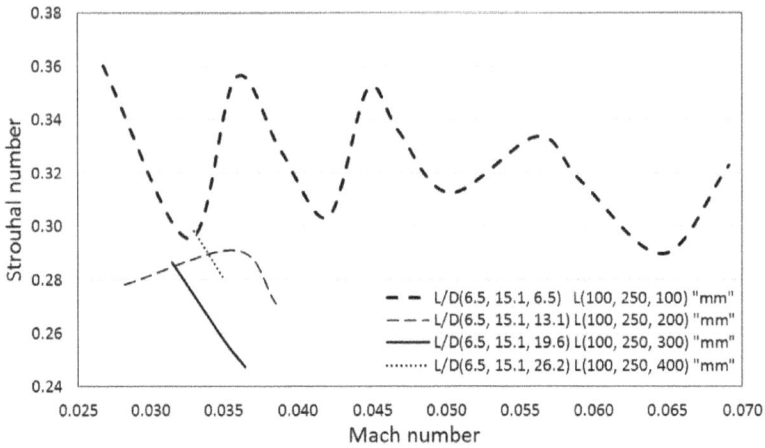

Figure 6. Strouhal number *St* plotted against Mach number *M* for $L_{up} = 100$ mm, $L_{corr} = 250$ mm, and $L_{dw} = 100$–400 mm.

Table 3. Mach number range, corresponding Strouhal number (*St*) range and maximum Strouhal number for each pipe configuration in Case a-I.

L_{up} (mm)	L_{corr} (mm)	L_{dw} (mm)	Mach Number Range	Corresponding *St* Range	Max. *St*
100	250	100	0.027–0.069	0.36–0.32	0.36
		200	0.028–0.039	0.28–0.27	0.29
		300	0.031–0.036	0.29–0.25	0.29
		400	0.033–0.035	0.30–0.28	0.30

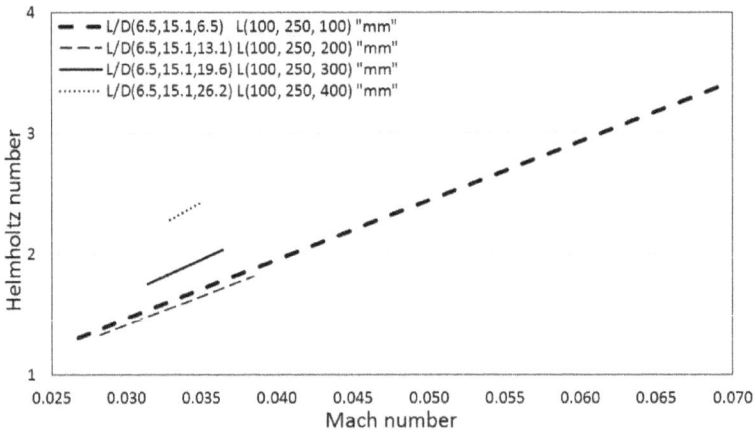

Figure 7. Helmholtz number *He* plotted against Mach number *M* for $L_{up} = 100$ mm, $L_{corr} = 250$ mm, and $L_{dw} = 100$–400 mm.

Table 4. Mach number and corresponding Helmholtz number (*He*) range for pipe systems in Case a-I.

L_{up} (mm)	L_{corr} (mm)	L_{dw} (mm)	Mach Number Range	Corresponding *He* Range
		100	0.027–0.069	1.31–3.38
100	250	200	0.028–0.039	1.33–1.81
		300	0.031–0.036	1.75–2.03
		400	0.033–0.035	2.28–2.42

4.2. Intermediate Corrugated Segment: Case a-II ($L_{corr}/D_{corr} = 21.2$)

The next pipe system is Case a-II, which has a corrugated segment of length $L_{corr} = 350$ mm, as shown in Figure 8. With $L_{up} = L_{dw} = 100$ mm, the minimum and maximum Mach numbers were estimated to be 0.023 and 0.078, respectively, corresponding to a whistling frequency range of 870–2600 Hz. For $L_{dw} = 200$ mm, the minimum and maximum Mach numbers were 0.025 and 0.069 with corresponding whistling frequencies of 750 and 2200 Hz, respectively. For $L_{dw} = 300$ mm, the onset of whistling occurred at $M = 0.025$ with a frequency of 850 Hz, and the pipe system whistled up to $M = 0.051$ with a frequency of 1500 Hz. Finally, for the downstream pipe with a length of 400 mm, whistling occurred from $M = 0.025$ to $M = 0.048$ with frequencies between 750 and 1550 Hz.

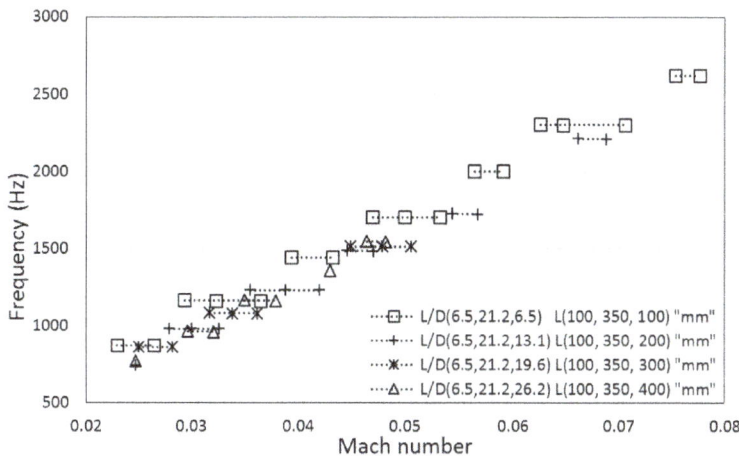

Figure 8. Frequency (Hz) plotted against Mach number M for $L_{up} = 100$ mm, $L_{corr} = 350$ mm, and $L_{dw} = 100$–400 mm.

For $L_{dw} = 100$ mm, the Strouhal number at a Mach number of 0.023, where whistling began, was 0.34, and at the maximum Mach number of 0.078, it was 0.30; the maximum Strouhal number within the whistling range was 0.35, as shown in Figure 9 and Table 5.

For the downstream pipe length of 200 mm, the range of Mach numbers was 0.025–0.069, corresponding to Strouhal numbers of 0.27 and 0.29, respectively, with a maximum Strouhal number of 0.31. For $L_{dw} = 300$ mm, the Strouhal number started at 0.31 and ended at 0.27 for $M = 0.025$ and 0.051, respectively, with a maximum Strouhal number of 0.31. Finally, for the downstream smooth pipe of length 400 mm, the starting and ending Strouhal numbers were 0.28 and 0.29 at Mach numbers of $M = 0.025$ and 0.048, respectively, with a maximum Strouhal number of 0.30.

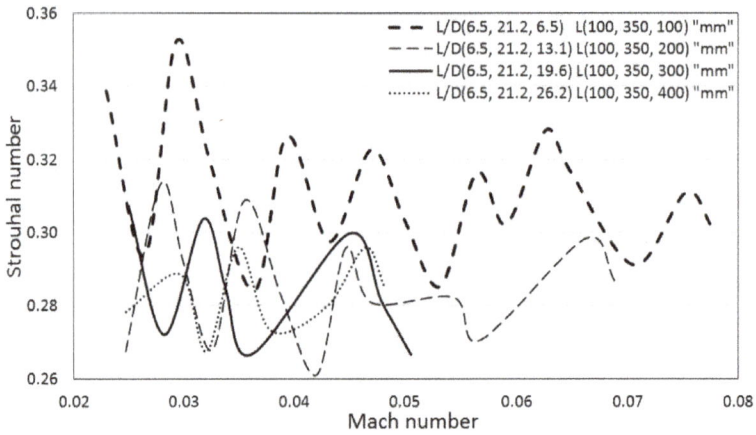

Figure 9. Strouhal number *St* plotted against Mach number *M* for $L_{up} = 100$ mm, $L_{corr} = 350$ mm, and $L_{dw} = 100$–400 mm.

Table 5. Mach number range, corresponding Strouhal number (*St*) range and maximum Strouhal number for each pipe configuration in Case a-II.

L_{up} (mm)	L_{corr} (mm)	L_{dw} (mm)	Mach Number Range	Corresponding *St* Range	Max. *St*
100	350	100	0.023–0.078	0.34–0.30	0.35
		200	0.025–0.069	0.27–0.29	0.31
		300	0.025–0.051	0.31–0.27	0.31
		400	0.025–0.048	0.28–0.29	0.30

As shown in Figure 10 and Table 6, for the smallest downstream pipe of length 100 mm, the minimum Helmholtz number *He* is 1.29 and occurs at $M = 0.023$, whereas the maximum *He* is 4.36 at $M = 0.078$. For $L_{dw} = 200$ mm, the minimum and maximum *He* are 1.32 and 3.68 and occur at $M = 0.025$ and 0.069, respectively. The minimum and maximum *He* for $L_{dw} = 300$ mm are 1.73 and 3.50 at $M = 0.025$ and 0.051, respectively. For the longest downstream smooth pipe of length 400 mm, the minimum *He* at $M = 0.025$ is reported to be 1.80, whereas the maximum *He* at $M = 0.048$ is found to be 3.51. Thus this case has shown the similar behavior as Case a-I; increasing downstream smooth pipe length consistently decreased the range of Helmholtz number.

Table 6. Mach number and corresponding Helmholtz number (*He*) range for pipe systems in Case a-II.

L_{up} (mm)	L_{corr} (mm)	L_{dw} (mm)	Mach Number Range	Corresponding *He* Range
100	350	100	0.023–0.078	1.29–4.36
		200	0.025–0.069	1.32–3.68
		300	0.025–0.051	1.73–3.50
		400	0.025–0.048	1.80–3.51

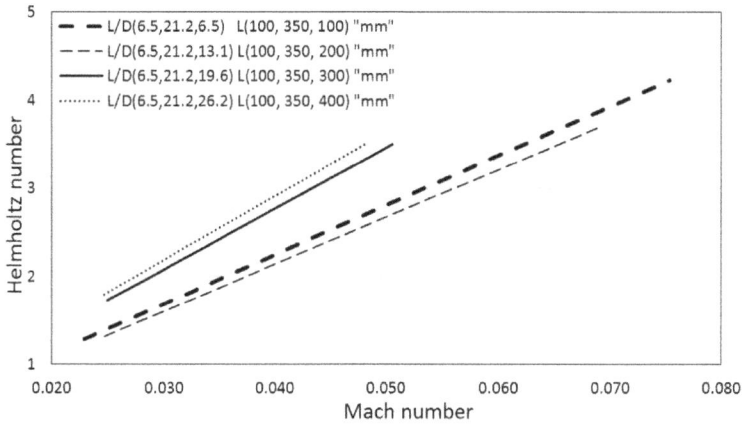

Figure 10. Helmholtz number *He* plotted against Mach number *M* for L_{up} = 100 mm, L_{corr} = 350 mm, and L_{dw} = 100–400 mm.

4.3. Longest Corrugated Segment: Case a-III (L_{corr}/D_{corr} = 27.3)

Pipe systems in Case a-III, which have the longest corrugated pipe considered in this study, L_{corr} = 450 mm, showed behavior similar to that in Cases a-I and a-II: increasing L_{dw} while maintaining L_{up} = 100 mm increased the frequency and Mach number ranges within which whistling occurred, as shown in Figure 11 and Table 7. For $L_{up} = L_{dw}$ = 100 mm, the system whistled from M = 0.026 to 0.084 with whistling frequencies between 1000 and 2700 Hz. For L_{dw} = 200 mm, the Mach number range was between M = 0.019 and 0.091, which correspond to frequencies of 650 and 2800 Hz, respectively. For L_{dw} = 300 mm, the whistling range lay between M = 0.026 and 0.090, and the corresponding frequencies lay between 800 and 2,550 Hz. Finally, for the downstream pipe of length 400 mm, whistling frequencies were between 700 and 2750 Hz for Mach numbers of M = 0.022 and 0.088, respectively.

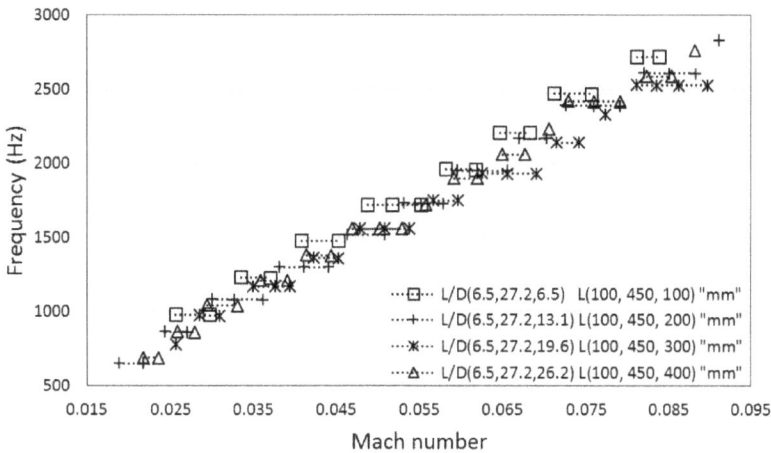

Figure 11. Frequency (Hz) plotted against Mach number M for L_{up} = 100 mm, L_{corr} = 450 mm, and L_{dw} = 100–400 mm.

Table 7. Mach number and corresponding frequency range (Hz) for pipe systems in Case a-III.

L_{up} (mm)	L_{corr} (mm)	L_{dw} (mm)	Mach Number Range	Corresponding Frequency Range (Hz)
		100	0.026–0.084	1000–2700
100	450	200	0.019–0.091	650–2800
		300	0.026–0.090	800–2550
		400	0.022–0.088	700–2750

For $L_{dw} = 100$ mm, the initial and final Strouhal numbers were 0.34 and 0.29 at $M = 0.026$ and 0.084, respectively, with a maximum Strouhal number of 0.34. For $L_{dw} = 200$ mm, the initial and final Strouhal numbers were 0.31 and 0.28 with corresponding Mach numbers of 0.019 and 0.091, respectively. The maximum Strouhal number for this case was estimated to be 0.32.

For $L_{dw} = 300$ mm, the Strouhal numbers began and ended at 0.27 and 0.25, corresponding to $M = 0.026$ and 0.090, respectively, with a maximum St of 0.30 occurring at multiple Mach numbers. Finally, for $L_{dw} = 400$ mm, the initial and final Strouhal numbers had the same value of 0.28 at the minimum and maximum Mach numbers of 0.022 and 0.088, with a maximum St of 0.32 in the whistling range, as shown in Figure 12 and Table 8. Figure 13 and Table 9 shows that for $L_{dw} = 100$ mm, the range of He was 1.75 to 5.74, which correspond to $M = 0.026$ and 0.084, respectively. For $L_{dw} = 200$ mm, the minimum and maximum He were estimated to be 1.30 and 6.31 at $M = 0.019$ and 0.091, respectively. The minimum and maximum He for the downstream length of $L_{dw} = 300$ mm were 1.79 and 6.28 at $M = 0.026$ and 0.090, respectively.

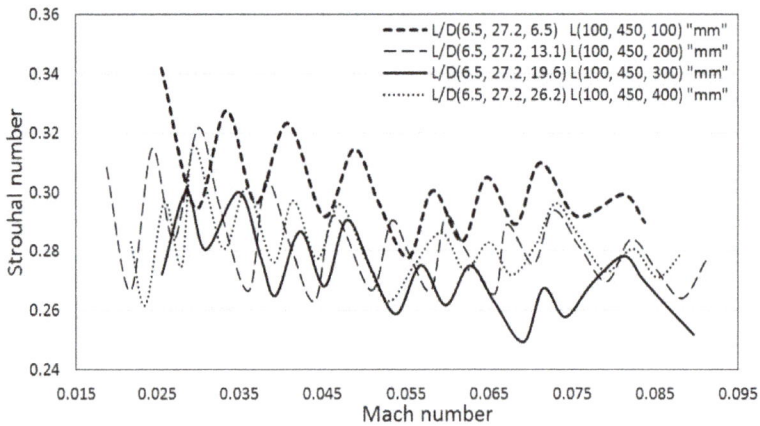

Figure 12. Strouhal number St plotted against Mach number M for $L_{up} = 100$ mm, $L_{corr} = 450$ mm, and $L_{dw} = 100$–400 mm.

Table 8. Mach number range, corresponding Strouhal number (St) range and maximum Strouhal number for each pipe configuration in Case a-III.

L_{up} (mm)	L_{corr} (mm)	L_{dw} (mm)	Mach Number Range	Corresponding St Range	Max. St
		100	0.026–0.084	0.34–0.29	0.34
100	450	200	0.019–0.091	0.31–0.28	0.32
		300	0.026–0.090	0.27–0.25	0.30
		400	0.022–0.088	0.28–0.28	0.32

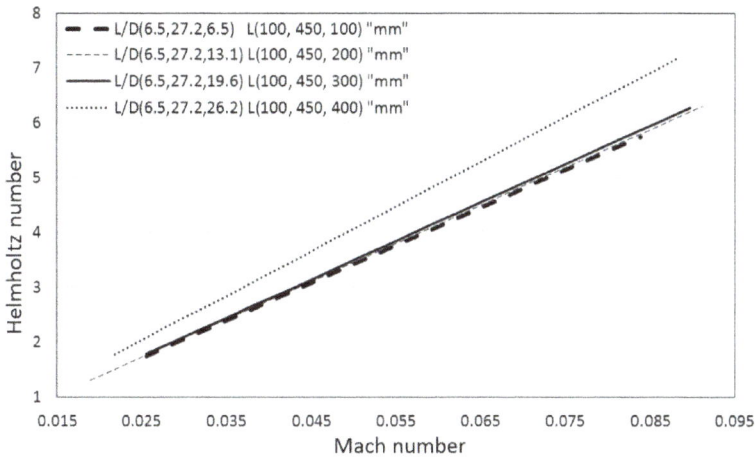

Figure 13. Helmholtz number *He* plotted against Mach number *M* for L_{up} = 100 mm, L_{corr} = 450 mm, and L_{dw} = 100–400 mm.

Table 9. Mach number and corresponding Helmholtz number (*He*) range for pipe systems in Case a-III.

L_{up} (mm)	L_{corr} (mm)	L_{dw} (mm)	Mach Number Range	Corresponding *He* Range
		100	0.026–0.084	1.75–5.74
100	450	200	0.019–0.091	1.30–6.31
		300	0.026–0.090	1.79–6.28
		400	0.022–0.088	1.77–7.18

For the longest downstream smooth pipe of length 400 mm, the minimum *He* at *M* = 0.022 was 1.77, whereas the maximum *He* at *M* = 0.088 was 7.18. Figure 13 demonstrates that increasing the length of the downstream smooth segment augments the range of Helmholtz numbers for a constant corrugated pipe length.

5. Effect of Upstream Smooth Pipe Length on the Whistling Behavior for Constant Corrugated Segment Length (Case b)

5.1. Shortest Corrugated Segment: Case b-I (L_{corr}/D_{corr} = 15.2)

In this section, we discuss the results of the cases in which L_{up} varied for the three corrugated segment lengths considered in this study while maintaining the downstream segment length at a fixed value of 100 mm. For L_{corr} = 250 mm with L_{up} = L_{dw} = 100 mm, the results were discussed in Section 4.1. For L_{up} = 200 mm, the whistling range decreased substantially from the case of L_{up} = L_{dw} = 100 mm, with narrow ranges of Mach numbers from *M* = 0.032 to 0.044 and frequencies from 1150 to 1450 Hz. For L_{up} = 300 mm, the Mach numbers ranged from *M* = 0.026 to 0.041 with corresponding whistling frequencies between 1000 and 1500 Hz.

However, for L_{up} = 400 mm, the Mach number range was found to broaden in comparison with L_{up} = 200 and 300 mm, although it was still very small compared with that of L_{up} = 100 mm. For this particular case, whistling occurred between *M* = 0.031 and 0.048 with frequencies ranging from 1100 to 1550 Hz. Again, in Figure 14 and Table 10, it is clear that the overall whistling range does not cover very high frequencies or Mach numbers because the length of the corrugated pipe used in this study is quite small. Evidently, fewer corrugations result in low-frequency whistling and smaller Mach number range. Varying the upstream and downstream lengths has a limited effect on enhancing the whistling range. For L_{up} = 100 mm, as mentioned in Case a-I, the initial and final Strouhal numbers were 0.36

and 0.32, corresponding to Mach numbers of $M = 0.027$ and 0.067, respectively, with a maximum Strouhal number of 0.36 in this range, as shown in Figure 15 and Table 11. For the upstream length of 200 mm, the Strouhal numbers were estimated to start and end at 0.33 and 0.30 for corresponding Mach numbers of $M = 0.032$ and 0.044, respectively, with a maximum St of 0.33. With the upstream smooth pipe of length 300 mm, the Strouhal number at the minimum and maximum Mach numbers of $M = 0.026$ and 0.041 were 0.34 and 0.32, respectively. The maximum St in the whistling range was 0.34. Finally, with the longest upstream smooth pipe with length $L_{up} = 400$ mm, the initial and final Strouhal numbers were 0.31 and 0.29 at $M = 0.031$ and 0.048, respectively, with a maximum of $St = 0.31$.

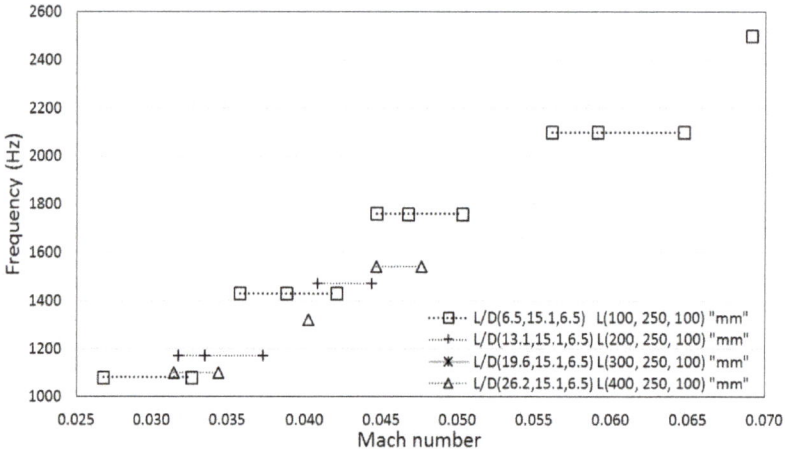

Figure 14. Frequency (Hz) plotted against Mach number M for $L_{up} = 100$–400 mm, $L_{corr} = 250$ mm, and $L_{dw} = 100$ mm.

Table 10. Mach number and corresponding frequency range (Hz) for pipe systems in Case b-I.

L_{up} (mm)	L_{corr} (mm)	L_{dw} (mm)	Mach Number Range	Corresponding Frequency Range (Hz)
100			0.027–0.069	1100–2500
200	250	100	0.032–0.044	1150–1450
300			0.026–0.041	1000–1500
400			0.031–0.048	1100–1550

Table 11. Mach number range, corresponding Strouhal number (St) range and maximum Strouhal number for each pipe configuration in Case b-I.

L_{up} (mm)	L_{corr} (mm)	L_{dw} (mm)	Mach Number Range	Corresponding St Range	Max. St
100			0.027–0.069	0.36–0.32	0.36
200	250	100	0.032–0.044	0.33–0.30	0.33
300			0.026–0.041	0.34–0.32	0.34
400			0.031–0.048	0.31–0.29	0.31

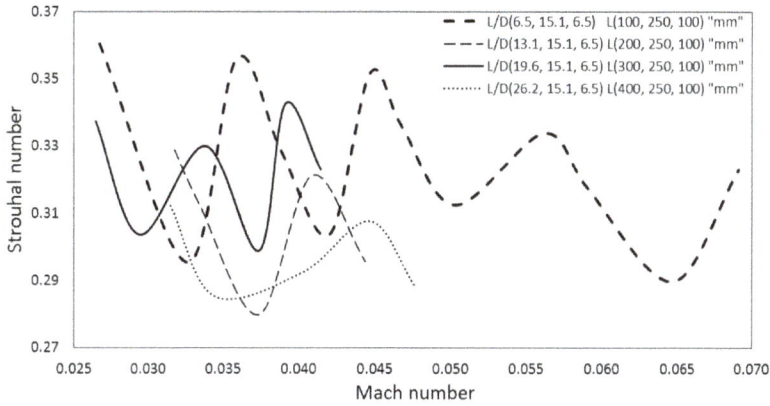

Figure 15. Strouhal number *St* plotted against Mach number *M* for L_{up} = 100–400 mm, L_{corr} = 250 mm, and L_{dw} = 100 mm.

Figure 16 and Table 12 show that, for the smallest upstream pipe of 100 mm, the minimum Helmholtz number *He* at *M* = 0.027 is 1.31, whereas the highest *He* at *M* = 0.069 is 3.38. For L_{up} = 200 mm, the minimum and maximum *He* are 1.73 and 2.41 at *M* = 0.032 and 0.044, respectively. The minimum and maximum *He* for L_{up} = 300 mm are 1.76 and 2.75 at *M* = 0.026 and 0.041, respectively. For the longest upstream smooth pipe of length 400 mm, the minimum *He* at *M* = 0.031 is 2.25, whereas the maximum *He* at *M* = 0.048 is 3.41. This particular case showed fluctuating behavior. It decreased at first but then increased again from the third configuration and continued to increase.

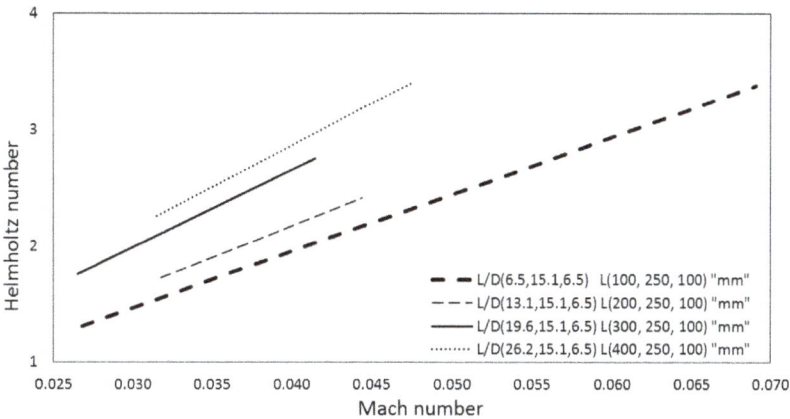

Figure 16. Helmholtz number *He* plotted against Mach number *M* for L_{up} = 100–400 mm, L_{corr} = 250 mm, and L_{dw} = 100 mm.

Table 12. Mach number and corresponding Helmholtz number (*He*) range for pipe systems in Case b-I.

L_{up} (mm)	L_{corr} (mm)	L_{dw} (mm)	Mach Number Range	Corresponding *He* Range
100			0.027–0.069	1.31–3.38
200	250	100	0.032–0.044	1.73–2.41
300			0.026–0.041	1.76–2.75
400			0.031–0.048	2.25–3.41

5.2. Intermediate Corrugated Segment: Case b-II ($L_{corr}/D_{corr} = 21.2$)

The next configuration to be discussed is the pipe system with a 350 mm corrugated pipe. This particular configuration showed an overall increasing trend regarding (the frequency and Mach number ranges) with increasing upstream pipe length, as shown in Figure 17 and Table 13. For upstream and downstream pipes with equal lengths of 100 mm, the lowest and highest Mach numbers were $M = 0.023$ and 0.078 with whistling frequencies ranging from 850 to 2600 Hz. For $L_{up} = 200$ mm, the minimum and maximum Mach numbers were $M = 0.023$ and 0.081 with frequencies of 750 and 2750 Hz, respectively. Surprisingly, for $L_{up} = 300$ mm, we observed a significant reduction in the Mach number range, extending from $M = 0.017$ to 0.056, but the frequency range had increased. Even at the lower modes, the pipe system whistled at higher intensities with frequencies ranging from 850 to 2600 Hz. Finally, the 400 mm upstream pipe followed the initial trend of increased whistling range from $M = 0.026$ to 0.081 with frequencies from 950 to 2700 Hz.

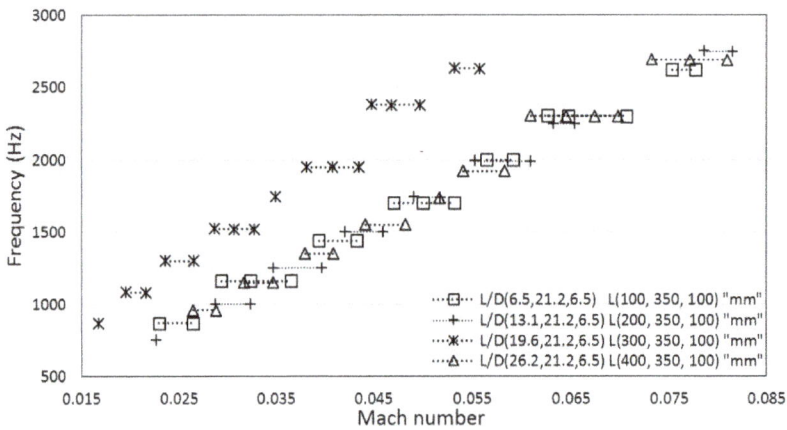

Figure 17. Frequency (Hz) plotted against Mach number M for $L_{up} = 100$–400 mm, $L_{corr} = 350$ mm, and $L_{dw} = 100$ mm.

Table 13. Mach number and corresponding frequency range (Hz) for pipe systems in Case b-II.

L_{up} (mm)	L_{corr} (mm)	L_{dw} (mm)	Mach Number Range	Corresponding Frequency Range (Hz)
100			0.023–0.078	850–2600
200	350	100	0.023–0.081	750–2750
300			0.017–0.056	850–2600
400			0.026–0.081	950–2700

When the upstream and downstream smooth pipe lengths had equal lengths of 100 mm for the corrugated length of 350 mm, the initial and final Strouhal numbers were 0.34 and 0.30 at $M = 0.023$ and 0.078, respectively, with a maximum St of 0.35. For $L_{up} = 200$ mm, the St at the minimum and maximum Mach numbers of $M = 0.023$ and 0.081 were 0.30 with 0.32, respectively, as the highest achieved value of Strouhal number for this configuration. As the upstream length increased to 300 mm, we estimated the St values to begin and end at 0.35 and 0.32 with corresponding Mach numbers of $M = 0.017$ and 0.056, respectively, and the maximum St was 0.37.

With further augmentation of the upstream length to 400 mm, the starting and ending St were 0.32 and 0.30 at $M = 0.026$ and 0.081, respectively, with a maximum St of 0.34 in this range, as shown in Figure 18 and Table 14.

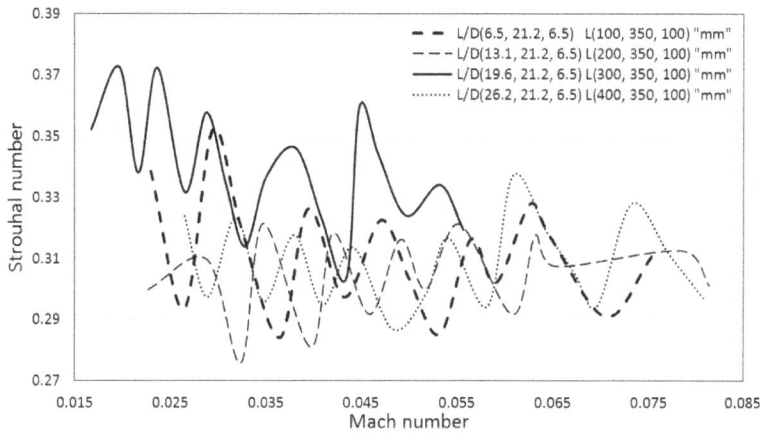

Figure 18. Strouhal number *St* plotted against Mach number *M* for L_{up} = 100–400 mm, L_{corr} = 350 mm, and L_{dw} = 100 mm.

Table 14. Mach number range, corresponding Strouhal number (*St*) range and maximum Strouhal number for each pipe configuration in Case b-II.

L_{up} (mm)	L_{corr} (mm)	L_{dw} (mm)	Mach Number Range	Corresponding *St* Range	Max. *St*
100			0.023–0.078	0.34–0.30	0.35
200	350	100	0.023–0.081	0.30–0.30	0.32
300			0.017–0.056	0.35–0.32	0.37
400			0.026–0.081	0.32–0.30	0.34

As shown in Figure 19 and Table 15, for the smallest upstream pipe of length 100 mm, the minimum Helmholtz number *He* was 1.29 at *M* = 0.023, whereas the maximum *He* was 4.36 at *M* = 0.078. For L_{up} = 200 mm, the minimum and maximum *He* were 1.31 and 4.71 at *M* = 0.023 and 0.081, respectively. The minimum and maximum *He* for L_{up} = 300 mm were 1.36 and 4.54 at *M* = 0.017 and 0.056, respectively. For the longest upstream smooth pipe of length 400 mm, the minimum *He* was 2.22 at *M* = 0.026, whereas the maximum *He* was 6.79 at *M* = 0.081. This particular trend suggests that the range of *He* with respect to *M* continuously increases as the length of upstream smooth pipe increases, except in the case of the 300 mm upstream pipe, which resulted in the reduction of *He* but started to whistle at a lower Mach number than all other configurations.

Table 15. Mach number and corresponding Helmholtz number (*He*) range for pipe systems in Case b-II.

L_{up} (mm)	L_{corr} (mm)	L_{dw} (mm)	Mach Number Range	Corresponding *He* Range
100			0.023–0.078	1.29–4.36
200	350	100	0.023–0.081	1.31–4.71
300			0.017–0.056	1.36–4.54
400			0.026–0.081	2.22–6.79

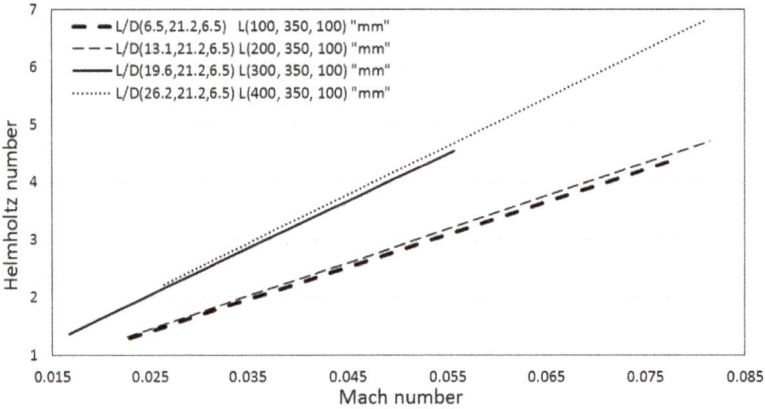

Figure 19. Helmholtz number He plotted against Mach number M for L_{up} = 100–400 mm, L_{corr} = 350 mm, and L_{dw} = 100 mm.

5.3. Longest Corrugated Segment: Case b-III (L_{corr}/D_{corr} = 27.3)

Pipe systems in Case b-III, which have the longest corrugated pipe considered in this study, L_{corr} = 450 mm, showed behavior similar to Cases b-I and b-II; increasing L_{up} while maintaining L_{dw} at 100 mm increased the range of whistling frequency and Mach number, as shown in Figure 20 and Table 16. For L_{up} = L_{dw} = 100 mm, the system whistled from M = 0.026 to 0.084 with whistling frequencies between 1000 and 2700 Hz. For L_{up} = 200 mm, the Mach number range was between M = 0.024 and 0.086, corresponding to frequencies of 850 and 2800 Hz, respectively. For L_{up} = 300 mm, the whistling range lay between M = 0.021 to 0.085, with a corresponding range of frequencies between 750 and 3100 Hz. Finally, for the upstream pipe of length 400 mm, whistling frequencies ranged from 700 to 2750 Hz at M = 0.022 and 0.093, respectively.

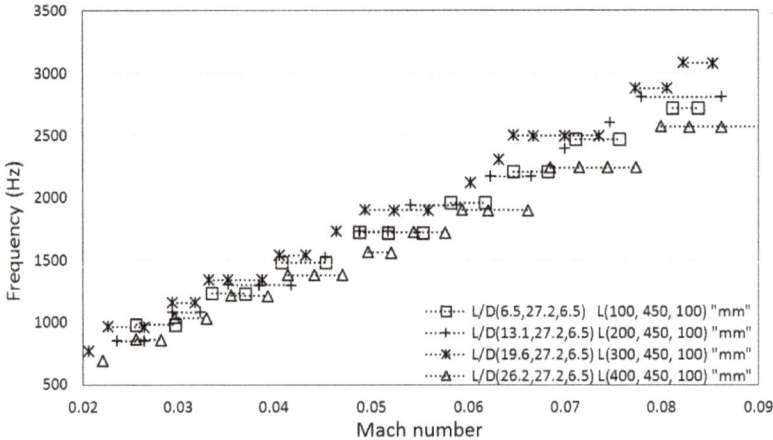

Figure 20. Frequency (Hz) plotted against Mach number M for L_{up} = 100–400 mm, L_{corr} = 450 mm, and L_{dw} = 100 mm.

Table 16. Mach number and corresponding frequency range (Hz) for pipe systems in Case b-III.

L_{up} (mm)	L_{corr} (mm)	L_{dw} (mm)	Mach Number Range	Corresponding Frequency Range (Hz)
100			0.026–0.084	1000–2700
200	450	100	0.024–0.086	850–2800
300			0.021–0.085	750–3100
400			0.022–0.093	700–2750

For L_{up} = 100 mm, the initial and final Strouhal numbers were 0.34 and 0.29 at M = 0.026 and 0.084, respectively, as shown in Figure 21 and Table 17, with a maximum of 0.34 in this range. For L_{up} = 200 mm, the Strouhal numbers were 0.32 and 0.29 at the initial and final Mach numbers of M = 0.024 and 0.086, respectively, with a maximum of 0.33 in this range.

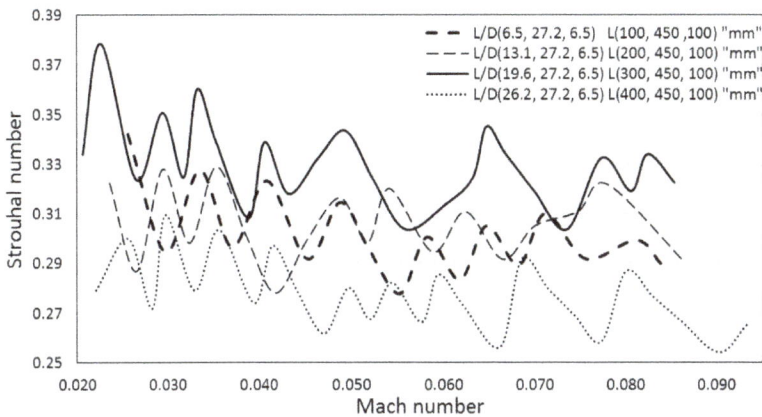

Figure 21. Strouhal number *St* plotted against Mach number M for L_{up} = 100 mm, L_{corr} = 450 mm, and L_{dw} = 100–400 mm.

Table 17. Mach number range, corresponding Strouhal number (*St*) range and maximum Strouhal number for each pipe configuration in Case b-III.

L_{up} (mm)	L_{corr} (mm)	L_{dw} (mm)	Mach Number Range	Corresponding *St* Range	Max. *St*
100			0.026–0.084	0.34–0.29	0.34
200	450	100	0.024–0.086	0.32–0.29	0.33
300			0.021–0.085	0.33–0.32	0.38
400			0.022–0.093	0.28–0.27	0.31

For the 300 mm upstream smooth pipe, the starting Strouhal number at Mach number M = 0.021 was 0.33 and the ending one was 0.32 at M = 0.085, with a maximum Strouhal number of 0.38 in this range; this was also the peak value of *St* for all cases considered in this study. For L_{up} = 400 mm, Strouhal numbers of 0.28 and 0.27 corresponded to the starting and ending Mach numbers of 0.022 and 0.093, respectively, with a maximum Strouhal number of 0.31 in this range.

As shown in Figure 22 and Table 18, for the smallest upstream pipe of length 100 mm, the minimum Helmholtz number *He* was 1.75 at M = 0.026, whereas the maximum *He* was 5.74 at M = 0.084. For L_{up} = 200 mm, the minimum and maximum *He* were 1.74 and 6.39 at M = 0.024 and 0.086, respectively. The minimum and maximum *He* for L_{up} = 300 mm were 1.79 and 7.40 at M = 0.021 and 0.085, respectively. For the longest upstream smooth pipe of length 400 mm, the minimum *He* was 1.79 at M = 0.022, whereas the maximum *He* was 7.59 at M = 0.093.

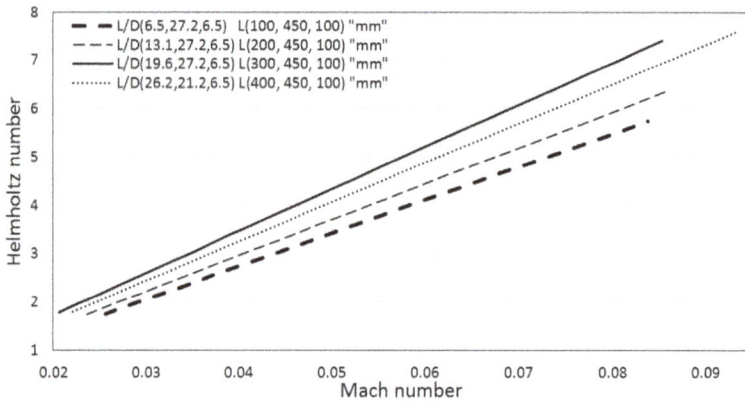

Figure 22. Helmholtz number *He* plotted against Mach number *M* for L_{up} = 100–400 mm, L_{corr} = 450 mm, and L_{dw} = 100 mm.

Table 18. Mach number and corresponding Helmholtz number (*He*) range for pipe systems in Case b-III.

L_{up} (mm)	L_{corr} (mm)	L_{dw} (mm)	Mach Number Range	Corresponding *He* Range
100			0.026–0.084	1.75–5.74
200	450	100	0.024–0.086	1.74–6.39
300			0.021–0.085	1.79–7.40
400			0.022–0.093	1.79–7.59

6. Acoustic Modes and Average Mode Gap

In this section, the whistling behaviors of various pipe configurations included in Cases a and b are discussed in terms of the acoustic modes and average mode gaps between successive modes. As shown in Tables 19 and 20, for each individual pipe system, there was significant variation in the excited acoustic mode numbers and the average mode gap between consecutive modes. For all configurations in Case a-I (see Table 19), increasing the length of the downstream smooth pipe resulted in a reduction in the number of excited acoustic modes from four modes (Modes 3–7) for L_{dw} = 100 mm to a single excited acoustic mode for L_{dw} = 300 and 400 mm. We could not predict which mode was excited in the last two pipe systems (L_{dw} = 300 and 400 mm) corresponding to Case a-I, because at least two whistling frequencies are required to estimate the excited mode numbers and average mode gap. Moreover, the average mode gap also decreased with increasing L_{dw}. For configurations in Case a-II, a similar but less abrupt decrement in the number of excited modes occurred. The whistling covered Modes 3–9 for L_{dw} = 100 mm, whereas for L_{dw} = 400 mm, the whistling was found to occur between Modes 4 and 8. For configurations in this case with L_{dw} = 200 and 300 mm, the pipe system did not whistle at Modes 8 and 6, respectively. The average mode gap decreased from 275 to 190 Hz for L_{dw} = 100 and 400 mm, respectively. For Case a-III, the behavior was very much in contrast to the previous cases. The number of excited acoustic modes increased for increasing downstream smooth segment length. For L_{dw} = 100 mm, the whistling range included Modes 4–11, whereas for L_{dw} = 400 mm, this range consisted of Modes 4–16 with all modes included in the whistling range. The average mode gap showed a similar response to prior cases. For L_{dw} = 100 mm, the average mode gap was 245 Hz, whereas for the longest downstream smooth pipe, L_{dw} = 400 mm, the value was estimated to be 175 Hz.

Table 19. Excited acoustic mode numbers, mode numbers within the whistling range at which there was no whistling, and average mode gap between two successive modes for all configurations included in Case a. In the table, MN and MG denote Acoustic Mode Numbers and Mode Gap (Hz), respectively.

Pipe Configurations		L_{up} (mm)	L_{corr} (mm)	L_{dw} (mm)	Excited AMN	Missing MN	Average MG
Case a	I	100	250	100	3–7	-	350
				200	3–4	-	290
				300	-	-	-
				400	-	-	-
Case a	II	100	350	100	3–9	-	275
				200	3–9	8	245
				300	4–7	6	220
				400	4–8	-	190
Case a	III	100	450	100	4–11	-	245
				200	3–13	-	220
				300	4–13	-	195
				400	4–16	-	175

Table 20. Excited acoustic mode numbers, mode numbers within the whistling range at which there was no whistling, and average mode gap between two successive modes for all configurations included in Case b.

Pipe Configurations		L_{up} (mm)	L_{corr} (mm)	L_{dw} (mm)	Excited AMN	Missing MN	Average MG [Hz]
Case b	I	100	250	100	3–7	-	350
		200			4 and 5	-	300
		300			4–6	-	250
		400			5–7	-	220
Case b	II	100	350	100	3–9	-	275
		200			3–11	10	250
		300			3–11	9	215
		400			5–14	11 and 13	190
Case b	III	100	450	100	4–11	-	245
		200			4–13	-	220
		300			3–15	13	195
		400			4–16	12 and 14	170

The pipe systems in Case b have consistently increasing upstream smooth pipes with lengths ranging 100–400 mm with downstream pipe lengths fixed at 100 mm. For all configurations in Case b-I, increasing the length of the upstream smooth pipe results in a reduction in the number of excited acoustic modes. This behavior is very similar to that in Case a-I. For L_{up} = 100 mm, the whistling covers acoustic Modes 3–7, whereas for L_{up} = 400 mm, the whistling range included Modes 5–7. The average mode gap between successive modes for L_{up} = 100 mm was 350 Hz, whereas for L_{up} = 400 mm, it was estimated to be 220 Hz. For Case b-II, the number of excited acoustic modes increased continuously with increasing upstream lengths. It increased from seven modes (Modes 3–9) for L_{up} = 100 mm to nine modes (Modes 5–14) for L_{up} = 400 mm. For Case b-II, we observed that in all remaining configurations, with the exception of the configuration with an upstream segment of length 100 mm, whistling did not occur at some modes within the covered mode ranges during the whistling regime. Configurations with L_{up} = 200 and 300 mm did not result in whistling at Modes 10 and 9, respectively, whereas Modes 11 and 13 were excluded in the case where L_{up} = 400 mm. The average mode gap decreased from 275 to 190 Hz with increasing L_{up} from 100 to 400 mm, respectively. For Case b-III, the number of excited modes continued to increase with increasing upstream length. The shortest upstream length of 100 mm covered Modes 4–11, whereas for the longest upstream length of 400 mm, the range was broadened to include modes 4 to 16. Mode 13 was excluded for the pipe system with L_{up} = 300 mm, whereas Modes 12 and 14 were excluded for the configuration with L_{up} = 400 mm.

The average mode gap began at 245 Hz and decreased to 170 Hz when increasing the upstream length from L_{up} = 100 to 400 mm. In both Cases a and b, the range of frequencies corresponding to the average mode gap showed a continuous downhill trend as the corrugated pipe length increased. The reason for the excluded modes reported for some configurations in both cases in not yet known.

7. Conclusions

This paper presents the results obtained from an experimental study in which both ends of a corrugated pipe were attached to smooth pipes. Three different corrugated pipe lengths (250, 350, and 450 mm) and four different smooth pipe lengths (100, 200, 300, and 400 mm) were included in the scope of our study. The configurations were divided into two cases, each of which was divided into three sub-cases. Case a included pipe systems where $L_{dw} \geq L_{up}$, whereas Case b included configurations where $L_{dw} \leq L_{up}$. Each corrugated pipe length was tested with each smooth pipe length, as was previously mentioned.

- Case a: For the 250 mm corrugated pipe, the whistling range decreased sharply as the downstream pipe length increased. The corrugated pipe of length 350 mm also showed a decreasing trend in its whistling range. For the 450 mm pipe, the behavior was completely different; the whistling remained in almost the same range of Mach numbers for all sub-cases.
- Case b: For the 250 mm corrugated pipe, the whistling range was again found to decrease with increasing smooth pipe length but not as abruptly as in Case a. For the 350 mm corrugated pipe, the Mach number range for the whistling consistently increased and reached a maximum for the longest smooth pipe. The number of excited acoustic modes also increased with increasing smooth pipe length. Finally, for L_{corr} = 450 mm, the pipe system behaved in a manner similar to the corrugated pipe with the same length in Case a (Case a-III). The overall Mach number range showed very little variation, and, for longer smooth pipes, the range showed a slight increment.

The average mode gap between successive modes continuously decreased for both cases as the smooth pipe increased in length. As the corrugation length increased, a greater number of modes were excited. The corrugated pipe with a length of 450 mm was not significantly affected by increasing the upstream or downstream smooth pipes. This may be because the lengths of the smooth pipe were small considering the number of cavities in the corrugated segment. Smaller corrugated pipes showed less whistling as a result of the lower number of corrugations, because each cavity acts as a source of sound, as reported in the literature. To more clearly observe the effect of the lengths of the upstream and downstream smooth pipes on longer corrugated segments, even longer smooth pipes should be investigated.

Author Contributions: H.-C.L. wrote and reviewed the manuscript. F.R. collected the data and made figures.

Funding: This research received no external funding.

Acknowledgments: This work was supported by "Human Resources Program in Energy Technology" of the Korea Institute of Energy Technology Evaluation and Planning (KETEP), granted financial resource from the Ministry of Trade, Industry & Energy, Republic of Korea (No. 20164030201230). In addition, this work was supported by the National Research Foundation of Korea (NRF) grant funded by the Korea government (MSIP) (No. 2016R1A2B1013820). This research was also supported by the Fire Fighting Safety & 119 Rescue Technology Research and Development Program funded by the Ministry of Public Safety and Security (MPSS-2015-79). This work was also supported by the China-Korea International Collaborative work (Project ID: SLDRCE15-04).

Conflicts of Interest: The authors declare no conflicts of interest.

References

1. Belfroid, S.P.; Shatto, D.P.; Peters, R.M.C.A.M. Flow induced pulsations caused by corrugated tubes. In Proceedings of the ASME Conference Proceedings, Seattle, DC, USA, 11–15 November 2007.
2. Goyder, H. Noise generation and propagation within corrugated pipes. *J. Press. Vessel Technol.* **2013**, *135*, 1–7. [CrossRef]
3. Crawford, F.S. Singing corrugated pipes. *Am. J. Phys.* **1974**, *42*, 278–288. [CrossRef]
4. Silverman, M.P.; Cushman, G.M. Voice of the dragon: The rotating corrugated resonator. *Eur. J. Phys.* **1989**, *10*, 298–304. [CrossRef]
5. Serafin, S.; Kojs, J. Computer models and compositional applications of plastic corrugated tubes. *Organ. Sound* **2005**, *10*, 67–73. [CrossRef]
6. Burstyn, W. Eine neue Pfeife (a new pipe). *Z. Tech. Phys.* **1922**, *3*, 179–180.
7. Cermak, P. Uber die Tonbildung bei Metallschlauchen mit eingedracktem Spiral gang (On the sound generation in flexible metal hoses with spiraling grooves). *Z. Tech. Phys.* **1922**, *23*, 394–397.
8. Gutin, L. On the sound field of a rotating propeller. *Z. Tech. Phys.* **1936**, *12*, 76–83.
9. Nakamura, Y.; Fukamachi, N. Sound generation in corrugated tubes. *Fluid Dyn. Res.* **1991**, *7*, 255–261. [CrossRef]
10. Petrie, A.M.; Huntley, I.D. The acoustic output produced by a steady airflow through a corrugated duct. *J. Sound Vib.* **1980**, *70*, 1–9. [CrossRef]
11. Kristiansen, U.R.; Wiik, G.A. Experiments on sound generation in corrugated pipes with flow. *J. Acoust. Soc. Am.* **2007**, *121*, 1337–1344. [CrossRef] [PubMed]
12. Kopev, V.F.; Mironov, M.A.; Solntseva, V.S. Aeroacoustic interaction in a corrugated duct. *Acoust. Phys.* **2008**, *54*, 197–203. [CrossRef]
13. Rockwell, D. Oscillations of impinging shear layers. *AIAA J.* **1983**, *21*, 645–664. [CrossRef]
14. Bruggeman, J.C.; Wijnands, P.J.; Gorter, J. Self-sustained low-frequency resonance in low-Mach-number gas flow through pipelines with side branch cavities. In Proceedings of the 10th Aeroacoustics Conference, Seattle, DC, USA, 9–11 July 1986.
15. Tonon, D.; Landry, B.; Belfroid, S.; Willems, J.; Hofman, G.; Hirschberg, A. Whistling of a pipe system with multiple side branches: Comparison with corrugated pipes. *J. Sound Vib.* **2010**, *329*, 1007–1024. [CrossRef]
16. Binnie, A.M. Self-induced waves in a conduit with corrugated walls II. Experiments with air in corrugated and finned tubes. *Proc. R. Soc. Lond.* **1961**, *262*, 179–191. [CrossRef]
17. Cadwell, L.H. Singing corrugated pipes revisited. *Am. J. Phys.* **1994**, *62*, 224–227. [CrossRef]
18. Elliott, J. Corrugated pipe flow. In *Lecture Notes on the Mathematics of Acoustics*; Wright, M.C.M., Ed.; Imperial College Press: London, UK, 2004; pp. 207–222.
19. Rudenko, O.; Nakiboglu, G.; Holten, A.; Hirschberg, A. On whistling of pipes with a corrugated segment: Experiment and theory. *J. Sound Vib.* **2013**, *332*, 7226–7242. [CrossRef]
20. Ziada, S.; Buhlmann, E. Flow induced vibration in long corrugated pipes. In Proceedings of the 5th International Conference on Flow-Induced Vibrations, Brighton, UK, 20–22 May 1991.
21. Debut, V.; Antunes, J.; Moreira, M. Experimental study of the flow-excited acoustical lock-in in a corrugated pipe. In Proceedings of the 14th International Conference on Sound and Vibration, Cairns, Australia, 9–12 July 2007.
22. Debut, V.; Antunes, J.; Moreira, M. Flow-acoustic interaction in corrugated pipes: Time domain simulation of experimental phenomena. In Proceedings of the 9th International Conference on Flow-Induced Vibrations, Prague, Czech Republic, 30 June–3 July 2008.
23. Kristiansen, U.R.; Mattei, P.O.; Pinhede, C.; Amielh, M. Experimental study of the influence of low frequency flow modulation on the whistling behavior of a corrugated pipe. *J. Acoust. Soc. Am.* **2011**, *130*, 1851–1855. [CrossRef] [PubMed]
24. Nakiboglu, G.; Belfroid, S.P.; Tonon, D.; Willems, J.F.; Hirschberg, A. A parametric study on the whistling of multiple side branch system as a model for corrugated pipes. In Proceedings of the ASME Conference Proceedings, Liverpool, UK, 11–15 October 2009.
25. Nakiboglu, G.; Belfroid, S.P.; Willems, J.; Hirschberg, A. Whistling behavior of periodic systems: Corrugated pipes and multiple side branch system. *Int. J. Mech. Sci.* **2010**, *52*, 1458–1470. [CrossRef]

26. Nakiboglu, G.; Belfroid, S.P.; Golliard, J.; Hirschberg, A. On the whistling of corrugated pipes: Effect of pipe length and flow profile. *J. Fluid Mech.* **2011**, *672*, 78–108. [CrossRef]
27. Nakiboglu, G.; Manders, H.B.; Hirschberg, A. Aeroacoustic power generated by a compact axisymmetric cavity: Prediction of self-sustained oscillation and influence of the depth. *J. Fluid Mech.* **2012**, *703*, 163–191. [CrossRef]
28. Howe, M.S. *Theory of Vortex Sound*; Cambridge University Press: Cambridge, UK, 2003.
29. Howe, M.S. The dissipation of sound at an edge. *J. Sound Vib.* **1980**, *70*, 407–411. [CrossRef]
30. Howe, M.S. Contributions to the theory of aerodynamic sound, with application to excess jet noise and the theory of the flute. *J. Fluid Mech.* **1975**, *71*, 625–673. [CrossRef]
31. Daugherty, R.; Franzini, J.; Finnemore, E. *Fluid Mechanics with Engineering Applications*, SI metric ed.; McGraw-Hill: New York, NY, USA, 1989.

energies

MDPI

Article

A Novel Adaptive Neuro-Control Approach for Permanent Magnet Synchronous Motor Speed Control

Qi Wang [1,2,*], Haitao Yu [1], Min Wang [3] and Xinbo Qi [2]

[1]   School of Electrical Engineering, Southeast University, Nanjing 210096, China; htyu@seu.edu.cn
[2]   Department of Automatic Control, Henan Institute of Technology, Xinxiang 453003, China; xbq.hait@outlook.com
[3]   School of Electrical Engineering, Zhengzhou University, Zhengzhou 450001, China; mwang.zz@outlook.com
[*]   Correspondence: qwang.seu@outlook.com

Received: 3 August 2018; Accepted: 30 August 2018; Published: 6 September 2018

Abstract: A speed controller for permanent magnet synchronous motors (PMSMs) under the field oriented control (FOC) method is discussed in this paper. First, a novel adaptive neuro-control approach, single artificial neuron goal representation heuristic dynamic programming (SAN-GrHDP) for speed regulation of PMSMs, is presented. For both current loops, PI controllers are adopted, respectively. Compared with the conventional single artificial neuron (SAN) control strategy, the proposed approach assumes an unknown mathematic model of the PMSM and guides the selection value of parameter K online. Besides, the proposed design can develop an internal reinforcement learning signal to guide the dynamic optimal control of the PMSM in the process. Finally, nonlinear optimal control simulations and experiments on the speed regulation of a PMSM are implemented in Matlab2016a and TMS320F28335, a 32-bit floating-point digital signal processor (DSP), respectively. To achieve a comparative study, the conventional SAN and SAN-GrHDP approaches are set up under identical conditions and parameters. Simulation and experiment results verify that the proposed controller can improve the speed control performance of PMSMs.

Keywords: permanent magnet synchronous motor (PMSM); single artificial neuron goal representation heuristic dynamic programming (SAN-GrHDP); single artificial neuron (SAN); reinforcement learning (RL); goal representation heuristic dynamic programming (GrHDP); adaptive dynamic programming (ADP)

1. Introduction

Permanent magnet synchronous motors (PMSMs) have many advantages, such as high power density, simple structure, small volume, high efficiency and reliability. PMSMs are widely used in numerical control machine tools, aerospace and industrial robotic manipulators [1]. A PMSM is a typical nonlinear and strongly coupled system, with unpredictable external disturbances, as well as internal parameter variations [2]. In recent years, various nonlinear control methods [3–11], such as fuzzy logic control, sliding mode control, neural network control, nonlinear optimal control, internal model control, adaptive control, have been used to meet the requirements of high reliability and performance in PMSM control [7–10]. The fuzzy logic control is successfully applied in the speed control of PMSMs [12,13]. However, the fuzzy control membership function is mainly based on expert experience, which is difficult to obtain. Sliding mode control is a preferred research topic, due to its insensitivity to variation of control object parameters and load disturbances [14,15]. Nevertheless, chattering phenomena exist in this control method. Meanwhile, nonlinear optimal control has been put forward as a new PMSM control method [16]. However, the parameters of the PMSM must be sufficiently accurate, and control results cannot adapt in time when the mechanical parameters of PMSM change. In [17], a novel control scheme combining the inverse system method and the internal model control for a bearingless permanent magnet synchronous motor (BPMSM) was proposed by Sun et al., although in order to regulate the tracking and disturbance rejection properties, the values of control parameter sets need to be adjusted separately [18].

Recently, adaptive dynamic programming (ADP) has attracted significantly increasing attention as a novel level reinforcement learning approach. It can solve the "curse of dimensionality" of conventional dynamic programming by approximately computing cost function [19]. ADP can be categorized into three classical structures [20]: the first is heuristic dynamic programming (HDP), the second is dual heuristic dynamic programming (DHP), and the last is globalized dual heuristic dynamic programming (GDHP). The main difference is that the critic network is used to approximate the value function J in HDP, while it is used to approximate the derivatives of value function J in DHP. GDHP incorporates the benefits of HDP and DHP, by approximating both value function J and its derivatives, respectively.

In paper [21], a novel hierarchical structure of ADP approach named goal representation heuristic dynamic programming (GrHDP) is proposed. Compared with the conventional ADP approach, the proposed approach has an additional reference network which can automatically build an internal reinforcement signal to facilitate the optimal learning, control effectively and efficiency [22]. This novel hierarchical ADP approach is of a superior learning performance over the traditional ones. The GrHDP approach is used in various fields of electrical engineering, such as power system stability control for a wind farm [23], power oscillation damping control for superconducting magnetic energy storage [24], and load frequency control for an islanded smart grid [25].

Meanwhile, the single artificial neuron (SAN) control approach has been used in many applications for its robust control in the presence of noise and uncertainties [26]. Generally speaking, traditional SAN control has been applied to engineering practices for a long time due to its good performance and easy implementation [27–29].

It has been pointed out that although the conventional SAN control approach can provide an online learning ability for the PMSM parameter variation, it may not provide a satisfactory property of load disturbance rejection. The reason is that the control effect of SAN mainly depends on the parameter K (neuron scale-up factor). The parameter K is very important to the control response performance. The selection of K is very difficult in traditional SAN control approaches. The control system will respond faster if the K value increases. However, the K value will lead to the instability of the system, if it is out of a certain range. Moreover, there is no profound theoretical background, which can be used to tune the parameter K for complicated systems with uncertainties and disturbance. It is a new idea to use machine learning to adjust the K value of SAN and make it applicable to PMSM control. At the same time, for the ultimate convergence, the action network weights of GrHDP approach usually need repeating online learning to achieve optimization solutions to the Bellman equation. So far, articles about ADP approaches mostly focus on the simulation stage [23,24,30–36].

To solve the above problems, in this paper, a novel neuro-control framework using GrHDP and SAN is proposed. Moreover, an application study on PMSM vector control system is also presented in this research. The main contributions of this paper are summarized as follows:

(1) A novel adaptive neuro-control controller, called single artificial neuron goal representation heuristic dynamic programming (SAN-GrHDP), based on SAN and GrHDP has been proposed in this paper. This framework, under which the parameter K in the SAN has been updated through a reference learning mechanism, can provide a sequential online control policy.
(2) The formula of SAN-GrHDP approach is derived, and the reinforcement signal and learning process are designed for the vector control of PMSM. Simulation studies have been carried out for the proposed approach. Simulation results demonstrate that the proposed controller has a higher potential of disturbance rejection, with much less speed fluctuation and shorter recovering time towards load disturbance.

(3) Moreover, comparative experiments of original SAN and SAN-GrHDP approaches are performed on the speed control of PMSM under the same conditions and parameters. The results of the experiments verify that SAN-GrHDP can better improve the control effect by interacting with the control object, and has much better robustness than SAN with load mutation and load disturbance.

The remainder of the paper is organized as follows. Section 2 describes the servo control system of a PMSM as well as the certain modeling of the speed controller used in this paper. Section 3 illustrates the details of the SAN-GrHDP controller, and the learning algorithm associated. In Section 4, the simulation of the speed control of the PMSM and the experimental setup based on SAN-GrHDP are presented. The results prove the effectiveness of the proposed SAN-GrHDP by comparing with the conventional SAN control approach. Finally, Section 5 presents our conclusions and a few future study directions.

2. Model of Permanent Magnet Synchronous Motor Control System

Assuming that magnetic circuit saturation, hysteresis eddy current losses are disregarded and the sinusoidal magnetic field is distributed in space, a surface-mounted PMSM is considered as the controlled object. In *d-q* coordinates, the model of a surface mounted PMSM can be expressed as follows [37,38]:

$$
\begin{pmatrix} i_d' \\ i_q' \\ \omega' \end{pmatrix} = \begin{pmatrix} -\dfrac{R_s}{L_d} & n_p\omega & 0 \\ -n_p\omega & -\dfrac{R_s}{L_d} & -\dfrac{n_p\psi_f}{L_q} \\ 0 & -\dfrac{1.5n_p\psi_f}{J} & -\dfrac{B}{J} \end{pmatrix} \begin{pmatrix} i_d \\ i_q \\ \omega \end{pmatrix} + \begin{pmatrix} \dfrac{u_d}{L_d} \\ \dfrac{u_q}{L_q} \\ -\dfrac{T_L}{J} \end{pmatrix}
\tag{1}
$$

where u_d and u_q are the stator *d*- and *q*-axes voltages, i_d and i_q are the stator *d*- and *q*-axes currents, L_d and L_q are the stator *d*- and *q*-axes inductances, n_p is the number of pole pairs, R_s is the stator resistance, ω is the rotor angular velocity, ψ_f is the flux linkage, T_L is the load torque, and B is the viscous friction coefficient.

The strategies of the vector control of PMSM have $i_d = 0$ control, power factor $\cos\phi = 1$ control, the maximum torque control, maximum output power control, flux weakening control and so on. The approach of $i_d = 0$ control which has many advantages such as small torque ripple and wide speed range, is the most simple strategy of vector control and used in this article. The field oriented control (FOC) diagram of PMSM system by $i_d = 0$ control approach is shown in Figure 1. There are three controllers in the diagram: one speed tracking loop controller and two current tracking loop controllers. The *d*- and *q*-axes currents i_d, i_q can be calculated from the two-phase static coordinate currents i_α, i_β of PMSM by the PARK transform. Similarly, the i_α, i_β currents can be obtained from the actual phase currents of PMSM through the CLARK transform. The rotor angular velocity ω and rotor position θ can be calculated from encoder. Usually, the reference current value i_q^* is determined by the speed loop controller output, and i_d^* is set to zero. Due to saturation phenomena of PMSMs, some values can depend on the operating point of the machine, such as rotor inductance and rotor resistance. This can affect the performance and the accuracy of the conventional controller. The SAN-GrHDP approach is a kind of machine learning algorithm (ADP approach). When motor parameters change, the controller can learn from a complex, uncertain environment (controlled plant) according to the optimal cost function, which is also the essence of ADP method [19]. Compared with the traditional control approach, the SAN-GrHDP can realize self-regulation by critic network and provide an online sequential control policy, not subject to the external load disturbances and parameter variations. This article mainly discusses the external load disturbances rejection capacity of proposed control strategy. The current-loop sampling period is 200 μs, and the speed-loop sampling time is ten times that of the current-loop. The current-loop controllers require faster response. Therefore, the inner current-loop controllers adopt the traditional PI controllers. Here, the task is to design a speed controller based on SAN-GrHDP approach.

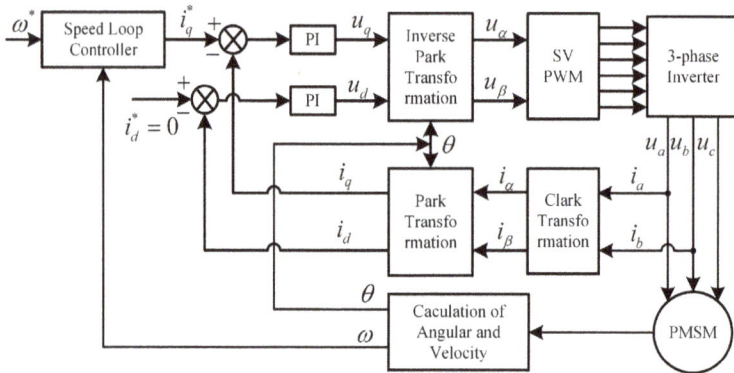

Figure 1. The field oriented control (FOC) diagram of permanent magnet synchronous motor (PMSM) system by $i_d = 0$ control approach.

3. Single Artificial Neuron Goal Representation Heuristic Dynamic Programming Controller

Like the conventional GrHDP approach [21,39,40], the proposed SAN-GrHDP controller also includes three approximate networks: an action network, a critic network, and a reference network. The critic network is set to approximate the cost-to-go function in Bellman equation by online learning. The reference network provides an adaptive internal reinforcement signal to facilitate the critic network to better approximate the value function. Compared with the classic ADP structure, GrHDP approach has an additional reference network to generate an internal goal-representation signal to facilitate learning and optimization. It provides an effective method for the intelligent system to achieve the goals by adaptive and automatic construction of internal goal representations [21]. This structure, due to the addition of reference network, also has some disadvantages, such as complex structure and high computation burden.

However, the action network of conventional GrHDP approach must be trained many times to ensure the convergence of weights. Because the action network is BP network, and it is difficult to use the conventional GrHDP approach for real-time control, especially in the field of PMSM speed control. In this article, the traditional GrHDP approach is improved and the action network is replaced by SAN control approach. Different from that of the conventional SAN control approach, the parameter K of the action network (SAN) is not fixed, and can be updated through interaction with controlled object in real time.

The schematic diagram of FOC by proposed SAN-GrHDP is shown in Figure 2. The ultimate objective for the SAN-GrHDP controller is still to solve the Bellman's optimal equation [20,22] as:

$$J^*(x, u) = \min_u \left(r(x, u) + \alpha J^*(x', u') \right) \tag{2}$$

so that the optimal control strategy can be achieved. Here the $J^*(x, u)$ is the immediate cost incurred by u at current time, the $J^*(x', u')$ is refer to the one-step future cost, the α is a discounted factor ($0 < \alpha < 1$), and the $r(x, u)$ is the external reinforcement signal.

Compared with conventional SAN control approach, the SAN-GrHDP approach has two additional networks (i.e., the reference network and the critic network). The reference network is related to the primary reinforcement signal $r(t)$, and generates the internal reinforcement signal $S(t)$ to facilitate the critic network to better approximate the value function. The critic network generates the cost function $J(t)$, according to $S(t)$.

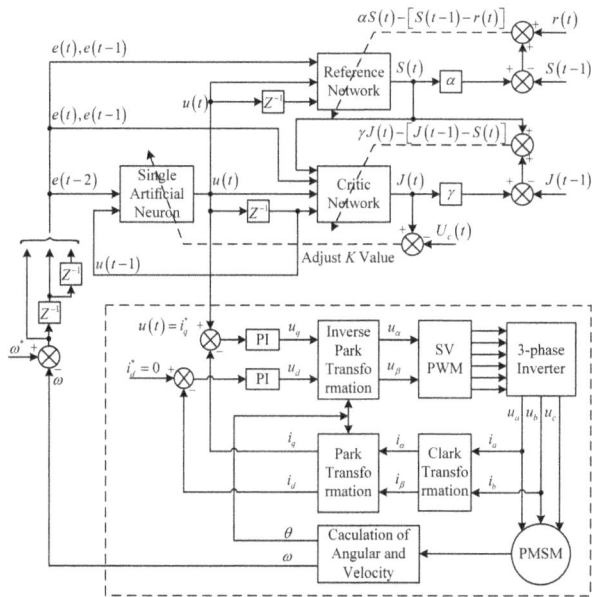

Figure 2. Schematic diagram of FOC by proposed single artificial neuron goal representation heuristic dynamic programming (SAN-GrHDP).

3.1. Learning and Adaptation of Reference Network

The structure of the reference network is shown in Figure 3. It can be seen that the reference network is designed with three-layer nonlinear architecture (including one hidden layer).

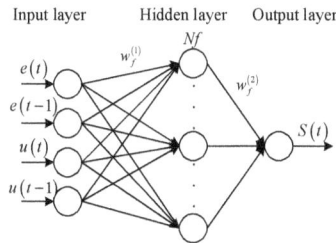

Figure 3. Schematic diagram of the reference network.

The feed-forward propagation formulas of the reference network are as follows:

$$S(t) = \sum_{i=1}^{Nf} w_{fi}^{(2)}(t) p_i(t) \; i = 1, \ldots, Nf \tag{3}$$

$$p_i(t) = \frac{1 - \exp^{-q_i(t)}}{1 + \exp^{-q_i(t)}} \; i = 1, \ldots, Nf \tag{4}$$

$$q_i(t) = \sum_{j=1}^{4} w_{f_{i,j}}^{(1)}(t) a_j(t) \; i = 1, \ldots, Nf \tag{5}$$

$$a(t) = [e(t), e(t-1), u(t), u(t-1)] \tag{6}$$

where $a(t)$ is the input vector of the reference network whose number is 4, including error value $e(t)$ at time t, error value $e(t-1)$ at time $t-1$, action value $u(t)$ at time t, and action value $u(t-1)$ at time $t-1$. $q_i(t)$ is the ith hidden node input of the reference network. $p_i(t)$ is the corresponding output of the hidden node. Nf is the total number of the hidden nodes. $S(t)$ is the output of the reference network.

We define the error function of the reference network as [25]:

$$e_f(t) = \alpha S(t) - [S(t-1) - r(t)] \tag{7}$$

and the objective function to be minimized as:

$$E_f(t) = \frac{1}{2}e_f^2(t) \tag{8}$$

To calculate the back propagation through the chain rule, the weights updating rules can be presented as follows [25]:

$\Delta w_f^{(2)}(t)$ (the weights adjustments of reference network for the hidden to the output layer):

$$\Delta w_{f_i}^{(2)}(t) = l_f(t)\left[-\frac{\partial E_f(t)}{\partial w_{f_i}^{(2)}(t)}\right] \tag{9}$$

$$\frac{\partial E_f(t)}{\partial w_{f_i}^{(2)}(t)} = \frac{\partial E_f(t)}{\partial e_f(t)}\frac{\partial e_f(t)}{\partial S(t)}\frac{\partial S(t)}{\partial w_{f_i}^{(2)}(t)} = \alpha e_f(t)p_i(t) \tag{10}$$

$\Delta w_f^{(1)}(t)$ (the weights adjustments of reference network for the input to the hidden layer):

$$\Delta w_{f_{i,j}}^{(1)}(t) = l_f(t)\left[-\frac{\partial E_f(t)}{\partial w_{f_{i,j}}^{(1)}(t)}\right] \tag{11}$$

$$\frac{\partial E_f(t)}{\partial w_{f_{i,j}}^{(1)}(t)} = \frac{\partial E_f(t)}{\partial e_f(t)}\frac{\partial e_f(t)}{\partial S(t)}\frac{\partial S(t)}{\partial p_i(t)}\frac{\partial p_i(t)}{\partial q_i(t)}\frac{\partial q_i(t)}{\partial w_{f_{i,j}}^{(1)}(t)} = \alpha e_f(t)w_{f_i}^{(2)}\left[\frac{1}{2}\left(1-p_i^2(t)\right)\right]a_j(t) \tag{12}$$

3.2. Learning and Adaptation of Critic Network

The structure of the critic network is shown in Figure 4. It is designed with a three-layer nonlinear architecture (with one hidden layer).

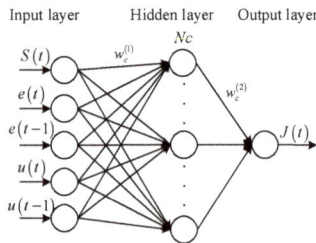

Figure 4. Schematic diagram of the critic network.

The feed-forward propagation formulas of the critic network are as follows:

$$J(t) = \sum_{l=1}^{Nc} w_{c_l}^{(2)}(t) y_l(t) \; l = 1, \ldots, Nc \tag{13}$$

$$y_l(t) = \frac{1 - \exp^{-z_l(t)}}{1 + \exp^{-z_l(t)}} \; l = 1, \ldots, Nc \tag{14}$$

$$z_l(t) = \sum_{k=1}^{5} w_{c_{l,k}}^{(1)}(t) c_k(t) \; l = 1, \ldots, Nc \tag{15}$$

$$c(t) = [S(t), e(t), e(t-1), u(t), u(t-1)] \tag{16}$$

where $c(t)$ is the input vector of the critic network which number is 5, including the internal reinforcement signal $S(t)$ (produced by reference network), error value $e(t)$ at time t, error value $e(t-1)$ at time $t-1$, action value $u(t)$ at time t and action value $u(t-1)$ at time $t-1$. $z_l(t)$ is the lth hidden node input of the critic network. $y_l(t)$ is the corresponding output of the hidden node. Nc is the total number of hidden nodes. $J(t)$ is the output of the critic network.

Define the error function of the critic network as [19,21]:

$$e_c(t) = \gamma J(t) - [J(t-1) - S(t)] \tag{17}$$

and the objective function to be minimized as:

$$E_c(t) = \frac{1}{2} e_c^2(t) \tag{18}$$

To calculate the backpropagation through the chain rule, the weights updating rules can be presented as follows [21]:

$\Delta w_c^{(2)}(t)$ (the weights adjustments of critic network for the hidden to the output layer):

$$\Delta w_{c_l}^{(2)}(t) = l_c(t) \left[-\frac{\partial E_c(t)}{\partial w_{c_l}^{(2)}(t)} \right] \tag{19}$$

$$\frac{\partial E_c(t)}{\partial w_{c_l}^{(2)}(t)} = \frac{\partial E_c(t)}{\partial e_c(t)} \frac{\partial e_c(t)}{\partial J(t)} \frac{\partial J(t)}{\partial w_{c_l}^{(2)}(t)} = \gamma e_c(t) y_l(t) \tag{20}$$

$\Delta w_c^{(1)}(t)$ (the weights adjustments of critic network for the input to the hidden layer):

$$\Delta w_{c_{l,k}}^{(1)}(t) = l_c(t) \left[-\frac{\partial E_c(t)}{\partial w_{c_{l,k}}^{(1)}(t)} \right] \tag{21}$$

$$\frac{\partial E_c(t)}{\partial w_{c_{l,k}}^{(1)}(t)} = \frac{\partial E_c(t)}{\partial e_c(t)} \frac{\partial e_c(t)}{\partial J(t)} \frac{\partial J(t)}{\partial y_l(t)} \frac{\partial y_l(t)}{\partial z_l(t)} \frac{\partial z_l(t)}{\partial w_{c_{l,k}}^{(1)}(t)} = \gamma e_c(t) w_{c_l}^{(2)} \left[\frac{1}{2} \left(1 - y_l^2(t) \right) \right] c_k(t) \tag{22}$$

3.3. Learning and Adaptation of Action Network

The structure of the action network (SAN) is shown in Figure 5. The SAN is employed as the controller, which is different from the traditional GrHDP (action network is BP network). The feed-forward propagation formulas of the SAN are introduced as follows [27]:

$$u(t) = u(t-1) + K(t)\Delta u(t) \tag{23}$$

$$\Delta u(t) = \sum_{i=1}^{2} w_i(t)x_i(t) \tag{24}$$

$$\begin{cases} w_1(t) = w_1(t-1) + \eta_P e(t)u(t)(e(t) + \Delta e(t)) \\ w_2(t) = w_2(t-1) + \eta_I e(t)u(t)(e(t) + \Delta e(t)) \end{cases} \tag{25}$$

$$\begin{cases} x_1(t) = e(t) - e(t-1) \\ x_2(t) = e(t) \end{cases} \tag{26}$$

where $u(t)$ is the output of the action network (SAN), which is applied to the controlled object directly. η_P, η_I are proportion, integral study rate respectively.

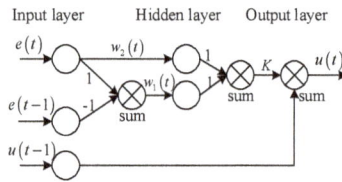

Figure 5. Schematic diagram of the action network (SAN).

The parameter K named neuron scale-up factor (where $K > 0$) is very important to the control response performance. The selection of K is very difficult of traditional SAN control approach. The control system will respond faster, if the K value is greater. However, the K value will lead to the instability of the system, if it is out of a certain range.

The key point of the SAN-GrHDP approach is to use the approximate function J from critic network to achieve the K value of optimization adjustment. Define "0" as the reinforcement signal for "success", and "−1" for "failure", so $U_c(t)$ is set to "0" for our following studies.

To calculate the backpropagation, the error function $e_a(t)$ is defined as follows [19]:

$$e_a(t) = J(t) - U_c(t) \tag{27}$$

and the objective function to be minimized as:

$$E_a(t) = \frac{1}{2}e_a^2(t) \tag{28}$$

For backward propagation, the error function of the reference network is not only related to the primary reinforcement signal $r(t)$, but also the internal reinforcement signal $S(t)$.

To calculate the backpropagation through the chain rule, the error function of the critic network involves the internal reinforcement signal $S(t)$. The signal $S(t)$ from reference network is related to the primary reinforcement signal $r(t)$. So the parameter K updating rules are composed of two parts: one is from the critic network path and the other is from the reference network path.

The detailed learning and adaptation formulas can be presented as follows:

$$\Delta K(t) = l_a(t)\left[-\frac{\partial E_a(t)}{\partial K(t)}\right] \tag{29}$$

$$\frac{\partial E_a(t)}{\partial K(t)} = \frac{\partial E_a(t)}{\partial J(t)}\frac{\partial J(t)}{\partial K(t)} + \frac{\partial E_a(t)}{\partial J(t)}\frac{\partial J(t)}{\partial S(t)}\frac{\partial S(t)}{\partial K(t)} \tag{30}$$

$$\begin{aligned} P_{a1} &= \frac{\partial E_a(t)}{\partial J(t)}\frac{\partial J(t)}{\partial K(t)} = \sum_{l=1}^{Nc} \frac{\partial E_a(t)}{\partial J(t)}\frac{\partial J(t)}{\partial y_l(t)}\frac{\partial y_l(t)}{\partial z_l(t)}\frac{\partial z_l(t)}{\partial u(t)}\frac{\partial u(t)}{\partial K(t)} \\ &= e_a(t) \cdot \sum_{l=1}^{Nc} w_{c_l}^{(2)}(t)\left[\frac{1}{2}\left(1 - y_l^2(t)\right)\right]w_{c_{l,A}}^{(1)}(t)\Delta u(t) \end{aligned} \tag{31}$$

$$P_{a2} = \frac{\partial E_a(t)}{\partial J(t)} \frac{\partial J(t)}{\partial S(t)} \frac{\partial S(t)}{\partial K(t)} = \sum_{l=1}^{Nc} \frac{\partial E_a(t)}{\partial J(t)} \frac{\partial J(t)}{\partial y_l(t)} \frac{\partial y_l(t)}{\partial z_l(t)} \frac{\partial z_l(t)}{\partial S(t)} \frac{\partial S(t)}{\partial K(t)}$$
$$= \sum_{l=1}^{Nc} \frac{\partial E_a(t)}{\partial J(t)} \frac{\partial J(t)}{\partial y_l(t)} \frac{\partial y_l(t)}{\partial z_l(t)} \frac{\partial z_l(t)}{\partial S(t)} \cdot \sum_{i=1}^{Nf} \frac{\partial S(t)}{\partial p_i(t)} \frac{\partial p_i(t)}{\partial q_i(t)} \frac{\partial q_i(t)}{\partial u(t)} \frac{\partial u(t)}{\partial K(t)}$$
$$= e_a(t) \cdot \sum_{l=1}^{Nc} w_{c_l}^{(2)}(t) \left[\tfrac{1}{2}\left(1 - y_l^2(t)\right) \right] w_{c_{l,1}}^{(1)}(t) \cdot \sum_{i=1}^{Nf} w_{f_i}^{(2)}(t) \left[\tfrac{1}{2}\left(1 - p_i^2(t)\right) \right] w_{f_{i,3}}^{(1)}(t) \Delta u(t)$$

(32)

where $l_a(t)$ is the learning rate of the parameter K. In the end, the gradient descent rule is selected as the tuning method of the parameter K, the formula is presented as follows:

$$K(t+1) = K(t) + \Delta K(t) \tag{33}$$

4. Simulation and Experiment Results

4.1. Reinforcement Signal Design of Speed Controller

The SAN-GrHDP controller is a real-time controller with immediate online learning from the surroundings, and its overall performance depends upon the design of the input, output and reinforcement signal.

The input signal of the controller is designed as follows:

$$\begin{cases} e(t) = \omega^*(t) - \omega(t) \\ e(t-1) = \omega^*(t-1) - \omega(t-1) \\ e(t-2) = \omega^*(t-2) - \omega(t-2) \end{cases} \tag{34}$$

where $\omega(t)$ is actual angular velocity of PMSM (obtained by the encoder) in time t, $\omega^*(t)$ is the reference angular velocity.

The output signal of the controller is i_q^*. The cost-to-go function (reinforcement signal) is designed as follows:

$$r(t) = 0.98 * e(t) + 0.02 * e(t-1) \tag{35}$$

Conventional controller designs are primarily based on on-linear analysis gear such as eigenvalue analysis, Bode diagrams, Nyquist diagrams and so on. In contrast, the SAN-GrHDP is based totally on online learning to regulate its parameters to reduce the reinforcement signal. Due to the similar approximation functionality of the neural network, it's far more liable to find the proper mapping among the input and output signals to withstand the disturbance of PMSM parameters. The critic network is used to approximate the cost-to-go function (reinforcement signal) $r(t)$ in the Bellman's optimal equation of dynamic programming [20]. The Bellman's optimal equation is shown in Equation (2). The reference network is integrated in the typical ADP structure to approximate an internal reinforcement signal $S(t)$. The internal reinforcement signal is used to interact with the operation of the critic network [21]. It can better facilitate the optimal learning and control over time to accomplish goals [30].

It is known that the initial parameters are significant for the performance of the SAN-GrHDP controller. Table 1 shows the parameters setting of the proposed SAN-GrHDP approach. Where, $l_a(t)$ is the learning rate of the action network, $l_f(t)$ is the learning rate of the reference network, and $l_c(t)$ is the learning rate of the critic network. The learning rate of the reference network is usually set same as the critic network. When these two learning rates are set too big, it will lead to instability of the controller. When these two learning rates are set too small, the convergence rate of controller is slow. When training offline, these two learning rates can be set bigger, and weights of these two networks can be obtained rapidly. After offline training, these two learning rates can be set a little bit lower, which can enhance the stability of the controller. The selection of K is very difficult in the traditional SAN control approach. The control system will respond faster, if the K value is greater. However, the K value will lead to the instability of the system, if it is out of a certain range. The rate of value K

variation is decided by the learning rate of the action network, which is usually set according to the experimental process. The α is the discount factor of the reference network, γ is the discount factor of the critic network. The discount factor determine how much the t moment affects the previous $t-1$ moment. If the discount factor is set too small, the effect of reinforcement learning signal at the current moment is small; otherwise, the effect is large. They are usually set between 0.95 and 0.99. The Nf is the hidden node number of the reference network. The Nc is the hidden node number of the critic network. Both the hidden node number of the critic network and the reference network are set to 8. The more layers, the better performance of controller. However, the more layers need a more powerful processor. According to experimental research, the quantity of layers is 8, so that computing speed of DSP28335 processor is acceptable. For a more detailed description of the process for setting the parameters of the ADP method readers may refer to relevant works [22].

Table 1. Parameters setting of the SAN-GrHDP approach.

Quantity	Symbol	Value
Learning rate of the action network	$l_a(t)$	0.5
Learning rate of the reference network	$l_f(t)$	0.03
Learning rate of the critic network	$l_c(t)$	0.03
Discount factor of the reference network	α	0.98
Discount factor of the critic network	γ	0.95
Hidden node number of the critic network	Nc	8
Hidden node number of the reference network	Nf	8

Using the ADP approaches with the characteristics of the interaction of the control object (vector control system of PMSM). Through the evaluation value J of critic network, the state variable feedback control object is calculated with the gradient descent rule, to guide the selection of SAN controller's K value, expressed as follows:

$$K(t+1) = K(t) + \Delta K(t) = K(t) + l_a(t)\left[-\frac{\partial E_a(t)}{\partial K(t)}\right] = K(t) + l_a(t)\left[-\frac{1}{2}\frac{\partial(J^2(t))}{\partial K(t)}\right] \tag{36}$$

The detailed learning and adaptation are shown in Equations (27)–(33).The selection of K value is used to promote the rapid convergence of the J value. The appropriate K value is selected and applied to the SAN (action network), and the optimal control value is output to vector control system of PMSM directly. The detailed calculating process is shown in Equations (23)–(26). The SAN-GrHDP optimal control output signal is q-axis current reference value i_q^* of vector control system of PMSM. The weights of the reference network and critic network in SAN-GrHDP approach are initialized randomly. For comparative studies, the parameters of SAN approach are set the same as the SAN-GrHDP approach.

4.2. Learning Process of Single Artificial Neuron Goal Representation Heuristic Dynamic Programming Speed Controller for Permanent Magnet Synchronous Motor

In the field of oriented control system of PMSM, speed difference is usually chosen as the input signal for the speed controller. In this SAN-GrHDP controller, previous control output is usually used as a supplementary signal input of the controller, so the controller input is of error value $e(t)$ at time t, error value $e(t-1)$ at time $t-1$, error value $e(t-2)$ at time $t-2$, previous control output value $u(t-1)$, and the controller output is $u(t)$. The optimization parameters of controller will be updated accordingly by online learning. The data flowchart is shown in Figure 6 and the algorithm training process is described as follows:

(1) Initialize the various parameters of the SAN-GrHDP, such as neural network learning rate, the initial weights values of neural network, discount factor and so on.

(2) Observe the differences of speed and obtain the control signal $u(t)$ that is q-axis current reference value for the control system of PMSM.

(3) Calculate the internal reinforcement learning signal $S(t)$, and the value function signal $J(t)$.

(4) Retrieve the previous time data $S(t-1)$ and $J(t-1)$, calculate the temporal difference errors and obtain the objective functions in reference network and critic network.

(5) Update the weights values of reference network, critic network and the K value of action network (SAN).

(6) Repeat from the second step when entering the $t+1$ step.

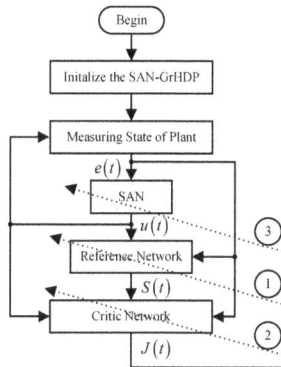

Figure 6. Flowchart of the SAN-GrHDP procedure.

4.3. Simulation and Experimental Results

The weights of the reference network and critic network in SAN-GrHDP approach are initialized randomly. For comparative studies, the parameters of SAN control approach are set the same as the SAN-GrHDP approach. To check the overall performance of the SAN-GrHDP control approach, simulation, and experiment on the speed control system of PMSM are carried out.

4.3.1. Simulation Results

To compare the disturbance rejection performance of both approaches, the comparative simulation of the proposed SAN-GrHDP control approach and the traditional SAN control approach are implemented on Simulink Matlab2016a (MathWorks, Natick, MA, USA). The parameters of the PMSM used in the simulation are listed in Table 2. The parameters of both current PI are the same: the proportional coefficient is 9, the integral coefficient is 3375. The saturation limit of the q-axis reference current i_q^* is ± 10 A. The initial load of PMSM is 0.2 N·m.

Table 2. Parameters setting of the PMSM.

Parameter	Symbol	Value
Rated Voltage	V	36 V
Rated Current	I	4.6 A
Maximum Current	I_{max}	13.8 A
Rated Power	P	100 W
Rated Torque	T	0.318 N·m
Stator Phase Resistance	R	0.375 Ohm
Motor Inertia	SI	0.0588 kg·m^2·10^{-4}
Pole Pairs	P_n	4 Pair
Q-axis Inductance	L_q	0.001 H
D-axis Inductance	L_d	0.001 H
Incremental Encoder Lines	N	2500PPR

Figure 7 shows that simulation responses under SAN and SAN-GrHDP approaches in the presence of load torque disturbance at 1300 rpm. Figure 8 shows that simulation responses under SAN and SAN-GrHDP approaches in the presence of load torque disturbance at 800 rpm. Figure 7a shows that the SAN-GrHDP-based controller gives the same settling time with a same overshoot compared with the SAN-based controller, in the case of 1300 rpm reference speed. Figure 8a shows that the SAN-GrHDP-based controller gives the same settling time with a same overshoot compared with the SAN-based controller, in the case of 800 rpm reference speed. It can also be seen that, when a load torque 0.5 N·m is applied at 0.1 s, the SAN-GrHDP approach has less speed fluctuation than the traditional SAN approach.

Figure 7b shows that the q-axis current response under SAN and SAN-GrHDP approaches in the presence of load torque disturbance at 1300 rpm. It shows that the q-axis current i_q is quite large at the moment of the start of PMSM. The i_q^* is much less than 10 A, which is the saturation limit of the output. As the speed is steady, the actual q-axis current i_q decreases down to reference q-axis current i_q^*. It can also be seen that, when a load torque 0.5 N·m is applied at 0.1 s, the actual q-axis current i_q of both approaches rise quickly under the sudden load disturbance impact. However, the SAN-GrHDP approach has less current fluctuation than the traditional SAN control approach. Figure 8b shows that the q-axis current response under SAN and SAN-GrHDP approaches in the presence of load torque disturbance at 800 rpm. It can be seen from the Figures 7b and 8b, when the same load torque 0.5 N·m is added suddenly at different speed, the q-axis current response at 1300 rpm is same as 800 rpm.

The evolution of the neural network parameters is presented in the SAN-GrHDP controller at 1300 rpm in Figure 7c–e. Figure 7c shows that the trajectory of the parameter K. At the load disturbance time (0.1 s), the neural network weights are adapting dramatically, which is constant with the full-size adjustments in the reinforcement signals, as shown in Figure 7d,e. The reason is that in spite of the load mutation, the system is converting according to the controller learning surroundings, so that it adapt its parameters to provide the most suitable control signal for the system again to achieve its normal working point. The evolution of the neural network parameters is presented in the SAN-GrHDP controller at 800 rpm in Figure 8c–e.

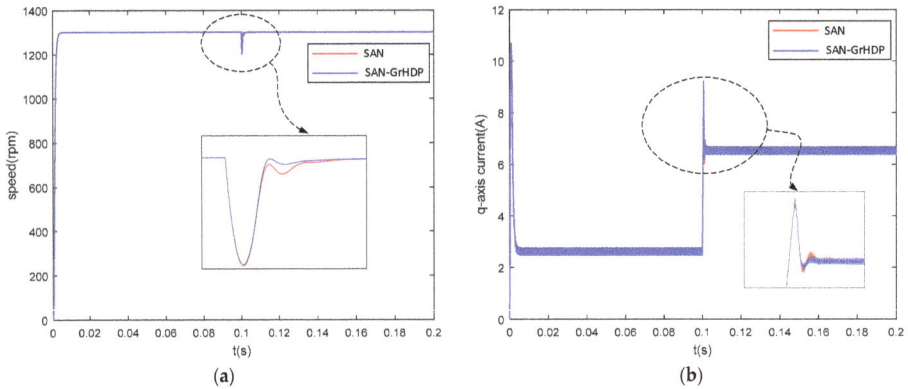

Figure 7. *Cont.*

(c)

(d)

(e)

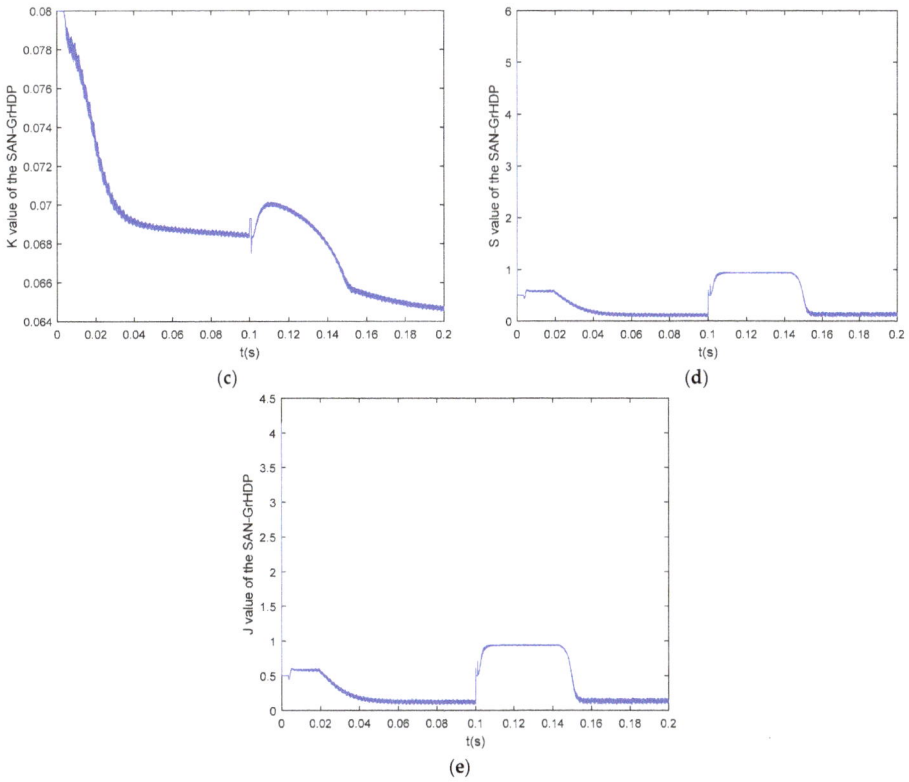

Figure 7. Simulation responses under SAN and SAN-GrHDP approaches in the presence of load torque disturbance at 1300 rpm. (**a**) Speed. (**b**) i_q. (**c**) K value of the SAN-GrHDP approach. (**d**) S value of the SAN-GrHDP approach. (**e**) J value of the SAN-GrHDP approach.

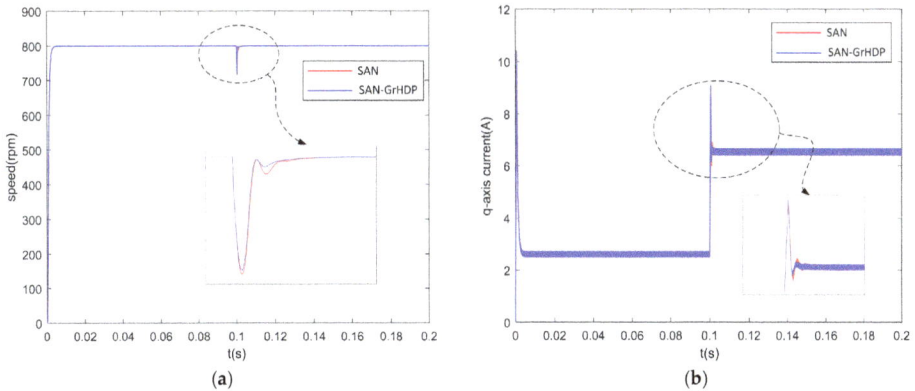

(a)

(b)

Figure 8. *Cont.*

(c)

(d)

(e)

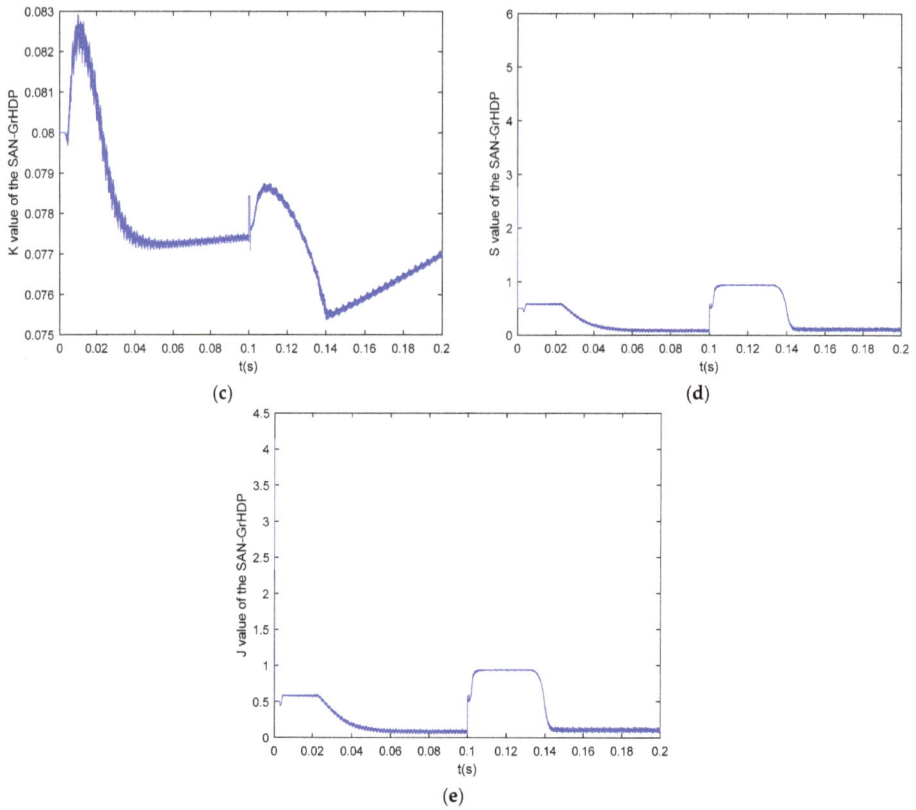

Figure 8. Simulation responses under SAN and SAN-GrHDP approaches in the presence of load torque disturbance at 800 rpm. (**a**) Speed. (**b**) i_q. (**c**) K value of the SAN-GrHDP approach. (**d**) S value of the SAN-GrHDP approach. (**e**) J value of the SAN-GrHDP approach.

4.3.2. Experimental Results

An experimental platform for a PMSM device is built to evaluate the overall performance of the proposed SAN-GrHDP control approach. The configuration and the experimental test setup are shown in Figures 9 and 10, respectively.

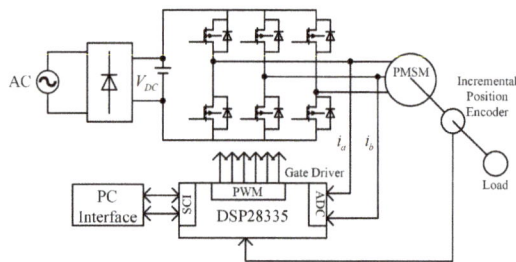

Figure 9. Configuration of the experimental system.

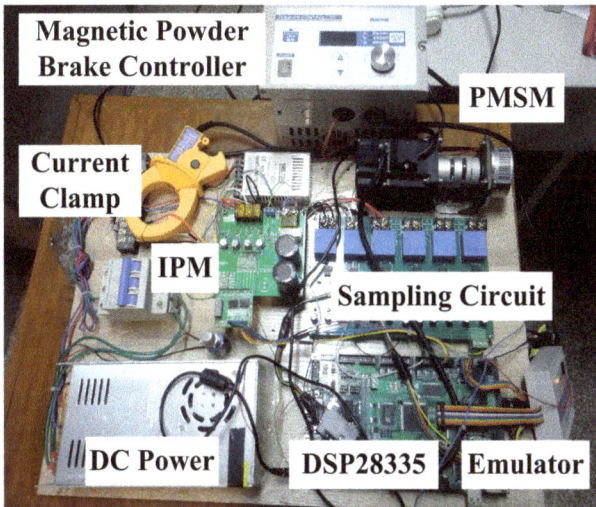

Figure 10. Experimental test setup.

All the control algorithms, which include the SVPWM technique, are implemented by using this system of the floating DSP TMS320F28335 with a clock frequency of one hundred and fifty MHz, the usage of a C-language. The current-loop sampling period is 200 μs, the speed loop sampling time is ten times that of the current loop. The saturation restriction of the q-axis reference current is ± 10 A. The PMSM is driven by using an intelligent power module (IPM) PS21965, which is designed by the Mitsubishi Company (Tokyo, Japan). The phase currents are measured by Hall sensors, converted to voltages by sampling resistances and AD7606 converter. The rotor speed and absolute rotor position can be measured by the incremental position encoder of 2500 lines. The speed and q-axis current signals are displayed on the oscilloscope, through a DAC converter (AD5344) output.

The parameters of both current PI units are the same: the proportional coefficient is 0.2, the integral coefficient is 0.006. The parameters of SAN are as follow: $\eta_p = 0.05$, $\eta_I = 0.05$. The initial value of scale-up factor $K = 0.01$. The parameters of SAN-GrHDP are shown in Table 1, the parameters of action network are same as SAN control approach.

Figure 11 shows the experimental response curves of speed and i_q with sudden load disturbance by SAN control approach at 1300 rpm. Figure 12 shows the experimental response curves of speed and i_q with the same sudden load disturbance by SAN-GrHDP approach at 1300 rpm. From Figure 11, it can be seen that the speed of SAN approach fluctuates greatly when load is added. It can be inferred from Figure 11 that the control effect of SAN can be improved with application of the machine learning (GrHDP) to tuning the K value. The proposed control strategy can quickly stabilize the speed when load is added. Figures 13 and 14 show the comparative experimental response curves with the SAN and proposed SAN-GrHDP approach at 800 rpm, respectively. The experimental results in Figures 13 and 14 are similar in Figures 11 and 12. From the experimental results, it can be seen that there are some differences from the results of simulation. The reason is that the PMSM model in simulation is ideal, and it has some disparities in practical application. In the process of experiment, the fluctuation error of speed is greater than the simulation result in steady state. The proposed SAN-GrHDP approach is a kind of machine learning algorithm (ADP). It can learned by itself according to the environmental characteristics. Therefore, the weights of neural networks in experiment are different from that of simulation. This is also the reason for disparities between simulation and experimental results. It is found that compared with the SAN control approach, the proposed SAN-GrHDP approach indicates a higher disturbance rejection potential, with much less speed fluctuation and shorter recovering time towards load disturbance.

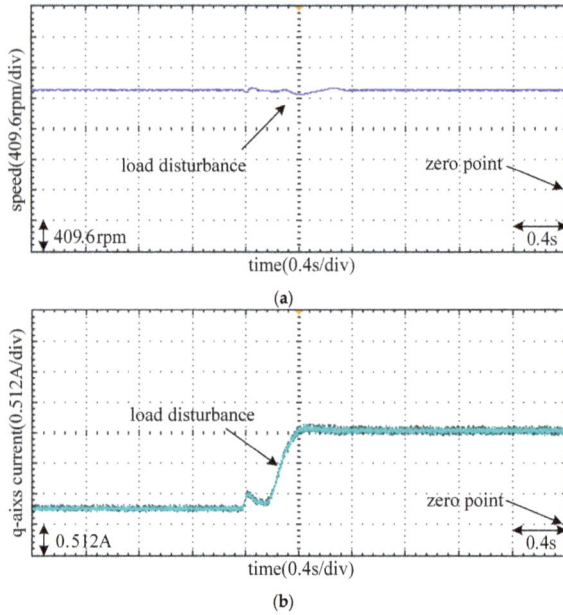

Figure 11. Experimental responses under SAN in the presence of load torque disturbance at 1300 rpm. (**a**) Speed; and (**b**) i_q.

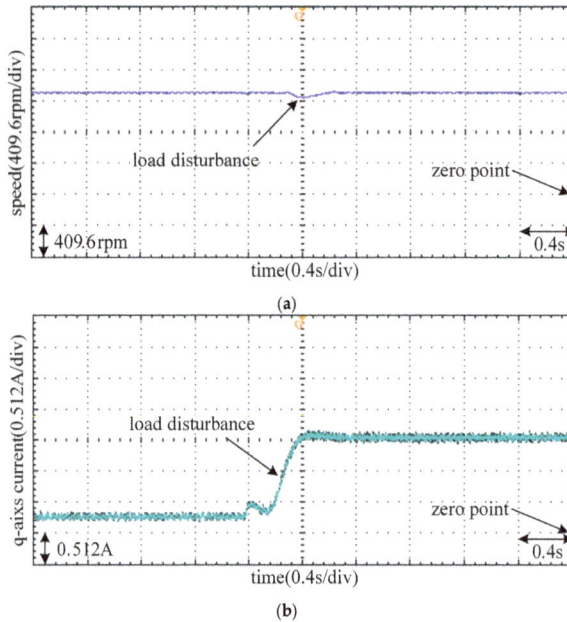

Figure 12. Experimental responses under SAN-GrHDP in the presence of load torque disturbance at 1300 rpm. (**a**) Speed; and (**b**) i_q.

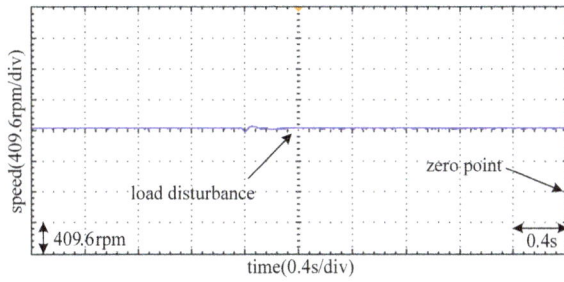

(a)

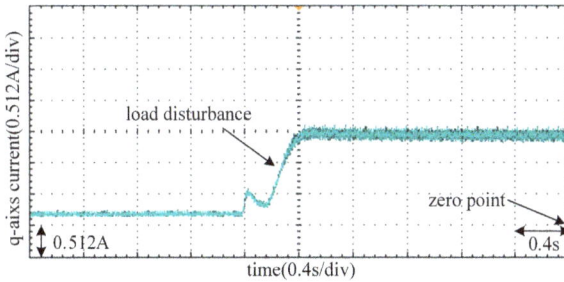

(b)

Figure 13. Experimental responses under SAN in the presence of load torque disturbance at 800 rpm. (**a**) Speed; and (**b**) i_q.

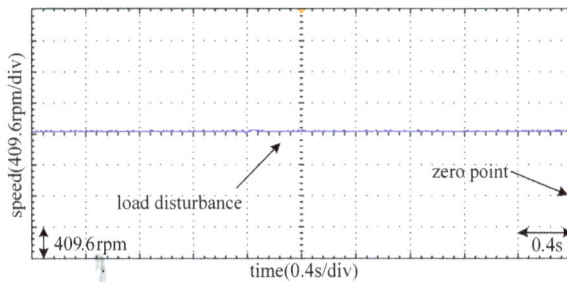

(a)

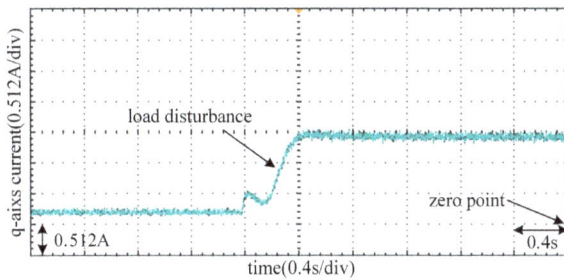

(b)

Figure 14. Experimental responses under SAN-GrHDP in the presence of load torque disturbance at 800 rpm. (**a**) Speed; and (**b**) i_q.

5. Conclusions

In order to improve the disturbance rejection capacity of PMSM closed-loop systems, the design and implementation of a novel adaptive speed controller for a PMSM was investigated in this paper. This controller was a composite reinforcement learning control approach which combines SAN and GrHDP collectively, namely SAN-GrHDP. The proposed control approach could develop an internal reinforcement learning signal to adjust the K value of the traditional SAN control approach, whenever external parameters varies.

From our simulation and experimental results, it could be concluded that the dual closed-loop structures of PMSM under the proposed SAN-GrHDP approach had a satisfying dynamic overall performance. The composite SAN-GrHDP approach was able to achieve a fulfilling performance with speedy temporary reaction, precise disturbance rejection capacity.

Because of the uncertainty of the network weights, most articles about the ADP approach put all the emphasis on the simulation part, instead of actual applications. In this article, the traditional GrHDP approach was improved and the action network was replaced by SAN. The stability of this proposed algorithm could be improved in practical applications by using SAN. The core idea of the proposed algorithm is machine learning. At this stage, there are still some unstable situations in practical applications, such as longer training time, complex structure, and so on. However, it is promising to apply it in actual control system to solve electric engineering problems.

Finally, perspectives on future research may be listed as follows. (1) The learning rate of the neural network can be chosen in an optimal way; (2) The success rate of the algorithm should be improved; (3) A rigorous stability analysis is required to show the convergence of SAN-GrHDP approach; (4) The critic and reference neural networks should be replaced by other mathematical models; (5) Some experimental designs of internal disturbance rejection capacity (PMSM parameter variations) of proposed SAN-GrHDP approach should be discussed in detail; (6) Only one input variable (PMSM speed) is taken into consideration in the proposed scheme. Therefore, further investigation can expand it to a more generalized case of multi-variables.

Author Contributions: Q.W. derived the algorithm, designed the simulation and experiment. M.W. formatted the manuscript and discussed the experimental results. H.Y. and X.Q. supervised the manuscript writing.

Funding: This research was funded by the National Natural Science Foundation of China 41576096 and the Foundation of Henan Educational Committee 17A120008.

Conflicts of Interest: The authors declare no conflict of interest.

Acronyms

PMSM	Permanent magnet synchronous motor
FOC	Field oriented control
SAN-GrHDP	Single artificial neuron goal representation heuristic dynamic programming
SAN	Single artificial neuron
DSP	Digital signal processor
RL	Reinforcement learning
ADP	Adaptive dynamic programming
GrHDP	Goal representation heuristic dynamic programming
HDP	Heuristic dynamic programming
DHP	Dual heuristic dynamic programming
GDHP	Globalized dual heuristic dynamic programming
BP	Back propagation

Energies **2018**, *11*, 2355

Constants

L_d	Stator d-axes inductance
L_q	Stator q-axes inductance
R_s	Stator resistance
ψ_f	Flux linkage
B	Viscous friction coefficient
n_p	Number of pole pairs
Nf	The hidden node number of the reference network
Nc	The hidden node number of the critic network

Variables

K	Neuron scale-up factor
u_d	Stator d-axes voltage
u_q	Stator q-axes voltage
i_d	Stator d-axes current
i_q	Stator q-axes current
T_L	Load torque
i_α	d-axes static coordinate current
i_β	q-axes static coordinate current
θ	Rotor position
i_q^*	q-axes reference current
α	Discounted factor of reference network $(0 < \alpha < 1)$
γ	Discounted factor of critic network $(0 < \gamma < 1)$
r	External reinforcement signal
S	Internal reinforcement signal
J	Cost function
a	Input vector of the reference network
u	Control signal
q_i	ith hidden node input of the reference network
p_i	ith hidden node output of the reference network
$\Delta w_f^{(2)}$	The weights adjustments of reference network for the hidden to the output layer
$\Delta w_f^{(1)}$	The weights adjustments of reference network for the input to the hidden layer
c	Input vector of the critic network
z_l	lth hidden node input of the critic network
y_l	lth hidden node output of the critic network
$\Delta w_c^{(2)}$	The weights adjustments of critic network for the hidden to the output layer
$\Delta w_c^{(1)}$	The weights adjustments of critic network for the input to the hidden layer
η_P	Proportion study rate of SAN
η_I	Integral study rate of SAN
l_a	Learning rate of the parameter K
ω	Actual angular velocity of PMSM
ω^*	Reference angular velocity of PMSM
l_f	Learning rate of the reference network
l_c	Learning rate of the critic network

References

1. Calvini, M.; Carpita, M.; Formentini, A.; Marchesoni, M. PSO-Based Self-Commissioning of Electrical Motor Drives. *IEEE Trans. Ind. Electron.* **2015**, *62*, 768–776. [CrossRef]
2. Li, S.; Liu, Z. Adaptive Speed Control for Permanent-Magnet Synchronous Motor System with Variations of Load Inertia. *IEEE Trans. Ind. Electron.* **2009**, *56*, 3050–3059. [CrossRef]
3. Jung, J.W.; Leu, V.Q.; Do, T.D.; Kim, E.K.; Choi, H.H. Adaptive PID Speed Control Design for Permanent Magnet Synchronous Motor Drives. *IEEE Trans. Power Electron.* **2015**, *30*, 900–908. [CrossRef]

4. Underwood, S.J.; Husain, I. Online Parameter Estimation and Adaptive Control of Permanent-Magnet Synchronous Machines. *IEEE Trans. Ind. Electron.* **2010**, *57*, 2435–2443. [CrossRef]

5. Liu, J.; Li, H.; Deng, Y. Torque Ripple Minimization of PMSM Based on Robust ILC via Adaptive Sliding Mode Control. *IEEE Trans. Power Electron.* **2018**, *33*, 3655–3671. [CrossRef]

6. Repecho, V.; Biel, D.; Arias, A. Fixed Switching Period Discrete-Time Sliding Mode Current Control of a PMSM. *IEEE Trans. Ind. Electron.* **2018**, *65*, 2039–2048. [CrossRef]

7. Joo, K.; Park, J.; Lee, J.; Joo, K.J.; Park, J.S.; Lee, J. Study on Reduced Cost of Non-Salient Machine System Using MTPA Angle Pre-Compensation Method Based on EEMF Sensorless Control. *Energies* **2018**, *11*, 1425. [CrossRef]

8. Park, J.; Wang, X.; Park, J.B.; Wang, X. Sensorless Direct Torque Control of Surface-Mounted Permanent Magnet Synchronous Motors with Nonlinear Kalman Filtering. *Energies* **2018**, *11*, 969. [CrossRef]

9. Su, D.; Zhang, C.; Dong, Y.; Su, D.; Zhang, C.; Dong, Y. An Improved Continuous-Time Model Predictive Control of Permanent Magnetic Synchronous Motors for a Wide-Speed Range. *Energies* **2017**, *10*, 2051. [CrossRef]

10. Yang, M.; Liu, Z.; Long, J.; Qu, W.; Xu, D.; Yang, M.; Liu, Z.; Long, J.; Qu, W.; Xu, D. An Algorithm for Online Inertia Identification and Load Torque Observation via Adaptive Kalman Observer-Recursive Least Squares. *Energies* **2018**, *11*, 778. [CrossRef]

11. Sun, X.; Chen, L.; Yang, Z.; Zhu, H. Speed-Sensorless Vector Control of a Bearingless Induction Motor with Artificial Neural Network Inverse Speed Observer. *IEEE/ASME Trans. Mechatron.* **2013**, *18*, 1357–1366. [CrossRef]

12. Chaoui, H.; Sicard, P. Adaptive Fuzzy Logic Control of Permanent Magnet Synchronous Machines with Nonlinear Friction. *IEEE Trans. Ind. Electron.* **2012**, *59*, 1123–1133. [CrossRef]

13. Barkat, S.; Tlemçani, A.; Nouri, H. Noninteracting Adaptive Control of PMSM Using Interval Type-2 Fuzzy Logic Systems. *IEEE Trans. Fuzzy Syst.* **2011**, *19*, 925–936. [CrossRef]

14. Lai, C.K.; Shyu, K.-K. A novel motor drive design for incremental motion system via sliding-mode control method. *IEEE Trans. Ind. Electron.* **2005**, *52*, 499–507. [CrossRef]

15. Liu, J.; Vazquez, S.; Wu, L.; Marquez, A.; Gao, H.; Franquelo, L.G. Extended State Observer-Based Sliding-Mode Control for Three-Phase Power Converters. *IEEE Trans. Ind. Electron.* **2017**, *64*, 22–31. [CrossRef]

16. Do, T.D.; Kwak, S.; Choi, H.H.; Jung, J.W. Suboptimal Control Scheme Design for Interior Permanent-Magnet Synchronous Motors: An SDRE-Based Approach. *IEEE Trans. Power Electron.* **2014**, *29*, 3020–3031. [CrossRef]

17. Sun, X.; Shi, Z.; Chen, L.; Yang, Z. Internal Model Control for a Bearingless Permanent Magnet Synchronous Motor Based on Inverse System Method. *IEEE Trans. Energy Convers.* **2016**, *31*, 1539–1548. [CrossRef]

18. Sun, X.; Chen, L.; Jiang, H.; Yang, Z.; Chen, J.; Zhang, W. High-Performance Control for a Bearingless Permanent-Magnet Synchronous Motor Using Neural Network Inverse Scheme Plus Internal Model Controllers. *IEEE Trans. Ind. Electron.* **2016**, *63*, 3479–3488. [CrossRef]

19. Si, J.; Wang, Y.-T. Online learning control by association and reinforcement. *IEEE Trans. Neural Netw.* **2001**, *12*, 264–276. [CrossRef] [PubMed]

20. Prokhorov, D.V.; Wunsch, D.C. Adaptive critic designs. *IEEE Trans. Neural Netw.* **1997**, *8*, 997–1007. [CrossRef] [PubMed]

21. He, H.; Ni, Z.; Fu, J. A three-network architecture for on-line learning and optimization based on adaptive dynamic programming. *Neurocomputing* **2012**, *78*, 3–13. [CrossRef]

22. He, H. Adaptive Dynamic Programming for Machine Intelligence. In *Self-Adaptive Systems for Machine Intelligence*; John Wiley & Sons, Inc.: Hoboken, NJ, USA, 2011; pp. 140–164, ISBN 978-1-118-02560-4.

23. Tang, Y.; He, H.; Wen, J.; Liu, J. Power System Stability Control for a Wind Farm Based on Adaptive Dynamic Programming. *IEEE Trans. Smart Grid* **2015**, *6*, 166–177. [CrossRef]

24. Tang, Y.; Mu, C.; He, H. SMES-Based Damping Controller Design Using Fuzzy-GrHDP Considering Transmission Delay. *IEEE Trans. Appl. Supercond.* **2016**, *26*, 1–6. [CrossRef]

25. Tang, Y.; Yang, J.; Yan, J.; He, H. Intelligent load frequency controller using GrADP for island smart grid with electric vehicles and renewable resources. *Neurocomputing* **2015**, *170*, 406–416. [CrossRef]

26. Mishra, S. Neural-network-based adaptive UPFC for improving transient stability performance of power system. *IEEE Trans. Neural Netw.* **2006**, *17*, 461–470. [CrossRef] [PubMed]

27. Cao, H.; Li, X. Thermal Management-Oriented Multivariable Robust Control of a kW-Scale Solid Oxide Fuel Cell Stand-Alone System. *IEEE Trans. Energy Convers.* **2016**, *31*, 596–605. [CrossRef]
28. Butt, C.B.; Rahman, M.A. Untrained Artificial Neuron-Based Speed Control of Interior Permanent-Magnet Motor Drives Over Extended Operating Speed Range. *IEEE Trans. Ind. Appl.* **2013**, *49*, 1146–1153. [CrossRef]
29. Ma, G.Y.; Chen, W.Y.; Cui, F.; Zhang, W.P.; Wu, X.S. Adaptive levitation control using single neuron for micromachined electrostatically suspended gyroscope. *Electron. Lett.* **2010**, *46*, 406–408. [CrossRef]
30. Zhong, X.; Ni, Z.; He, H. A Theoretical Foundation of Goal Representation Heuristic Dynamic Programming. *IEEE Trans. Neural Netw. Learn. Syst.* **2016**, *27*, 2513–2525. [CrossRef] [PubMed]
31. Liu, D.; Zhang, Y.; Zhang, H. A self-learning call admission control scheme for CDMA cellular networks. *IEEE Trans. Neural Netw.* **2005**, *16*, 1219–1228. [CrossRef] [PubMed]
32. Lin, W.-S.; Yang, P.-C. Adaptive critic motion control design of autonomous wheeled mobile robot by dual heuristic programming. *Automatica* **2008**, *44*, 2716–2723. [CrossRef]
33. Liu, Y.J.; Gao, Y.; Tong, S.; Li, Y. Fuzzy Approximation-Based Adaptive Backstepping Optimal Control for a Class of Nonlinear Discrete-Time Systems with Dead-Zone. *IEEE Trans. Fuzzy Syst.* **2016**, *24*, 16–28. [CrossRef]
34. Nguyen, T.L. Adaptive dynamic programming-based design of integrated neural network structure for cooperative control of multiple MIMO nonlinear systems. *Neurocomputing* **2017**, *237*, 12–24. [CrossRef]
35. Mu, C.; Ni, Z.; Sun, C.; He, H. Data-Driven Tracking Control with Adaptive Dynamic Programming for a Class of Continuous-Time Nonlinear Systems. *IEEE Trans. Cybern.* **2017**, *47*, 1460–1470. [CrossRef] [PubMed]
36. Tang, Y.; He, H.; Ni, Z.; Zhong, X.; Zhao, D.; Xu, X. Fuzzy-Based Goal Representation Adaptive Dynamic Programming. *IEEE Trans. Fuzzy Syst.* **2016**, *24*, 1159–1175. [CrossRef]
37. Li, S.; Zhou, M.; Yu, X. Design and Implementation of Terminal Sliding Mode Control Method for PMSM Speed Regulation System. *IEEE Trans. Ind. Inf.* **2013**, *9*, 1879–1891. [CrossRef]
38. Krause, P.C.; Wasynczuk, O.; Sudhoff, S.D. *Analysis of Electric Machinery*; Wiley-IEEE Press: New York, NY, USA, 1995; ISBN 978-0-7803-1101-5.
39. Ni, Z.; He, H.; Wen, J.; Xu, X. Goal Representation Heuristic Dynamic Programming on Maze Navigation. *IEEE Trans. Neural Netw. Learn. Syst.* **2013**, *24*, 2038–2050. [CrossRef] [PubMed]
40. Shen, Y.; Chen, W.; Yao, W.; Liao, S.; Wen, J. Supplementary Damping Control of VSC-HVDC for Interarea Oscillation Using Goal Representation Heuristic Dynamic Programming. In Proceedings of the 12th IET International Conference on AC and DC Power Transmission (ACDC 2016), Beijing, China, 28–29 May 2016. [CrossRef]

energies

MDPI

Article

Full-Speed Range Encoderless Control for Salient-Pole PMSM with a Novel Full-Order SMO

Yuanlin Wang [1], Xiaocan Wang [1,*], Wei Xie [2] and Manfeng Dou [1]

[1] The School of Automation, Northwestern Polytechnical University, Xi'an 710072, China;
    yuanlin.wang@nwpu.edu.cn (Y.W.); doumf@nwpu.edu.cn (M.D.)
[2] Quanzhou Institute of Equipment Manufacturing, Haixi Institutes, Chinese Academy of Sciences,
    Jinjiang 362200, China; xiewei.life@gmail.com
*   Correspondence: xiaocan.wang@nwpu.edu.cn; Tel.: +86-186-0296-8247

Received: 15 August 2018 ; Accepted: 30 August 2018 ; Published: 13 September 2018

Abstract: For salient-pole permanent magnet synchronous motor (PMSM), the amplitude of extended back electromotive force (EEMF) is determined by rotor speed, stator current and its derivative value. Theoretically, even at extremely low speed, the back EEMF can be detected if the current in q-axis is changing. However, it is difficult to detect the EEMF precisely due to the current at low speed. In this paper, novel full-order multi-input and multi-output discrete-time sliding mode observer (SMO) is built to detect the rotor position. With the proposed rotor position estimation technique, the motor can start up from standstill and reverse between positive and negative directions without a position sensor. The proposed method was evaluated by experiment.

Keywords: sliding mode observer (SMO); permanent magnet synchronous motor (PMSM); extended back electromotive force (EEMF); position sensorless

1. Introduction

Permanent magnet synchronous motor (PMSM) has many benefits, such as high efficiency, high power density, and good dynamic performance, which has been widely used in various kinds of industrial and domestic applications [1,2].

As is well known, rotor position is required in high-performance control of PMSM, which is usually obtained by using an external dedicated sensor. However, the position sensor may increase cost, weight, volume, and complexity; reduce reliability; and restrict the application area [3,4].

To detect the rotor position information directly from the model of PMSM, various kinds of strategies have been proposed up to date, such as voltage model based methods [5,6], Kalman filter based methods [7,8], and state observers based methods [9,10]. Among them, sliding mode observer (SMO) is a very promising option [11–13].

The SMO-based rotor position estimation algorithm of salient-pole PMSM is more complicated than that of non-salient pole PMSM [14]. In $\alpha\beta$ coordinate system, the state equations of salient-pole PMSM are coupled with each other. The amplitude of the extended back electromotive force (EEMF) is determined by rotor speed, stator current and its derivative value. It is a challenge to estimate the EEMF accurately [12,13].

To obtain the rotor position of salient-pole PMSM, some SMO-based methods have been proposed. To facilitate digital control applications, a discrete-time SMO is constructed in [11], and a kind of position extraction algorithm is proposed to mitigate the oscillations. In [14], a rotor position estimation method based on extended flux model is proposed, while a discrete-time SMO and a position compensator are designed. A full-order discrete-time SMO-based position sensorless control method is introduced in [15], where the modeling uncertainties and external disturbances are considered.

However, in these studies, two single input and single output (SISO) SMO are built in α-axis and β-axis, respectively, while the effect of coupling is neglected. A signum function or a sigmoid function is used as switching function, which cannot guarantee the convergence in the boundary layer [16]. During load (torque and/or speed) variations, it is a challenge to estimate the EEMF accurately [13]. Due to unwanted chattering, a filter is required to achieve desired back EEMF signal, which may cause phase shift and estimation error in the rotor position.

In this paper, an alternative rotor position estimation strategy for salient-pole PMSM is proposed. To improve the estimation results, the transient state of back EEMF is considered [17]. A fourth-order state equation of salient-pole PMSM in $\alpha\beta$ coordinate system is established; the state vector consists of currents and back EEMF. As the state vector is four-dimensional and input vector is two-dimensional, a novel multiple input and multiple output (MIMO) sliding mode observer is built for the system. To facilitate digital control applications, the sliding mode observer is studied in the discrete-time field. Pole placement technique [18,19] is used to design the switching surface; desired dynamic characteristics can be achieved through eigenvalue placement. To force the state trajectories reach and subsequently remain on the eventual sliding surface with a good movement quality, free hierarchical law is adopted as switching scheme, and discretized reaching law [20] is used to design the quasi-sliding mode and reaching process. Reaching law approach has many merits, such as guaranteeing robustness, reducing chattering, and revealing the motion mechanism of the system [21].

This paper is organized as follows: the SMO-based position sensorless control strategies for salient-pole PMSM are introduced in Section 1. In Section 2, a full-order state equation of salient-pole PMSM is built. In Section 3, a full-order SMO is proposed to detect the back EEMF and rotor position. In Section 4, the experimental results of the proposed position sensorless control are given. The paper is concluded in Section 5.

2. Full-Order State Equation of Salient-Pole PMSM

The model of salient-pole PMSM is shown in Figure 1. A, B and C are the three phase windings, $\alpha\beta$ represents the stationary reference frame, and dq means the rotating reference frame.

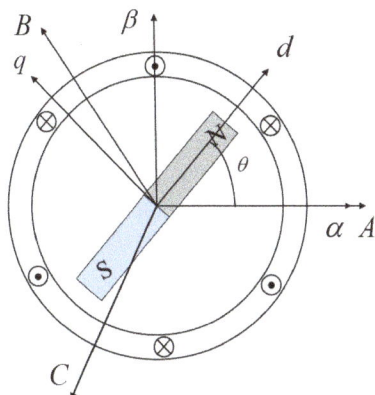

Figure 1. Illustration of the salient-pole permanent magnet synchronous motor (PMSM) model.

The motor equation in *dq* coordinate frame is expressed as Equation (1) [11].

$$
\begin{bmatrix} u_d \\ u_q \end{bmatrix} = \begin{bmatrix} R + PL_d, & -\omega L_q \\ \omega L_q, & R + PL_d \end{bmatrix} \begin{bmatrix} i_d \\ i_q \end{bmatrix}
$$
$$
+ \begin{bmatrix} 0, & 0 \\ \omega(L_d - L_q), & -P(L_d - L_q) \end{bmatrix} \begin{bmatrix} i_d \\ i_q \end{bmatrix} \tag{1}
$$
$$
+ \begin{bmatrix} 0 \\ \omega \psi_{PM} \end{bmatrix}
$$

where P is a derivative operator; R_s is stator resistance; ω is electrical rotor speed; θ is electrical rotor angle; ψ_{PM} is PM flux linkage; L_d, L_q are stator inductances; u_d, u_q are stator voltages; and i_d, i_q are stator currents.

To facilitate the rotor position estimation, inverse Park transformation is used to transform Equation (1) into $\alpha\beta$ coordinate frame, as shown in Equation (2).

$$
\begin{bmatrix} u_\alpha \\ u_\beta \end{bmatrix} = \begin{bmatrix} R + PL_d, & \omega(L_d - L_q) \\ -\omega(L_d - L_q), & R + PL_d \end{bmatrix} \begin{bmatrix} i_\alpha \\ i_\beta \end{bmatrix}
$$
$$
+ \{(L_d - L_q)(\omega i_d - Pi_q) + \omega \psi_{PM}\} \begin{bmatrix} -\sin\theta \\ \cos\theta \end{bmatrix} \tag{2}
$$

As is shown in Equation (2), the state equations in α-axis and β-axis couples with each other. The second term in the right side of Equation (2) is the EEMF; the amplitude of the EEMF is determined by rotor speed, stator current and its derivative value.

The differential of i_q exists in the EEMF. Even the motor is standstill, only if the current i_q changes, the EEMF is not zero. This property is useful for the motor to start up from zero speed and reverse from one direction to the other.

Let η denotes the term $(L_d - L_q)(\omega i_d - Pi_q) + \omega \psi_{PM}$, then the current model of PMSM is shown as Equation (3).

$$
\begin{bmatrix} Pi_\alpha \\ Pi_\beta \end{bmatrix} = \frac{1}{L_d} \begin{bmatrix} -R, & -\omega(L_d - L_q) \\ \omega(L_d - L_q), & -R \end{bmatrix} \begin{bmatrix} i_\alpha \\ i_\beta \end{bmatrix}
$$
$$
+ \frac{1}{L_d} \begin{bmatrix} u_\alpha \\ u_\beta \end{bmatrix} - \frac{1}{L_d} \begin{bmatrix} e_\alpha \\ e_\beta \end{bmatrix} \tag{3}
$$

e_α and e_β in Equation (3) are EEMF, which can be expressed as Equation (4). In conventional second-order SMO-based encoderless control methods, the derivatives of the EMF terms are assumed to be zero ($de/dt = 0$), so the dynamic performance is limited [17,22].

$$
\begin{bmatrix} e_\alpha \\ e_\beta \end{bmatrix} = \eta \begin{bmatrix} -\sin\theta \\ \cos\theta \end{bmatrix} \tag{4}
$$

The differential equation of EEMF is shown as Equation (5) [23].

$$
\begin{bmatrix} Pe_\alpha \\ Pe_\beta \end{bmatrix} = \omega \begin{bmatrix} -e_\beta \\ e_\alpha \end{bmatrix} \tag{5}
$$

Combining Equations (3) and (5), a full-order state equation of PMSM is shown in Equation (6).

$$\begin{bmatrix} P e_{\alpha\beta} \\ P i_{\alpha\beta} \end{bmatrix} = \begin{bmatrix} A_{11} & O \\ A_{21} & A_{22} \end{bmatrix} \begin{bmatrix} e_{\alpha\beta} \\ i_{\alpha\beta} \end{bmatrix} + \begin{bmatrix} O \\ B_1 \end{bmatrix} u_{\alpha\beta} \tag{6}$$

where,

$$i_{\alpha\beta} = [i_\alpha, i_\beta]^T, e_{\alpha\beta} = [e_\alpha, e_\beta]^T, u_{\alpha\beta} = [u_\alpha, u_\beta]^T$$
$$A_{11} = \omega J_2, \quad A_{21} = -B_1 = -\frac{1}{L_d} I_2,$$
$$A_{22} = -\frac{R}{L_d} I_2 + \omega \frac{L_d - L_q}{L_d} J_2,$$
$$I_2 = \begin{bmatrix} 1 & 0 \\ 0 & 1 \end{bmatrix}, J_2 = \begin{bmatrix} 0 & -1 \\ 1 & 0 \end{bmatrix}, O = \begin{bmatrix} 0 & 0 \\ 0 & 0 \end{bmatrix}$$

3. Full-Order Sliding Mode Observer Design

Based on motor state Equation (6), estimated state equation is shown in Equation (7).

$$\begin{bmatrix} P \hat{e}_{\alpha\beta} \\ P \hat{i}_{\alpha\beta} \end{bmatrix} = \begin{bmatrix} A_{11} & O \\ A_{21} & A_{22} \end{bmatrix} \begin{bmatrix} \hat{e}_{\alpha\beta} \\ \hat{i}_{\alpha\beta} \end{bmatrix}$$
$$+ \begin{bmatrix} O \\ B_1 \end{bmatrix} u_{\alpha\beta} + \begin{bmatrix} M \\ N \end{bmatrix} z_{\alpha\beta} \tag{7}$$

where $\hat{i}_{\alpha\beta}$ are estimated currents; $\hat{e}_{\alpha\beta}$ are estimated EEMF; z_α and z_β are inputs of the estimator, $z_\alpha = f(\hat{i}_\alpha - i_\alpha)$; $z_\beta = f(\hat{i}_\beta - i_\beta)$; $M = mI_2$; and $N = nI_2$.

Subtracting Equation (6) from Equation (7), estimation errors are shown in Equation (8).

$$\begin{bmatrix} P \bar{e}_{\alpha\beta} \\ P \bar{i}_{\alpha\beta} \end{bmatrix} = \begin{bmatrix} A_{11} & O \\ A_{21} & A_{22} \end{bmatrix} \begin{bmatrix} \bar{e}_{\alpha\beta} \\ \bar{i}_{\alpha\beta} \end{bmatrix} + \begin{bmatrix} M \\ N \end{bmatrix} z_{\alpha\beta} \tag{8}$$

where $\bar{e}_{\alpha\beta} = \hat{e}_{\alpha\beta} - e_{\alpha\beta}$ and $\bar{i}_{\alpha\beta} = \hat{i}_{\alpha\beta} - i_{\alpha\beta}$.

Equation (8) is a Multi-Input and Multi-Output (MIMO) system, which can be expressed as Equation (9).

$$\dot{x}_0 = A x_0 + B z \tag{9}$$

where $x_0 \in R^4$ is a state vector, and $z \in R^2$ is a input vector.

A linear transformation shown as Equation (10) is used to transform Equation (9) into a regular form that has reduced-order, simpler computation, and equivalent dynamics [24].

$$x = T x_0$$

$$T = \begin{bmatrix} I_2 & -MN^{-1} \\ 0 & I_2 \end{bmatrix} \tag{10}$$

After transformation, the regular form of the sliding mode observer is shown as Equation (11).

$$\dot{x} = \tilde{A} x + \tilde{B} z \tag{11}$$

where,

$$\tilde{A} = TAT^{-1} = \begin{bmatrix} \tilde{A}_{11} & \tilde{A}_{12} \\ \tilde{A}_{21} & \tilde{A}_{22} \end{bmatrix}, \tilde{B} = TB = \begin{bmatrix} 0 \\ M \end{bmatrix},$$
$$\tilde{A}_{11} = A_{11} - \frac{m}{n} A_{21}, \tilde{A}_{21} = A_{21}, \tilde{A}_{22} = \frac{m}{n} A_{21} + A_{22}$$
$$\tilde{A}_{12} = (A_{11} - \frac{m}{n} A_{21}) \frac{m}{n} I_2 - \frac{m}{n} A_{22}$$

To facilitate digital processor applications, the SMO is studied in discrete-time field. The discrete-time form of the SMO in Equation (11) is expressed as Equation (12).

$$x(k+1) = Dx(k) + Ez(k) \tag{12}$$

For the Multiple Input and Multiple Output (MIMO) system, a SMO is designed to estimate the currents and back EEMF in α-axis and β-axis. The scheme is shown as Figure 2.

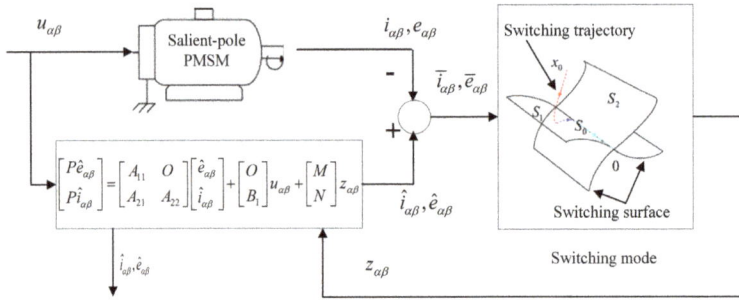

Figure 2. Sliding mode observer diagram.

The design of the sliding mode control involves two parts: switching surfaces and switching trajectory, as shown in Figure 2. The switching surfaces are designed to ensure the system has desired dynamic characteristics. System state trajectories should reach and remain on the eventual switching surface [19,21].

The dynamics of the system only depends on switching surfaces and is not influenced by system structure and parameter uncertainties [13].

Linear switching surfaces are used in the variable structure control. There are two inputs, so two switching surfaces are designed, which are shown as S_1 and S_2 in Figure 2.

$$S_i(k) = \tilde{C}x(k) \tag{13}$$

where $i = 1, 2, \tilde{C} \in R^{2 \times 4}$

The eventual switching surface is S_0, which is shown as Equation (14).

$$S_0 = S_1 \cap S_2 \tag{14}$$

When the sliding mode is enforced in the switching surface, the system's dynamic characteristics are determined by sliding eigenvalues [19].

In the following, the matrix $\tilde{C}$ is expressed as $\tilde{C} = [\tilde{C}_1, \tilde{C}_2]$, $\tilde{C}_1 \in R^{2 \times 2}$ and $\tilde{C}_2 \in R^{2 \times 2}$ are unknown.

As the design of switching surface is not affected by $\tilde{C}_2$, it can take any arbitrary value[18,19]. To simplify the switching function, $\tilde{C}_2$ is set as a unit matrix I_2.

Equation (12) can be expressed as Equation (15), and the switching function Equation (13) can be expressed as Equation (16).

$$\begin{aligned} x_1(k+1) &= D_{11}x_1(k) + D_{12}x_2(k) \\ x_2(k+1) &= D_{21}x_1(k) + D_{22}x_1(k) + Ez(k) \end{aligned} \tag{15}$$

$$S(k) = \tilde{C}x(k) = \tilde{C}_1 x_1(k) + \tilde{C}_2 x_2(k) \tag{16}$$

Equation (16) can be transformed to Equation (17).

$$\begin{aligned} x_1(k) &= x_1(k) \\ x_2(k) &= \tilde{C}_2^{-1}S(k) - \tilde{C}_2^{-1}C_1x_1(k) \end{aligned} \tag{17}$$

Substituting Equation (17) into Equation (15):

$$x_1(k+1) = (D_{11} - D_{12}\tilde{C}_2^{-1}\tilde{C}_1)x_1(k) + D_{12}\tilde{C}_2^{-1}S(k) \tag{18}$$

$$\begin{aligned} S(k+1) &= [\tilde{C}_1D_{11} + \tilde{C}_2D_{21}]x_1(k) \\ &\quad - [(\tilde{C}_1D_{12} + \tilde{C}_2D_{22})\tilde{C}_2^{-1}\tilde{C}_1]x_1(k) \\ &\quad + (\tilde{C}_1D_{12} + \tilde{C}_2D_{22})\tilde{C}_2^{-1}S(k) + \tilde{C}_2Ez(k) \end{aligned} \tag{19}$$

When the switching trajectory arrives at the switching surface, $S(k) \approx 0$. Substituting $S(k) = 0$ into Equation (18):

$$\begin{aligned} x_1(k+1) &= (D_{11} - D_{12}F)x_1(k) \\ F &= \tilde{C}_2^{-1}\tilde{C}_1, \quad F \in R^{2\times2} \end{aligned} \tag{20}$$

Assume the desired poles of the system are λ_1 and λ_2.

$$\det(\lambda I_2 - (D_{11} - D_{12}F)) = (\lambda - \lambda_1)(\lambda - \lambda_2) \tag{21}$$

Matrix F can be calculated based on Equation (21), and $\tilde{C}_1$ can be obtained according to F and Equation (20). The switching functions are achieved by substituting $\tilde{C}_1$ and $\tilde{C}_2$ into Equation (13).

In this control system, there are three switching surfaces (S_0, S_1, S_2). Free-order switching scheme is adopted, as shown in Figure 2.

Discrete-time reaching law is used to design the sliding mode trajectory, which is shown as Equation (22).

$$\begin{aligned} S(k+1) - S(k) &= -qT_sS(k) - \varepsilon T_s\text{sgn}(S(k)) \\ \varepsilon &> 0, \quad q > 0, \quad 1 - qT_s > 0 \end{aligned} \tag{22}$$

where $\varepsilon \in R^{2\times2}$ and $q \in R^{2\times2}$ are diagonal matrices.

According to Equations (12) and (13), $S(k+1)$ can be expressed as Equation (23).

$$S(k+1) = \tilde{C}x(k+1) = \tilde{C}Dx(k) + \tilde{C}Ez(k) \tag{23}$$

Combining Equations (22) and (23), inputs of the system $z(k)$ are shown as Equation (24).

$$z(k) = (\tilde{C}E)^{-1}[\tilde{C}(I_4 - D)x(k) - \varepsilon T_s\text{sgn}(S(k)) - qT_s\tilde{C}x(k)] \tag{24}$$

Substituting Equation (24) into Equation (12), the sliding mode trajectory is shown as Equation (25).

$$\begin{aligned} x(k+1) &= (D + E(\tilde{C}E)^{-1}\tilde{C}(I_2 - D))x(k) \\ &\quad - (E(\tilde{C}E)^{-1}qT_s\tilde{C})x(k) \\ &\quad - E(\tilde{C}E)^{-1}\varepsilon T_s\text{sgn}(S(k)) \end{aligned} \tag{25}$$

Substituting Equation (24) into the discrete-time form of Equation (7), state variables of the system, including currents (i_α, i_β) and EEMF signals (e_α, e_β), can be achieved.

An angle-tracking observer (ATO) is used to estimate rotor angle and speed from estimated EEMF signals [22,25]. Because the amplitude of EEMF changes with speed, normalization is adopted in the observer. The position error is shown as Equation (26).

$$\xi_N = -\frac{1}{\sqrt{\hat{e}_\alpha^2 + \hat{e}_\beta^2}}(\hat{e}_\alpha \cos\hat{\theta}_e + \hat{e}_\beta \sin\hat{\theta}_e) \tag{26}$$

The scheme of the ATO is shown as Figure 3, and rotor position is achieved by Equation (27).

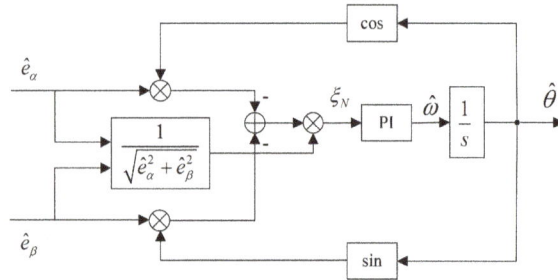

Figure 3. Angle-tracking observer.

$$\hat{\theta}_e = \frac{1}{s}(K_p + \frac{K_i}{s})\xi_N \tag{27}$$

where K_p and K_i are parameters of the PI regulator in Figure 3.

4. Experimental Results

The proposed encoderless method is shown as Figure 4. A conventional full-order discrete-time SMO-based position sensorless control method [15] and the proposed method are compared under the same condition. When $Sw = 1$, the proposed method is used, otherwise the conventional method is used. The comparison includes computation time, speed variation and torque variation. To show the effectiveness of the proposed method in low speed range, speed reversal experiment and startup experiment are implemented.

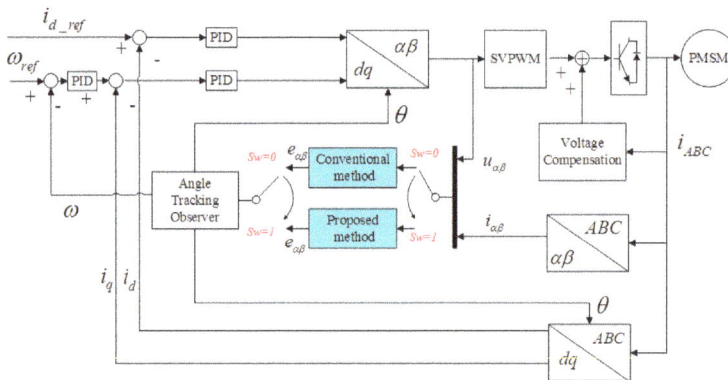

Figure 4. Rotor position and speed estimation diagram.

The test bench is shown as Figure 5. The inverter is a specially designed two-level three phase voltage source inverter, the type of the MOSFETS (IXFR180N10, IXYS corporation, Leiden, Netherlands) is IXFR180N10, and the current sensor is T60404-N4646-X100. The parameters of the salient-pole PMSM are shown in Table 1. The ratio between amplitude of back EEMF and speed is very low, which is a severe condition for back EEMF based position sensorless control methods when working in low speed range.

A DC motor is mechanically coupled with the salient-pole PMSM to produce the load torque, an adjustable resistor that connected to the terminal of the DC motor is used to change the load torque.

An absolute encoder is used to measure the actual position used for comparison. Both the switching frequency and sampling frequency are 10 kHz.

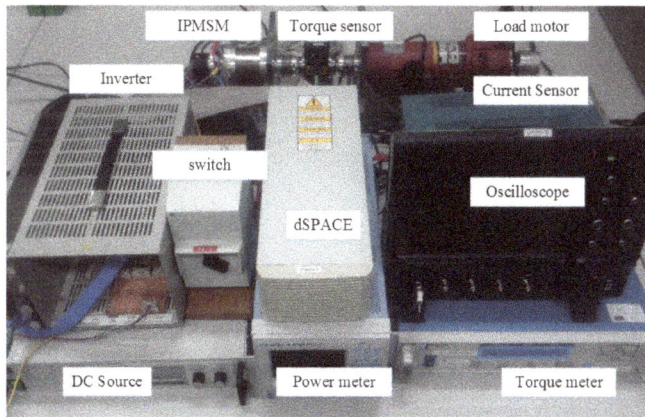

Figure 5. The test bench.

Table 1. Parameters of the tested IPMSM.

Parameter	Value
Rated torque	2 Nm
Rated current/voltage (rms)	50 A/13 V
Number of pole pairs	5
d/q-axis inductance	0.05/0.095 mH
Resistance	18 mΩ
Rated speed	2000 rpm
The moment of inertia	0.00187 kg·m^2
PM flux linkage	0.00707 Vs

In the figures shown in the Experimental Results, "Red" represents the reference signals, "Black" means measured signals, "Green" represents the signals that are achieved by using convention method and "Blue" denotes the signals that are obtained by using the proposed method.

4.1. Computation Load Comparison

To evaluate the computational load of the two methods, computation time are compared. In this experiment, the turnaround time is used as a criterion, which can be read directly from the control desk of dSAPCE. The turnaround time includes the communication time, data conversion time, code implementation time and data saving time. Except for code implementation time, the other times of the two methods are the same.

The turnaround time of the two strategies is shown in Table 2. Compared with the conventional method, the time increase of the proposed method is 3.4% of the sampling period.

Table 2. Turnaround time comparison of the conventional method and the proposed method.

Time	Value
Conventional method t_1 (μs)	8.9
Proposed method t_2 (μs)	12.3
Increased time $t_2 - t_1$ (μs)	3.4
Sampling period Ts (μs)	100
Increased time percentage $(t_2 - t_1)/T_s$	3.4%

4.2. Speed Variation Comparison

In this experiment, the motor accelerates from 100 rpm to 2000 rpm and then decelerates to 100 rpm without load torque. To make a fair comparison, the switching surface of the proposed method is adjusted to make the estimation error between the proposed method and conventional method is approximately equal at 100 rpm.

During this process, the reference speed, measured speed, estimated speed, speed estimation error and electrical position estimation error are given in Figure 6. The experimental results of the conventional method are shown in Figure 6a, while the experimental results of the proposed method are shown in Figure 6b.

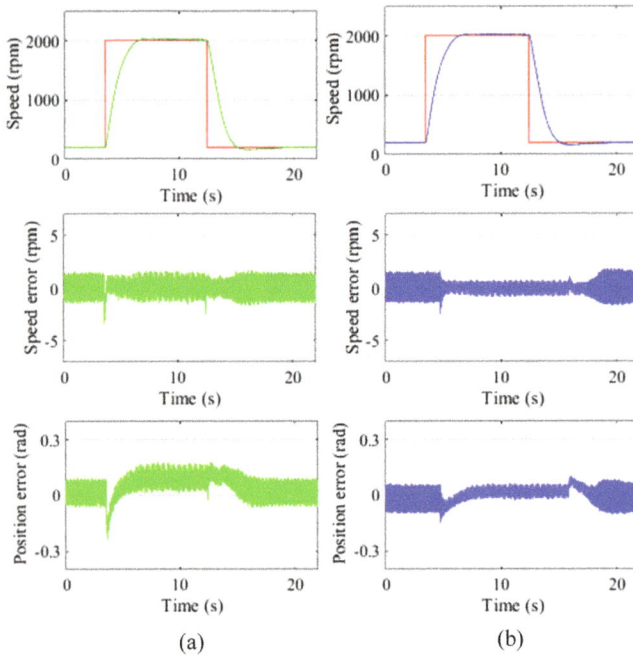

Figure 6. The speed changes from 100 rpm (0.05 p.u.) to 2000 rpm (1 p.u.): (**a**) conventional method; and (**b**) proposed method. "Red Line" represents the reference signals, "Green Line" represents the signals that are achieved by using convention method and "Blue Line" denotes the signals that are obtained by using the proposed method.

The experimental results show that, at 100 rpm, the speed estimation errors of the two methods are almost the same. With the increase of speed, the speed estimation error of the proposed method is smaller than that of the conventional method. At 2000 rpm, the speed estimation error of the proposed method is about 50% of the conventional method.

At 100 rpm, the electrical rotor position estimation errors of the two methods are similar. During the speed variation process, the maximum electrical rotor position estimation error of the conventional method is −0.2 rad, and the maximum electrical rotor position estimation error of the proposed method is −0.1 rad. At 2000 rpm, the phase lag of the conventional method is 0.1 rad, and the phase lag of the proposed method is zero.

4.3. Torque Variation Comparison

In this experiment, the motor operates under speed control; the speed is 1500 rpm (0.75 p.u.). In the beginning, the load torque is 0.2 Nm (0.1 p.u.), a load torque 1.2 Nm (0.6 p.u.) is provided to the motor as a disturbance, and then the load torque is reduced to 0.2 Nm (0.1 p.u.). During the torque variation process, the electromagnetic torque, load torque, measured speed, estimated speed, electrical position estimation error and speed estimation error are shown as Figure 7.

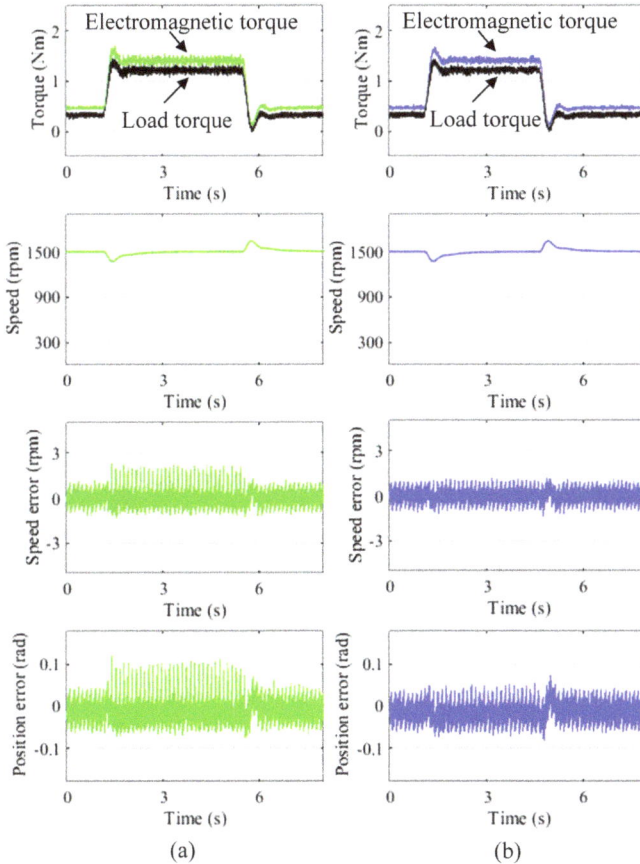

Figure 7. The load torque changes from 0.2 Nm (0.1 p.u.) to 1.2 Nm (0.6 p.u.): (**a**) conventional method; and (**b**) proposed method. "Black Line" means measured signals, "Green Line" represents the signals that are achieved by using convention method and "Blue Line" denotes the signals that are obtained by using the proposed method.

The experimental results show that, during the torque disturbance process, both the electrical rotor position and speed estimation errors of the proposed method are significantly lower than those of the conventional method.

4.4. Speed Reversal

With the conventional full-order discrete-time SMO, the motor can not reverse from one direction to the other without position sensor in this experiment. Therefore, in this section, only the experimental results of the proposed method are shown.

The motor reverses between 200 rpm (0.1 p.u.) and −200 rpm (−0.1 p.u.). The reference speed, measured speed, estimated speed, measured rotor position, estimated rotor position, currents in *d*-axis and *q*-axis, speed estimation error and electrical position estimation error are shown in Figure 8.

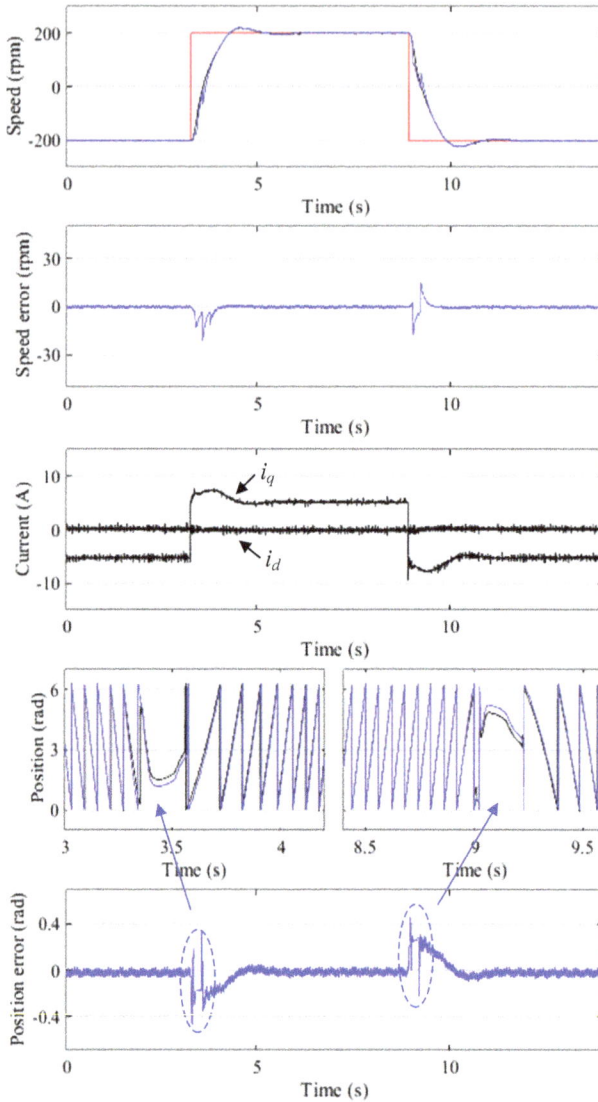

Figure 8. The speed changes between 200 rpm (0.1 p.u.) and −200 rpm (−0.1 p.u.). "Red Line" represents the reference signals, "Black" means measured signals and "Blue Line" denotes the signals that are obtained by using the proposed method.

The experimental results show that the motor can reverse successfully from one direction to the other. During the reversal process, the maximum electrical rotor position and speed estimation errors occur around zero speed. The maximum electrical rotor position estimation error is 0.4 rad, and the maximum rotor speed estimation error is 15 rpm.

4.5. Start up from Standstill

With the conventional full-order discrete-time SMO, the motor cannot start up from standstill in this experiment. Therefore, in this section, only the experimental results of the proposed method are shown.

The initial rotor position is achieved by initial rotor position Estimation [26]. During the startup process, the reference speed, measured speed, estimated speed, measured rotor position, estimated rotor position, phase current, electrical position estimation error and speed estimation error are shown in Figure 9.

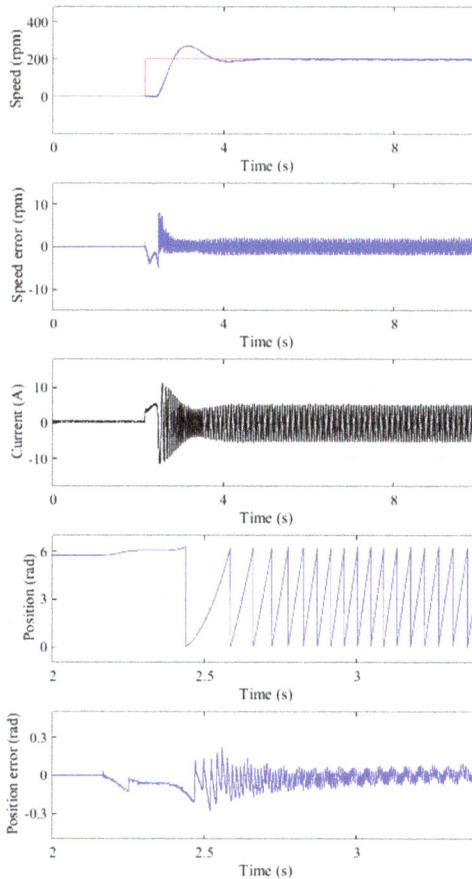

Figure 9. The motor starts up from zero speed to 200 rpm (0.1 p.u.).

The experimental results show that the motor can start up successfully from standstill. During the startup process, the maximum speed estimation error is 8 rpm and the maximum electrical rotor position estimation error is −0.27 rad.

5. Conclusions

This paper proposes a novel position sensorless control strategy for PMSM considering saliency. A novel full-order SMO is built to estimate the rotor position. The effectiveness of the proposed method is validated on a low voltage salient-pole PMSM.

The performance of a conventional full-order SMO-based position sensorless control method and the proposed method are compared under the same condition. The computational burden of the proposed method is higher than that of the conventional method. In the test bench, the computation time increase is 3.4% of the sampling period. With the rapid development of fast microprocessors, the computational time increase can be ignored.

During speed variation and torque variation process, the performance of the proposed method is obviously better than that of the conventional method. The rotor speed estimation error and position estimation error of the proposed method are about 50% of the conventional method. During speed variations, there is no phase lag in the proposed method. Based on the proposed method, the motor can reverse between positive and negative directions and start up from standstill without a position sensor.

The rotor position is estimated based on the the the differential of i_q, so the proposed method can be used for salient-pole PMSM with different load at stand still and low speed. However, due to the restriction of the test bench, it is incapable of producing satisfied load torque at low speed. In the next step, a new test bench will be built to repeat the experiments under heavy mechanical load at zero speed and low speed.

The proposed position sensorless control method can be used for salient-pole PMSMs in electrical car, robot joints, etc., where startup and low speed operation of PMSM are required.

Author Contributions: Y.W. and M.D. designed the position sensorless control method for PMSM. X.W. and W.X. performed the experiments and analyzed the data.

Funding: This research was funded by "the Fundamental Research Funds for the Central Universities".

Conflicts of Interest: The authors declare no conflict of interest.

References

1. Fortino, M.; Víctor, M.; Juvenal, R. Robust speed control of permanent magnet synchronous motors using two-degrees-of-freedom control. *IEEE Trans. Ind. Electron.* **2018**, *65*, 6099–6108. [CrossRef]
2. Zhang, Y.; Xu, D.; Huang, L. Generalized multiple-vector-based model predictive control for PMSM drives. *IEEE Trans. Ind. Electron.* **2018**, *65*, 9356–9366. [CrossRef]
3. Wei, Q.; Xi, Z.; Jin, F.; Hua, B.; Lu, D.; Cheng, B. Using high-control-bandwidth fpga and sic inverters to enhance high-frequency injection sensorless control in interior permanent magnet synchronous machine. *IEEE Access* **2018**, *58*, 2169–3536. [CrossRef]
4. Xu, P.L.; Zhu, Z.Q. Novel square-wave signal injection method using zero-sequence voltage for sensorless control of pmsm drives. *IEEE Trans. Ind. Electron.* **2016**, *63*, 7444–7454. [CrossRef]
5. Ichikawa, S.; Tomita, M.; Doki, S.; Okuma, S. Sensorless control of permanent-magnet synchronous motors using online parameter identification based on system identification theory. *IEEE Trans. Ind. Electron.* **2016**, *53*, 363–372. [CrossRef]
6. Genduso, F.; Miceli, R.; Rando, C.; Galluzzo, G.R. Back EMF sensorless-control algorithm for high-dynamic performance PMSM. *IEEE Trans. Ind. Electron.* **2010**, *57*, 2092–2100. [CrossRef]
7. Bolognani, S.; Tubiana, L.; Zigliotto, M. Extended kalman filter tuning in sensorless pmsm drives. *IEEE Trans. Ind. Appl.* **2013**, *39*, 1741–1747. [CrossRef]
8. Fuentes, E.; Kennel, R. Sensorless-predictive torque control of the pmsm using a reduced order extended kalman filter. *Symp. Sensorless Control Electron. Drives (SLED)* **2011**, *39*, 123–128. [CrossRef]
9. Junggi, L.; Jinseok, H.; Kwanghee, N.; Ortega, R.; Praly, L.; Astolfi, A. Sensorless control of surface-mount permanent-magnet synchronous motors based on a nonlinear observer. *IEEE Trans. Power Electron.* **2010**, *25*, 290–297. [CrossRef]
10. Foo, G.H.B.; Rahman, M.F. Direct torque control of an ipmsynchronous motor drive at very low speed using a sliding-mode stator flux observer. *IEEE Trans. Power Electron.* **2010**, *25*, 933–942. [CrossRef]
11. Zhao, Y.; Qiao, W.; Wu, L. Position extraction from a discrete sliding-mode observer for sensorless control of ipmsms. *IEEE Int. Symp. Ind. Electron.* **2012**, *25*, 725–730. [CrossRef]
12. Foo, G.; Rahman, M.F. Sensorless sliding-mode mtpa control of an ipm synchronous motor drive using a sliding-mode observer and hf signal injection. *IEEE Trans. Ind. Electron.* **2010**, *57*, 1270–1278. [CrossRef]

13. Zhao, Y.; Qiao, W.; Wu, L. An adaptive quasi-sliding-mode rotor position observer-based sensorless control for interior permanent magnet synchronous machines. *IEEE Trans. Power Electron.* **2013**, *28*, 5618–5629. [CrossRef]

14. Yue, Z.; Zhe, Z.; Wei, Q.; Long, W. An extended flux model-based rotor position estimator for sensorless control of salient-pole permanentmagnet synchronous machines. *IEEE Trans. Power Electron.* **2015**, *30*, 4412–4422. [CrossRef]

15. Guoqiang, Z.; Gaolin, W.; Ronggang, N.; Dianguo, X. Active flux based full-order discrete-time sliding mode observer for position sensorless ipmsm drives. In Proceedings of the 2014 17th International Conference on Electrical Machines and Systems (ICEMS), Hangzhou, China, 22–25 Octorber 2014. [CrossRef]

16. Feng, Y.; Zheng, J.; Yu, X.; Truong, N. V. Hybrid terminal slidingmode observer design method for a permanent-magnet synchronous motor control system. *IEEE Trans. Ind. Appl.* **2009**, *56*, 3424–3431. [CrossRef]

17. Kim, M.; Sul, S.K. An enhanced sensorless control method for pmsm in rapid accelerating operation. In Proceedings of the 2010 International Power Electronics Conference-ECCE ASIA, Sapporo, Japan, 21–24 June 2010. [CrossRef]

18. Ackermann, J.; Utkin, V. Sliding mode control design based on ackermann's formula. *IEEE Trans. Autom. Control* **2009**, *43*, 234–237. [CrossRef]

19. Ramon, G.; Luis, G.; Miguel, C.; Jaume, M.; Helena, M. Variable structure control in natural frame for three-phase grid-connected inverters with LCL filter. *IEEE Trans. Power Electron.* **2018**, *33*, 4512–4522. [CrossRef]

20. Zhu, Q.; Wang, T.; Jiang, M.; Wang, Y. A new design scheme for discrete-time variable structure control systems. In Proceedings of the 2009 International Conference on Mechatronics and Automation, Changchun, China , 9–12 August 2009. [CrossRef]

21. Ma, H.; Wu, J.; Xiong, Z. A novel exponential reaching law of discrete-time sliding-mode control. *IEEE Trans. Ind. Electron.* **2017**, *64*, 3840–3850. [CrossRef]

22. Wang, G.; Li, Z.; Zhang, G.; Yu, Y.; Xu, D. Quadrature PLL-based high-order sliding-mode observer for IPMSM sensorless control with online MTPA control strategy. *IEEE Trans. Ind. Electron.* **2013**, *28*, 214–224. [CrossRef]

23. Comanescu, M. Cascaded EMF and speed sliding mode observer for the nonsalient PMSM. In Proceedings of the IECON 2010-36th Annual Conference on IEEE Industrial Electronics Society, Glendale, AZ, USA, 7–10 November 2010. [CrossRef]

24. DeCarlo, R.A.; Zak, S.H.; Matthews, G.P. Variable structure control of nonlinear multivariable systems: A tutorial. *Proc. IEEE* **2013**, *76*, 212–232. [CrossRef]

25. Hoseinnezhad, R.; Harding, P. A novel hybrid angle tracking observer for resolver to digital conversion. In Proceedings of the 44th IEEE Conference on Decision and Control, Seville, Spain, 15 December 2005. [CrossRef]

26. Wu, X.; Feng, Y.; Liu, X.; Huang, S.; Yuan, X.; Gao, J.; Zheng, J. Initial rotor position detection for sensorless interior pmsm with square-wave voltage injection. *IEEE Trans. Magn.* **2017**, *53*, 1–4. [CrossRef]

Article

Vibration Characteristics Analysis of Planetary Gears with a Multi-Clearance Coupling in Space Mechanism

Huibo Zhang [1,2], Chaoqun Qi [1,2], Jizhuang Fan [3,*], Shijie Dai [1,2,*] and Bindi You [4]

1 School of Mechanical Engineering, Hebei University of Technology, Tianjin 300130, China; zhanghb@hebut.edu.cn (H.Z.); 201621202009@stu.hebut.edu.cn (C.Q.)
2 Hebei Key Laboratory of Robot Perception and Human-Robot Interaction, Tianjin 300130, China
3 State Key Laboratory of Robotics and System, Harbin Institute of Technology, Harbin 150080, China
4 School of Naval Architecture and Ocean Engineering, Harbin Institute of Technology at Weihai, Weihai 264209, China; youbindi@hithw.edu.cn
* Correspondence: fanjizhuang@hit.edu.cn (J.F.); dsj@hebut.edu.cn (S.D.); Tel.: +86-138-3619-5902 (J.F.); +86-139-2051-7176 (S.D.)

Received: 24 September 2018; Accepted: 29 September 2018; Published: 9 October 2018

Abstract: Multi-clearance is the main cause for the performance and reliability decline of complicated mechanical systems. The increased clearance could induce contacts and impacts in joints, and consequently affect control accuracy. A nonlinear dynamic model of planetary gears with multi-clearance coupling is proposed in the current study to investigate the mechanism of influence of clearance on the dynamic performance. In addition, the coupling relationship between radial clearance and backlash is integrated into the multi-body system dynamics. The vibration characteristics of planetary gears with the changes of rotational velocity, clearance size and inertia load are explored. The numerical simulation results show that there are complex coupling relations in planetary gear systems, due to the multi-clearance coupling. The phenomenon of system resonance may occur with the changes of rotational velocities and clearances' sizes. Multi-clearance coupling can significantly increase the resonant response of planetary gear systems in empty-load or light-load states.

Keywords: space mechanism; multi-clearance; nonlinear dynamic model; planetary gears; vibration characteristics

1. Introduction

With the increasing requirements for the performance of space mechanisms, the transmission joints, as the core components, have become the focus of attention. Planetary gears are widely used in aerospace applications, which include space manipulators, satellite antenna drive mechanisms and other spacecraft mechanisms [1]. The transmission system of the Canadian Manipulator and the European Manipulator on the International Space Station both use planetary gears for their advantages of compactness, high transmission ratio and low gear noise [2]. However, clearance in planetary gears are inevitable due to manufacturing and assembly errors, fatigue and wear. Moreover, those clearances cause contacts and impacts between gear sets, and consequently affect the control accuracy [3–7]. With the development of space technology, the high precision and high reliability of spacecraft mechanisms are increasingly required, thus research on the vibration characteristics of planetary gear systems is receiving more and more attention.

In theoretical research, Kahraman has presented a series of nonlinear time-varying dynamic models of planetary gears since 1994 [8]. The tooth surface wear model was established, and the influences of tooth surface wear on the dynamic responses of planetary gear systems were analyzed [9]. On this basis, the initial model considered gear manufacturing errors, assembly errors, backlash, time-varying meshing stiffness and other problems [10–13]. Then, a new method for measuring the average load of plane and radial orbit of the Sun with strain gauges and probes was presented. The influence rules of the number of planetary gears, lubes temperature and surface roughness on the efficiency of transmission were analyzed by experiments [14,15]. Ambarisha and Parker [16] used a finite element method to analyze the vibration mode of planetary gear systems. In an early dynamic model the influence of bearing stiffness on the static characteristics of planetary gear systems with manufacturing errors was first analyzed [17]. The effects of meshing stiffness, tooth profile modification, meshing phase, contact ratio and other factors on the suppression of system vibration and noise were researched [18,19]. Then, lumped-parameter and finite element models with bearing clearance, tooth separation, and gear mesh stiffness variation are developed to investigate the nonlinear dynamic behavior of planetary gear systems [20]. Based on this model, Ericson and Parker [21] discovered that the natural frequencies of modes with significant planet bearing deflection are particularly sensitive to torque. Cooley et al. [22] summarized the calculation methods of meshing stiffness into two categories, namely the average slope method and the local slope of the force–deflection curve. Pappalardo and Guida [23] proposed a new methodology to address the problems of suppressing structural vibrations and attenuating contact forces in nonlinear mechanical systems. Then a new computational algorithm for the numerical solution of the adjoint equations for the nonlinear optimal control problem was introduced [24]. Ouyang et al. [25] formulated the eigenstructure assignment as an inverse eigenvalue problem within the frame of constrained nonlinear integer programming, which solves the precision problem of discrete optimal solutions. Palermo et al. [26] have presented a contact element for global dynamic simulations of gear assemblies using multibody modeling that enables to take into account real-case parameters in a scalable way. To achieve high calculation efficiency, Shweiki et al. [27] used finite element (FE) simulations to establish an angle-dependent stiffness function and then stored the results in a lookup table, which is then interpolated during the dynamic simulation. Regarding modeling methods, Vivet et al. [28] modeled the local contact deformation based on Hertz theory, and proposed a multibody approach to tooth contact analysis. Wei et al. [29] established a comprehensive, fully coupled, dynamic modeling method by applying a virtual equivalent shaft element to overcome the lack of fidelity of the lumped parameter models and the high computational cost of finite element models. In addition, it was proposed that different calculation methods lead to different vibration models. To obtain a satisfactory space manipulator positioning control accuracy, a nonlinear dynamic model of the manipulator joint with planetary gear train transmission is developed by considering time-variant joint stiffness, backlash and reduction ratio [30]. To reflect the nonlinear behavior of the space manipulator's joint, factors such as backlash clearance, gear tooth profile error, and time-variant meshing stiffness were considered in the modeling process [31]. Marques et al. [32] presented a new formulation to model spatial revolute joints with radial and axial clearance. Pan et al. [2] established a planetary gear dynamics model for the space manipulator joint considering the nonlinear factors, including gear tooth flexibility, meshing damping, backlash, meshing error, etc.

So far, studies about planetary gear transmission systems have become rather mature in methods and contents. However, the spacecraft mechanisms typically operate at low velocity and under light load conditions. The effect of clearance on the vibration characteristics of the mechanism in microgravity are significantly larger than in the ground environment. Previous studies on the clearance of gear systems are mainly concentrated on backlash, while much less attention has been paid to radial clearance and multi-clearance coupling.

A refined dynamic model of planetary gear transmission joint is proposed in this paper to analyze the influence of multi-clearance coupling on the vibration characteristics of the system. Backlash, radial clearance and time-varying meshing stiffness are considered in the dynamic model. Planetary gear transmission joints with multi-clearance coupling are used as a numerical example to investigate the vibration characteristics of the system, and clearance size, rotational velocity and load magnitude are also analyzed separately.

This paper is organized as follows: Section 2 introduces the establishment process of dynamics model in planetary gear driven joint, and obtains the model of single gear and planetary gear with multi-clearance coupling. In Section 3, the accuracy of dynamic model with multi-clearance coupling is verified by numerical simulation. Then, the vibration characteristics of planetary gears with the changes of rotational velocity, clearance size and inertia load are explored. Finally, Section 4 includes a summary and the conclusions of the paper.

2. Dynamic Modeling of Planetary Gear Drive Joint

The 2K-H planetary gear reducer (shown in Figure 1), as an example of a typical joint system, is the current study object. The system is composed of four parts: sun gear (s), ring gear (r), planet carrier (c), and a certain number of planetary gears (p). Each planetary gear is fixed to the planet carrier through bearing, and can be freely rotated relative to the carrier. In order to establish the joint dynamics model of the planetary gear drive, the following assumptions are made:

(1) Each gear in joints is considered to be a rigid gear, neglecting the plastic deformation during contact collision;
(2) The elastic deformation of drive shaft is neglected, and the only effect of radial run-out caused by bearing clearance on the dynamic characteristics of system is considered;
(3) The joint system is assumed to be a planar system. In other words, the radial vibration at the bearing is equivalent to two-degree-of-freedom translational motion in the gear rotation plane, and the torsional vibration of gear is equivalent to single-degree-of-freedom rotation in the plane of rotation.

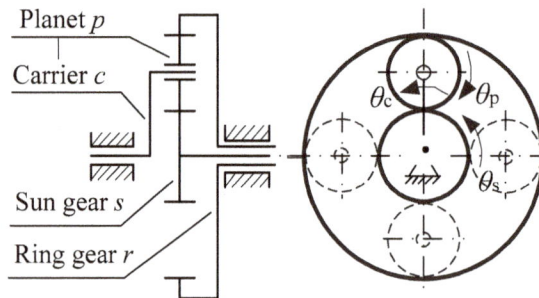

Figure 1. 2K-H planetary gear.

2.1. Multi-Clearance Coupling Model

In a gear system, it is known that a radial clearance exists in bearings, and a backlash occurs between the teeth, so vibrations will occur in the clearance during the course of movement. Then the radial vibration at bearing lead to the change of gear's actual center distance, and the backlash will dynamically change. The appearance of dynamic backlash will affect the torsional vibration and even the whole vibration characteristics of the gear system. Therefore, the coupling clearance model with bearing radial clearance and backlash is established as shown in Figure 2.

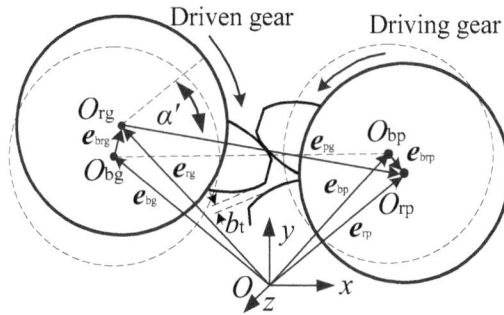

Figure 2. Multi-clearance coupling model.

Figure 2 shows the multi-clearance coupling model of a single-pair gear pair. Here, gear p represent the driving gear and gear g represents the driven gear. O_{bp} and O_{bg} are defined as the bearing centers of the driving and driven gear, respectively. It is noteworthy that O_{rp} (the rotation center of the driving gear) and O_{rg} (the rotation center of the driven gear) will deviate from the corresponding bearing center, due to the influence of the bearing's radial clearance. Thus, the vectors of above four points in the global inertia coordinate are defined as e_{bp}, e_{bg}, e_{rp} and e_{rg}, respectively. Then e_{brp} and e_{brg} define the radial clearance vectors which is the displacement error of diving and driven gear relative to bearing, respectively. Thus, the equations are written as:

$$e_{brp} = e_{rp} - e_{bp} \tag{1}$$

$$e_{brg} = e_{rg} - e_{bg} \tag{2}$$

The relative displacement relation between gear shaft and bearing in the radial direction can be represented as follows:

$$\delta_{ri} = |e_{bri}| - c_{ri}(i = p, g) \tag{3}$$

where c_{ri} (i = p, g) represents bearing radial clearance of driving or driven gear.

Thus, the function $f_r(\delta_{ri})$ can be used to describe the state of gear shaft and bearing during the collision:

$$f_r(\delta_{ri}) = \begin{cases} \delta_{ri} & \delta_{ri} > 0 \\ 0 & \delta_{ri} = 0 \\ 0 & \delta_{ri} < 0 \end{cases}, \tag{4}$$

In Equation (4), the first condition represents the gear shaft and bearing without collision, while the second and third conditions represent a critical contact (impact force is zero) and a contact condition, respectively. In Figure 2, the actual center distance e_{pg} between gears has changed due to the radial offset of the gear:

$$e_{pg} = e_{rp} - e_{rg} \tag{5}$$

When the gear standard center distance is A_0, and the initial engagement angle is α_0. With the change of center distance, the actual meshing angle α' of gear varies as the following equation:

$$\alpha' = \arccos\left(\frac{A_0}{|e_{pg}|}\cos(\alpha_0)\right), \tag{6}$$

According to the involute geometry, taking the initial backlash b_0 into consideration, the dynamic backlash b_t can be expressed as:

$$b_t = 2A_0\cos(\alpha_0)(\text{inv}(\alpha') - \text{inv}(\alpha_0)) + b_0, \tag{7}$$

and in the global coordinate system, the relative meshing displacement of driving and driven gear can be expressed as:

$$g_t = R'_p \cdot \theta_p - R'_g \cdot \theta_g + (e_{brpx} - e_{brgx}) \sin \alpha' + (e_{brpy} - e_{brgy}) \cos \alpha', \tag{8}$$

where R_p' and R_g' are the radius of driving and driven gear reference circle respectively. α' is the actual pressure angle. θ_p and θ_g are the angular displacement of driving and driven gear, respectively. e_{brpx} and e_{brpy} are components of clearance vector e_{brp} in the x-axis and y-axis directions, respectively. e_{brgx} and e_{brgy} are components of clearance vector e_{brg} in the x-axis and y-axis direction, respectively.

Finally, the backlash function f_g (g_t) is used to describe the motion state during gear meshing as shown in Equation (9):

$$f_g(g_t) = \begin{cases} g_t & g_t \geq 0 \\ 0 & -b_t < g_t < 0 \\ g_t + b_t & g_t \leq -b_t \end{cases}, \tag{9}$$

In Equation (9), the first condition represents normal gear engagement, while the second and third conditions represent the occurrence of separation phenomenon of gear pair and double-sided impacts between a gear pair.

2.2. System Dynamics Model

In Figure 3, there are n_p number of planetary gears. Each gear body j (j = s, p_i, c) is modeled as a rigid gear radius R_j, angular displacement θ_j, and mass moment of inertia I_j. External torques T_s and T_c represent the input and output values. Between the planetary gear p_i and gear n (s or r), K_{Gnpi}, D_{Gnpi} and b_{npi} are the periodically time-varying meshing stiffness, the viscous damper coefficient and backlash, respectively. α is theoretically pressure angle. K_{Rpi} and D_{Rpi} are the nonlinear stiffness and damping of the radial contact collision model at the planetary gear bearing, respectively. θ_{spi} is the initial phase angle of planetary gear relative to sun gear. R_r is the radius of the ring gear. K_{Rc} and D_{Rc} are the nonlinear stiffness and the damping of the carrier output shaft and bearing radial contact collision model, respectively.

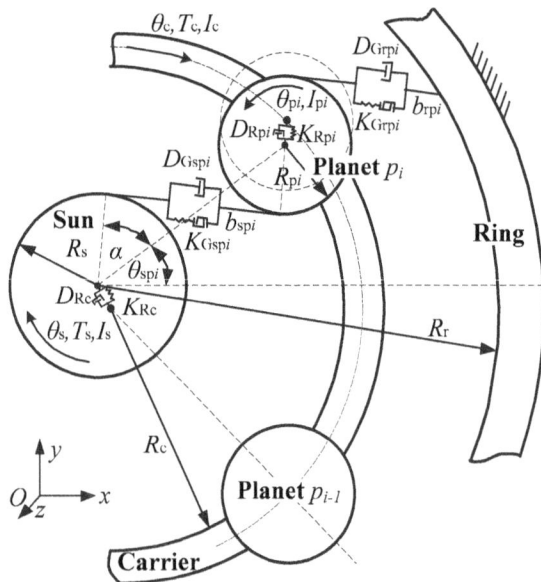

Figure 3. Joint dynamics model of planetary gear transmission.

The equations of motion of planetary system are written as follows:

$$I_s \ddot{\theta}_s(t) + R_s \sum_{i=1}^{n_p} \left\{ D_{Gspi} \cdot \dot{g}_{tspi}(t) + K_{Gspi} \cdot f_{tspi}[g_{tspi}(t)] \right\} = T_s(t), \tag{10}$$

$$I_{cp}\ddot{\theta}_c(t) + I_p \sum_{i=1}^{n_p} \ddot{\theta}_{cpi}(t) - R_s \sum_{i=1}^{n_p} \left\{ D_{Gspi} \cdot \dot{g}_{tspi}(t) + K_{Gspi} \cdot f_{tspi}[g_{tspi}(t)] \right\} - R_r \sum_{i=1}^{n_p} \left\{ D_{Grpi} \cdot \dot{g}_{trpi}(t) + K_{Grpi} \cdot f_{trpi}[g_{trpi}(t)] \right\} = T_c(t), \tag{11}$$

$$m_c \ddot{\delta}_{cx}(t) + \sum_{i=1}^{n_p} \left\{ D_{Rpi} \cdot \dot{\delta}_{pix}(t) + K_{Rpi} \cdot f_{rpix}[\delta_{pix}(t)] \right\} - D_{Rc} \cdot \dot{\delta}_{cx}(t) - K_{Rc} \cdot f_{rcx}[\delta_{cx}(t)] = 0, \tag{12}$$

$$m_c \ddot{\delta}_{cy}(t) + \sum_{i=1}^{n_p} \left\{ D_{Rpi} \cdot \dot{\delta}_{piy}(t) + K_{Rpi} \cdot f_{rpiy}[\delta_{piy}(t)] \right\} - D_{Rc} \cdot \dot{\delta}_{cy}(t) - K_{Rc} \cdot f_{rcy}[\delta_{cy}(t)] = 0, \tag{13}$$

$$I_p \ddot{\theta}_{cpi}(t) + I_{cp}\ddot{\theta}_c(t) - R_p D_{Gspi} \cdot \dot{g}_{tspi}(t) - R_p K_{Gspi} \cdot f_{tspi}[g_{tspi}(t)] + R_p D_{Grpi} \cdot \dot{g}_{trpi}(t) + R_p K_{Grpi} \cdot f_{trpi}[g_{trpi}(t)] = 0, \tag{14}$$

$$m_{pi}\ddot{\delta}_{pix}(t) - D_{Rpi} \cdot \dot{\delta}_{pix}(t) - K_{Rpi} \cdot f_{rpix}[\delta_{pix}(t)] = 0, \tag{15}$$

$$m_{pi}\ddot{\delta}_{piy}(t) - D_{Rpi} \cdot \dot{\delta}_{piy}(t) - K_{Rpi} \cdot f_{rpiy}[\delta_{piy}(t)] = 0, \tag{16}$$

In these equations, $i = 1, 2, \ldots, n_p$, and in Figure 3, $\theta_{cpi}(t) = \theta_{pi}(t) - \theta_c(t)$ represents rotation angle of planetary gear with respect to carrier. Here, $I_{cp} = I_c + n_p(I_p + R_c^2 m_p)$ is equivalent mass moment of inertia of carrier.

In these equations, the radial collision stiffness K_{Rk} ($k = c, p_i$) could be obtained by collision experiment of two spheres, it is defined as:

$$K_{Rk} = \frac{4}{3\pi \left(\frac{1-v_{bk}^2}{\pi E_{bk}} + \frac{1-v_{jk}^2}{\pi E_{jk}} \right)} \left[\frac{R_{bk} R_{jk}}{R_{bk} - R_{jk}} \right]^{1/2}, \tag{17}$$

where, v_{ik} and E_{ik} ($i = b, j$) are Poisson's ratio and elastic modulus of center element i, respectively. The variable b represents bearing, and j represents shaft.

The damper coefficients of radial collision, namely, D_{Rc} and D_{Rpi} are defined as:

$$D_{Rk} = \frac{3K_{Rk}(1 - c_{ek}^2)\delta_k^m}{4\dot{\delta}_k^{(-)}}, \tag{18}$$

where c_{ek} is the coefficient of restitution, $\dot{\delta}_k^{(-)}$ is the initial relative velocity at the collision location.

The time-varying meshing stiffness K_{Gh} ($h = spi, rpi$) between gear teeth is given by:

$$K_{Gh}(t) = k_{Gmk} + k_{Gah} \cos(\omega_{Gh} t + \varphi_{Gh}), \tag{19}$$

where, k_{Gmk} is average meshing stiffness, k_{Gah} is time-varying meshing stiffness, ω_{Gh} is gear meshing frequency. φ_{Gh} is initial phase of variable stiffness ($\varphi_{Gh} = 0$ in general).

D_{Gspi} and D_{Grpi} are the non-linear dampers of gear meshing. In order to prevent the discontinuity of damping force during the meshing process and avoid the phenomenon in which the impact force of linear damping model is not equal to zero in critical contact state, the nonlinear damping force can be expressed as:

$$D_{Gh} = \begin{cases} D_{Gmh} & f_{gh}(g_{th}) \geq d_h \\ D_{Gmh} - \beta_h^2(3 - 2\beta_h) & 0 \leq f_{gh}(g_{th}) < d_h \\ 0 & f_{gh}(g_{th}) < 0 \end{cases}, \tag{20}$$

where D_{Gmh} is the maximum damping coefficient; d_h is the maximum embedding depth and values as 0.1 mm, and β_h is defined as $\beta_h = f_{gh}(g_{th})/d_h$.

3. Results

It is known that planetary gears are used in space manipulators and drive mechanisms, primarily for rotating motion of space mechanisms. As the kinematic accuracy and reliability requirements are extremely high in space mechanisms, the vibrations within joint, which are most likely to have a significant impact on the operational stability of the entire spacecraft, are not negligible. Therefore, in this section, the influence of multi-clearance coupling effects on vibration characteristics of a planetary gear drive is studied by analyzing the vibration characteristics of the planetary gear joint and the influence of velocity, clearance and load on its vibration characteristics. In this paper, the Newmark method is used to solve the dynamic model. This algorithm is unconditionally stable under certain parameters and is helpful to solve high-dimensional nonlinear differential algebraic equations. The radial and torsional acceleration curves of planetary gear train are obtained by simulation, and then the corresponding vibration spectrum is obtained by Fast Fourier Transform (FFT). To ensure the versatility of modeling and analysis methods, here, the main parameters of numerical simulation are referred to Kahraman's simulation case [11], as shown in Table 1.

Table 1. Parameters of the planetary gear drive joint.

Parameter	Sun	Planet	Ring
Number of teeth	34	18	70
Module (mm)	1.5	1.5	1.5
Pressure angle (deg)	21.3	21.3	21.3
Circular tooth thickness (mm)	1.895	2.585	1.844
Face width (mm)	30	30	30

In order to get theoretical velocity of planetary gear and planet carrier, the general formula for planetary gear ratio calculation is as follows:

$$i_{sr}^c = \frac{n_s^c}{n_r^c} = \frac{n_s - n_c}{n_r - n_c} = (-1)^m \frac{z_p \cdot z_r}{z_s \cdot z_p},$$

(21)

where, i_{sr}^c is the transmission ratio of the sun gear and the ring gear in inverted gear train. n_s, n_r and n_c represent the velocities of sun gear, ring gear and planet carrier respectively. And m is the number of external meshing between sun gear and ring gear in the inverted gear train. z_s, z_r and z_p represent the number of teeth of sun gear, ring gear and planetary gear, respectively, and the values are shown in Table 1. Here, due to the ring gear is fixed, n_c is taken as zero. To verify the reliability of dynamic model and calculation results, a verification simulation experiment was performed. When the ring gear and frame is relatively static, the sun gear velocity is taken as 30°/s, and the theoretical value of planet carrier velocity is taken as w_c = 9.808°/s. Under the same condition, the numerical results of planet carrier velocity is w'_c = 9.809°/s. It can be seen that the motion law of dynamic model in this paper is realistic and the accuracy of dynamic model is verified.

3.1. Analysis of Coupling Vibration of Transmission Joint

As the multi-clearance coupling relation in dynamic model of planetary gear drive joint is considered, there is a complex coupling vibration relation inside the transmission joint. Therefore, the next step is finding the coupling law of internal vibration of transmission joint by spectrum analysis. Here, assuming that the connection between sun gear and drive motor is ideal, there are four clearances in the drive joint, namely, the backlash between sun gear and planetary gear, taking b_{sp} = 200 μm; the backlash exists between planet and ring gear, taking b_{rp} = 200 μm. The radial clearance between the planetary gear shaft and bearing of the planet carrier, c_{cp} = 200 μm; The radial clearance between planet carrier shaft and output bearing, c_c = 200 μm. As the drive velocity of space mechanism is slow, the drive velocity of sun gear is taken as 30°/s.

Figures 4–7 show the vibration frequency spectra of the planetary gear in the case of an empty inertial load. Figure 4 shows the radial vibration frequency spectrum of the planet carrier's bearing, in which the abscissa represents frequency and the ordinate represents the amplitude, reflecting the radial vibration characteristics of the planet carrier relative to the bearing. It can be seen from the figure that the radial vibration frequency spectrum of the planet carrier mainly contains the frequency f_r of radial clearance vibration, and the two times and three times the frequency of f_r. Among them, the frequency of radial clearance is the vibration frequency of the rotating shaft and bearing in a continuous contact state, and in this case, $f_r = 0.87$ Hz. In addition, this figure shows a higher harmonics of f_r cause by the mutual coupling between the radial vibration of planet carrier and planet.

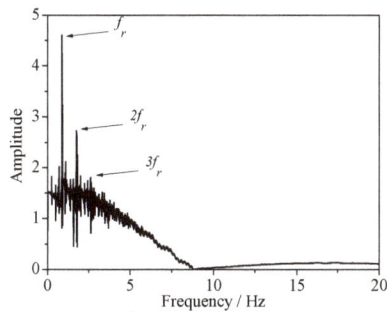

Figure 4. Radial vibration frequency spectrum of planet carrier bearing.

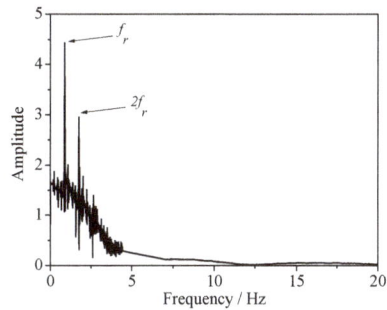

Figure 5. Radial vibration frequency spectrum of planetary gear bearing.

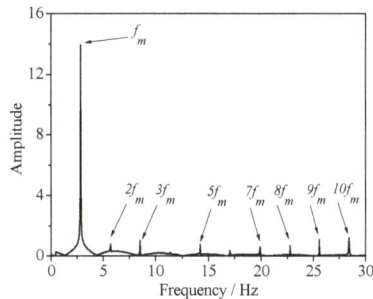

Figure 6. The frequency spectrum of torsional vibration of planet carrier.

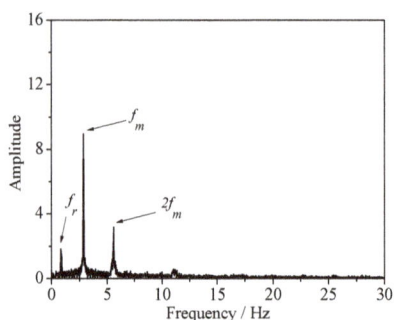

Figure 7. The frequency spectrum of torsional vibration of planetary gear.

Figure 5 shows the radial vibration frequency of the planetary gear relative to the bearing. The frequency f_r of the radial clearance vibration is also included in the frequency spectrum for the planet shaft and the bearings are also in a continuous contact state at low velocity. Meanwhile, since the planetary gears are mounted on the planet carrier, the radial vibration of the planet carrier not only directly affect the radial vibration of planet as a whole, but also be affected by the radial vibration of the planetary gear bearings. In addition, due to the mutual radial vibration coupling between the planet carrier and planet, the radial vibration of the planet also shows two times the frequency f_r.

Figure 6 shows the frequency spectrum of torsional vibration of the planet carrier, which reflects the vibration characteristics in the direction of rotation when the planet carrier rotates at the output velocity. The rotary motion of the planet carrier is driven by meshing among the planet, sun gear and ring gear, so its torsional vibration characteristics are mainly related to the meshing frequency f_m among gears. It can be seen from the figure that the torsional vibration frequency spectrum of the planet carrier includes the gear meshing frequency f_m, $2f_m$, $3f_m$, and so on, where the amplitude of meshing frequency f_m is the largest, and it is worth mentioning that a higher harmonics occurs due to the influence of time-varying meshing stiffness and backlash.

Figure 7 shows the torsional vibration frequency spectrum of the planetary gear, which not only contains the gear meshing frequencies f_m and $2f_m$, but also reflects the frequency f_r of the radial clearance vibration. It shows that radial vibration of planetary gears causes a change in backlash, resulting in dynamic backlash fluctuations that then influence the torsional vibration characteristics of planetary gears. It can be seen that although the radial and torsional vibrations of multi-clearance gear systems are not in the same direction, there is still a coupling relationship between them.

3.2. Effect of Velocity on Joint Vibration Characteristics

According to the analysis presented in the previous section, it can be seen that planet carrier is coupled with the planetary gear vibrations, and the radial and torsional vibration of the planetary gear itself also has a coupled relation. These coupling vibrations may cause resonance phenomena under certain conditions. Since the space mechanisms are operated under environmental loads and in diverse working states, the rotational velocities of transmission joints are quite different in order to accomplish different tasks like attitude-adjusting, tracking, pointing and so on, so this section analyzes the influence of rotational velocity on the vibration characteristics of joints. It reflects the variation of the vibration peak values of the system steady-state response at different velocities. It is assumed that the radial clearance of the joint interior and backlash are taken as 200 μm. In empty-load state, the velocity range is from 0°/s to 30°/s. The amplitude-frequency response curves of planetary gears are shown in Figures 8–11.

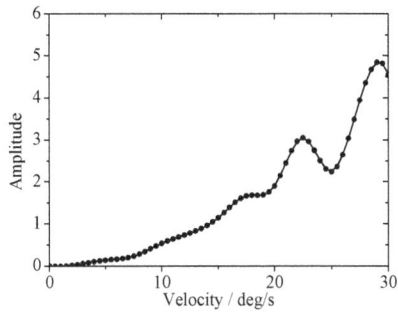

Figure 8. Radial vibration amplitude of the planet carrier at different rotational velocities.

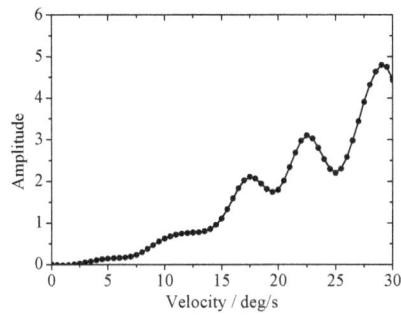

Figure 9. Radial vibration amplitude of planetary gears at different rotational velocities.

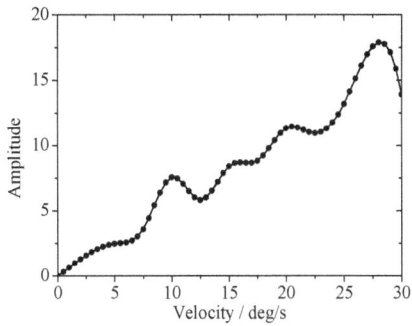

Figure 10. The amplitude of torsional vibration of planet carrier at different rotational velocities.

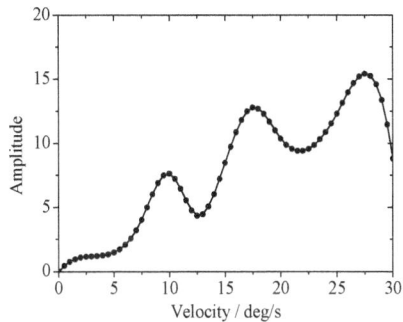

Figure 11. The amplitude of torsional vibration of planetary gears at different rotational velocities.

Figures 8 and 9 show the amplitude-response curves of the radial vibration of the planet carrier and planetary gear at different driving velocities, respectively. These graphs display the change law of the amplitude of frequency f_r at different driving velocities. By comparing the two graphs, it can be found that vibration peaks occur at specific velocities. Both graphs show the vibration peaks' value are nearly equal at the same velocities. It shows that the radial vibration of the planet carrier is mainly caused by the radial vibration of planet. Compared to Figure 8, there is a new vibration peak at a driving velocity of 17.5°/s in Figure 9. It can be inferred that this phenomenon is caused by the coupling effect of the radial vibration of the planet carrier and planet.

Figures 10 and 11 show the amplitude-response curves of the torsional vibration of the planet carrier and planet at different driving velocities, respectively. It shows the change law of the amplitude of gear meshing frequency f_m in the torsional vibration of planet carrier and planet at different driving velocities. As the curve shows, both graphs have vibration peaks when the driving velocity is close to 10°/s and 27.5°/s, and the same phenomenon can be found in Figure 11. It can be proved that the coupling relation of torsional vibration between the planet carrier and the planet makes the resonance phenomenon occur in the system when the velocities are close to 10°/s and 27.5°/s. Similar to Figure 9, a vibration peak also can be found when the driving velocity near 17.5°/s in Figure 11. It can be inferred that this phenomenon is caused by the coupling effect between the radial vibration and the torsional vibration of planet, when the driving velocity of the planet drive joint is close to 17.5°/s.

3.3. Influence of Clearance Size on Joint Vibration Characteristics

From the dynamic modeling process, it can be seen that the clearance size is also one of the major factors which have an influence on the vibration characteristics of joint systems. On the basis of the above analysis, it can be concluded that the coupling relation of vibrations between the planet carrier and planet gear is a significant factor. The radial vibration of planetary gears also has a coupling relation with their torsional vibration. Hence, this section will analyze the influence rule of clearance size on the coupling vibration of joint, providing a theoretical basis for the design of the joint system and reasonable selection of the clearance.

First, we will analyze the influence of radial clearance size of the planet carrier and planet on the coupling relation of radial vibration, taking the driving velocity as 29°/s, and the range of radial clearance of planet carrier and planet as 10 to 200 μm. The radial vibration response curves of planetary gears at different radial clearances are obtained, as shown in Figure 12.

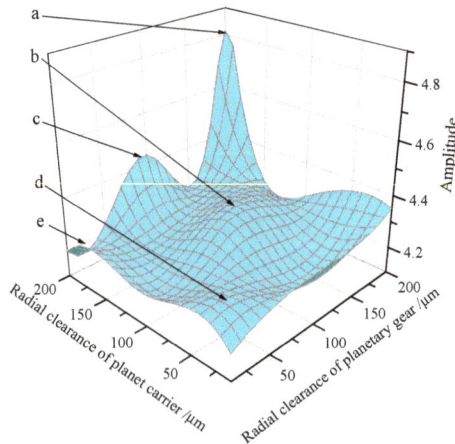

Figure 12. Radial vibration amplitude of planetary gears under different radial clearance.

In Figure 12, the x and y axes represent the radial clearance of the planet carrier and planet, respectively, and z axis represents the amplitude of the frequency of the radial vibration. There are totally five major vibration peaks in Figure 12. The a-point's vibration peak is highest when cc = 200 μm and ccp = 200 μm, and other vibration peaks are represented by b, c, d and e. It can be concluded that the size of radial clearance has a significant effect on the radial coupling vibration, and the system will resonate when the radial clearance displays a specific combination.

To study the influence of backlash size on the coupling relation of torsional vibration, we take the torsion vibration response curve of planet carrier as an example here. The driving velocity of the torsional vibration peak is taken as 27.5°/s according to the analysis presented in the previous section. It is considered that $b_{sp} = b_{rp}$, and its variation range is 10 to 200 μm. The numerical calculation results are shown in Figure 12.

In Figure 13, the horizontal axis represents backlash, and vertical axis represents the amplitude of meshing frequency f_m in torsional vibration of the planet carrier. It must also be noted that a vibration peak appears when the backlash is taken as 30, 100 and 180 μm. It can be seen that the effect of backlash size on the torsional vibration characteristics is very significant.

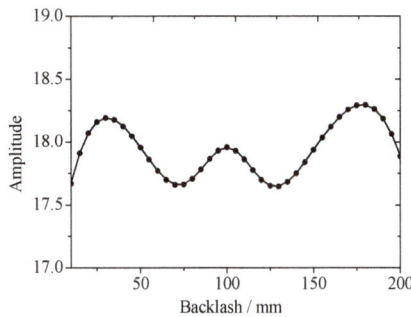

Figure 13. The amplitude of torsional vibration of planet carrier under different backlash.

Through the above numerical calculation, it can be found that there is a coupling relation between the planetary gear's radial and the torsional vibration. In other words, it indirectly verifies that the coupling between radial clearance and backlash is correct in the dynamic model. On the basis of the vibration characteristics of planet itself, this section aims to further study the influence of radial clearance and backlash's size on the coupling vibration.

According to the analysis of Figures 9 and 11, the driving velocity is taken as 17.5°/s for the radial and torsional vibration of planet shows a resonance phenomenon at this velocity, and the ranges of radial clearance and backlash of the planet are 10 to 200 μm, the amplitude variation rule of frequency f_r of planets with different radial clearance and the backlash sizes is shown in Figure 14.

In Figure 14, x-axis and y-axis respectively indicate the sizes of radial clearance and backlash of planet, and the z-axis represents the frequency's amplitude of the radial vibration of the planet. There are five vibration peaks in the figure, and the a-point vibration peak is highest at $c_{cp} = 40$ μm and $b_{sp} = b_{rp} = 40$ μm. It can be concluded that the sizes of the radial clearance and backlash have a significant effect on the coupling between the radial and torsional vibration of planet, and a large amplitude of vibration peak will occur when the clearance are in a specific combination.

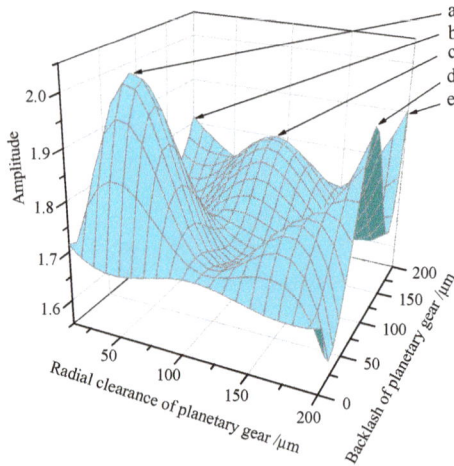

Figure 14. Radial vibration amplitude of planetary gears under different radial and backlashes.

3.4. Influence of Load on Joint Vibration Characteristics

A typical spacecraft, such as a large space manipulator, will bear a certain inertial load during the execution of object-grabbing or object-transferring. The end of the satellite biaxial drive mechanism also has an antenna reflector, a camera or other loads. On the one hand these inertial loads are indispensable for the spacecraft; on the other hand, the vibration characteristics of joints will be affected during the course of motion due to the presence of inertial loads. Taking the single joint as an example, this section analyzes the influences of different inertial loads on the vibration characteristics of planetary gears. Here, the variation range of inertial load is taken as 0 to 100 kg·m^2, the driving velocity is taken as 30°/s, both the radial clearance and the backlash are taken as 200 μm, and the numerical calculation results are shown in Figures 15–18.

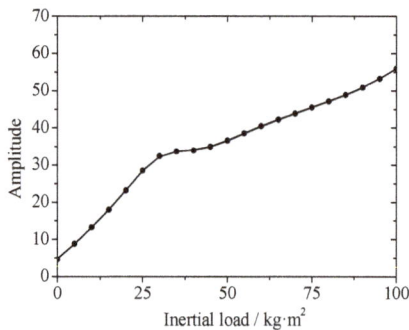

Figure 15. Radial vibration amplitude of planet carrier under different inertial loads.

Figure 15 shows the change law of the frequency f_r's amplitude in the radial vibration of the planet carrier at different inertial loads. Since the inertial load is directly working on the planet carrier, with the increase of inertial load, the radial vibration amplitude of the planet carrier increases and the radial vibration will be more intense. This phenomenon is caused by the increase of inertial load, and the radial vibration of planet carrier can be alleviated only by reducing the inertial load.

Figure 16 shows the change law of planetary gear radial vibration with different inertial loads. It can be seen from Figure 16 that the radial vibration amplitude of the planetary gear is reducing with the increasing inertia load. The reason is that the planet carrier is in a free vibration state in the radial direction without the inertial load, and four planetary gears mounted on the carrier are vibrating with the vibration of the planet carrier. When the system bears an inertial load, the joints are subjected to inertial forces in the radial direction, but the four evenly arranged planetary gears which are in the circumferential direction of planet carrier will dynamically balance the radial forces. In this condition, the changes of all planetary gear's radial force are quite stable, and then the radial vibration of planetary gears will be reduced.

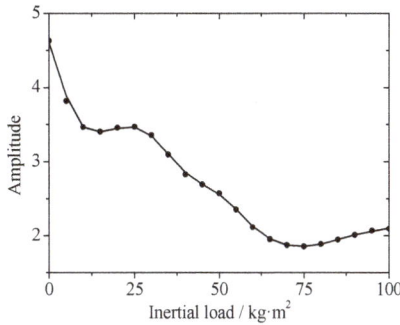

Figure 16. Radial vibration amplitude of planetary gear under different inertial loads.

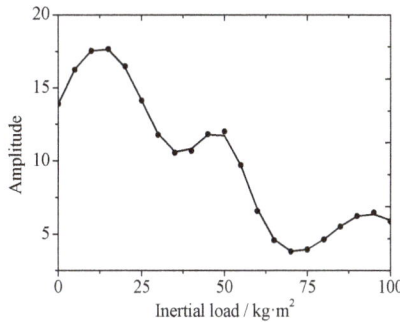

Figure 17. The amplitude of torsional vibration of planet carrier under different inertial loads.

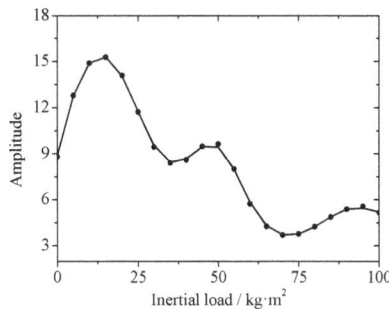

Figure 18. The amplitude of torsional vibration of planetary gears under different inertial loads.

A similar phenomenon to the one seen in Figure 16 also could be seen in Figures 17 and 18, in which the torsional vibrations of planet carrier and planetary gear are reduced as the inertia load increases. One reason is that the increase of inertia load results in the raise of gear meshing force and fewer off tooth and tooth back impacts. Another reason is that the dynamic backlash's changes are weakening at any time due to the weakness of the radial vibration of the planetary gear, and the gear meshing process is more stable.

4. Conclusions

This paper sets up a new nonlinear dynamic model of planetary gear joints, in which the radial clearance of bearings, dynamic backlash and time-varying meshing stiffness are considered. The effects of multi-clearance coupling on the vibration characteristics of driving joint planetary gears are analyzed. In this paper, a coupling effect of the internal vibration of the transmission joint is found in the frequency domain. The effects of velocity, clearance size and load on the joint vibration characteristics of planetary gear drives are analysed. The conclusion of the current study can be summarized as follows:

- At some specific velocities, resonance may occur between the vibration of the planet carrier and the planetary gear, and there is a coupling relation between the radial and torsional vibrations of the planetary gear itself;
- The clearance may cause nonlinearities in the system dynamic and changes in the vibration amplitude. Vibration peaks occur when the radial vibration of the planet carrier and planetary gears have specific radial clearance sizes. The change of backlash size causes a torsional vibration peak between the planet carrier and planetary gear. The radial and torsional vibration of planetary gears also produce a vibration peak with certain radial clearance and backlash combinations;
- Due to the special structure of planetary gears, the radial vibration amplitude of the output will increase when the load is heavy, but the amplitude of torsional vibration will decrease. Therefore, when the inertia load is heavy, the system's rotation accuracy and stability are good, but when in empty-load or light-load states, the operating strategies also need to be adjusted to ensure operational reliability.

An inherent characteristic analysis of planetary gears with multi-clearance coupling will be developed by modal analysis in future work.

Author Contributions: Conceptualization, H.Z.; Data curation, J.F.; Formal analysis, C.Q. and S.D.; Funding acquisition, J.F.; Investigation, C.Q.; Project administration, S.D. and B.Y.; Resources, S.D.; Validation, J.F. and B.Y.; Writing–original draft, H.Z. and C.Q.; Writing–review & editing, H.Z.

Funding: This research was supported by State Key Laboratory of Robotics and System (HIT) (Grant No. SKLRS-2017-KF-15), Science and Technology on Space Intelligent Control Laboratory (Grant No. ZDSYS-2017-08), National Natural Science Foundation of China (NSFC) (Grant No. 51575126; 51605133; 51705127), and Graduate Innovative Ability Training Project of Hebei Province (Grant No. CXZZSS2018025).

Acknowledgments: The authors would like to thank the research grant from Spacecraft Dynamics Design and Simulation Lab (SDDSL) of HIT.

Conflicts of Interest: The authors declare no conflict of interest.

References

1. Chen, Z.; Shao, Y. Dynamic simulation of planetary gear with tooth root crack in ring gear. *Eng. Fail. Anal.* **2013**, *31*, 8–18. [CrossRef]
2. Pan, B.; Yu, D.; Sun, J. Research on dynamic modeling and analysis of joint in large space manipulator. *J. Asrton.* **2010**, *31*, 2448–2455. [CrossRef]
3. Parenti-Castelli, V.; Venanzi, S. Clearance influence analysis on mechanisms. *Mech. Mach. Theory* **2005**, *40*, 1316–1329. [CrossRef]
4. Bai, Z.; Zhao, Y.; Wang, X. Wear analysis of revolute joints with clearance in multibody systems. *Sci. China Phys. Mech. Astron.* **2013**, *56*, 1581–1590. [CrossRef]
5. Megahed, S.M.; Haroun, A.F. Analysis of the dynamic behavioral performance of mechanical systems with multi-clearance joints. *J. Comput. Nonlinear Dyn.* **2012**, *7*, 354–360. [CrossRef]
6. Zhang, H.; Tian, J.; Zhou, J.; Zhao, Y.; You, B. Dynamic and experimental investigation of gear-rotor system with multiple clearances coupled. *Chin. J. Mech. Eng.* **2017**, *53*, 29–37. [CrossRef]

7. Chaker, A.; Mlika, A.; Laribi, M.A.; Romdhane, L.; Zeghloul, S. Clearance and manufacturing errors' effects on the accuracy of the 3-RCC spherical parallel manipulator. *Eur. J. Mech. A-Solids* **2013**, *37*, 86–95. [CrossRef]
8. Kahraman, A. Load sharing characteristics of planetary transmission. *Mech. Mach. Theory* **1994**, *29*, 1151–1165. [CrossRef]
9. Yuksel, C.; Kahraman, A. Dynamic tooth loads of planetary gear sets having tooth profile wear. *Mech. Mach. Theory* **2004**, *39*, 695–715. [CrossRef]
10. Al-Shyyab, A.; Kahraman, A. Non-linear dynamic analysis of a multi-mesh gear train using multi-term harmonic balance method: Sub-harmonic motions. *J. Sound Vib.* **2005**, *284*, 151–172. [CrossRef]
11. Al-Shyyab, A.; Kahraman, A. Non-linear dynamic model for planetary gear sets. *Proc. Inst. Mech. Eng. K-J. Multi-Body Dyn.* **2007**, *221*, 567–576. [CrossRef]
12. Ligata, H.; Kahraman, A.; Singh, A. An experimental study of the influence of manufacturing errors on the planetary gear stresses and planetary load sharing. *J. Mech. Des.* **2008**, *130*, 1–9. [CrossRef]
13. Inalpolat, M.; Kahraman, A. A dynamic model to predict modulation sidebands of a planetary gear set having manufacturing errors. *J. Sound Vib.* **2010**, *329*, 371–393. [CrossRef]
14. Hotait, M.A.; Kahraman, A. Experiments on the relationship between the dynamic transmission error and the dynamic stress factor of spur gear pairs. *Mech. Mach. Theory* **2013**, *70*, 116–128. [CrossRef]
15. Li, S.; Kahraman, A. A tribo-dynamic model of a spur gear pair. *J. Sound Vib.* **2013**, *332*, 4963–4978. [CrossRef]
16. Ambarisha, V.K.; Parker, R.G. Nonlinear dynamics of planetary gears using analytical and finite element models. *J. Sound Vib.* **2007**, *302*, 577–595. [CrossRef]
17. Gill-Jeong, C.; Parker, R.G. Influence of bearing stiffness on the static properties of a planetary gear system with manufacturing errors. *KSME Int. J.* **2004**, *18*, 1978–1988. [CrossRef]
18. Canchi, S.V.; Parker, R.G. Effect of ring-planet mesh phasing and contact ratio on the parametric instabilities of a planetary gear ring. *J. Mech. Des.* **2007**, *130*, 014501. [CrossRef]
19. Liu, G.; Parker, R.G. Nonlinear dynamics of idler gear systems. *Nonlinear Dyn.* **2008**, *53*, 345–367. [CrossRef]
20. Guo, Y.; Parker, R.G. Dynamic analysis of planetary gears with bearing clearance. *J. Comput. Nonlinear Dyn.* **2012**, *7*, 041002. [CrossRef]
21. Ericson, T.M.; Parker, R.G. Experimental measurement of the effects of torque on the dynamic behavior and system parameters of planetary gears. *Mech. Mach. Theory* **2014**, *74*, 370–389. [CrossRef]
22. Cooley, C.G.; Liu, C.; Dai, X.; Parker, R.G. Gear tooth mesh stiffness: A comparison of calculation approaches. *Mech. Mach. Theory* **2016**, *105*, 540–553. [CrossRef]
23. Pappalardo, C.M.; Guida, D. Control of nonlinear vibrations using the adjoint method. *Meccanica* **2017**, *52*, 2503–2526. [CrossRef]
24. Pappalardo, C.M.; Guida, D. Adjoint-based optimization procedure for active vibration control of nonlinear mechanical systems. *J. Dyn. Syst. Meas. Control* **2017**, *139*, 081010. [CrossRef]
25. Ouyang, H.; Richiedei, D.; Trevisani, A.; Zanardo, G. Discrete mass and stiffness modifications for the inverse eigenstructure assignment in vibrating systems: Theory and experimental validation. *Int. J. Mech. Sci.* **2012**, *64*, 211–220. [CrossRef]
26. Palermo, A.; Mundo, D.; Hadjit, R.; Desmet, W. Multibody element for spur and helical gear meshing based on detailed three-dimensional contact calculations. *Mech. Mach. Theory* **2013**, *62*, 13–30. [CrossRef]
27. Shweiki, S.; Palermo, A.; Mundo, D. A study on the dynamic behaviour of lightweight gears. *Shock Vib.* **2017**, *2017*, 1–12. [CrossRef]
28. Vivet, M.; Mundo, D.; Tamarozzi, T.; Desmet, W. An analytical model for accurate and numerically efficient tooth contact analysis under load, applied to face-milled spiral bevel gears. *Mech. Mach. Theory* **2018**, *130*, 137–156. [CrossRef]
29. Wei, J.; Zhang, A.; Qin, D.; Lim, T.; Shu, R.; Lin, X.; Meng, F. A coupling dynamics analysis method for a multistage planetary gear system. *Mech. Mach. Theory* **2017**, *110*, 27–49. [CrossRef]
30. Yang, T.; Yan, S.; Han, Z. Nonlinear model of space manipulator joint considering time-variant stiffness and backlash. *J. Sound Vib.* **2015**, *341*, 246–259. [CrossRef]

31. Yang, T.; Yan, S.; Wei, M.; Han, Z. Joint dynamic analysis of space manipulator with planetary gear train transmission. *Robotica* **2016**, *34*, 1042–1058. [CrossRef]

32. Marques, F.; Isaac, F.; Dourado, N.; Flores, P. An enhanced formulation to model spatial revolute joints with radial and axial clearances. *Mech. Mach. Theory* **2017**, *116*, 123–144. [CrossRef]

energies   MDPI

Article

Pulse-Current Sources for Plasma Accelerators

Alexei Shurupov [1], Alexander Kozlov [1], Mikhail Shurupov [1], Valentina Zavalova [1,*],
Anatoly Zhitlukhin [2], Vitalliy Bakhtin [2], Nikolai Umrikhin [2] and Alexei Es'kov [2]

[1]   Joint Institute for High Temperatures of the Russian Academy of Sciences (JIHT RAS),
      125412 Moscow, Russia; shurupov@fites.ru (A.S.); Kozlov@fites.ru (A.K.); m.a.shurupov@yandex.ru (M.S.)
[2]   State Research Center of the Troitsk Institute for Innovation & Fusion Research (SRC RF TRINITI),
      142190 Moscow, Russia; zhitlukh@triniti.ru (A.Z.); bakhtin@triniti.ru (V.B.); umrikhin@triniti.ru (N.U.);
      eskov@triniti.ru (A.E.)
*   Correspondence: zavalova@fites.ru; Tel.: +7-916-153-4243

Received: 21 September 2018; Accepted: 1 November 2018; Published: 7 November 2018

Abstract: The pulse source for plasma-accelerators supply operates under the conditions of nonlinear growth of load inductance, which complicates the matching of the source and the load. This article presents experimental studies of the use of both traditional pulse-energy sources based on capacitive storage and alternative ones based on explosive magnetic generators (EMG). It is shown that the EMG with the special device of the current-pulse formation more effectively matches with such a plasma load as the pulse plasma-accelerator (PPA). This device allows a wide range to manage the current-pulse formation in a variable load and, consequently, to optimize the operation of the power source for the specific plasma load. A mathematical model describing the principle of operation of this device in EMG on inductive load was developed. The key adjustable parameters are the current into the load, the residual inductance of the EMG, and the sample time of the specified inductance and the final current in the load. The device was successfully tested in experiments with the operation on both one and two accelerators connected in parallel. In the experiments, the optimal mode of device operation was found in which the total energy inputted to a pair of accelerators in one pulse reached 0.55 MJ, and the maximum current reached about 3.5 MA. A comparison with the results of experiments performed with capacitive sources of the same level of stored energy is given. The experiments confirmed not only the principal possibility of using EMG with a special device of current-pulse formation for operation with plasma loads in the MJ energy range but also showed the advantages of its application with specific types of plasma load.

Keywords: explosion-magnetic generator; plasma accelerator; current-pulse formation

1. Introduction

Research of the physical and technical basis of sources for the generation of super-power electromagnetic fields and current pulses with the help of explosion-magnetic generators (EMG), outlined in classical works [1,2], found a continuation in new problems for plasma loads. Among the common problems we can note the following: pulsed plasma accelerator (PPA) [3–5], plasma focus [6], modeling and study of the dynamics of plasma jets in the ionosphere [7], high-current plasma interrupters [8,9], the generation of X-ray radiation [10], and studies of erosion of metal samples under the influence of intense plasma flows created by the PPA [11]. All of these areas have special requirements for the energy source, forming a pulsed current to generate a plasma flow with the necessary parameters.

Capacitive storage devices are traditionally used as energy sources in these tasks. They are very practical to use, but they do not match well with plasma loads, especially in the mega-ampere range of plasma accelerator currents. This is due to the feature of changing the power source during operation of the PPA [5]. The rapid acceleration of the plasma leads to a significant change in the inductance of the load, which requires an increase in the power supply to maintain large current amplitude. Capacitive storage devices have the highest power at the highest voltage, and therefore in the process of operation on the PPA only their power falls. However, EMG have their maximum power at the end of their operations, which allows them to keep and even increase their current in the load despite a significant change in the inductance of the load. This EMG feature allows effective matching with the non-linear load of plasma, and this will be confirmed in this article also.

The experience of using an alternative source for PPA, based on EMG, is described in article [5]. As is known from the early estimates [12], the highest efficiency of energy transfer from the HMG directly to the inductive load is when the load inductance is significantly less than the generator inductance. However, both the inductance of the generator and the inductance of the load change during operation on the plasma load. Moreover, if the inductance of the generator is reduced and we can control the law of its output at the design stage, the plasma load increases nonlinearly. In this article, we propose a mathematical model that allows one to predict the current in the load at the design stage of the EMG. The model takes into account the output of the EMG inductance to a specific load for the selected electrical circuit. The model calculates the dynamics of inductance change in EMG, taking into account changes in the geometry of both the generator's coil and the liner. A liner means a tube, usually made of copper, filled with an explosive. With the end explosion of the charge, the liner expands in the form of a cone, cutting off parts of the generator coil over time. As the liner contacts move along the EMG spiral, the magnetic flux is compressed and displaced into the load. Because the change in the inductance of the plasma load is nonlinear, the role of experimental studies is very important. The paper [5] presents the results of research and development of the device for forming a current pulse consisting of a solid-state circuit breaker and explosive breaker keys with an adjustable delay of their operation. The perspectives of such a scheme due to the wide possibilities to manage the keys and the movement of the liner were noted as well. In the present work, research was continued, and as a result the value of the supplied energy from the EMG to the load reached more than 500 kJ. The proposed design of the generator, optimized for specific load, allows us to effectively pump the magnetic flux into the plasma load. This avoids premature falling of the current amplitude. The idea of the operation of the device built into the EMG design is to connect the load circuit and disconnect the EMG circuit with an adjustable delay. Selecting the appropriate delay allows one to control the residual inductance of the EMG at the switching moment and optimize the shape of the current pulse in the load. The optimization of the device operation consists in the pre-selected mode of change (output) of the inductance of the EMG and the choice of the time diagram for the solid-state circuit closer and the explosive breaker. Generator residual inductance output is the main phase of the power supply to the load. This phase provides an increase in electrical power in the load circuit at a significant change in its inductance. In addition, to increase the power of the generator, the copper liner was equipped with a new type of explosives, which narrowed the front of the current pulse.

The article presents the comparative oscillograms of currents and voltages for a pulsed source based on capacitive storage and inductive pulse sources based on EMG when PPA operates in two modes. The modes were different by the connection time of the load. A series of experiments to control the shape of the output pulse of the current EMG and the matching of the latter with the work of a specific plasma load was carried out. Modes of efficient energy transfer to the load have been obtained.

The principles of operation of the helical EMG and PPA are not considered in this article. Here, we consider EMG as a pulsed-current source based on nonlinear inductive energy storage, and PPA as a nonlinear load for EMG with a growing inductance during operation, which is typical for a wide class of plasma loads.

The structure of the article is as follows. After the introduction, a chapter with the results of the study of the work of the plasma accelerator with a capacitive power source is presented. Then, the results of modeling and experimental studies of an alternative source based on an explosion-magnetic generator with built-in devices for current-pulse formation are presented.

After that, the modes of operation of this device were considered. In conclusion, the main results of the research are shortly presented, the mode of the most efficient power supply of PPA is noted, and the methods of application of the developed EMG are indicated.

2. Experimental Study of the PPA Work with Use of Capacitive Power Source

The scheme of direct connection of the capacitive storage to the plasma accelerator is a typical LCR circuit with a switch-solid-state discharger. The parameters of applied capacitors are as follows: capacity is 1 mF, and the operating voltage is up to 25 kV. The voltage was measured by means of a divider built into the input collector of the load; the current was measured by the Rogowski coil built into the switching discharger.

Characteristic oscillograms of the PPA from the capacitor power source are shown in Figure 1. The current and voltage distributions on the load and the change in load inductance are plotted on the same time scale. As can be seen from the presented oscillograms, when the current reaches a value of about 1.5–2 MA, the load inductance begins to increase sharply, which leads to a significant increase in the voltage to 20 kV or more. The time voltage on the capacitor battery does not exceed 10 kV. It is obvious that the residual voltage of the battery is not able to provide the discharge circuit with the magnetic flux necessary to maintain the current. The results of the described situation are in sharp contrast to the current amplitude when the load is rebuilt and the deviation of the current waveform from the sinusoidal shape. Actually, the experimental task is to achieve the maximum value of the current before the sharp increase of the system inductance, after which the parasitic inductance of the circuit works as energy storage. It is not possible to significantly increase or even maintain the current in the system with a capacitor battery after this moment.

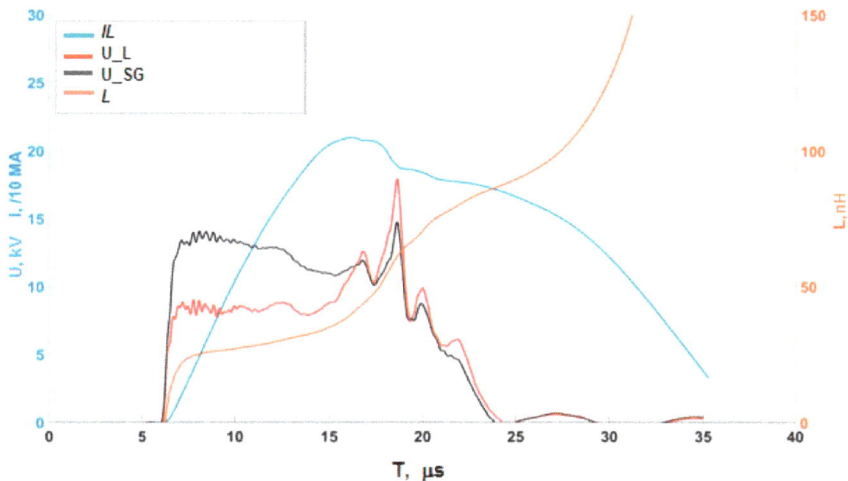

Figure 1. Oscillograms of current in the load (I_L—blue solid line), and voltage on the PPA (U_L—red line) and on the discharger (U_SG—black line) for left axis with respect to the changing inductance of the load (*L*—brown line) for right axis.

Thus, when PPA works with use of capacitive power source and its inductance increases sharply, this leads to the appearance of "features" on the current and voltage oscillograms. These features on the oscillograms at the beginning and at the maximum current are what distinguishes them from the classical distributions when the capacitor is operating at a constant inductive load. One of the main parameters that determine the efficiency of such systems is the maximum achievable current amplitude before the appearance of this feature, after which the current in the system falls. In this scheme, with about 270 kJ stored in the capacitive energy storage, about 25% of the energy was transferred to the load.

3. Operation of PPA with Use EMG Power Source

The equivalent electric circuit for operation of EMG on the load is presented in Figure 2. The initial magnetic flux in the generator is powered from the initial energy source E_0, representing the capacitive storage. The energy of explosives goes to the expansion of the liner, which in turn compresses the magnetic flux inside the helical EMG.

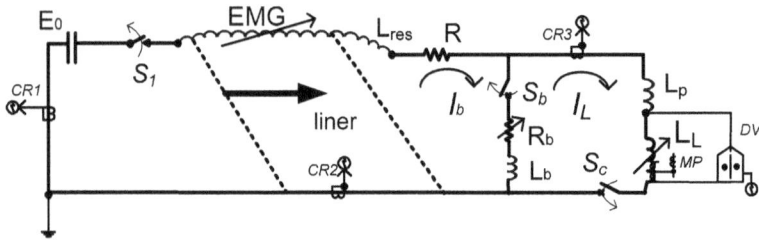

Figure 2. Equivalent electric circuit of load operation with EMG. Here, E_0 is the source of initial energy, R—the ohmic resistance in the EMG circuit including all of the ohmic losses of the magnetic flux, L_{res} is the residual inductance of EMG at the time of load connection, L_b is the inductance of the breaker circuit, S_b—breaker key, S_c—closer key, L_L is the inductance of the load, L_p is the parasitic inductance in the load circuit (inductance of the EMG energy output device), and R_b is the active resistance of the breaker. CR1-CR3—Rogowski coils, DV—the voltage divider, and MP—magnetic probes.

The main parameters of the circuit are: the initial energy E_0, the initial and residual inductance of the EMG, ohmic resistance in the EMG circuit and the circuit of the breaker, parasitic inductance in the circuit of the circuit and the load circuit, and inductance load. Parasitic inductance is determined by the leading cables, and it should be minimized. This scheme also corresponds to experimental setup when the EMG operates on a single accelerator. The scheme demonstrates the location of the main elements of the circuit, including the primary current sensors—Rogowski coils (RC), the voltage divider (VD), and magnetic probes (MP).

Switching device for the current-pulse formation in the load includes the following keys: the solid-state circuit closer (S_c) and the assembly with explosive breaker (S_b). The switching keys do not work at the same time. S_b is triggered with a delay with respect to S_c in accordance with the time diagram. This delay determines the current in the load and the residual inductance of EMG in the circuit after the switching.

Capacitive storage of 130 μF with voltage up to 30 kV was used as a source of initial energy. EMG initial inductance was approximately 10 μH. Currents in the experiments were calculated by numerical integration of data from Rogowski coils while taking into account the sensitivity of each coil. The recording of signals from the Rogowski coils is made on oscilloscopes PICOSCOPE 5000 series. Voltage measurements were made using voltage dividers DV (1:3000), designed for pulse signals.

3.1. Results of Numerical Modeling of the Current Pulse with EMG Source

For theoretical estimates, a simplified electrical circuit of the EMG operation on a variable load was taken, which is shown in Figure 2.

The calculation of the generator is performed in two stages. First, the calculation of the dynamics of the inductance output of the EMG was performed with use the open numerical complex of Finite Element Method Magnetics (FEMM 4.2). This complex allows two-dimensional, steady-state calculations of magnetic fields for the specified design of the EMG entered into the program at a constant frequency by the finite element method. The possibility of using scripts for geometry transformation made it possible to calculate the high-frequency inductance for each position of the liner. The dynamics of the liner movement was set manually. The final dynamics of the inductance output is shown Figure 3 and used in the calculation of the electrical circuit of the generator to the load.

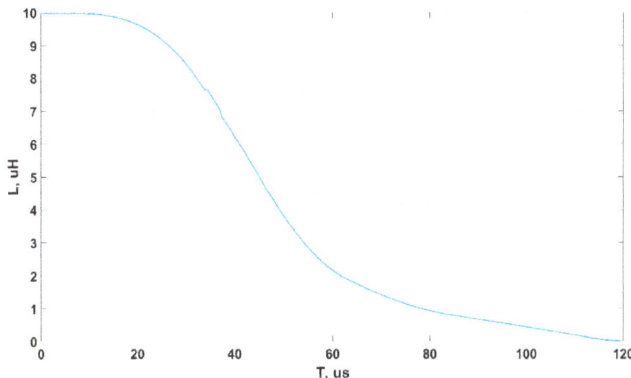

Figure 3. Dynamics of EMG inductance output at the operation on a load.

Before connecting the load, the EMG is shorted through the explosive breaker S_b; the current goes only along one circuit—1, and it greatly simplifies the equations for its calculation. After connecting the load, the current flows through both circuits I_b and I_L, as long as the resistance of the explosive breaker dynamically changes in the first circuit and it is triggered. Dynamics of changes of this resistance determine the process of switching current to the load. The dependence of the breaker resistance on time was approximated from experimental data by a function of the following form:

$$R_b = A \times exp(Bt) + C \times exp(Dt) \tag{1}$$

The Equation (1) describes well the experimentally measured resistance dynamics of the developed breaker in the EMG circuit. It agrees with experimental data with high accuracy at the coefficient's selection, as seen in Figure 4. However, it should be noted that at high cut-off currents (2.5–3 MA for our design), when the aluminum foil in the explosive breaker is heated by the EMG current to melting temperatures, this function becomes inadequate. In this case, an increase in the geometric dimensions of the explosive breaker is required. Nevertheless, this function of resistance increases the explosive breaker (see Figure 4), and the dynamics of the inductance output (see Figure 3) allow carrying out of a package of preliminary calculations to determine the output parameters of the EMG current. These calculations were carried out only on the concentrated load. The equations describing the dynamics of current switching to the load are given below:

$$\begin{cases} I_{emg} = I_b + I_L \\ L_L \frac{dI_L}{dt} = L_b \frac{dI_b}{dt} + R_b I_b \\ -I_{emg} \frac{dL_{emg}}{dt} - L_{emg} \frac{dI_{emg}}{dt} - I_{emg} \times \left(R - (1-f) \frac{dL_{emg}}{dt} \right) = L_b \frac{dI_b}{dt} + R_b I_b \end{cases} \tag{2}$$

in which f is the coefficient of EMG perfection, and $f = \ln\left(\frac{I_L}{I_{emg}}\right) / \ln\left(\frac{L_{res}+L_L}{L_L}\right)$, which reflects the degree of magnetic flux saving during EMG operation. The typical value of f for the EMG is 0.85, [2]. I_{emg} is the current through the EMG, I_b is the current through the circuit breaker, I_L is the current through the load, and L_{emg} is inductance of EMG.

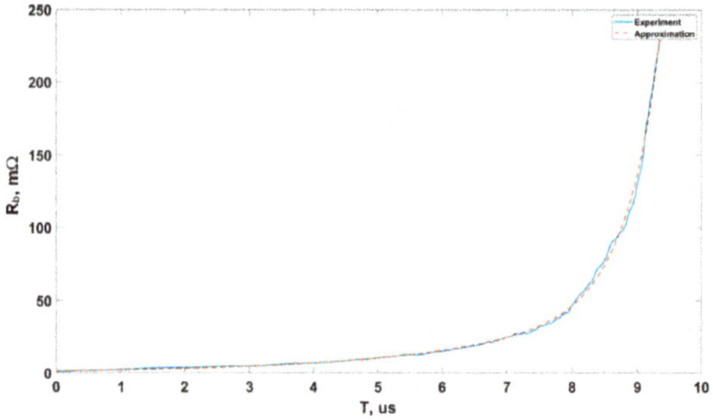

Figure 4. Experimental and calculated (dashed) dependence of the resistance of the circuit breaker from time to time.

The system (2) can be reduced to the following system of differential Equations (3) suitable for the calculation of the current switching process:

$$
\begin{cases}
\frac{dI_b}{dt} = \dfrac{-I_{emg} \times \left(R + f \times \frac{dL_{emg}}{dt}\right) + \left(\frac{L_{emg}}{L_n} + 1\right) \times R_b I_b}{L_b + \left(\frac{L_b}{L_L} + 1\right) \times L_{emg}} \\
\frac{dI_L}{dt} = \frac{L_b}{L_L}\frac{dI_b}{dt} + \frac{R_b I_b}{L_L}
\end{cases}
\tag{3}
$$

The constructed model allows varying the parameters of the load, EMG, and the explosive breaker in a wide range, which is necessary to optimize the design in order to obtain the necessary current pulse in the load. The specific characteristics of various subassemblies are specified according to the experimental data. The results of calculations of the output parameters of the current pulse of EMG, of the load, and of the load voltage for the developed designs are presented below in Figure 5.

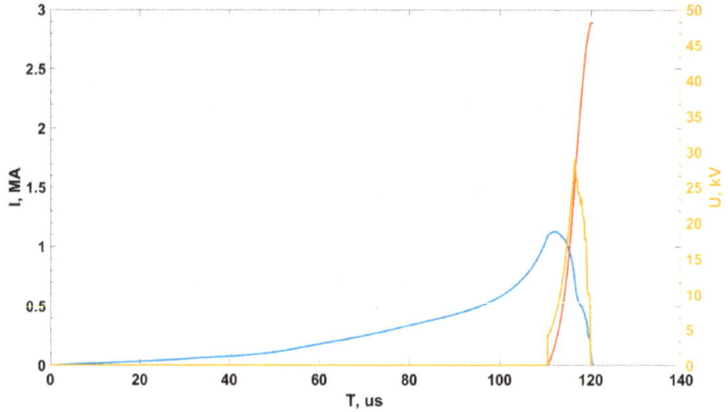

Figure 5. Calculated pulses of current in the load and load voltage during operation of the EMG. Blue—I_b, current EMG; red—I_L, load current; and yellow—U_L, voltage on the load.

The presented calculations refer to a constant inductance of the load of 50 nH, while the plasma load is variable to a large extent, and at the end of the accelerator operation, its inductance becomes at least doubled. This leads to a change in the voltage shape, which depends on the dynamics of the inductance in the load. This calculation allows us to reliably estimate the dependences, especially for currents, which have been confirmed experimentally many times. That is why this model is convenient to use for verification of various designs in order to obtain an adequate representation of the current pulse in the load, when the input parameters change in a wide range.

3.2. Results of Experimental Studies with Use EMG Power Source

To maintain the current rise in the load, it is necessary to pump the magnetic flux at an increasing rate, which requires increasing voltage and power of the power source. It is obvious that the capacitive storage does not have this characteristic. The EMG with a special current-pulse-forming device was developed to solve this problem. General view of the EMG design is shown in Figure 6a,b.

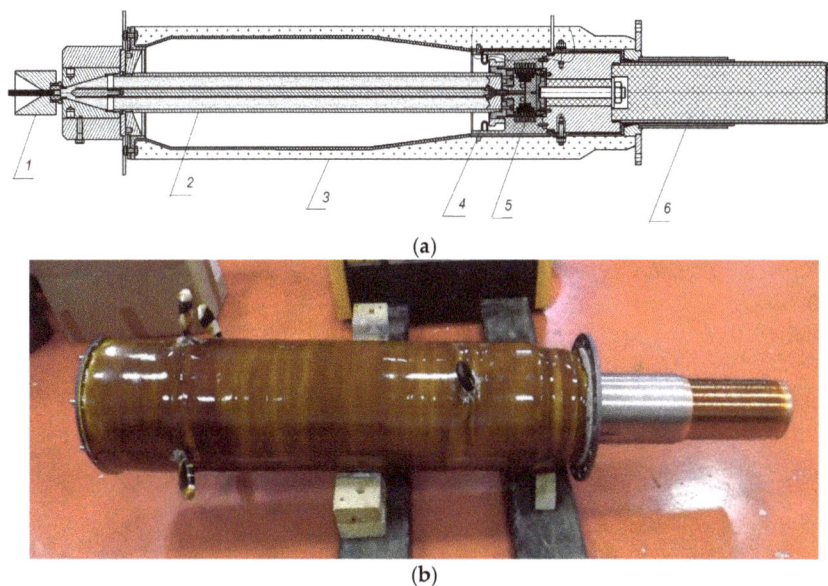

(a)

(b)

Figure 6. (**a**) General view of the EMG. 1—initiator unit, 2—copper liner, 3—helical EMG, 4—solid-state closer of circuit S_c, 5—explosive breaker-S_b, and 6—current collector. (**b**) Photos of EMG.

This version of the EMG is a classic helical generator combined with two switches in one design. The principle of operation of the developed design is as follows. In the first phase of operation of the device, the spiral coil of EMG (element #3, Figure 6a) is powered by initial energy source E_0 and the initial magnetic flux is formed. After that, the main charge of the liner is detonated, and the EMG most of the time works on the explosive breaker (element #5) in a short-circuited mode, converting the explosive energy into electrical energy. During the next stage, the load is connected in parallel with the circuit breaker using a solid-state circuit closer (element #4). After connecting the load, the explosive breaker (5) is activated with delay and switches the main current to the load. Selection of residual inductance of EMG is the final stage of the described design. Depending on the selected mode of synchronization, inductance was from 50 to 500 nH at the time of operation start of the explosive breaker.

The increasing power at the working on the load in the final stage is a key aspect of the described concept of operation of the EMG, which is provided by the continually-increasing derivative of the inductance ($\frac{dL}{dt}$). In addition, it is important to note that this design has considerable flexibility, because it has many adjustable parameters. These parameters include the moments of connection of the load and the operation of the explosive breaker, and the residual inductance and its sampling time. The general time diagram of the EMG is determined by the speed of expansion of the liner, which is regulated by the type of used explosives and the geometry of the liner. The moment of the load connection is determined by the contactor position. The moment of explosive breaker operation is determined by its own response time and the detonation line delay between the main charge of the explosive and the charge in the breaker. The time of the residual inductance transformation of generator at the last stage of work is regulated by adjusting the angle of the spiral slope of EMG in the last section to the angle of the liner opening and the liner speed. The detonation of explosive is carried out from one detonator and does not require additional detonation lines to synchronize the switching devices. The variability of the generator parameters allows optimizing EMG operation for a wide range of non-linear plasma loads. The developed design of the generator is equipped with a special coaxial current output of energy. It allows use of explosive protective cameras for EMG, while the load is located outside the chamber and can be fully saved.

A key moment in the development of design EMG was the need to hold a lot of charge inside the liner is in the range of 5–6 kg, because the explosion chamber "Titan", with a maximum permissible weight of 8 kg, was used in experiments in laboratory conditions. Copper liners with diameters from 90 to 95 mm with charges made from bulk explosives were used in a preliminary series of experiments. Also, two types of the main spiral of the EMG were used. The spiral with a diameter of 200 mm was used in the first startups for testing the design of the generator. After testing the design of the first series generators, the transition to more powerful ones with a spiral diameter of up to 300 mm was carried out. The used types of charges provided the transverse speed of the liner from 1 to 1.5 km/s, which corresponds to the angle (from the axis) of liner opening from 7 to 9 degrees. These speeds allowed reaching of the sampling time of generator residual inductance at the level of 10–18 μs. Another advantage of the use of EMG is their constructive ability to minimize the parasitic inductance between the current source and the load due to the use of the parallel circuit of the load. Electrical scheme of EMG with a load in the form of two PPA connected in parallel is shown in Figure 7.

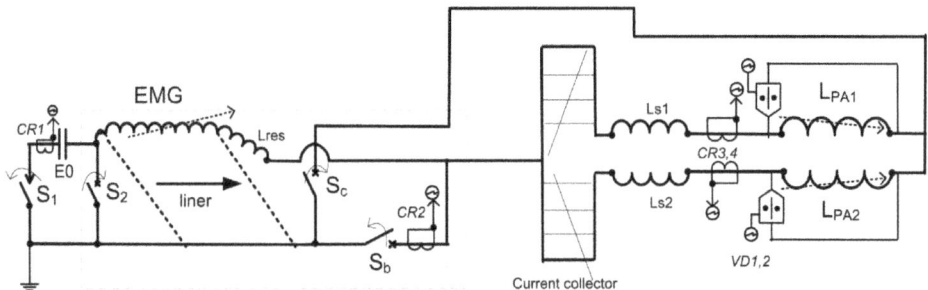

Figure 7. Schematic diagram of the connection of the pulse-current source–EMG load in the form of two PPA connected in parallel. L_{PA1}, L_{PA2} and L_{s1}, L_{s2}—inductances of PPA1, PPA2, and parasitic inductances, respectively; E_0 is the initial source of energy, S—contactors: S_c—solid-state closer and S_b—explosive breaker, CR1-4—Rogowski coils, and VD1-2—voltage dividers.

Both loads are connected in parallel through the collector. Current collector in this case is an element of the electrical circuit, which is used to connect the current output of the EMG with the load by means of coaxial cables. This design allows reduction of the parasitic inductance in the circuit and increase of the power source without increasing its dimensions.

Two main modes of operation of the device for the current-pulse formation of the EMG in the load in the form of PPA were investigated during the experimental series. The modes differed mainly by the synchronization of the switching assemblies of the generator, S_c, and S_b.

3.2.1. The First Operation Mode of the Current-Pulse Formation Devices

The first mode is characterized by later operation of the explosive breaker with respect to the circuit closer and, accordingly, the larger sample of the main inductance of the EMG. In this synchronization mode, the current is switched to the load at a sufficiently high current >2 MA, and the residual inductance of the EMG after switching is practically absent. In the described mode, the residual inductance transformation increases the current in the load after switching slightly.

Illustration of the first mode in the form of current distributions in EMG, PPA, and voltage on PPA is presented in Figure 8. It corresponds to the scheme in Figure 1.

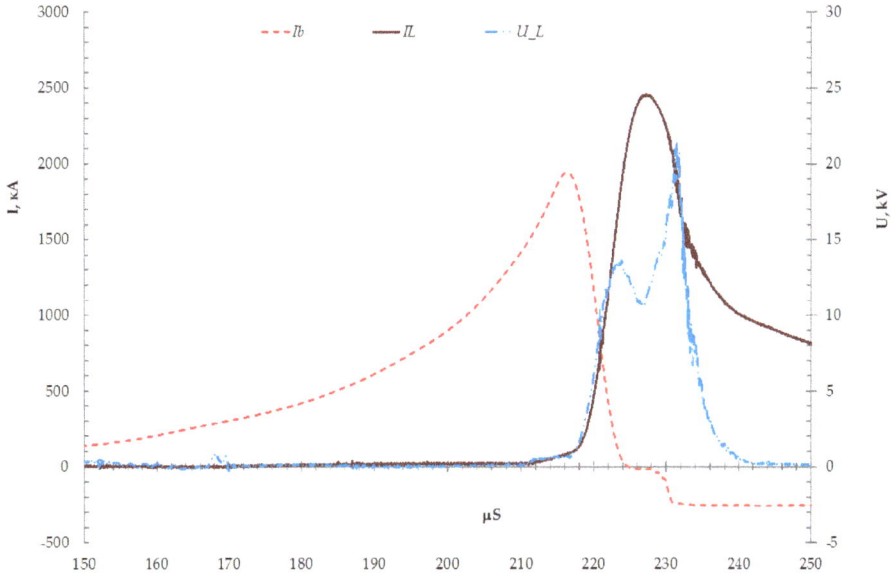

Figure 8. Current distributions in EMG, PPA, and voltage on PPA.

In fact, this mode repeats the task of fast acceleration of the current in the capacitor battery, but it significantly increases the derivative of the current and the current in the load. The time of current switching in the experiments was 3–4 μs at the maximum current amplitude in the load of ~2.5 MA. For comparison, the current is increased to 1.8–1.9 MA for about 10 μs with use the capacitor power source. Reducing the parasitic inductance of the system plays a significant role in increasing the derivative current in the load. Note that the parasitic inductance of the bus arrangement and the switching discharger is at the level of 35–40 nH in the system with a capacitor power source, while the passive (parasitic) inductance can be kept at the level of 15–20 nH in the system with EMG. Comparison of experimental and calculated results (see Figure 1) showed that the calculations correctly describe the processes; the accuracy of the coincidence values is about 15%–20%.

3.2.2. The Second Operation Mode of the Current-Pulse Formation Devices

The second mode of EMG operation of the developed design is the mode with an explicit displacement of the residual magnetic flux into the load after switching-over the current into the load. This mode was made according to the scheme with connected of two accelerators. In this mode, the load on the explosive breaker is significantly reduced due to the fact that the switching is at a current of ~1–1.5 MA. After switching, the residual inductance of EMG remains significant (about 180–200 nH) and its sample takes 5–6 μs. At the same time, it is possible to match the process of pumping the residual magnetic flux into the load with the dynamics of increasing its inductance. The experimental results are presented in Figures 9 and 10.

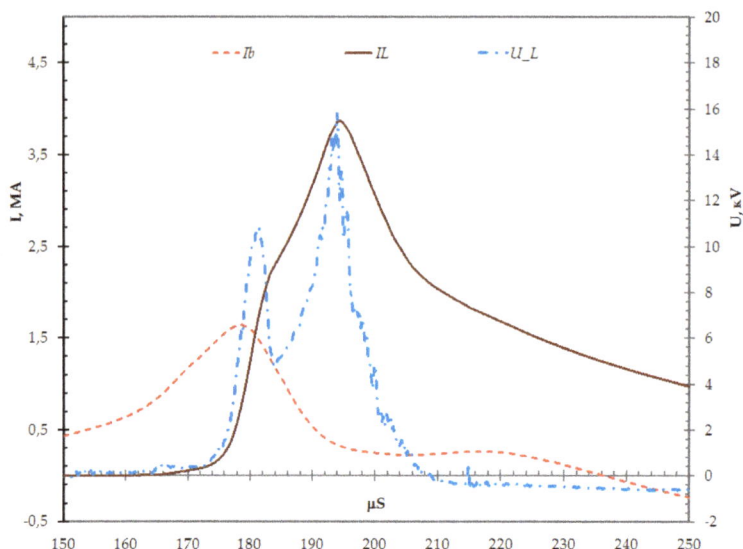

Figure 9. Oscillograms of current and voltage pulses.

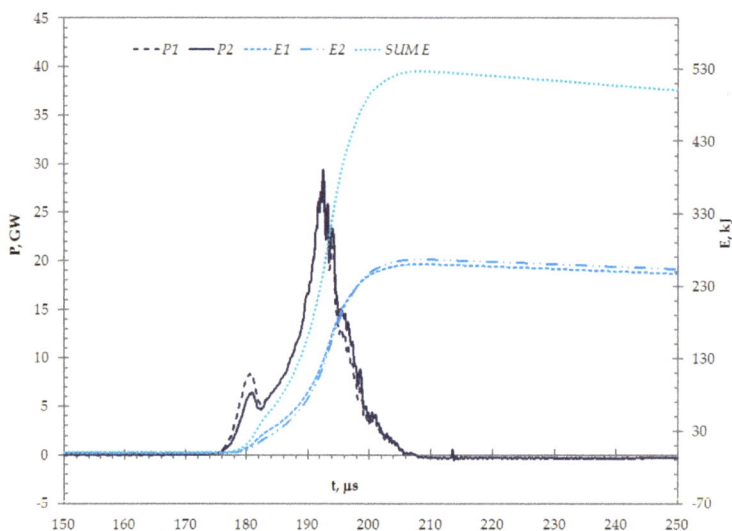

Figure 10. Power and energy inputted for each branch of the load.

The oscillograms of current and voltage pulses during the operation of the EMG on two plasma accelerators (PA), connected in parallel, are shown in Figure 9, and correspond to the schematic diagram in Figure 7. The voltages were measured on the current collectors of each PPA. A series of experiments allowed one to sequentially choose the best conditions for matching the time diagram of the EMG with the dynamics of the load. In this mode, the current rise in the load is observed at the transformation of the residual inductance of EMG after switching the current to the load. The maximum current reaches of 3.8 MA. Moreover, the main phase of PPA operation takes place with the increasing power of EMG. The peak power value is 27 GW. The energy transmitted from the EMG to the load reached 0.55 MJ, as shown in Figure 10.

4. Conclusions

Research of the matching pulse sources, in the form of capacitive storage and an explosive-magnetic generator, with a plasma load, showed:

1. For the operation with plasma loads in the mega joule energy range, the use of explosive magnetic generators with a special device of current-pulse formation increases the efficiency of energy transfer to the load in comparison with the capacitive power source. This device includes the following switching keys: a solid-state closer and an explosive breaker of electrical circuit with adjustable response delays.

2. The use of effective-mode current-switching in the PPA, with the residual inductance of the EMG into the load circuit, allowed transferring to the load 0.55 MJ at a maximum current—up to 3.8 MA. For this start-up, the energy inputted directly into the plasma flow was approximately 250 kJ (45%). This is a significant progress compared to the results outlined in [5]. At that time, EMG managed to transfer to the plasma flow about 85 kJ.

3. The developed computational model of EMG with the load and the switching keys has been successfully applied in the generator design and modeling of its operation. A comparison of the shape of the pulse current and the maximum value of the calculated and experimental results showed that they correspond with an accuracy of 15%–20%.

4. The proposed design of EMG can be used for any pulse plasma loads. The produce of a specific generator should be accompanied by preliminary calculations of the time diagram of switching the keys of the current-pulse formation device.

Author Contributions: Conceptualization, A.S.; Data curation, M.S.; Formal analysis, A.S.; Funding acquisition, A.S. and A.Z.; Investigation, M.S., A.K., V.B., N.U. and A.E.; Methodology, A.Z.; Project administration, V.Z.; Resources, A.K. and V.B.; Software, M.S.; Supervision, A.S.; Writing—original draft, V.Z.

Funding: This research received no external funding.

Conflicts of Interest: The authors declare no conflict of interest.

References

1. Knoepfel, H. *Pulsed High Magnetic Fields*; North-Holland: Amsterdam, The Netherlands, 1970.
2. Fortov, V.E. (Ed.) *Explosive Generators of Powerful Pulses of Electric Current*; Nauka: Moscow, Russia, 2002. (In Russian)
3. Grabovskii, E.V.; Bakhtin, V.P.; Zhitlukhin, A.M.; Lototskii, A.P.; Toporkov, D.A.; Umrikhin, N.M.; Efremov, N.M.; Krylov, M.K.; Khomutinnikov, G.N.; Sulimin, Y.N.; et al. Operation of a magnetic pulse compressor with electrodynamic acceleration of a liner. *Tech. Phys. Russ. J. Appl. Phys.* **2014**, *59*, 1072–1081. [CrossRef]
4. Putrik, A.B.; Klimov, N.S.; Gasparyan, Y.M.; Yaroshevskaya, A.D.; Kovalenko, D.V. Plasma-facing material erosion products formed under iter-like transient loads at QSPA-T plasma gun facility. *Fusion Sci. Technol.* **2014**, *66*, 70–76. [CrossRef]
5. Fortov, V.E.; Shurupov, A.V.; Zhitlukhin, A.M.; Cherkovets, V.E.; Dudin, S.V.; Mintsev, V.B.; Ushnurtsev, A.E.; Kozlov, A.V; Leontev, A.A.; Shurupova, N.P. Magnetocumulative generator as the power supply for pulsed plasma accelerator. *High Temp.* **2010**, *48*, 1–6. [CrossRef]
6. Demidov, V.A.; Kazakov, S.A. Helical magneto-cumulative generator for plasma focus powering. *IEEE Trans. Plasma Sci.* **2010**, *38*, 1758–1761. [CrossRef]
7. Delamere, P.; Stenbaek, N.; Nielsen, H.; Pfaff, R.; Erlandson, R.E.; Meng, C.I.; Zetzer, J.I.; Kiselev, Y.; Gavrilov, B.G. Dynamics of the Active Plasma Experiment North Star Artificial Plasma Jet. *J. Spacecr. Rocket.* **2004**, *41*, 503–508. [CrossRef]
8. Selemir, V.D.; Demidov, V.A.; Ermolovich, V.F.; Spirov, G.M.; Repin, P.B.; Pikulin, I.V.; Volkov, A.A.; Orlov, A.P.; Boriskin, A.S.; Tatsenko, O.M.; et al. Generator of soft X-ray emission by Z-pinches powered from helical explosive magneto-cumulative generators. *Plasma Phys. Rep.* **2007**, *33*, 381–390. [CrossRef]

9. Bryxin, V.A.; Chlenov, A.M.; Fedorov, A.A.; Ivaschenko, D.M.; Kamensky, V.; Kochergin, V.V.; Metelev, A.P.; Mikulin, I.N.; Vyskubov, V.P. UIN-10, High-Power Direct Acting Pulse Electron Accelerator for Radiation Investigations. In Proceedings of the 15th International Conference on High-Power Particle Beams, S-Petersburg, Russia, 18–23 July 2004; p. 299.
10. Drozdov, Y.M.; Duday, P.V.; Zimenkov, A.A.; Ivanov, V.A.; Ivanovskii, A.V.; Ablesimov, V.E.; Andrianov, A.V.; Bazanov, A.A.; Glybin, A.M.; Dolin, Y.N.; et al. Development of a plasma focus neutron source powered by an explosive magnetic generator. *J. Appl. Mech. Tech. Phys.* **2015**, *56*, 77–85. [CrossRef]
11. Poznyak, I.M.; Safonov, V.M.; Zybenko, V.Y. *Movement of Melt Metal Layer under Conditions Typical Transient Events in ITER*; Problems of Atomic Science and Technology, Series: Thermonuclear Fusion; Coordination Center "Controlled Thermonuclear Fusion-International Projects": Moscow, Russia, 2016; Volume 39, pp. 15–21. ISSN 0202-3822. (In Russian)
12. Gerasimov, L.S. Agreement of EMG with inductive load. *J. Tech. Phys.* **1974**, *44*, 1973–1979.

energies

MDPI

Article

Influences of the Load of Suspension Point in the *z* Direction and Rigid Body Oscillation on Steel Catenary Riser Displacement and Frequency Under Wave Action

Bo Zhu [1,*], Weiping Huang [1], Xinglong Yao [1], Juan Liu [2] and Xiaoyan Fu [3]

[1] Shangdong Province Key Laboratory of Ocean Engineering, Ocean University of China, Qingdao 266071, China; wphuang@ouc.edu.cn (W.H.); 17864272321@163.com (X.Y.)
[2] Institute of Civil Engineering, Agriculture University of Qingdao, Qingdao 266009, China; qianqian070712@163.com
[3] College of oceanography, Hohai University, Nanjing 210098, China; fuxiaoyan1988@163.com
* Correspondence: beiji_dongjie@hotmail.com

Received: 15 November 2018; Accepted: 10 January 2019; Published: 16 January 2019

Abstract: The rigid body swing is an important problem for steel catenary risers (SCRs). In addition to many other important issues, the transverse flow direction response is studied in this paper. By extending the load terms of the large deflection slender beam equation, the load of suspension point in the z direction, Morison and rigid body swing are superimposed on the beam equation. On the basis of the above work, a Cable3d subroutine is written to complete the task. Then the structural response is simulated and verified by the Lissajous phenomenon and spectral phase analysis. On the basis of verification, the response is analyzed from an angle of three-dimensional space and the influence coefficient is adopted to evaluate the effect of rigid body swing. The importance of loads is determined by spectral analysis. Phase curve and the change of vibration direction are analyzed by higher orders of frequency. The results show the verification of Lissajous and spectral phase analysis are feasible. The analysis of the spatial response shows the vibration direction of the 140th node is in the same direction as the rigid body swing vector, so the interaction is relatively of more intensity and the influence coefficient is relatively larger. This influence interval of rigid body swing displacement statistical analysis is −0.02 to 0.02 and the effect is weak. The spectrum analysis indicates there is no resonance between the main load and the bending vibration, and the analysis also shows the main influence load of the transverse flow response in this paper is the top load in the z direction. According to phase analysis, the load has a high order effect on the spectral phase curve of the structure. This paper has drawn a conclusion that rigid body swing has limited effect on transverse flow response, however, it has a relatively strong impact on the middle region of the riser, so it plays an influential role on the safety of the riser to some extent. The key point for this paper is to provide qualitative standards for the verification of rigid body swing through Lissajous graphs, which are central factors to promote the development of rigid body swing. It is hoped that the above research can provide some reasonable suggestions for the transverse flow response simulation of the steel catenary riser.

Keywords: steel catenary riser; rigid body rotation; wave; the load of suspension point in the *z* direction; Cable3D

1. Introduction

The development of offshore oil and gas exploration from shallow water to deep sea is promising at present. In recent years, with the rapid development of deep sea resource exploration, more stringent and specific standards are put forward. In oil production system, there are many types of development modes, and the mode consisting of platforms, risers, and underwater trees is a relatively common one. In this type of mode, the steel catenary riser is a key device. It connects the top platform to the bottom tree [1]. The environmental impact on the steel catenary riser increases with the increase of water depth. It has a wide range of application depths and significant economic advantages over the other risers. For example, the top tension riser (TTR) lacks adaptability to platforms and the development cost of flexible risers is higher. Researchers worldwide have conducted extensive studies on SCRs, but for waves and the load of suspension point in the z direction, the following problems remain an urgent topic: first, the loads of wave and suspension point are perpendicular to each other, but the frequencies are very close. The result is close to Lissajous' Figures. Second, the wave load is also perpendicular to the rotating plane of rigid body swing. This requires studies from three dimensions to one dimension. This paper focuses on the change of structural response caused by waves, a SCR's top load in the z direction and rigid body swing.

Zhu and Gao [2] studied the influence of a free rotating impeller on the vortex-induced vibration (VIV) response of a riser. The results showed that the reduction rate and energy extraction objective could be achieved at the same time. Qiu et al. [3] introduced the drag crisis phenomenon caused by unsteady shear layer separation when the Reynolds number ranges from 2×10^5 to 5×10^5. Experiments and numerical simulations were conducted and adopted by the 27th league ITTC committee for the simulation. The report showed the LES method has relatively more advantages for simulating the drag crisis phenomenon. Li et al. [4] studied the stress wave transmission in the riser. The multi-signal complex exponential method was used to solve this problem. The dominant forms for the downstream and cross direction responses are a standing and travelling wave, respectively. Teixeira and Morooka [5] adopted semi-empirical methods to calculate VIV responses, considering the energy balance in the method. The results showed it is in good agreement with physical experiments. Xu et al. [6] proposed a nonlinear wake model of VIV for the prediction of the fatigue life of marine risers. Their results agreed well with the experiments. Cabrera-Miranda and Paik [7] analyzed models associated with in-situ marine meteorological data. At the same time, a model related to input variables was also adopted to calculate the load probability distribution. The calculations show the load model associated with the influence variables is more suitable. Do and Lucey [8] analyzed and researched the Lyapunov direct method. The design of an active control system at the top and bottom of the riser was proposed. The system was designed for tensile, non- shear and in-plane deformation risers. Zhang et al. [9] adopted a finite element method to research vortex-induced vibrations under axial harmonic load conditions. Liu [10] coupled a rigid body swing and bending vibration model to simulate the response. The results showed the wave response and rigid body swing decreases with depth. Yao [11] adopted the Cable3d program to simulate the structural response. The calculations show the influence of rigid body swing is between 10% and 20% under the action of wave loads, and the influence is positively correlated with the swing vector S. Komachi et al. [12] analyzed the wake oscillator and Newmark-Beta method. VIV response and three dimensional dynamic response and stress fatigue ratio were simulated for the system. Domala and Sharma [13] researched the vortex-induced vibration responses of three types of riser, including a top tensioned riser, steel catenary riser and SCR with platforms. Hong and Shah [14] adopted the Euler Bernoulli method to set up a model for riser movement. The model considers load and ship motion, and theory and test results were compared for verification. Alfosail et al. [15] proposed a new state space method. The method involves solving the vibration frequency and critical buckling limit of the riser. For verification, the results were in good agreement with other methods. Yang et al. [16] established a new model with Kelvin Voigt elasticity for simulation of three dimensional nonlinear dynamic responses. The results showed the coefficient has obvious effects on the natural frequency, maximum displacement and stress. O 'Halloran et al. [17] researched the characteristics of flexible risers with tests. The results showed friction and damage characteristics have obvious effects on risers. Wang et âl. [18] studied the HHT method for gas entering into the riser. Hu et al. [19] adopted the Keller box method to studied a model with different riser lengths during installation. The displacement of risers in the top zone and subsea trees were researched. The results show the method could solve this kind of installation problem.

In recent years, research has mainly focused on the response characteristics of risers. The influence of rigid body swing and other vibration models is important for the response, however, direct simulation of rigid body swing takes a long time and other programs cannot directly simulate this phenomenon. Therefore, research on the characteristics of transverse flow response is still insufficient. The interaction between the platform and the rigid body swing needs further study.

From the perspective of overall movement, the response and frequency along water depth are worth studying. The response characteristics of key nodes such as the top suspension point and bottom contact point are important. The influence of rigid body swing is also worth researching. From the perspective of the FFT analysis of risers, there are few studies on the phase curve. The load frequency in the phase curve is characterized by the phase cliff-break or rapid change. It can be used to distinguish the load frequency of the structure. The phase curve, should also not be neglected for the research.

This paper takes the Petrobras Marlim oil and gas field as the simulation model. The development mode includes the FPSO, steel catenary riser and underwater tree. Then a simplified Cable3D SCR model [10,11] is established for Petrobras Marlim oil and gas field. The response is calculated under the action of linear waves and cross-flow loads on the platform. Ther rigid body swing without VIV is superimposed on the model. The rigid body swing is worthy of attention because of its effect on the transverse response. Then the characteristics of SCR with and without rigid body swing are analyzed. During the calculation and evaluation, the following problems need to be solved and studied:

(1)   The structure is affected by the oscillating effect of rigid body swing. At present, commercial finite element programs cannot calculate it directly. The effect can be taken into account by a certain coefficient in engineering, but the coefficient still needs to be calculated and analyzed.

(2)   The actual SCR structure is subject to relatively more loads perpendicular to each other. VIV, represented by vertical lift and drag, is studied in depth. However, there are few studies for waves and the top platform, which are perpendicular to each other. Whether the Lissajous' Figure can be used as the mathematical validation is worthy of attention.

(3)   In the plane, the Lissajous' Figures with wave and load of suspension point in the z direction are presented. In space, the plane load has two characteristics: in-plane wave load and out-of-plane rigid body swing.

(4)   The structural response is affected by waves, top loads in the z direction, rigid body swing and natural vibration frequencies, but there is no simple formula for calculating the load frequency, especially the frequency of rigid body swing.

The response characteristics of steel catenary risers under rigid body swing and platform are studied. Through the above work, it is hoped to provide some reasonable suggestions for the research.

2. SCR Numerical Simulation Model

2.1. General Description of the Numerical Model

In this paper, a model consisting of a SCR large deflection slender beam module and a wave—load of suspension point in the z direction—rigid body swing module is developed. Based on Cable3D, the model can simulate wave loads, platform linear harmonic loads and SCR rigid body swing.

The Cable3D model [10] is developed to solve the complex response of steel catenary riser. Riser-platform, fluid-solid and pipe-soil interactions and the effect of rigid body swing are researched by the Cable3d software. The load model q as shown in Equation (4) is usually substituted by the subterm of SCR. These load subterms q integrated for the new complex phenomenon allow Cable3D to be further developed and extended. The integration model for simulation is shown in Figure 1. In this paper, a SCR large deflection slender beam model is adopted to solve the structure response.

It is subjected to wave loads and loads in the z direction of the suspension point. The rigid body swing is simulated by using the SCR rigid body swing model.

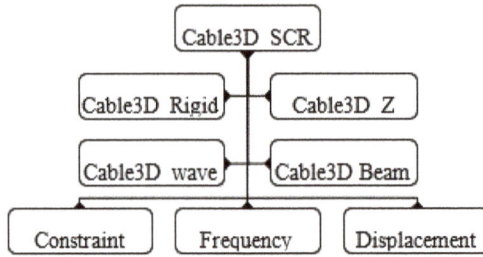

Figure 1. Process of the structural model.

2.2. Basic Control Equation of SCR Motion

The basic governing equations [10] of the riser motion range from the beam balance equation to the riser vibration equation. In this process, the control equation is obtained from the expressions of load and mass terms. The basic theory of SCR provides a basic solving model for solutions. Then the wave, load of suspension point in the z direction and rigid body swing can be solved in the equation. The calculation of load such as rigid body swing usually has an iterative process. The following steps are detailed. Figure 2 shows the coordinate system of the riser. The shape of beam is represented by a vector S which is a function of the length r and time t.

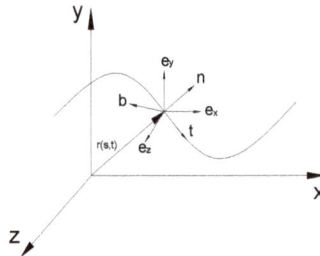

Figure 2. Coordinate system of the beam.

According to the conservation, moment, linear and angular of momentum theorem, the equilibrium equation can be obtained:

$$\rho \frac{d^2 r}{dt^2}(s,t) - q = \frac{dF}{ds} \tag{1}$$

$$\lambda \frac{dr}{ds}(s,t) - B \frac{d^4 r}{ds^4}(s,t) = F \tag{2}$$

where F is the internal force of the beam section, q is the distributed external force per unit length on the beam, ρ is the mass of beam per unit length, λ is the Lagrangian operator and B is the beam bending stiffness.

After substituting the equation F', the motion equation of the large-deflection slender beam is obtained. In this process, The Bernoulli-Euler theory is applied:

$$\rho \frac{d^2 r}{dt^2}(s,t) = \lambda \frac{d^2 r}{ds^2}(s,t) - B \frac{d^4 r}{ds^4}(s,t) + q \tag{3}$$

After substituting the equation for the term q, the vibration control equation of the riser is obtained. In this process, the load calculation formula and vector transformation matrix are applied:

$$M\frac{d^2r}{dt^2}(s,t) = \lambda\frac{d^2r}{ds^2}(s,t) - B\frac{d^4r}{ds^4}(s,t) + q \tag{4}$$

where M is the mass matrix, B is the stiffness matrix, q is the load matrix and λ is the Lagrangian operator. The above formula is the basis of the structural numerical simulation. The mass, load and Lagrange operators of the structure form the basic solution and the basis model of SCR is composed of gravity, inertial force, drag force and F-K force.

2.3. SCR Rigid Body Rotation Submodel

The motion equation model of rigid body swing is established to simulate the phenomena of the steel catenary riser. Because the rigid body swing model cannot be directly simulated in commercial softwares, Cable3D is relatively superior to other commercial softwares in rigid motion simulations. The model adopts the theorem of momentum to study the moment balance of the riser. The response is solved through coupling the load term with the vibration equation. Then the result is obtained for the structural response of the rigid body swing in the rotation plane. The practical significance of the model is in providing a solution for the rigid body swing phenomenon. Figure 3 shows the rigid body swing system caused by loads such as deep water waves, flows and others. Figure 4 shows a real system of FMC Technologies. The rotation vector is represented by S which is a function of the node coordinates.

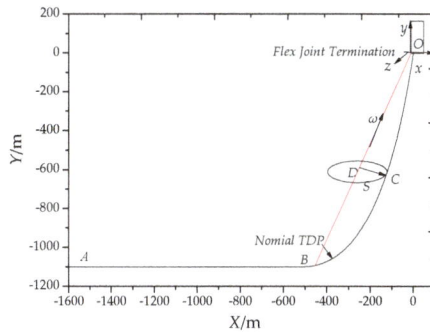

Figure 3. Rigid body swing system.

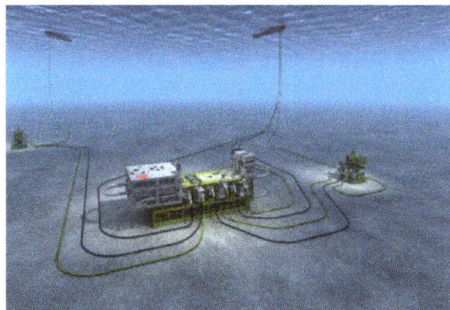

Figure 4. Petrobras Marlim oil and gas fields.

According to the moment of momentum theorem, the oscillating equation of rigid body swing is:

$$(m + m_a)s^2\frac{d^2a_r}{dt^2} + c_a s^2\frac{da_r}{dt} = q_r \tag{5}$$

$$q_r = q_z\sqrt{s_1^2 + s_2^2} + q_x\omega_2 s_3 - mg\omega_1 s a_r \tag{6}$$

Substituting Equation (5) into Equation (4) and rearranging yields:

$$M\frac{d^2r}{dt^2} = \lambda\frac{d^2r}{ds^2} - B\frac{d^4r}{ds^4} + q_f + q_r + mg - (m + m_a)\frac{d^2r_r}{dt^2} - c_a\frac{dr_r}{dt} \tag{7}$$

$$M\frac{d^2r}{dt^2} = \lambda\frac{d^2r}{ds^2} - B\frac{d^4r}{ds^4} + q + mg - (m + m_a)\frac{d^2r_r}{dt^2} - c_a\frac{dr_r}{dt} \tag{8}$$

where m, m_a are the mass and additional mass per unit length of riser, c_a is the additional damping coefficient per unit length. q_x and q_z are the environmental load acting on per unit length. a_r, $\dot{a}_r$, $\ddot{a}_r$ are the angular displacement, angular velocity, angular acceleration of rigid body rotation, s_1, s_2 and s_3 are the projections of the radius S onto the axes x, y, z and ω_1, ω_2 and ω_3 are the projections of the vector onto the axes x, y, z.

It is expressed as the coordinate component form:

$$(m + m_a)\left(\frac{d^2u_b}{dt^2} + \frac{d^2u_r}{dt^2}\right) + (c + c_a)\frac{du_b}{dt} = q_x - c_a\frac{du_r}{dt} - ku_b \tag{9}$$

$$(m + m_a)\left(\frac{d^2v_b}{dt^2} + \frac{d^2v_r}{dt^2}\right) + (c + c_a)\frac{dv_b}{dt} = q_y - mg - c_a\frac{dv_r}{dt} - kv_b \tag{10}$$

$$(m + m_a)\left(\frac{d^2w_b}{dt^2} + \frac{d^2w_r}{dt^2}\right) + (c + c_a)\frac{dw_b}{dt} = q_z - c_a\frac{dw_r}{dt} - kw_b \tag{11}$$

where u_b, v_b, w_b, $\dot{u}_b, \dot{v}_b, \dot{w}_b$, $\ddot{u}_b, \ddot{v}_b, \ddot{w}_b$ are the projection of displacement, velocity and acceleration on x, y, z and u_r, v_r, w_r, $\dot{u}_r, \dot{v}_r, \dot{w}_r$, $\ddot{u}_r, \ddot{v}_r, \ddot{w}_r$ are the projection of displacement, velocity and acceleration on of rigid body swing projected on x, y, z.

$$\vec{v}_r = \dot{a}_r(\vec{c} \times \vec{s}), \ \vec{v}_r = \dot{a}_r(\vec{c} \times \vec{s}), \ \vec{a}_r = \ddot{a}_r(\vec{c} \times \vec{s}) \tag{12}$$

Expand it to the following expression:

$$\dot{u}_r = \frac{da_r}{dt}(\omega_2 s_3 - \omega_3 s_2), \ \dot{v}_r = \frac{da_r}{dt}(\omega_3 s_1 - \omega_1 s_3) \ \dot{w}_r = \frac{da_r}{dt}(\omega_1 s_2 - \omega_2 s_1) \tag{13}$$

$$\ddot{u}_r = \frac{d^2a_r}{dt^2}(\omega_2 s_3 - \omega_3 s_2) \ \ddot{v}_r = \frac{d^2a_r}{dt^2}(\omega_3 s_1 - \omega_1 s_3) \ \ddot{w}_r = \frac{d^2a_r}{dt^2}(\omega_1 s_2 - \omega_2 s_1) \tag{14}$$

2.4. SCR Wave Load Submodel

The wave load submodel is to solve the wave load equation in deep sea with riser moving relative to the water particle. The model uses Morison's equation to study the relative load of the riser. The wave is calculated by a linear wave in deep water. The response of linear waves in deep water decreases as the depth increases. It is worth noting that this decreasing trend has a great influence on the trend of overall response. The loads such as waves and suspension point load in the z direction have the same trend.

Velocity potential is:

$$\phi_0 = \frac{gA}{\sigma}e^{kz}\sin(kx - \sigma t) \tag{15}$$

Horizontal velocity is:

$$u_x = A\sigma e^{kz} \cos\theta \tag{16}$$

Horizontal acceleration is:

$$\frac{\partial u_x}{\partial t} = A\sigma^2 e^{kz} \sin\theta \tag{17}$$

In the wave model, the motion equation of SCR is:

$$f_H = \frac{1}{2}C_D\rho A(u_x - \dot{x})|u_x - \dot{x}| + C_{M}\rho\frac{\pi D^2}{4}\frac{\partial u_x}{\partial t} - C_{m}\rho\frac{\pi D^2}{4}\ddot{x} \tag{18}$$

where ϕ_0 is the velocity potential, u_x is the horizontal velocity of a water particle, $\partial u_x/\partial t$ is the horizontal acceleration, A is the wave amplitude, σ is the wave circular frequency, k is the wave number, ρ is the density, $\dot{x}$ is the velocity of the riser, D is the diameter, C_D is the drag force coefficient, C_M is the mass coefficient of mass and Cm is the additional mass coefficient.

2.5. SCR Load Submodel of Suspension Point in the z Direction

The main goal of load model in the z direction is to simulate the transverse load of the platform. The top load in the z direction is:

$$F_z = k_1 A \cos\omega_s t \tag{19}$$

Considering the interaction between waves and solids, the top load submodel is:

$$A_r = A - r \tag{20}$$

$$Q_z = k_1 A_r \cos\omega_s t \tag{21}$$

where A is the load amplitude, k_1 is the adjustment coefficient, usually 1, and ω_s is the load frequency.

The top load submodel is an equation to solve the simulation of cross flow direction load of platform. It is worth noting that the two loads appear perpendicular to each other in space, and as the wave load moves along the x direction, and the top load is in the z direction. Two harmonic load oscillations perpendicular to each other will form a stable elliptic curve if the frequencies are close to each other. If the frequency is expressed as an integer ratio, it is called Lissajous' Figures [20,21].

Lissajous' Figures provide theoretical support for analysis and verification of the load response perpendicular to each other. Equation (18) and Equation (5) can be substituted into Equation (4), and the control Equation (22) can be obtained:

$$M\frac{d^2r}{dt^2} = \lambda\frac{d^2r}{ds^2} - B\frac{d^4r}{ds^4} + q + mg - (m + m_a)\frac{d^2r_r}{dt^2} - c_a\frac{dr_r}{dt} + f_H + Q_z \tag{22}$$

where Q_z is the top load in the z direction. For Equation (22), the effect of load frequency on the equation is the frequency input on the right side. The load term is related to frequency, amplitude and phase. Then the input of different frequencies, amplitudes and phases makes the response complex and random.

2.6. Boundary Constraints and Iteration Conditions

According to the hypothesis that small deformations are allowed, the following formula is obtained:

$$\frac{dr}{dt}(s,t) \cdot \frac{dr}{dt}(s,t) = (1 + \varepsilon)^2 \tag{23}$$

The convergence criteria for the above solution is that the number of iterations is less than the allowed maximum number, or the difference between the two iterations is within the error range:

$$n > N_{max} \text{ or } eps_n < \left| \frac{r_n - r_{n-1}}{r_n} \right| \tag{24}$$

where n is the number of iterations, N_{max} is the allowed maximum number of iterations, and eps_n is the iterative error. Besides point A adopts an anchoring constraint, and the top suspension point O restrains the x and y directions. The load of the suspension point in the z direction (transverse flow direction) is added to the input file.

2.7. Summary of the Numerical Model

In this section, SCR and load numerical model are introduced. The SCR numerical model is a large deflection flexible beam model. The load numerical model introduces rigid body swing, load of suspension point in the z direction and wave model.

The Cable3D_Vswing program calculates the response under top load in the z direction, wave and rigid body swing. The difference between RT_CABLE [10] and Vswing is whether the top load in the z direction is considered or not. It is a key factor of the top load in the z direction for Vswing. The basic process mode of platform load includes the call of load function and the external input file.

This article adopts the method of external input file. Its biggest advantage is that it reduces the difficulty of load function fit. In addition, this paper promotes vector calculation from two-dimensional to three-dimensional, and this is a small improvement over the original RT_CABLE program.

3. The parameters and Verification for SCR Structure Simulation

The response of the SCR model in the cross-flow direction has been studied previously [10,11]. After the wave in the x direction and load of suspension point in the z direction are applied to the structure it appears that the loads are perpendicular to each other in space. The influence of response is worth studying after superposition of the rigid body swing.

The structure is affected by the mutually perpendicular load effect and rigid body swing. The response changes as the water depth increases and the influence of rigid body swing is of concern. Flow and fluid-solid coupling are not considered in this paper.

In this section, the Cable3D program and the Cable3D_Vswing program are respectively used for our simulation [11]. The Cable3D_Vswing program is a modified version of the load term Qforce. The coupled program improves the model of wave, load of suspension point in the z direction and rigid body swing. Cable3D calculates the structural response under the influence of wave load in the x direction and top suspension point load in the z direction. Cable3D_Vswing is used to calculate the response under the influence of waves, load of the suspension point in the z direction and rigid body swing. The difference is whether the influence of rigid body swing is considered in the response.

3.1. The Parameters of the SCR Structure

The density of crude oil in the riser is 865 kg/m^3 and the density of sea water outside the pipe is 1025 kg/m^3. The deep water linear wave is selected for calculation. The wave height is 3.5 m, period is 8.60 s, and the frequency is 0.11622 Hz. Petrobras Marlim, similar to the simulated structure, with a water depth of 1330 m, is connected to the undersea tree by pipelines. The development mode including SPAR, SCR and underwater tree is designed, with a SPAR draft of 153.169 m. It can accommodate eight top-tensioned risers. The SCR riser is suspended outside the soft cabin.

The design depth is 1100 m, the length of SCR is 2500 m, and the anchor point is 1800 m. The SCR adopts a five-layer structure from inside to outside. The different layers take corresponding roles respectively. The actual structure is exposed to dangerous conditions such as random response of the platform, waves and so on. In this paper, relatively simple conditions are selected for our simulation.

The linear model is adopted for the seabed soil, without consideration of flow. Table 1 shows the layers of the SCR bonded riser. Table 2 lists the parameters of the bearing layer, and Table 3 shows the top linear load parameters under different conditions.

Table 1. Structure layers of the SCR bonded riser.

Structure	Function	Material	Density (kg/m^3)
Bearing layer	Load bearing, etc.	API X-65	7850
Anticorrosive coating	Prevent osmotic corrosion	Epoxy resin	1440
Bonding layer	Interlayer bonding	Polypropylene	980
Thermal insulation layer	Cold oil not be transported	Foam Polypropylene	800
Protective layer	Prevent seabed damage, etc.	Polyethylene	900

Table 2. Parameters of the SCR bearing layer [10].

Parameters	Value	Unit	Parameters	Value	Unit
The outer diameter	0.355	m	The moment of inertia	0.36123×10^{-3}	m^4
The inner diameter	0.305	m	The outer diameter area	0.99315×10^{-1}	m^2
Modulus elasticity	207.0	Gpa	The inner diameter area	0.72966×10^{-1}	m^2
Minimum yield strength	408.0	MPa	Unit length mass	0.2960	kg/m^3
Poisson's ratio	0.3	/	Unit length buoyancy	$0.2137\,4 \times 10^4$	N/m

Table 3. Parameters of the load of the suspension point in the z direction [11].

Cdt	Load	Direction	Amplitude/m	Frequency/Hz	Wave	ZDL	RBS
1	Cab1	Transverse flow	3.0	0.093	✓	✓	
2	Cab 2	Transverse flow	2.0	0.101	✓	✓	
3	Cab 3	Transverse flow	1.0	0.111	✓	✓	
4	Csw1	Transverse flow	3.0	0.093	✓	✓	✓
5	Csw 2	Transverse flow	2.0	0.101	✓	✓	✓
6	Csw 3	Transverse flow	1.0	0.111	✓	✓	✓

RBS is a shortened form of Rigid Body Swing. ZDL is a shortened form of load of suspension point in the z direction. Four hundred 400 nodes, 399 beam elements and the three-time Hermite function are used to calculate the node response of SCR. The quadratic Hermite function is used to calculate different matrices. The matrices include the Lagrangian operator, mass, stiffness and external load matrix. The spring stiffness of the structural suspension point O and fixed point A is 0.1×10^{12} Pa. The friction coefficient of the riser and seabed interaction is 0.2. The relaxation factor is 0.8 and the iteration error is 0.1×10^{-5}. The number of iteration steps is 3600 and the calculation step is 0.1 s. Lift coefficient is 0.7, and drag force coefficient is 1.2. The mass coefficient is 1.0, and the hydrodynamic parameter is 0.355.

The wave load and the load of suspension point in the z direction make the structural response complex and random. Due to the action of water flow and waves, structural response of rigid body swing is complex and random. However, most finite element programs cannot directly calculate the rigid body swing. Therefore, in this paper the influence of rigid body swing is considered by a coefficient. The effect can be taken into account through multiplying the response under the main load by a coefficient.

In addition, the linear model is used for the load of suspension point in the z direction, which is suitable for better sea conditions, but not for worse ones. A linear model is adopted for waves, which is suitable for deep water, but not for inshore areas.

3.2. SCR Structural Lissajous Phenomenon and Verification

In real sea conditions, the load on the structure presents random characteristics and the response of random loads is shown in the perpendicular directions. This makes the response of the structure more intense or violent with random features, but there are also cases where the structure is subjected to simple loads. For example, the structure is subjected to waves, which adopt a deep water linear wave model, and the structure is also subjected to steady flow and the relatively stable movement of the platform, or if the structure is suspended on a fixed platform, the linear wave and steady flow in deep water models can be applied. In some of the above cases, the Lissajous phenomenon may also exist.

The Lissajous phenomenon of the SCR structure [20,21] is generated or defined by simple harmonic oscillations. The oscillations are perpendicular to each other. A stable curve will form when the frequency is close to the integral ratio. In this paper, the wave load is along the x direction and the load of the suspension point is in the z direction, so they are perpendicular to each other in space.

The response curve of the structure caused by wave load lacks check rules or standards. Then the Lissajous curve can play a certain auxiliary role in the verification. The check condition is that loads are perpendicular to each other, and the frequency ratio is integer ratio. The reference [20,21] can provide the curve to realize the verification:

$$x = A_1 \cos(2\pi n_1 t + \varphi_1) \quad y = A_2 \cos(2\pi n_2 t + \varphi_2) \tag{25}$$

$$2\pi n_1 t + \varphi_1 + 2k\pi = \pm\arccos\frac{x}{A_1} \quad 2\pi n_2 t + \varphi_2 + 2m\pi = \pm\arccos\frac{y}{A_2} \tag{26}$$

$\frac{n_1}{n_2} = \frac{m_1}{m_2}$ (m_1, m_2-prime numbers), and the trajectory equation is:

$$\cos\left(m_1\arccos\frac{y}{A_2} \pm m_2\arccos\frac{x}{A_1}\right) = \cos(m_1\varphi_2 - m_2\varphi_1) \tag{27}$$

The shape is determined by the amplitude ratio A_1/A_2, frequency ratio m_1/m_2 and $\cos(m_1\varphi_2 - m_2\varphi_1)$.

In this paper, the wave load frequency in the x direction is 0.11622 Hz and the load frequency of the suspension point in the z direction in the condition C1 is 0.093 Hz. After the data has been processed, the top load in the z direction is 0.09 Hz and the wave is 0.12 Hz. The common factor is 0.03, and the frequency ratio is 3:4.

From the viewpoint of load, the structure is subjected to the x direction wave load and the load of the suspension point in the z direction. The wave and the top load in the z direction are perpendicular to each other, and the frequency ratio ranges from 3:4 to 4:5. Then the wave and top load in the z direction can be simply checked and compared with the Lissajous curve.

For working condition 1, this paper adopts the Cable3D calculation results to compare with Lissajous's figure. The z-x plane graphic response of the 10th–200th structure is studied. The structural response is shown in Figure 5. The response of 10th and 140th structure is similar to that of the second picture in 3:4 of Figure 6. The 80th and 200th response is similar to that of the fourth picture in 3:4 of Figure 6. These figures are also more similar to a 4:5 ratio.

On the whole, the structural response conforms to the basic characteristics of a Lissajous curve. The z-x plane response of the structure presents a zonal track characteristic. It is consistent with the characteristics of the Lissajous curve with high phase. From the 10th to 200th response, the zonal track decays gradually from a more complete distribution into a disordered one.

The amplitudes of the response in the x direction and z direction decrease with depth. The amplitude is positively correlated with the weakening of the wave and load of the suspension point in the z direction with increasing depth. The degree of deformation is strengthened or intense. The similarity degree of the graph and standard curve is also weakened. This represents an increase in the complexity of the bottom motion as the water depth increases. After the superimposition of rigid body swing, the basic features of the graph remain unchanged. The graph is shifted downward in the *x* axis and to the right in the *z* axis. From the above phenomenon, the calculation of the structure can be verified.

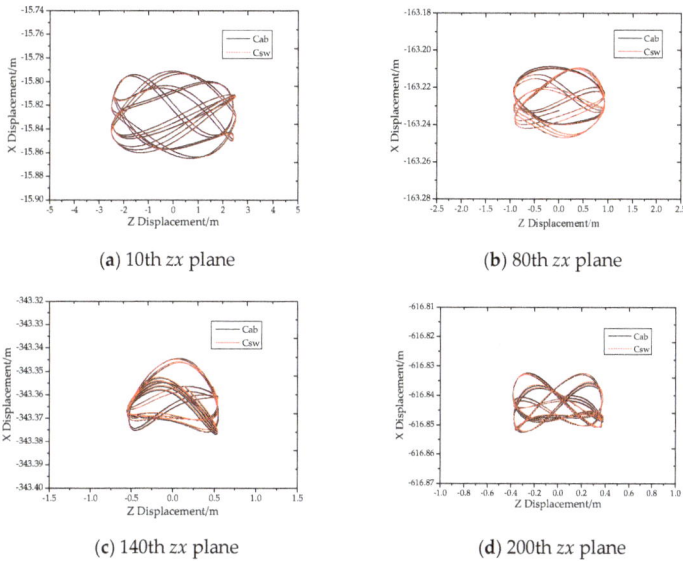

(**a**) 10th *zx* plane

(**b**) 80th *zx* plane

(**c**) 140th *zx* plane

(**d**) 200th *zx* plane

Figure 5. Lissajous phenomenon of 10th–200th in the *zx* plane.

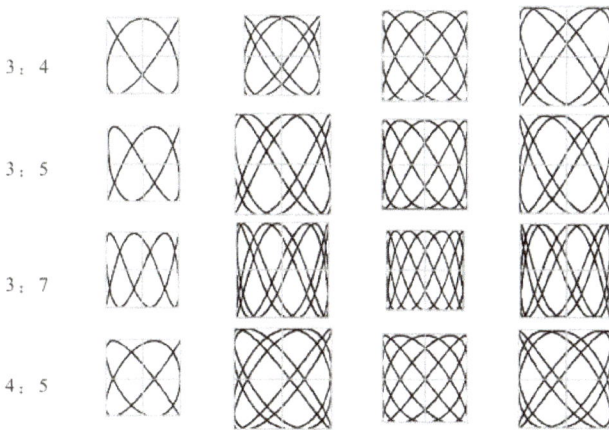

Figure 6. Lissajous verification table.

3.3. Structural Load Frequency Analysis and Verification

The response of a structure is usually a linear or nonlinear superposition of multiple load responses. Under the action of load, linear and nonlinear response curves are produced. The analysis of the load frequency can be helpful to the analysis of the main response. In this paper, the load frequency analysis is mainly to carry out with a FFT transformation which looks for the main frequency of the structure.

The solution method and data of load frequency are different for verification. Under load excitation, the frequency of the structure subjected to wave load in the x direction is 0.11622 Hz. Under the action of the load of suspension point in the z direction, the structure frequency is 0.093 Hz–0.111 Hz. The first frequency of bending vibration is solved by the Vandiver method to 0.016 Hz. The vibration frequency of the rigid body swing is solved by Rayleigh method to 0.029 Hz [11].

The Rayleigh method is shown in Equations (28)–(30):

$$T = \frac{1}{2}\sum_i m_i x_i^2 = \frac{1}{2}(\frac{1}{3}m_1 l^2 + \frac{1}{3}m_a l^2 + m_2 l^2)\omega^2 A^2 \cos^2\theta \tag{28}$$

$$V = \frac{1}{2}\sum_i k_i x_i^2 = \frac{1}{2}(kl^2)A^2 \sin^2\theta = \frac{1}{2}(\frac{1}{2}m_1 gl + \frac{1}{2}m_a gl + m_2 gl)A^2 \sin^2\theta \tag{29}$$

Because $T_{max} = V_{max}$, then the frequency can be derived:

$$\omega = \sqrt{\frac{1/2m_1 gl + 1/2m_a gl + m_2 gl}{1/3m_1 l^2 + 1/3m_a l^2 + m_2 l^2}} \tag{30}$$

Bending vibration method for solving natural vibration frequency is:

$$\omega = (n\pi)^2 \sqrt{\frac{EI}{ml^4}}, n = 1, 2, \cdots, \infty \tag{31}$$

The response spectrum of the SCR structure is calculated by the FFT method and the response graphs of the amplitude and phase of the structure are obtained. In the previous FFT transformation, less attention is paid to the phase, which makes the analysis of the main frequency of structure inadequate. This paper will supplement some advice for these analyses. The numerical comparison of frequencies is as follows: wave frequency > load frequency of suspension point in the z direction > rigid body swing frequency > structure bending frequency (1st).

The frequency comparison can determine the relationship between the load frequency and the natural frequency and it can determine whether resonance is generated. These data can also provide theoretical values for comparison and verification of the FFT frequencies. In addition, the FFT amplitude corresponding to structural frequency can be used to analyze the influence of various loads. The main frequency of the structure, which is obtained from the phase curve, is shown in Section 4.4.2. The frequency of the FFT analysis is in good agreement with the results of theoretical formula. Table 4 compares structural results of formula, FFT and literature [11] methods.

Table 4. Load excitations of structure.

Frequency		Theory	FFT	Literature [11] 0.27 m/s	Literature [11] 0.30 m/s
WAVE (Hz)		0.11622	0.11600	/	/
ZDL (Hz)		0.09300	0.09299	/	/
RBS(Hz)		0.02900	0.02400	0.02800	0.02860
BV (Hz)	1st	0.01600	/	0.01550	0.01560
	2nd	0.06400	0.06800	/	/

RBS is the Rigid Body swing frequency and BV is the second-order frequency for Bending Vibration. ZDL is load frequency of the suspension point in the z direction. WAVE is the wave frequency.

3.4. The Summary of Verification

The verification of rigid body swing is very difficult. In the existing literature [11], there is an experimental method for verification. Due to its high cost and low effect of rigid body swing, it is sometimes difficult to capture the results satisfyingly. On the basis of the above, the *z-x* plane curve can be used for verification. Under the action of a specific frequency and amplitude, a curve can be formed to check by comparing the results with the Lissajous figures.

4. Response Analysis of the SCR Structure

4.1. General Description of Response

In the existing literature [22], response research mainly focuses on the relationship between amplitude and frequency. Among them, the amplitude mainly studies whether the structural response changes suddenly and whether the vibration amplitude exceeds the limit or not. Frequency mainly studies whether the load frequency generates resonance with the natural frequency.

The basic formula for the displacement solution [10] is:

$$\gamma_{ikm} M_{njm} \ddot{u}_{kj} + \alpha_{ikm} B_m u_{kn} + \beta_{ikm} \widetilde{\lambda}_m u_{kn} = \mu_{im} q_{mn} + f_{in} \tag{32}$$

The above formula is a set of equations for the model of a large deflection slender beam. It is a basic equation to solve static and dynamic displacementz. The SCR static displacement equation is a simplified version of Equation (32). The simplified formula is an expression of space, independent of time, which is the basis of the calculation.

The static basic equation of a steel catenary riser is:

$$\alpha_{ikm} B_m u_{kn} + \beta_{ikm} \widetilde{\lambda}_m u_{kn} = \mu_{im} q_{mn} + f_{in} \tag{33}$$

where u is the displacement and $\widetilde{\lambda}$ is the Lagrange operator.

The basic dynamic equation of a steel catenary riser is:

$$\gamma_{ikm} M_{njm}^{(K)} \ddot{u}_{kj}^{(K)} + \alpha_{ikm} B_m u_{kn}^{(K)} + \beta_{ikm} \widetilde{\lambda}_m^{(K)} u_{kn}^{(K)} = \mu_{im} q_{mn}^{(K)} + f_{in}^{(K)} \tag{34}$$

The initial displacement, velocity, acceleration and Lagrange operator can be obtained by the static Equation (33). $M^{(K)}$ and $q^{(K)}$ can be obtained from the above Equation (34). The SCR displacement response calculation is based on the above solution.

4.2. SCR Three-Dimensional and Two-Dimensional Response Analysis

In this paper, SCR response analysis is carried out from three-dimensional analysis to one-dimensional analysis. The response characteristics of structural space are the main contents or emphases of this paper. The process from three dimensions to one represents the feature from integral, plane to one dimension. Under the action of wave and rigid body swing, the effect of load of suspension point in the *z* direction on structure has been studied. Figure 7 and Table 5 show the static location of different nodes.

The responses of the 10th, 80th, 140th and 200th nodes of the structure can be obtained. The three-dimensional figure without rigid body swing is obtained as shown in Figure 8 and the *y-z* plane diagram with and without rigid body swing is shown in Figure 9.

The vertical coordinate is the *y*-axis which is connected to the *z*-axis, and the *x*-axis is perpendicular to the *y-z* plane in Figure 8, which shows the three-dimensional response of non-rigid body swing under wave and load of suspension point in the *z* direction. Figure 9 shows a figure with or without rigid body swing in the *y-z* plane.

There is no obvious difference in shape or overall pattern between the two groups. This indicates that the influence of rigid body swing is weak. The *y-z* plane is a projection in the *y* (swaying or vertical) and z (cross-flow) plane of the structure. There is no significant difference in the plane of vertical and cross flow direction (the rigid body swing plane). This means that when the structure is subjected to the load of the suspension point in the z direction there is no significant difference in the rotation plane. The influence of rigid body swing on the plane is weak. This point can be understood from the amplitude of rigid body swing and top load in the z direction. The top load in the z direction is applied by 3 m (platform response) with significant amplitude relative to the rigid body swing.

The time curve in the z direction is expressed graphically and gradually in the following Section 4.3. Under different conditions, the vibration decreases gradually in the transverse direction, however, it increases gradually in the vertical depth. The oscillation with a graphics like eight in the plane gradually decreases.

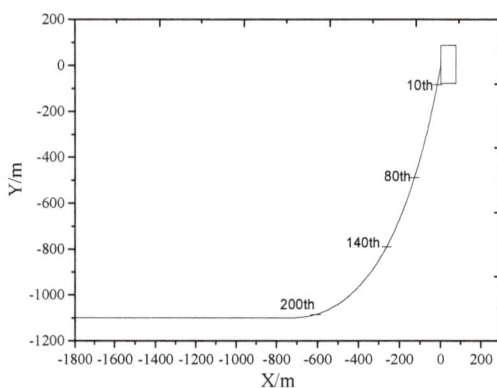

Figure 7. The static location.

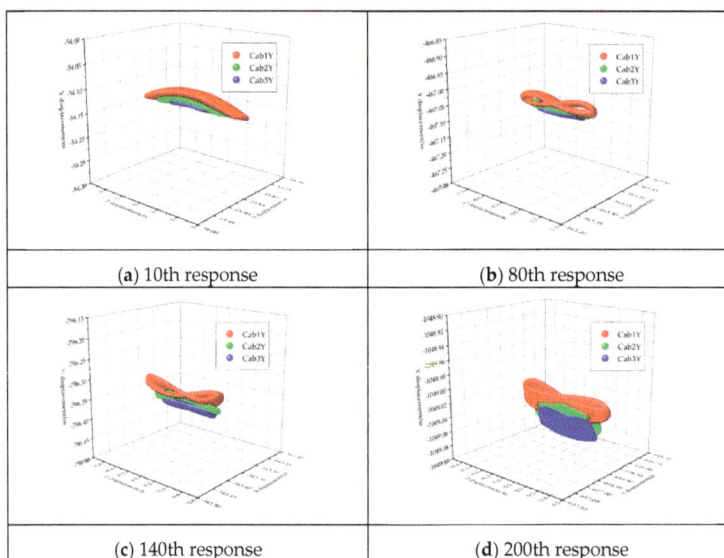

(a) 10th response	(b) 80th response
(c) 140th response	(d) 200th response

Figure 8. 10th–200th three dimensional response of no rigid body swing.

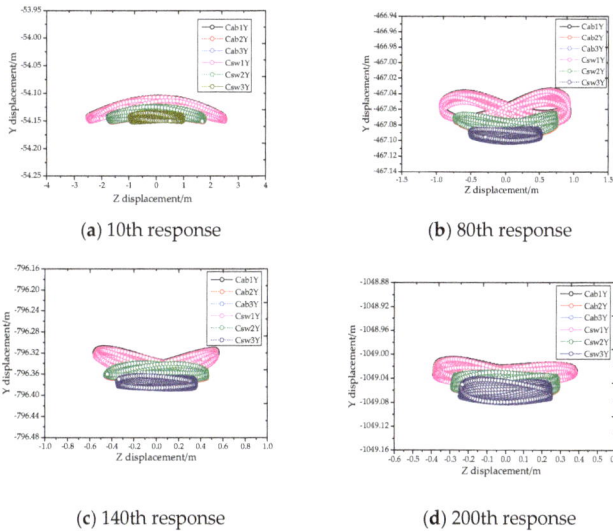

(a) 10th response

(b) 80th response

(c) 140th response

(d) 200th response

Figure 9. 10th–200th two-dimensional response of a/no rigid body swing in the *yz* plane.

Table 5. The location of nodes.

	X/m	Y/m
10	−15.8273	−54.1376
80	−163.2674	−467.0530
140	−343.3629	−796.3470
200	−616.8364	−1049.0293

In terms of the responses of different nodes, the response of the *y-z* plane becomes more complicated as the node number increases. The overall response is that the 10th's *y-z* plane oscillation gradually develops into the 200th's *z-x* plane oscillation. There is also a rotation around the *z* axis to the *z-x* plane. The swing mode: 10th's vibration is a narrow amplitude oscillation in the *y-z* plane. The 80th vibration is a wide amplitude oscillation with a graphic form shaped like eight. The 140th vibration is relatively narrow range oscillation in the *y-z* plane with an eight-like graph. Tthere is a rotation around the *z-x* axis. The 200th vibration is a narrow range oscillation in the *y-z* plane with a graph like an eight. There is also a rotation around the *z* axis, and *z-x* plane gradually oscillates with a graph like an eight.

The amplitude of the swing gradually decreases. The maximum deviations from the center of movement of the 10th, 80th, 140th and 200th node gradually decrease. The structural response also decreases after the superposition of rigid body swing in condition 1 and the rigid body swing decreases the motion of the structure in the *y-z* plane. For condition 2 and 3, the maximum influence of rigid body swing is 1.95% for the 140th node. The difference and amplitude are shown in Table 6 below.

The motion of the structure in the *y-z* plane presents a complicated state with the increase of water depth. The amplitude of oscillation attenuation of the top node is the greatest. The activity is the most intense near the top suspension point region of the 10th–80th nodes. As the water depth or the condition increases, the amplitude decreases, as shown in Figures 10 and 11. The figures help to understand the response from a three-dimensional perspective.

Table 6. Amplitude and difference of *yz* plane swing in different conditions.

Cdt	Node	Cab (m)	Csw (m)	Diff
1	10th	2.53294	2.53254	−0.016%
1	80th	0.93671	0.93484	−0.200%
1	140th	0.55687	0.55431	−0.460%
1	200th	0.38464	0.38395	−0.179%
2	10th	1.76859	1.76816	−0.024%
2	80th	0.74119	0.74525	0.548%
2	140th	0.45918	0.45993	0.163%
2	200th	0.28813	0.28781	−0.111%
3	10th	0.97038	0.97032	−0.006%
3	80th	0.51709	0.51871	0.313%
3	140th	0.35021	0.35704	1.950%
3	200th	0.24737	0.24684	−0.214%

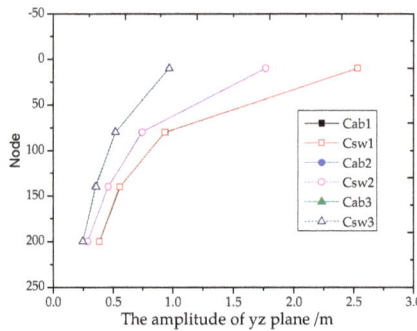

Figure 10. Amplitude change with water depth.

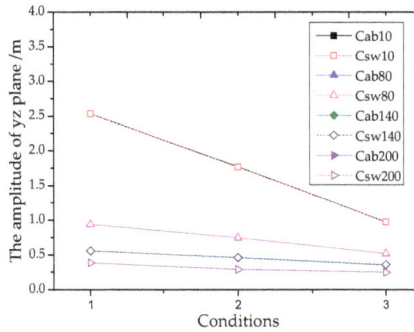

Figure 11. Amplitude change with conditions.

4.3. Analysis of the z-Direction Response of the SCR Structure

4.3.1. SCR Response Analysis

For SCR, the above response characteristics in the *y-z* plane are the response characteristics of a rigid body swing plane. The calculation in the *z* direction is from an angle perpendicular to the *x-y* plane (riser bending plane). It is an indispensable part of the three-dimensional motion response. Its calculation is very important for the vibration system.

The calculation of the cross-flow response of thestructure includes VIV, cross-flow load, rigid body swing and so on. There are plenty of studies on structures with bending vibrations and vortex-induced vibrations, while the response research of the rigid body swing in the cross-flow direction is relatively less common. The responses of the 10th, 80th, 140th and 200th nodes in condition 1 are shown in Figure 12. The formula of the coefficients is as follows:

$$R_{cab} = R_{wave} + R_z \quad R_{csw} = R_{wave} + R_z + R_{rbs} \tag{35}$$

$$Inc = (R_{csw} - R_{cab})/R_{cab} \tag{36}$$

$$Rdt = \begin{cases} (R_n - R_{next})/R_{10th}, n = 10th, 80th, 140th \\ (R_{200th} - 0)/R_{10th}, n = 200th. \end{cases} \tag{37}$$

where R_{cab} is the structural response under waves and the load of the suspension point in the z direction, R_{csw} is the structural response under the action of waves, top load in the z direction and rigid body swing, R_{wavew} is the structural response under wave action, R_{rbs} is the structural response under rigid body swing, R_z is the structural response under the top load in the z direction, Inc is the growth rate and Rdt the relative reduction.

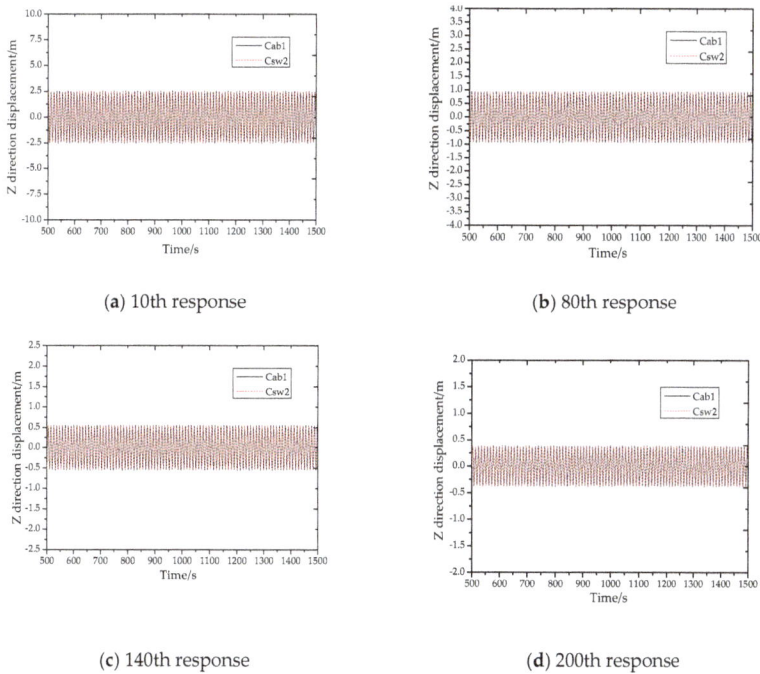

(a) 10th response

(b) 80th response

(c) 140th response

(d) 200th response

Figure 12. 10th–200th z-direction time history response in condition 1.

Figure 12 shows the results of condition 1. The overall response of the structure is observed by selecting four nodes. A linear harmonic load from the z direction is imposed on the structure. The cross-flow extreme, balance position and standard deviation are shown in Tables 6 and 7. They are on behalf of the vibration amplitude and intensity. This shows that the cross-flow maximum of structure decreases with depth under wave action and load of the suspension point in the z direction. The minimum value and standard deviation of the structural vibration are also decreasing. The main reason is that the wave velocity, etc. decreases with depth. As the water depth increases, the response of other particles produced by the top z-direction excitation decreases, therefore, the wave and top load in the z direction weaken the structural response.

Under the action of waves, load of the suspension point in the z direction and rigid body swing, the conclusion is similar to that of non-rigid body swing. The cross-flow response of condition 1 is shown in Tables 7 and 8. As the depth increases, the cross-flow response decreases under the wave, top load in the z direction and rigid body swing influence. After superimposition of rigid body swing, the response's trend is similar to that under the action of waves and the top load in the z direction only. Then the structure is relatively weakly affected by rigid body swing.

The rigid body swing, wave and top action are considered as simply a linear superposition. The maximum growth rates of 10th–200th nodes are −0.01%, −0.53%, −1.12% and −0.26%, respectively, as shown in Tables 7 and 8. By comparing condition 2 and 3, there is maximum superposition of 1.78%. The response of the touch down point and suspension point are weakened under the rigid body swing. For the 140th node the influence of vibration is in a range of −1.5% to 2.0%.

Table 7. Maximum, minimum and increase of node response.

Cdt	Node	Cab_{max} (m)	Csw_{max} (m)	Inc_{max}	Cab_{min} (m)	Csw_{min} (m)	Inc_{min}
1	10th	2.53031	2.52998	−0.01%	−2.52323	−2.52298	−0.01%
1	80th	0.93399	0.92900	−0.53%	−0.93500	−0.93754	0.27%
1	140th	0.55306	0.54688	−1.12%	−0.55430	−0.55152	−0.50%
1	200th	0.38398	0.38300	−0.26%	−0.38312	−0.38239	−0.19%
2	10th	1.76981	1.76936	−0.03%	−1.76465	−1.76436	−0.02%
2	80th	0.74551	0.74471	−0.11%	−0.73664	−0.74101	0.59%
2	140th	0.45820	0.46017	0.43%	−0.45952	−0.45950	0.00%
2	200th	0.28697	0.28663	−0.12%	−0.28717	−0.28764	0.16%
3	10th	0.96836	0.96843	0.01%	−0.96587	−0.96594	0.01%
3	80th	0.51933	0.52159	0.44%	−0.50926	−0.51324	0.78%
3	140th	0.35072	0.35698	1.78%	−0.34172	−0.34872	2.05%
3	200th	0.24683	0.24670	−0.05%	−0.24590	−0.24638	0.20%

Table 8. Equilibrium and standard deviation of node response.

Cdt	Node	Cab_{mean} (m)	Csw_{mean} (m)	Cab_{std} (m)	Csw_{std} (m)
1	10th	0.00315	0.00315	1.71816	1.71816
1	80th	−0.00238	−0.00238	0.62457	0.62443
1	140th	0.00111	0.00112	0.37026	0.37051
1	200th	0.00144	0.00149	0.25996	0.25951
2	10th	1.00×10^{-4}	9.87×10^{-5}	1.1907	1.19067
2	80th	6.71×10^{-4}	6.68×10^{-4}	0.49967	0.4997
2	140th	5.90×10^{-4}	5.86×10^{-4}	0.31063	0.31095
2	200th	4.23×10^{-4}	4.29×10^{-4}	0.19729	0.19767
3	10th	4.77×10^{-4}	4.72×10^{-4}	0.63321	0.63322
3	80th	4.28×10^{-4}	3.98×10^{-4}	0.34035	0.34057
3	140th	4.75×10^{-4}	4.23×10^{-4}	0.23391	0.23431
3	200th	4.34×10^{-4}	3.81×10^{-4}	0.17002	0.16968

Vibration in a certain range represents the phase difference between wave, linear load of the suspension point in the z direction and rigid body swing. However, the effect of rigid body swing first increases and then decreases with depth. It is positively correlated to vector S. The rest of the conditions are similar, as shown in Tables 7 and 8.

Figure 13 shows the variation of the 10th–200th nodes with different conditions. The results show the response of structure decreases when the amplitude of the top load in the z direction decreases. With the increase of water depth, the displacement of each node decreases relative to the response of the top node. The attenuation of the 10th–80th nodes is more intense, and the attenuation is weaker after the 80th, as shown in Tables 9 and 10.

(**a**) 10th response

(**b**) 80th response

(**c**) 140th response

(**d**) 200th response

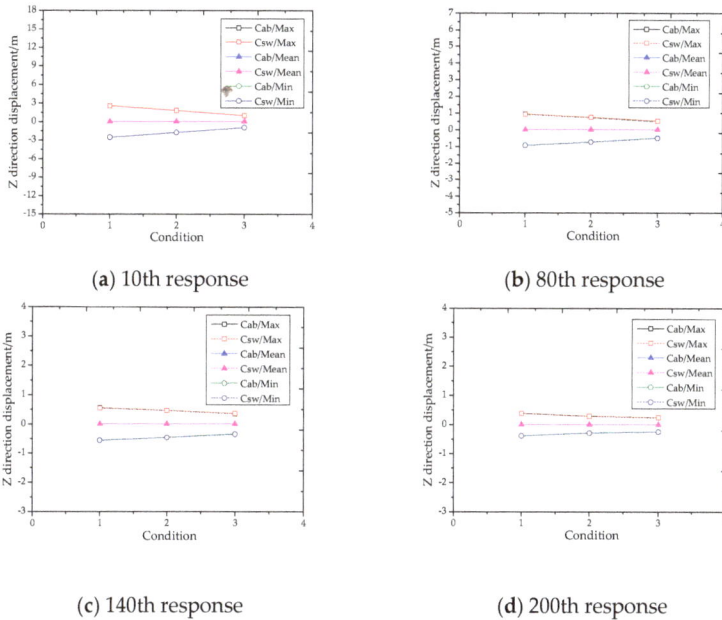

Figure 13. 10th–200th z-direction response changing with conditions.

Table 9. Reduction, difference and normalized values of node's maximum response.

Cdt	Node	Cab_{max} (m)	Csw_{max} (m)	Rdt_{cab}	Rdt_{csw}	Diff	Nml_{cab}	Nml_{csw}
1	10th	2.52716	2.52683	62.95%	63.14%	0.19%	1.00	1.00
1	80th	0.93637	0.93138	15.21%	15.26%	0.05%	1.00	1.00
1	140th	0.55195	0.54576	6.70%	6.50%	−0.20%	1.00	1.00
1	200th	0.38254	0.38151	15.14%	15.10%	−0.04%	1.00	1.00
2	10th	1.76971	1.76926	57.91%	57.95%	0.04%	0.70	0.70
2	80th	0.74484	0.74404	16.23%	16.08%	−0.15%	0.79	0.79
2	140th	0.45761	0.45958	9.67%	9.80%	0.13%	0.83	0.83
2	200th	0.28655	0.28620	16.19%	16.18%	−0.01%	0.75	0.75
3	10th	0.96788	0.96796	46.39%	46.16%	−0.23%	0.38	0.38
3	80th	0.51890	0.52119	17.42%	17.01%	−0.41%	0.55	0.55
3	140th	0.35025	0.35656	10.73%	11.39%	0.66%	0.62	0.63
3	200th	0.24640	0.24632	25.46%	25.45%	−0.01%	0.64	0.64

The influence coefficient ranges from −0.02 to 0.02, as shown in Figure 14a. The influences of rigid body swing on different extreme conditions are statistically analyzed. The maximum is 0.66% at the 140th node. To some extent, if the rigid body swing cannot be calculated, it is appropriate to use a response times 1.02 for the effect of rigid body swing. Figure 14b–d show that with the increase of nodes in each condition, the top region displays the maximum response and the displacement gradually weakens for conditions 1–3. This trend is similar to the change trend of wave loads. The oscillation of the load of suspension point in the z direction decreases with the depth of the water. Tthe pendulum of Equation (8) can explain the trend in a simple way. This tendency of decrease with depth can also be used as part of the verification for calculations.

Table 10. Reduction, difference and normalized values of node's minimum response.

Cdt	Node	Cab$_{min}$ (m)	Csw$_{min}$ (m)	Rdt$_{cab}$	Rdt$_{csw}$	Diff	Nml$_{cab}$	Nml$_{csw}$
1	10th	−2.52323	−2.52298	62.94%	62.84%	−0.10%	1.00	1.00
1	80th	−0.9350	−0.93754	15.09%	15.30%	0.21%	1.00	1.00
1	140th	−0.5543	−0.55152	6.78%	6.70%	−0.08%	1.00	1.00
1	200th	−0.38312	−0.38239	15.18%	15.16%	−0.03%	1.00	1.00
2	10th	−1.76465	−1.76436	58.26%	58.00%	−0.25%	0.70	0.70
2	80th	−0.73664	−0.74101	15.70%	15.96%	0.25%	0.79	0.79
2	140th	−0.45952	−0.45950	9.77%	9.74%	−0.03%	0.83	0.83
2	200th	−0.28717	−0.28764	16.27%	16.30%	0.03%	0.75	0.75
3	10th	−0.96587	−0.96594	47.27%	46.87%	−0.41%	0.38	0.38
3	80th	−0.50926	−0.51324	17.35%	17.03%	−0.31%	0.54	0.55
3	140th	−0.34172	−0.34872	9.92%	10.59%	0.67%	0.62	0.63
3	200th	−0.24590	−0.24638	25.46%	25.51%	0.05%	0.64	0.64

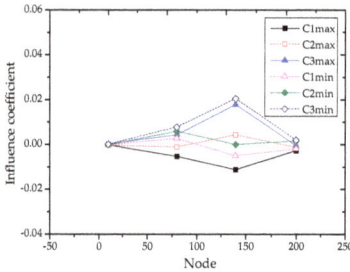

(a) influence coefficient

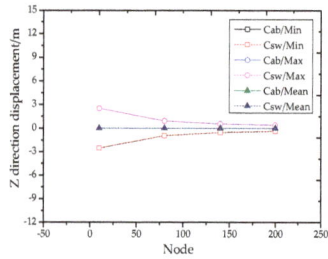

(b) z response in condition 1

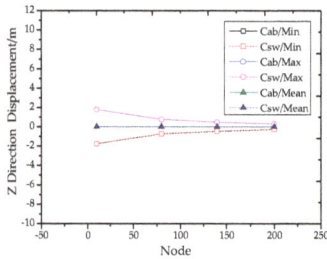

(c) z response in condition 2

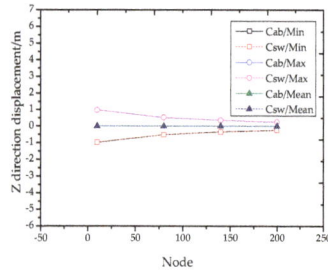

(d) z response in condition 3

Figure 14. z-Direction response in conditions 1–3 changing with the nodes.

The normalization value of Tables 9 and 10 can directly obtain the response coefficient of each condition relative to condition 1. This coefficient can explain why the response variation is more influenced by condition amplitude. The structural response decreases by 20–30% when the amplitude of structural condition decreases by 50%.

4.3.2. Response Analysis of the Top Suspension Point of the SCR Structure

The top suspension point of the SCR has a larger response amplitude in this paper. It belongs to the structural danger area, which should be studied further [22]. The 10th node presents the Lissajous phenomenon in the z-x plane when the frequency ratio is 3:4 in Figure 6. Figures 8 and 9 show that the oscillation in the y-z plane is in a narrow range. The maximum fluctuation standard deviation and reduction with the depth is shown in Tables 9 and 10. The reason is that there is a maximum of wave and top load in the z direction in the top suspension point area. In addition, the response decreases fastest in that region, then it becomes more dangerous.

On the above basis, this paper carried out the spectral analysis, and the results are shown in Table 11 and Figure 15. The response presents a unimodal state. The spectral peak corresponds to load frequency of suspension point in the z direction. The structural response presents a forced motion. As the frequency increases, the spectral peak of the structure decreases. The main reason is that the amplitude of the motion is decreasing. The spectral peak has increased in condition 1. However, it decreases in condition 2–3 after superimposition of the rigid body swing, as shown in Table 11.

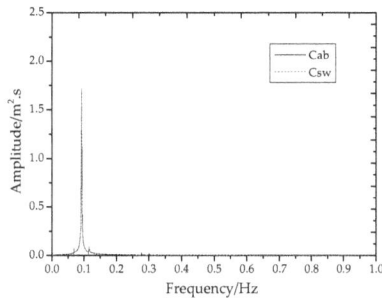

Figure 15. 10th spectrum of condition 1.

Table 11. Spectrum analysis of node 10.

C	Node	X (Hz)	Cab/Y (m²·s)	Csw/Y (m²·s)
1	10th	0.092991	1.710155	1.710178
2	10th	0.100990	1.676353	1.676343
3	10th	0.110989	0.885023	0.885041

4.4. Spectral Analysis

4.4.1. Basic Theory of FFT Calculation

FFT [11] is a transformation from the response signal with time to the response signal with frequency. FFT is also used to calculate the spectral curve of the structure in this paper. In this process, the main frequency and influence frequency can be obtained by the spectral analysis. In addition, the response of the structure includes amplitude and phase. The main frequency and the influence frequency can be obtained by the amplitude curve of the structure. They can also be obtained by the phase curve of the structure. In this paper, the phase curve is more accurate to obtain them:

$$F_n = \sum_{i=0}^{N-1} x_i e^{\frac{-2\pi j}{N} Ni} \tag{38}$$

$$Phase = atc\tan(I_m / R_e) \tag{39}$$

$$Amplitude = \begin{cases} \sqrt{R_e^2 + I_m^2}/n, i = 0, n/2 \\ 2 * \sqrt{R_e^2 + I_m^2}/n, other \end{cases} \tag{40}$$

where F_n is the discrete transform (DFT), N is the calculated length and x_i is the sequence. The phase and amplitude of the single side spectrum ($i = 0 - n/2$) are shown in the formula. The window function is calculated by rectangle. Spectrum analysis is rearranged to reduce leakage and the normalization method of the spectral density is the Mean Square Amplitude (MSA) method.

4.4.2. Analysis of Structural Main Frequency and the Influence Frequency

The frequency of the structure can be obtained by spectral analysis. The variation of the main frequency with water depth is studied. The influences of rigid body swing and the load of the suspension point in the z direction on the main frequency are analyzed. The main frequency is the dotted line position of the amplitude-phase diagram for the 10th-200th nodes in Figure 16. RBS is a shortened form of the rigid body swing. BV is a shortened form of the bending vibration. ZDL is a shortened form of the top load in the z direction. WAVE is a shortened form of wave action. These frequencies have been expressed in Section 3.2.

The amplitude frequency curve of the structural response is difficult to resolve due to the small impact. However, good resolution is obtained in the phase frequency curve. This can be used to judge and check the influence frequency. From the amplitude and phase curve of the structure, the main frequency of the structure is the load frequency of the suspension point in the z direction. The main frequency shows a cliff-edge or rapid change of the phase on the phase curve, but it turns into a peak on the amplitude curve.

Structural frequencies and amplitudes are shown in Tables 12 and 13. The FFT calculation frequency is close to the theoretical value. The importance of load can be obtained from a percentage comparison relative to the total frequency's amplitude, as shown in Table 13. From the top to the bottom, the bending vibration and waves have limited effects on the frequency. Rigid body swing has little effect on the frequency and increases with water depth, as shown in Table 13.

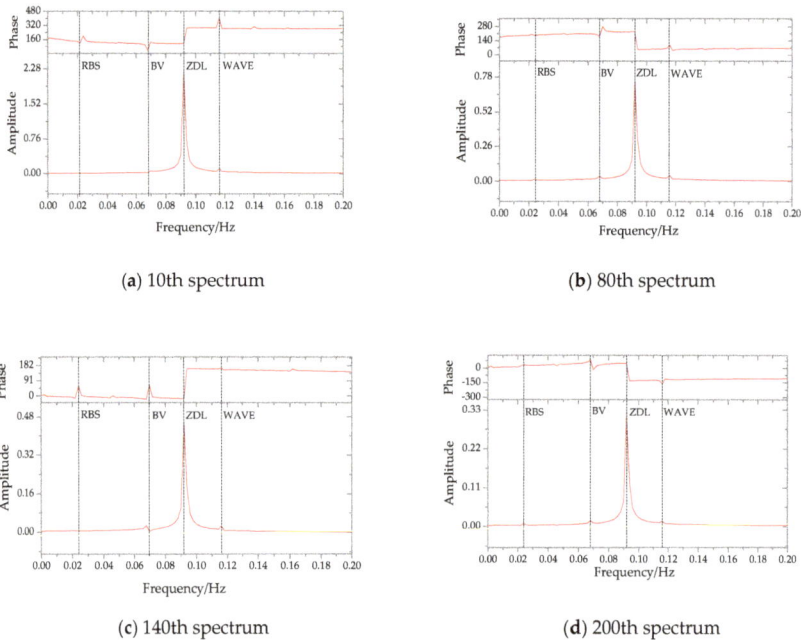

(**a**) 10th spectrum

(**b**) 80th spectrum

(**c**) 140th spectrum

(**d**) 200th spectrum

Figure 16. 10th–200th spectrum of condition 1.

Table 12. Structural frequencies and amplitudes of different loads.

Node	Cdt	RBS X (Hz)	RBS Y (m²·s)	BV X (Hz)	BV Y (m²·s)	ZDL X (Hz)	ZDL Y (m²·s)	WAVE X (Hz)	WAVE Y (m²·s)
10th	Cab	0.024	0.011804	0.0681	0.021044	0.0921	2.238628	0.1161	0.107761
	Csw	0.024	0.012534	0.0681	0.020742	0.0921	2.223184	0.1161	0.105901
80th	Cab	0.024	0.016019	0.0680	0.035739	0.0920	0.770012	0.1160	0.041897
	Csw	0.024	0.016045	0.0681	0.035919	0.0921	0.769957	0.1161	0.038988
140th	Cab	0.024	0.005458	0.0681	0.027404	0.0920	0.457074	0.1161	0.025370
	Csw	0.024	0.005445	0.0681	0.027346	0.0920	0.457155	0.1161	0.029915
200th	Cab	0.024	0.007584	0.0680	0.015419	0.0920	0.320702	0.1160	0.013523
	Csw	0.024	0.007606	0.0680	0.015786	0.0920	0.320106	0.1160	0.014074
Theory		0.0290	/	0.0640	/	0.0930	/	0.1162	

Table 13. Percentage relative to total frequency's amplitude of different loads.

Node	Cdt	RBS	BV	ZDL	WAVE
10th	Cab	0.496%	0.884%	94.090%	4.529%
	Csw	0.531%	0.878%	94.109%	4.483%
80th	Cab	1.855%	4.138%	89.156%	4.851%
	Csw	1.864%	4.172%	89.435%	4.529%
140th	Cab	1.059%	5.318%	88.700%	4.923%
	Csw	1.047%	5.260%	87.938%	5.754%
200th	Cab	2.123%	4.316%	89.775%	3.786%
	Csw	2.127%	4.415%	89.522%	3.936%

4.4.3. Analysis of the Structural Main Frequency

The frequency response of the 10th–200th nodes in Figure 17 is obtained through the spectrum analysis of condition 1. Tables 14 and 15 are the main frequency of the structure in different conditions. In terms of the main frequency, the *x* values are 0.092991 Hz, 0.100990 Hz and 0.110989 Hz. The numerical value is close to the load frequency of the suspension point in the *z* direction of the structure. The frequency of the external load is the same as the response frequency, which is a forced motion. After the superposition of rigid body swing, the main frequency of the structure is still same. It means that the rigid body swing has no obvious influence on the main frequency.

From the main frequency amplitude of the structure, the Y value (spectral amplitude) is studied. The spectral amplitude decreases as the node number increases. The maximum reduction is 40–60% in the top area, as shown in Table 14. As the water depth increases, it decreases first and then increases after reaching the minimum value at the 140th node. The effect does not change significantly after superimposition of rigid body swing. After superimposition of rigid body swing, the maximum impact is about −0.2% in bottom area, as shown in Table 15. At the 140th node, main frequency decreases the least with depth. The rigid body swing makes the main frequency amplitude of structure increase slightly. Then the main frequency amplitude of the structure is normalized. The amplitude of structure decreases gradually with the conditions, as shown in Table 15. The main frequency amplitude represents the energy of the response of the structure. The main frequency is strongly affected by the decrease of amplitude of conditions.

(**a**) 10th spectrum amplitude

(**b**) 80th spectrum amplitude

(**c**) 140th spectrum amplitude

(**d**) 200th spectrum amplitude

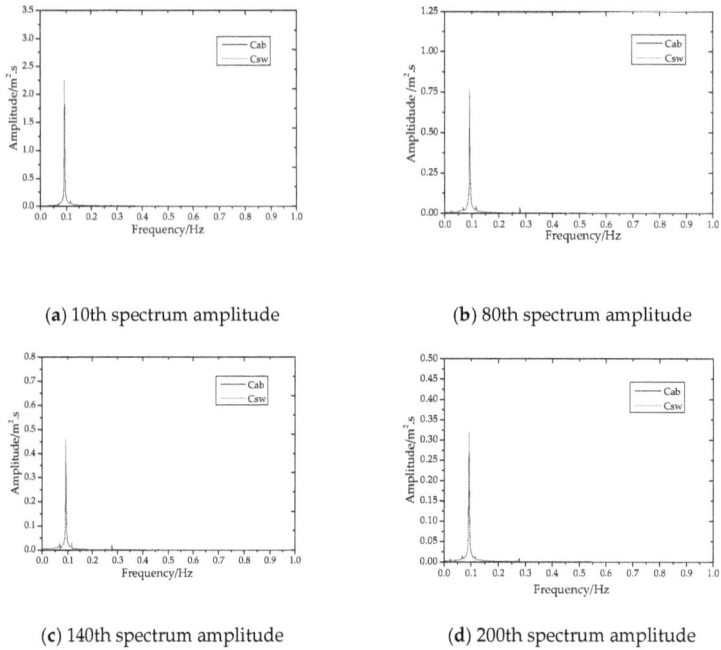

Figure 17. 10th–200th spectrum amplitude of condition 1.

Table 14. Main frequency and reduction of with/no rigid body swing along water depth.

Cdt	Node	Cab_x (Hz)	Cab_y (m$^2 \cdot$s)	Csw_x (Hz)	Csw_y (m$^2 \cdot$s)	Cab_{rdt}	Csw_{rdt}	$Diff_{rdt}$
1	10th	0.092991	2.238628	0.092991	2.223184	65.60%	65.37%	−0.24%
1	80th	0.092991	0.770012	0.092991	0.769957	13.98%	14.07%	0.09%
1	140th	0.092991	0.457074	0.092991	0.457155	6.09%	6.16%	0.07%
1	200th	0.092991	0.320702	0.092991	0.320106	14.33%	14.40%	0.07%
2	10th	0.100990	1.209967	0.100990	1.209961	62.08%	62.09%	0.01%
2	80th	0.100990	0.458852	0.100990	0.458673	11.75%	11.73%	−0.02%
2	140th	0.100990	0.316691	0.100990	0.316708	9.46%	9.43%	−0.03%
2	200th	0.100990	0.202235	0.100990	0.202626	16.71%	16.75%	0.03%
3	10th	0.110989	0.597511	0.110989	0.597504	46.20%	46.24%	0.04%
3	80th	0.110989	0.321454	0.110989	0.321218	16.88%	16.87%	−0.01%
3	140th	0.110989	0.220580	0.110989	0.220394	10.02%	10.04%	0.03%
3	200th	0.110989	0.160731	0.110989	0.160395	26.90%	26.84%	−0.06%

Table 15. Difference and normalized values of spectrum.

Cdt	Node	X (Hz)	Cab_y (m²·s)	Csw_y (m²·s)	Diff	Cab_{nml}	Csw_{nml}
1	10th	0.092991	2.238628	2.223184	−0.69%	1.00	1.00
2	10th	0.100990	1.209967	1.209961	0.00%	0.54	0.54
3	10th	0.110989	0.597511	0.597504	0.00%	0.27	0.27
1	80th	0.092991	0.770012	0.769957	−0.01%	1.00	1.00
2	80th	0.100990	0.458852	0.458673	−0.04%	0.60	0.60
3	80th	0.110989	0.321454	0.321218	−0.07%	0.42	0.42
1	140th	0.092991	0.457074	0.457155	0.02%	1.00	1.00
2	140th	0.100990	0.316691	0.316708	0.01%	0.69	0.69
3	140th	0.110989	0.22058	0.220394	−0.08%	0.48	0.48
1	200th	0.092991	0.320702	0.320106	−0.19%	1.00	1.00
2	200th	0.100990	0.202235	0.202626	0.19%	0.63	0.63
3	200th	0.110989	0.160731	0.160395	−0.21%	0.50	0.50

4.4.4. Analysis of the Structural Phase Response

The main frequency and other frequencies of the structure are usually obtained by a frequency amplitude graph. The phase diagram has a relatively obvious relations with the frequency which is a good complement to the amplitude analysis in this paper.

The influence of structural frequency can be obtained by analyzing the phase of the 10th–200th nodes in Figure 18 in working condition 1. Cliffs or rapid changes occur in the structures at frequencies of 0.1 Hz, 0.3 Hz and 0.4 Hz. This means the structures are affected by obvious loads such as waves here. 0.1 Hz is the effect of the wave load. The response frequencies 0.3 Hz and 0.4 Hz are the structure's natural vibration frequency influenced by the term n^2. Due to the effect oft the term n^2, the loads of waves and the suspension point in the z direction and rigid body swing have higher order effects. Specifically speaking, they have second and third order effects at 0.3 Hz and 0.4 Hz.

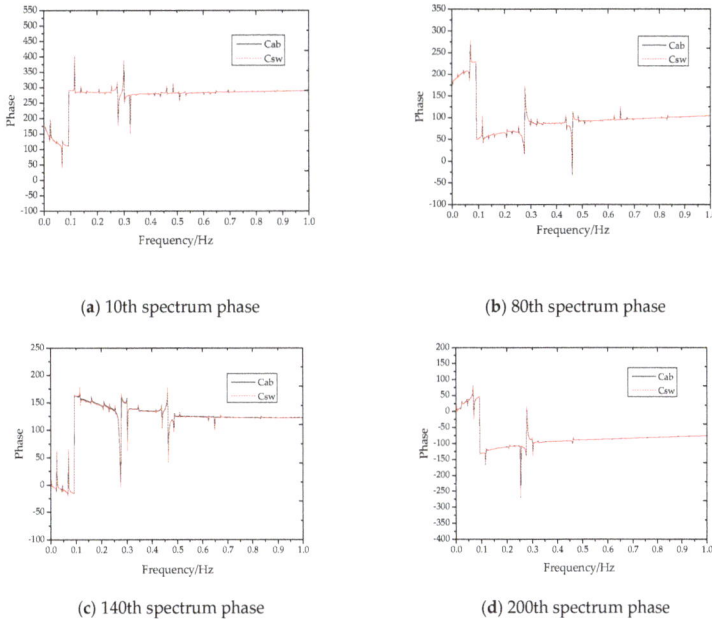

(a) 10th spectrum phase

(b) 80th spectrum phase

(c) 140th spectrum phase

(d) 200th spectrum phase

Figure 18. 10th–200th spectrum phase of condition 1.

4.5. The Summary of Response Analysis

In terms of the overall spatial motion, the structure is affected by the mutual loads perpendicular to each other. The response is gradually complicated along with the water depth. The characteristics of oscillations with or without rigid body are studied in this paper. The main frequency and other frequencies are obtained by a frequency amplitude graph. The phase diagram has a relatively obvious resolution to frequency and it is a good complement to the response analysis.

5. Conclusions

Based on the analysis of vibration theory, a SCR model is simulated and analyzed for the rigid body swing's influence. The characteristics of oscillations caused by water depth and with rigid body swing are researched in this paper. The spectrum amplitude and phase are adopted for analysis of main frequency. The main conclusions of this work are as follows:

(1) Program for rigid body swingares few and their iteration is relatively complex. For structural design it is relatively appropriate to consider the influence of rigid body swing by multiplying the response of the main load by a coefficient. According to the statistics of multiple conditions described in this paper, the influence coefficient range of displacement is −0.02–0.02 and the influence coefficient range of the main frequency amplitude is −0.007–0.002.

(2) Waves are perpendicular to the rigid body swing and load of the suspension point in the z direction in this paper. The analysis of the Lissajous curve for qualitative verification and spectral phase for quantitative verification together can provide a relatively good verification for the rigid body swing.

(3) In terms of overall spatial movement, the response is gradually complicated along the depth. The motion response varies from the yz plane at the 10th node to the zx plane at the 200th.There is a rotation around the z axis in this process. There are phase differences between rigid body swing, wave and the load of the suspension point in the z direction. The structural response is related to the conditions and its trend is relatively uncertain. As an influencing term, the effect of rigid body swing with depth displays a positive linear correlation with the diameter vector S. The response and main frequency have different attenuations at different nodes with depth. The response of the topside 10th–80th region has a sharp decrease, while the response of the 80th-bottom region has a smaller decrease. The response of nodes shows a strong positive correlation with the load of suspension point in the z direction. And it is a forced motion. For the response in top region, it poses a threat to the SCR as a whole for its large amplitude, fast attenuation and strong intensity plane vibration.

(4) In terms of frequency analysis, the result of the rigid body swing frequency solved by the Rayleigh method and the natural frequency of the bending vibration solved by the simply supported beam method are close to that of the FFT transformation. This can relatively meet the requirements of verification for further research. After a spectrum analysis, the results reveal that the load frequency of the suspension point in the z direction is the main frequency. In addition, the main load frequencies of the structure do not resonate with the frequency of the bending vibration. For phase analysis, the trajectory curve under load changes, and the phase curve will show a cliff or rapid change. On the contrary, this kind of variation can distinguish structural load frequencies well. In terms of deficiencies and prospects of this work, there is no real platform load and nonlinear calculation such as random wave. It is still necessary to continue to research the influence in a more realistic environment.

Author Contributions: This study was carried out in collaboration between all authors. B.Z. worked for the writing, analysis, methodology and data; W.H. designed the methodology; X.Y. and J.L. contributed analysis tools and data; X.F. performed the language review and editing.

Funding: The research received no external funding.

Energies **2019**, *12*, 273

Acknowledgments: Thanks to the support by National natural science foundation of China. (51239008, 51739010, 51679223, 51409232) And thanks to National high-tech research and development program (863) (2013AA09A218). Natural science foundation of SHANDONG province (ZR2016GM06) has provided help. Thanks to the help provided by professor ZHANG JUN from the university of Texas A&M in the United States. Thanks to all the help.

Conflicts of Interest: The authors declare no conflicts of interest.

References

1. Huang, W.P.; Li, H.J. A New type of deepwater riser in offshore oil & gas production:the steel catenary, SCR. *Period. Ocean Univ. China* **2006**, *36*, 775–780.

2. Zhu, H.J.; Gao, Y. Vortex induced vibration response and energy harvesting of a marine riser attached by a free-to-rotate impeller. *Energy* **2017**, *134*, 532–544. [CrossRef]

3. Qiu, W.; Lee, D.Y.; Lie, H.; Rousset, J.M.; Mikami, T.; Sphaier, S.; Tao, L.B.; Wang, X.F.; Magarovskii, V. Numerical benchmark studies on drag and lift coefficients of a marine riser at high Reynolds numbers. *Appl. Ocean Res.* **2017**, *69*, 245–251. [CrossRef]

4. Li, H.J.; Wang, C.; Liu, F.S.; Hu, S.J. Stress wave propagation analysis on vortex-induced vibration of marine risers. *China Ocean Eng.* **2017**, *31*, 30–36. [CrossRef]

5. Teixeira, D.C.; Morooka, C.K. A time domain procedure to predict vortex-induced vibration response of marine risers. *Ocean Eng.* **2017**, *142*, 419–432. [CrossRef]

6. Xu, J.; Wang, D.S.; Huang, H.; Duan, M.L.; Gu, J.J.; An, C. A vortex-induced vibration model for the fatigue analysis of a marine drilling riser. *Ships Offshore Struct.* **2017**, *12*, S280–S287. [CrossRef]

7. Cabrera-Miranda, J.M.; Paik, J.K. On the probabilistic distribution of loads on a marine riser. *Ocean Eng.* **2017**, *134*, 105–118. [CrossRef]

8. Do, K.D.; Lucey, A.D. Boundary stabilization of extensible and unshearable marine risers with large in-plane deflection. *Automatica* **2017**, *77*, 279–292. [CrossRef]

9. Zhang, X.D.; Gou, R.Y.; Yang, W.W.; Chang, X.P. Vortex-induced vibration dynamics of a flexible fluid-conveying marine riser subjected to axial harmonic tension. *J. Braz. Soc. Mech. Sci. Eng.* **2018**, *40*, 8. [CrossRef]

10. Liu, J. Study of Out-Of-Plane Motion of SCRs with Rigid Swing. Ph.D. Thesis, China Ocean University, Qingdao, China, 2013.

11. Yao, X.L. On the Dynamic Response and Fatigue of Steel Catenary Riser with Rigid Swinging. Ph.D. Thesis, China Ocean University, Qingdao, China, 2018.

12. Komachiy, M.S.; Tabeshpour, M.R. Wake and structure model for simulation of cross-flow/in-line vortex induced vibration of marine risers. *J. Vibroeng.* **2018**, *20*, 152–164.

13. Domala, V.; Sharma, R. An experimental study on vortex-induced vibration response of marine riser with and without semi subm-ersible. *Proc. Inst. Mech. Eng. Part M J. Eng. Marit. Environ.* **2018**, *232*, 176–198.

14. Hong, K.S.; Shah, U.H. Vortex-induced vibrations and control of marine risers: A review. *Ocean Eng.* **2018**, *152*, 300–315. [CrossRef]

15. Alfosail, F.K.; Nayfeh, A.H.; Younis, M.I. A state space approach for the eigenvalue problem of marine risers. *Meccanica* **2018**, *53*, 747–757. [CrossRef]

16. Yang, W.W.; Ai, Z.J.; Zhang, X.D.; Gou, R.Y.; Chang, X.P. Nonlinear three-dimensional dynamics of a marine viscoelastic riser subjected to uniform flow. *Ocean Eng.* **2018**, *149*, 38–52. [CrossRef]

17. O'Halloran, S.M.; Harte, A.M.; Shipway, P.H.; Leen, S.B. An experimental study on the key fretting variables for flexible marine risers. *Tribol Int.* **2018**, *117*, 141–151. [CrossRef]

18. Wang, X.H.; Guan, Z.C.; Xu, Y.Q.; Tian, Y. Signal analysis of acoustic gas influx detection method at the bottom of marine riser in deepwater drilling. *J. Process Control.* **2018**, *66*, 23–38. [CrossRef]

19. Hu, Y.L.; Cao, J.J.; Yao, B.h.; Zeng, Z.; Lian, L. Dynamic behaviors of a marine riser with variable length during the installation of a subsea production tree. *J. Marine Sci. Technol.* **2018**, *23*, 378–388. [CrossRef]

20. Yang, J.X. Study on the properties of Lissajous' Figures. *J. Xihua Univ. Nat. Sci.* **2008**, *27*, 98–100, 125.

21. Yang, J.X. Study on the synthesized Path of Two Simple Harmonic Vibrations with one Vertical to another. *J. Xihua Univ. Nat. Sci.* **2008**, *2*, 5, 76–78, 82.
22. Li, J.W.; Tang, Y.G.; Li, Y. Influence of areodynamic imbalance on an offshore wind turbine during pitch controller fault. *Ocean Eng.* **2017**, *35*, 37–43.

MDPI

St. Alban-Anlage 66

4052 Basel

Switzerland

Tel. +41 61 683 77 34

Fax +41 61 302 89 18

www.mdpi.com

Energies Editorial Office

E-mail: energies@mdpi.com

www.mdpi.com/journal/energies

www.ingramcontent.com/pod-product-compliance
Lightning Source LLC
Chambersburg PA
CBHW051704210326
41597CB00032B/5368